MATHEMATICS RESEARCH DEVELOPMENTS

FOCUS ON ARTIFICIAL NEURAL NETWORKS

Mathematics Research Developments

Additional books in this series can be found on Nova's website under the Series tab.

Additional E-books in this series can be found on Nova's website under the E-books tab.

Engineering Tools, Techniques and Tables

Additional books in this series can be found on Nova's website under the Series tab.

Additional E-books in this series can be found on Nova's website under the E-books tab.

MATHEMATICS RESEARCH DEVELOPMENTS

FOCUS ON ARTIFICIAL NEURAL NETWORKS

JOHN A. FLORES
EDITOR

Nova Science Publishers, Inc.
New York

LIBRARY OF CONGRESS CATALOGING-IN-PUBLICATION DATA

Focus on artificial neural networks / editor, John A. Flores.
p. cm.
Includes index.
ISBN 978-1-61324-285-8 (hardcover)
1. Neural networks (Computer science) I. Flores, John A.
QA76.87.F623 2011
006.3'2--dc23

2011012975

Published by Nova Science Publishers, Inc. † New York

CONTENTS

Preface		**vii**
Chapter 1	Application of Artificial Neural Networks (ANNs) in Development of Pharmaceutical Microemulsions *Ljiljana Djekic, Svetlana Ibric and Marija Primorac*	**1**
Chapter 2	Investgations of Application of Artificial Neural Network for Flow Shop Scheduling Problems *T. Radha Ramanan*	**29**
Chapter 3	Artificial Neural Networks in Environmental Sciences and Chemical Engineering *F. G. Martins, D. J. D. Gonçalves and J. Peres*	**55**
Chapter 4	Establishing Productivity Indices for Wheat in the Argentine Pampas by an Artificial Neural Network Approach *R. Alvarez and J. De Paepe*	**75**
Chapter 5	Design of Artificial Neural Network Predictors in Mechanical Systems Problems *İkbal Eski, Eyüp Sabri Topaland Şahin Yildirim*	**97**
Chapter 6	Massive-Training Artificial Neural Networks for Supervised Enhancement/Suppression of Lesions/Patterns in Medical Images *Kenji Suzuki*	**129**
Chapter 7	An Inverse Neural Network Model of Disc Brake Performance at Elevated Temperatures *Dragan Aleksendrić*	**151**
Chapter 8	Artificial Neural Networks; Definition, Properties and Misuses *Erkam Guresen and Gulgun Kayakutlu*	**171**
Chapter 9	Evidences of New Biophysical Properties of Microtubules *Rita Pizzi, Giuliano Strini, Silvia Fiorentini' Valeria Pappalardo and Massimo Pregnolato*	**191**

Chapter 10 Forecasting Stream Temperature Using Adaptive Neuron-Fuzzy Logic and Artificial Neural Network Models **209**
Goloka Behari Sahoo

Chapter 11 Neural Network Applications in Modern Induction Machine Control Systems **231**
Dinko Vukadinović and Mateo Bašić

Chapter 12 Wavelet Neural Networks: A Recent Strategy For Processing Complex Signals Applications to Chemistry **257**
Juan Manuel Gutiérrez Roberto Muñoz and Manel del Valle

Chapter 13 Robustness Verification of Artificial Neural Network Predictors in a Purpose-Built Data Compression Scheme **277**
Rajasvaran Logeswaran

Chapter 14 Intelligent Inverse Kinematics Solution for Serial Manipulators Passing through Singular Configurations with Performance Prediction Network **299**
Ali T. Hasan and H. M. A. A. Al-Assadi

Chapter 15 Using Artificial Neural Networks for Continuously Decreasing Time Series Data Forecasting **323**
Mebruk Mohammed, Kunio Watanabe and Shinji Takeuchi

Chapter 16 Application of Artificial Neural Networks in Enzyme Technology **341**
Mohd Basyaruddin Abdul Rahman, Naz Chaibakhsh, Mahiran Basri and Abu Bakar Salleh

Chapter 17 Development of an ANN Model for Runoff Prediction **355**
A. Bandyopadhyay and A. Bhadra

Chapter 18 Artificial Neural Networks Concept: Tools to Simulate, Predict and Control Processes **375**
Abdoul-Fatah Kanta

Index **399**

PREFACE

Chapter 1 – An artificial neural network (ANN) is an intelligent non-linear mapping system built to loosely simulate the functions of the human brain. An ANN model consists of many nodes and their connections. Its capacity is characterized by the structure, transfer function and learning algorithms. Because of their model independence, non-linearity, flexibility, and superior data fitting and prediction ability, ANNs have gained interest in the pharmaceutical field in the past decade. The present chapter highlights the potential of ANNs in the development of pharmaceutical microemulsions. Although microemulsions are currently of interest to the pharmaceutical scientist as promising drug delivery vehicles, the formulation of such unique and complex colloidal systems requires a great experimental effort due to a diverse range of possible colloidal systems as well as coarse dispersions, beside microemulsions, which may be formed in water–oil–tensides systems, depending on temperature and physico-chemical properties and concentrations of constituents. The determination of the region of existence of microemulsions, as the collection of numerous potential pharmaceutical formulations, requires complex and time consuming phase behaviour investigations. Therefore, there is a growing interest of researchers for *in silico* development of ANN models for prediction and/or optimization of the phase behaviour of microemulsion-forming systems using as inputs the data extracted from the phase diagrams already published in the literature or those collected by constructing the phase diagrams using the limited number of experiments. This chapter will be mainly focused on the recent results of the investigations conducted to estimate the applicability of ANN in evaluation of the phase behaviour of microemulsion-forming systems employing the complex mixtures of novel pharmaceutically acceptable nonionic surfactants.

Chapter 2 – The objective of this chapter is to present the research findings, of the author, that primarily use Artificial Neural Network (ANN) as a tool to find an improved solution for the performance measure(s) taken under consideration. The following studies are undertaken to investigate the applicability of ANN:

A bicriterian approach considering makespan and total flow time as performance measures to flow shop scheduling problem applying ANN with competitive network structure is made as a first attempt. With this objective, the architecture is constructed in two stages, viz. initial learning stage and implementation stage. In the initial learning stage the nodes of the network learns the scheduling incrementally and implements the same in the implementation stage. A number of problems are solved for different combinations of jobs and machines by varying jobs from 5 to 30 in steps of 5 and by varying machines from 5 to

30 in steps of 5. A total of 180 problems are solved by taking 5 problems in each set. The work is then extended to seek solutions for multicriteria flow shop scheduling considering makespan, earliness and lateness as performance measures. The result of the ANN is discussed in comparison with particle swarm optimization (PSO).

The next part of the study is modeled with the back propagation network of ANN and tested for seeking solutions to makespan as a performance measure. The results of ANN is sought to be further improved with improvement heuristics, Genetic algorithm (GA) and Simulated Annealing (SA). The problems are also tested against Taillard's benchmark problems (1993). The work aims at obtaining improved solutions by initializing SA and GA with a good starting solution provided by ANN. El-Bouri et al. (2005) show that neural sequences exhibit the potential to lead neighborhood search methods to lower local optima. This aspect is investigated in the study by making a comparison of the performance of the perturbation search and a non-perturbation search when starting from ANN initial solutions. The results show that neural sequences when made a perturbation, exhibit the potential to lead neighborhood search methods to lower local optima.

Chapter 3 – Artificial neural networks have been used for a long time in a wide range of fields inside Environmental Sciences and Chemical Engineering. The main reason for this extensive utilization is the ability of this technique to model easily the complexity of the systems related with these fields, keeping most of the valuable original information about each system.

The feedforward artificial neural networks are the most commonly used topology due to the inherent simple architecture, the diversity of the available training algorithms, and the good performances. Besides feedforward artificial neural networks, the self organizing maps, or also called Kohonen neural networks, have as well relevant applications.

In Environmental Sciences, the most relevant applications appear in modelling for both environmental and biological processes. In Chemical Engineering, artificial neural networks have been applied mainly in: i) modelling; ii) control; and iii) development of software sensors.

This chapter compiles several applications that have been published recently concerning the subjects referred above. A special attention is given to the relevance of the cases, the procedures/techniques, and the ability to be extrapolated to other applications.

Chapter 4 – The Pampas of Argentina is a vast fertile plain that covers approximately 60 Mha and is considered as one of the most suitable regions for grain production worldwide. Wheat production represents a main national agricultural activity in this region. Usually, regression techniques have been used in order to generate wheat yield models, at regional and subregional scales. In a whole regional analysis, using these techniques, climate and soil properties explained 64% of the spatial and interannual variability of wheat yield. Recently, an artificial neural network (ANN) approach was developed for wheat yield estimation in the region. In this chapter the authors compared the performance of multiple regression methods with the ANN approach as wheat yield estimation tools and propose developing productivity indexes by the latter technique. The ANN approach was able to generate a better explicative model than regression, with a lower RMSE. It could explain 76% of the interannual wheat yield variability with positive effects of harvest year, soil available water holding capacity, soil organic carbon, photothermal quotient and the ratio rainfall/crop potential evapotranspiration. Considering that the input variables required to run the ANN can be available 40-60 days before crop harvest, the model has a yield forecasting utility. The results of the ANN model can be used for estimating climate and soil productivity. A climate productivity index developed assessed the effect of the climate scenario and its changes on crop yield. A soil productivity index was also elaborated which represents the capacity to produce a certain amount of harvest grain per hectare, depending on soil characteristics. These indices are tools for characterizing climatic regions and for identifying productivity capabilities of soils at regional scale. The methodology developed can be applied in other cropping areas of the World and for different crops.

Chapter 5 – Due to nonlinearity of the mechanical systems, it is necessary to use adaptive predictors for analysing system parameters. Neural networks could be used as an alternative to overcome such problems. In this chapter, two approaches of mechanical systems are presented for CAD-CAM systems and vehicle suspension systems. In the first approach, surface roughness prediction studies on end milling operations are usually based on three main parameters composed of cutting speed, feed rate and depth of cut. The step-over ratio is usually neglected without investigating it. The aim of this study is to discover the role of the step-over ratio in surface roughness prediction studies in flat end milling operations. In realising this, machining experiments are performed under various cutting conditions by using sample specimens. The surface roughnesses of these specimens are measured. Two Artificial neural networks (ANN) structures were constructed. First of them was arranged with considering, and the second without considering the step-over ratio. ANN structures were trained and tested by using the measured data for predicting surface roughness. Average RMS error of the ANN model with considering step-over ratio is 0.04 and without considering stepover ratio is 0.26. The first model proved capable of prediction of average surface roughness (Ra) with a good accuracy and the second model revealed remarkable deviations from the experimental values.

Other approach is consisted of analyzes effects of vibrations on comfort and road holding capability of vehicles as observed in the variations of suspension springs, road roughness etc. Also, design of non-linear experimental car suspension system for ride qualities using neural networks is presented. Proposed active suspension system has been found more effective in vibration isolation of car body than linear active suspension system. Proposed neural network predictor could be used in vehicle's suspension vibration analysis.

The results of both approaches improved that ANN structure has superior performance at adapting large disturbances of mechanical systems.

Chapter 6 – Medical imaging is an indispensable tool for patients' healthcare in modern medicine. Machine learning plays an important role in the medical imaging field, including medical image processing, medical image analysis, computer-aided diagnosis, organ/lesion segmentation, lesion classification, functional brain mapping, and image-guided therapy, because objects in medical images such as lesions, structures, and anatomy often cannot be modeled accurately by simple equations; thus, tasks in medical imaging require some form of "learning from examples." Pattern enhancement (or suppression: enhancement of specific patterns means suppression of other patterns) is one of the fundamental tasks in medical image processing and analysis. When a doctor diagnoses lesions in medical images, his/her tasks are detection, extraction, segmentation, classification, and measurement of lesions. If we can enhance a specific pattern such as a lesion of interest in a medical image accurately, those tasks are almost complete. What is left to do is merely thresholding of the enhanced lesion. For the tasks of detection and measurement, calculation of the centroid of and the area in the thresholded region may be needed. Thus, enhancement (or suppression) of patterns is one of the fundamental tasks. In this chapter, the basic principles and applications of supervised enhancement/suppression filters based on machine learning, called massive-training artificial neural networks (MTANN), for medical image processing/analysis are presented.

Chapter 7 – The demands imposed on a braking system, under wide range of operating conditions, are high and manifold. Improvement and control of automotive braking systems' performance, under different operating conditions, is complicated by the fact that braking process has stochastic nature. The stochastic nature of braking process is determined by braking phenomena induced in the contact of friction pair (brake disc and disc pad) during braking. Consequently, the overall braking system's performance has been also affected especially at high brake interface temperatures. Temperature sensitivity of motor vehicles brakes has always been an important aspect of their smooth and reliable functioning. It is particularly related to front brakes that absorb a major amount (up to 80%) of the vehicle total kinetic energy. The friction heat generated during braking application easily raises temperature at the friction interface beyond the glass transition temperature of the binder resin and often rises above decomposition temperature. The gas evolution at the braking interfaces because of pyrolysis and thermal degradation of the material results in the friction force decreasing. At such high temperatures, friction force suffers from a loss of effectiveness. This loss of effectiveness (brake fading) cannot be easily predicted due to subsequent thermo-mechanical deformation of disc and disc pad (friction material) which modifies the contact profile and pressure distribution, altering the frictional heat. The instability of the brake's performance after a certain number of brake applications is common and depends on braking regimes represented by application pressure, initial speed, and brake interface temperature. Therefore, the most important issue is related to investigation of possibilities for control of brake performance, especially at elevated temperatures, in order to be stabilized and kept on some level. The control of motor vehicle brakes performance needs a model of how braking regimes, before all application pressure, affecting their performance for the specific friction pair characteristics. Analytical models of brakes performance are difficult, even impossible to be obtained due to complex and highly nonlinear phenomena involved during braking. That is why, in this chapter artificial neural network abilities have been used for modelling of the disc brake performance (braking torque) against synergy of influences of application pressure, initial speed, and brake interface temperature. Based on that, an inverse model of the disc brake performance has been developed able to predict the value of brake's application pressure, which, for current values of brake interface temperature and initial speed, provides wanted braking torque. Consequently, the brake's application pressure could be adjusted to keep the disc brake performance (braking torque) on some wanted level and prevent its decreasing during braking at elevated temperatures.

Chapter 8 – There are no such clear and good definitions of ANNs in the literature. Many of the definitions refer to the figures instead of well explaining the ANNs. That is why many weighted graphs (as in shortest path problem networks) fit the definition of ANN. This study aims to give a clear definition that will differentiate ANN and graphs (or networks) by referring to biological neural networks. Although there is no input choice limitation or prior assumption in ANN, sometimes researchers compare ANN achievements with the results of other methods using different input data and make comments on these results. This study also gives examples from literature to misuses, unfair comparisons and evaluates the underlying reasons which will guide researchers.

Chapter 9 – Microtubules (MTs) are cylindrical polymers of the protein tubulin, are key constituents of all eukaryotic cells cytoskeleton and are involved in key cellular functions. Among them MTs are claimed to be involved as sub-cellular information or quantum information communication systems. MTs are the closest biological equivalent to the well known carbon nanotubes (CNTs) material. The authors evaluated some biophysical properties of MTs through two specific physical measures of resonance and birefringence, on the assumption that when tubulin and MTs show different biophysical behaviours, this should be due to the special structural properties of MTs. The MTs, as well as CNTs, may behave as oscillators, this could make them superreactive receivers able to amplify radio wave signals. Our experimental approach verified the existence of mechanical resonance in MTs at a frequency of 1510 MHz. The analysis of the results of birefringence experiment highlights that the MTs react to electromagnetic fields in a different way than tubulin.

Chapter 10 – All biological processes in water are temperature dependent. The plunging depth of stream water and its associated pollutant load into a lake/reservoir depend on stream water temperature. Lack of detailed datasets and knowledge on physical processes of the stream system limits the use of a phenomenological model to estimate stream temperature. Rather, empirical models have been used as viable alternatives. In this study, models using artificial neural networks (ANN) were examined to forecast the stream water temperature from available solar radiation and air temperature data. Observed time series data were non-linear and non-Gaussian, thus the method of time delay was applied to form the new dataset that closely represented the inherent system dynamics. Mutual information function indicates that optimum time lag was approximately 3 days. Micro-genetic algorithms were used to optimize the ANN geometry and internal parameters. Results of optimized ANN models showed that the prediction performance of four layer back propagation neural network was highest to those of other models when data are presented to the model with one-day to three-day time lag. Air temperature was found to be the most important variable in stream temperature forecasting; however, the prediction performance efficiency was somewhat higher if short wave radiation was included.

Chapter 11 – This chapter shows an overview of neural network applications in modern induction machine control systems. Induction motors have been used as the workhorse in industry for a long time due to their being easy to build, highly robust, and having generally satisfactory efficiency. In addition, induction generators play an important role in renewable energy systems such as energy systems with variable-speed wind turbines. The induction machine is a nonlinear multivariable dynamic system with parameters that vary with temperature, frequency, saturation and operating point. Considering that neural networks are capable of handling time varying nonlinearities due to their own nonlinear nature, they are suitable for application in induction machine systems. In this chapter, the use of artificial neural networks for identification and control of induction machine systems will be presented. An overview of neural network applications in induction machine control systems will be the focus:

1. Drive feedback signal estimation,
2. Inverter control,
3. Identification of machine parameters,
4. Neural network based approaches for the efficiency improvement in induction machine systems,

5. Neural network implementations by digital signal processors and ASIC chips.

Chapter 12 – In the last three decades, Artificial Neural Networks (ANNs) have gained increasing attention due to their wide and important applications in different areas of knowledge as adaptive tool for processing data. ANNs are, unlike traditional statistical techniques, capable of identifying and simulating non-linear relationships without any a priori assumptions about the data's distribution properties. Furthermore, their abilities to learn, remember and compare, make them useful processing tools for many data interpretation tasks in many fields, for example in chemical systems or in the analytical field.

Nevertheless, the development of new analytical instruments producing readouts of higher dimensionality and the need to cope with each time larger experimental data sets have demanded for new approaches in data treatment. All this has lead to the development of advanced experimental designs and data processing methodologies based on novel computing paradigms, in order to tackle problems in areas such as calibration systems, pattern recognition, resolution and recovery of pure-components from overlapped spectra or mixtures.

This chapter describes the nature and function of Wavelet Neural Networks (WNNs), with clear advantages in topics such as feature selection, signal pre-processing, data meaning and optimization tasks in the treatment of chemical data. The chapter focuses on the last applications of WNNs in analytical chemistry as one of its most creative contributions from theoretical developments in mathematical science and artificial intelligence. Specifically, recent contributions from our laboratory showing their performance in voltammetric electronic tongue applications will be outlined and commented.

Chapter 13 – Artificial Neural Networks (ANN) are reputed to be error tolerant due to their massively parallel architecture, where the performance of faulty components maybe compensated by other parts of the network. However, most researchers take this for granted and do not verify the fault tolerance capabilities of their purpose-built ANN systems. This article reports on the robustness performance of various ANN architectures to the influences of noise and network failure in a block-adaptive predictor scheme developed to compress numeric telemetry data from remote sensors. Various single and multilayered feedforward and recurrent ANN architectures are tested as the predictor. For real-time adaptability, yet preventing network rigidity due to over-training, the ANN are retrained at the block level by segmenting the incoming data, providing good adaptability to even significantly varying input patterns. The results prove that while some ANN architectures in the proposed scheme do indeed provide better robustness as compared to classical schemes, this is not necessarily true for other architectures. The findings and discussions provided would be useful in determining the suitability of ANN architectures in future implementations that require sustainable robustness to influences such noise and network failures.

Chapter 14 – This chapter is devoted to the application of Artificial Neural Networks (ANN) to the solution of the *Inverse Kinematics* (IK) problem for serial robot manipulators, in this chapter two networks were trained and compared to examine the effect of considering the Jacobian Matrix to the efficiency of the IK solution.

Offline smooth geometric paths in the joint space of the manipulator are obtained through trajectory planning process to give the desired trajectory of the end effector of the manipulator in a free of obstacles workspace. Some of the obtained data sets were used in the training phase while the remaining data sets were used in the testing phase.

Even though it is very difficult in practice, data used in this study were recorded experimentally from sensors fixed on robot's joints to overcome the effect of kinematics uncertainties presence in the real world such as ill-defined linkage parameters, links flexibility and backlashes in gear train

The generality and efficiency of the proposed algorithm are demonstrated through simulation of a general six DOF serial robot manipulator, finally the obtained results were verified experimentally.

Chapter 15 – Data preprocessing is an issue that is often recommended to create more uniform data to facilitate ANN learning, meet transfer function requirements, and avoid computation problems. In ANN typical transfer functions, such as the sigmoid logistic function, or the hyperbolic tangent function, cannot distinguish between two very large values, because both yield identical threshold output values of 1.0. It is then necessary to normalize (preprocess) the inputs and outputs of a network. Usually normalization is carried out using the minimum and maximum values obtained in the in-sample (calibration) data. Such a network will result in absurd output, if the out-of-sample (test) data contain values that are beyond the in-sample data range. This ultimately limits the application of ANN in forecasting continuously increasing or decreasing time series data.

This study will present a novel and successful application of ANN, which is trained by the error backpropagation algorithm, in the context of forecasting beyond in-sample data range. The emphasis here is on continuously decreasing hydraulic pressure data forecasting that are observed at Mizunami underground research laboratory construction site, Japan. The ANN utilizes the sigmoid logistic function in its hidden and output layers.

Chapter 16 – Enzymes are protein molecules that speed up biochemical reactions without being consumed, so act as biocatalysts that help make or break the covalent bonds (Alberts, 1998). Enzyme technology is the technological concepts that enable application of enzymes in production processes to achieve sustainable industrial development. This field is leading to discovery, development and purification of enzymes, and their application in different industry sectors (van Beilen and Li, 2002). Custom design of enzyme activity for desired industrial applications, process control and bioparameter estimation are major goals in enzymatic process development.

Mathematical modeling and simulation is a powerful approach for understanding the complexity and nonlinear behavior of biological systems and identifying natural laws describing their behavior (Meng et al. 2004). Computational Intelligence (CI) techniques have been successfully applied to solve problems in the identification and control of biological systems (do Carmo Nicoletti and Jain, 2009). Artificial Neural Networks (ANNs), in particular, provide an adequate approach in estimating variables from incomplete information and handling nonlinear dynamic systems like enzymatic processes.

One of the major problems of ANNs is the cost of model development due to requiring relatively extensive training data (Montague and Morris, 1994). It is also difficult to interpret the network, and convergence to a solution is slow and depends on the network's structure (do Carmo Nicoletti and Jain, 2009). In order to overcome these limitations, Design of Experiments (DOE) has been introduced as a better methodology than the common trial and error techniques for generating the ANN's training data (Balestrassi et al., 2009).

This chapter reviews some applications of ANNs in enzyme technology. Some practical considerations including utilization of DOE for training the neural networks in enzymatic processes have also been introduced.

Chapter 17 – Over the years, several hydrological models ranging from empirical relationships to physically based models have been developed for prediction of runoff. The physically based models are better as they relate physical processes but at the same time their data requirement is also high. Therefore, there is a need to look for alternative methods for prediction of runoff using readily available information such as rainfall. Artificial Neural Network (ANN) is an information processing system that is composed of many nonlinear and densely interconnected processing elements or neurons. Feed forward multilayer neural networks are widely used as predictors in several fields of applications. The purpose of this study is to demonstrate development of an ANN model using both steepest descent and Levenberg-Marquardt optimization training algorithms and to investigate its potential for accurate runoff estimation. Different ANN networks were trained and tested to predict the daily runoff for Kangsabati reservoir catchment. The networks were selected using one, two, and three hidden layers. The network models were trained for seven years data and tested for one year data for different sizes of architecture. Training was conducted using both steepest descent and Levenberg- Marquardt Back Propagation where the input and output were presented to the neural network as a series of learning patterns. Results indicated that the neural networks trained with Levenberg-Marquardt Back Propagation converged much faster than simple steepest descent back propagation. Further, the ANN models performance improved with increase in number of hidden neurons as well as with increase in number of hidden layers up to a certain point 15-20-20-1 (best network architecture), after which the performance deteriorated.

Chapter 18 – Artificial neural network (ANN) is a powerful statistical procedure permitting to relate the parameter of a given problem to its desired result by considering a complex network of artificial neurons. This concept is based on model which offers the possibility to develop such a global and integrated approach, without providing any physical explanation for the relationships that have to be validated from a physical point of view.

The design of neural networks structures is an important problem for ANN applications which is difficult to solve theoretically. The definition of optimal network architecture for any particular problem is quite difficult and remains an open problem. The contribution of this paper is the description and the implementation of a formal neural networks concept.

In: Focus on Artificial Neural Networks
Editor: John A. Flores, pp. 1-28
ISBN: 978-1-61324-285-8

Chapter 1

APPLICATION OF ARTIFICIAL NEURAL NETWORKS (ANNS) IN DEVELOPMENT OF PHARMACEUTICAL MICROEMULSIONS

Ljiljana Djekic, Svetlana Ibric and Marija Primorac
Department of Pharmaceutical Technology and Cosmetology,
Faculty of Pharmacy, Vojvode Stepe, Belgrade, Serbia and Montenegro

1. INTRODUCTION

An artificial neural network (ANN) is an intelligent non-linear mapping system built to loosely simulate the functions of the human brain. An ANN model consists of many nodes and their connections. Its capacity is characterized by the structure, transfer function and learning algorithms. Because of their model independence, non-linearity, flexibility, and superior data fitting and prediction ability, ANNs have gained interest in the pharmaceutical field in the past decade. The present chapter highlights the potential of ANNs in the development of pharmaceutical microemulsions. Although microemulsions are currently of interest to the pharmaceutical scientist as promising drug delivery vehicles, the formulation of such unique and complex colloidal systems requires a great experimental effort due to a diverse range of possible colloidal systems as well as coarse dispersions, beside microemulsions, which may be formed in water–oil–tensides systems, depending on temperature and physico-chemical properties and concentrations of constituents. The determination of the region of existence of microemulsions, as the collection of numerous potential pharmaceutical formulations, requires complex and time consuming phase behaviour investigations. Therefore, there is a growing interest of researchers for *in silico* development of ANN models for prediction and/or optimization of the phase behaviour of microemulsion-forming systems using as inputs the data extracted from the phase diagrams already published in the literature or those collected by constructing the phase diagrams using the limited number of experiments. This chapter will be mainly focused on the recent results of the investigations conducted to estimate the applicability of ANN in evaluation of the phase behaviour of microemulsion-forming systems employing the complex mixtures of novel pharmaceutically acceptable nonionic surfactants.

2. Artificial Neural Networks (ANNs)

Rigorous regulations in pharmaceutical industry urge for more sophisticated tools that could be used for designing and characterizing dosage forms. It is of great importance to be fully aware of all the factors impacting the process of dosage form manufacturing and, if possible, predict the intensity of these impacts on product characteristics. Computer programs based on artificial intelligence concepts are proving to be distinctive utilities for this purpose.

Artificial neural network (ANN), usually called "neural network" (NN), is a mathematical model or computational model that tries to simulate the structure and/or functional aspects of biological neural networks. It consists of an interconnected group of artificial neurons and processes information using a connectionist approach to computation. In most cases an ANN is an adaptive system that changes its structure based on external or internal information that flows through the network during the learning phase. Modern neural networks are non-linear statistical data modeling tools. They are usually used to model complex relationships between inputs and outputs or to find patterns in data (Bishop, 1995).

Like humans, neural networks learn directly from input data. The learning algorithms take two main forms. Unsupervised learning, where the network is presented with input data and learns to recognize patterns in the data, is useful for organizing amounts of data into a smaller number of clusters. For supervised learning, which is analogous to “teaching” the network, the network is presented with a series of matching input and output examples, and it learns the relationships connecting the inputs to the outputs. Supervised learning has proved most useful for pharmaceutical formulation, where the goal is to determine cause-and-effect links between inputs (ingredients and processing conditions) and outputs (measured properties) (Rowe et al., 1996).

The basic component of the neural network is the neuron, a simple mathematical processing unit that takes one or more inputs and produces an output. For each neuron, every input has an associated weight that defines its relative importance, and the neuron simply computes the weighted sum of all the outputs and calculates an output. This is then modified by means of a transformation function (sometimes called a transfer or activation function) before being forwarded to another neuron. This simple processing unit is known as a perceptron, a feed-forward system in which the transfer of data is in the forward direction, from inputs to outputs, only. A neural network consists of many neurons organized into a structure called the network architecture. Although there are many possible network architectures, one of the most popular and successful is the multilayer perceptron (MLP) network. This consists of identical neurons all interconnected and organized in layers, with those in one layer connected to those in the next layer so that the outputs in one layer become the inputs in the subsequent layer. Data flow into the network via the input layer, pass through one or more hidden layers, and finally exit via the output layer (Figure 1). In theory, any number of hidden layers may be added, but in practice multiple layers are necessary only for those applications with extensive nonlinear behavior, and they result in extended computation time. It is generally accepted that the performance of a well-designed MLP model is comparable with that achieved by classic statistical techniques.

Unlike conventional computer programs, which are explicitly programmed, supervised neural networks are "trained" with previous examples. The network is presented with example data, and the weights of inputs feeding into each neuron are adjusted iteratively until the output for a specific network is close to the desired output. The method used to adjust the weights is generally called back propagation, because the size of the error is fed back into the calculation for the weight changes. There are a number of possible back propagation algorithms, most with adjustable parameters designed to increase the rate and degree of convergence between the calculated and the desired (actual) outputs. Although training can be a relatively slow process, especially if there are large amounts of data, once trained, neural networks are inherently fast in execution.

The utility of artificial neural network models lies in the fact that they can be used to infer a function from observations. This is particularly useful in applications where the complexity of the data or task makes the design of such a function by hand impractical.

The tasks to which artificial neural networks are applied tend to fall within the following broad categories:

- Function approximation, or regression analysis, including time series prediction, fitness approximation and modeling.
- Classification, including pattern and sequence recognition, novelty detection and sequential decision making.
- Data processing, including filtering, clustering, blind source separation and compression.

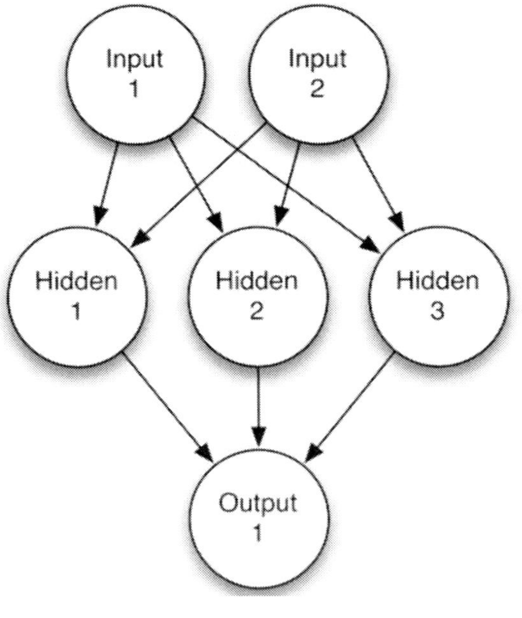

Figure 1. Diagram of a multilayer perceptron with one hidden layer.

Artificial neural networks (ANN) have been introduced into the field of pharmaceutical technology in 1991 by Hussain et al. (1991) and coworkers and gained interest in several pharmaceutical applications (Bourquin et al., 1998a, Bourquin et al., 1998b, Bourquin et al., 1998c, Murtoniemi et al., 1994 and Turkoglu et al., 1999). Ever since, they received great attention, especially when it was realized how powerful tools these networks can be. Authors (Chen et al., 1999) have used ANNs in the design of controlled release formulations. Varying formulation variables were used as inputs and in vitro cumulative percentages of drug released were used as outputs. Other researchers (Zupancic Bozic et al., 1997) have developed an ANN model to optimize diclofenac sodium sustained release matrix tablets. Trained model was employed to predict release profiles and to optimize the formulation composition. A generalized regression neural network (GRNN) was used in the design of extended-release aspirin tablets (Ibric et al., 2002). There are many other examples of applications of ANN in pharmaceutical technology, cited in Sun et al. (2003). Among the many possible ANN architectures, the multi-layer perceptron (MLP) network is one of the most widely used (Peh et al., 2000, Reis et al., 2004 and Rowe and Roberts, 1998). It has been shown that many artificial intelligence systems, especially neural networks, can be applied to the fundamental investigations of the effects of formulation and process variables on the delivery system (Sun et al., 2003).

Genetic programming, generally regarded as a subset of genetic algorithms (GA), having been widely popularized only in the 1990s primarily by Koza (Koza, 1992). It has had limited use in pharmaceutical formulation, but it shows great promise since it has the learning capabilities similar to that of neural networks but the transparency associated with a straightforward mathematical expression. In genetic programming, each solutions is a 'tree', in which each tree node has an operator function and each terminal node is an operand. These trees provide an alternative way of representing equations. An initial population of solutions is assumed, and as with other evolutionary methods, the fitness of each member is assessed. The population then evolves allowing crossover (whereby parts of trees are swapped) and mutation. The evolution is biased so that the fittest solutions are emphasized in successive generations, leading to increased improvement in the fit of the model to the training data.

In the same way as for other genetic algorithms, a criterion of fitness needs to be defined. The simplest criterion would simply minimize the mean-squared error between the calculated and actual values, but this could result in an overly complex, and potentially over-fitted, model. Therefore, it is often appropriate to use a model assessment criterion (such as Structural Risk Minimization) to penalize those solutions whose added complexity does not return significant new knowledge.

Genetic programming currently suffers from the disadvantage that it is time consuming, and its application is less well understood in the formulation domain than are neural networks. Nonetheless they are attractive possibilities for future work, because they can produce 'transparent' models.

3. Microemulsions

Microemulsions are thermodynamically stable and optically isotropic transparent colloidal systems consisting of water, oil and appropriate amphiphiles (surfactant, usually in combination with a cosurfactant). They form spontaneously when admixing the appropriate quantities of the components applying mild stirring and/or heating. The formation of microemulsions are related to ultra-low interfacial tension ($\leq 10^{-3}$ mN/m) which corresponds with the relatively high concentrations of tensides. Furthermore, such extreme lowering of the tension on a water-oil interface usually requires an introduction of additional component which acts as a cosurfactant. In general, microemulsions are clear low viscous liquids. However, on a microscopical level, they are heterogeneous and three type of microstructure were identified: water-in-oil (w/o) microemulsions, oil-in-water (o/w) microemulsions and bicontinuous microstructure. The w/o and o/w structures of microemulsions usually describes as the droplet types of microemulsions where the droplets of one phase (oil or water) are sorounded by the monomolecular film of a surfactant (or surfactant/cosurfactant) molecules and dispersed within the other phase. The diameter of the droplets ranges from 10 – 100 nm. For the third type of the microstructure, both phases are assumed as the continuous while a surfactant (or surfactant/cosurfactant) molecules form flexible, continuous monomolecular film on the interface. The type of the structure and long term stability of microemulsions at given temperature and pressure are determined mainly by physico-chemical properties and concentrations of the constituents (Fanun, 2009). Therefore, the main characteristics of microemulsions such as ease of preparation, good stability and the large interface area representing an additional phase suitable for solubilisation of different substances, increase their relevance for various applications including drug delivery. Microemulsions are currently of interest to the pharmaceutical scientist as promising vehicles with a great potential in improvement of bioavailability of numerous drugs applied orally, on/via skin, on eye etc. The observed drug delivery potential ascribes primary to large solubilisation capacity, although a significant number of components of microemulsions may affect the biological membranes and act as absorption enhancers (Bagwe et al., 2001; Gupta and Moulik, 2008; Lawrence and Rees, 2000; Malmstein, 1999; Spernath and Aserin, 2006, Fanun, 2009).

The particular interes is the formulation of microemulsions using nontoxic, biocompatible, pharmaceutically acceptable oils (e.g medium-chain triglycerides, partial glycerides (glyceryl monocaprylocaprate (Capmul® MCM), glyceryl monostearate (Geleol™, Imwitor® 191, Cutina™ GMS, or Tegin™), glyceryl distearate (Precirol™ ATO 5), glyceryl monooleate (Peceol™), glyceryl monolinoleate (Maisine™ 35-1), or glyceryl dibehenate (Compritol® 888 ATO), fatty acid esters (isopropyl myristate, isopropyl palmitate, isostearyl isostearate, ethyl oleate, cetearil octanoate), fatty acids (oleic acid), alcohols (octanol, decanol)), surfactants (polyoxylglycerides (Labrasol®, Labrafil®-s, or Gelucire®-s), ethoxylated glycerides derived from castor oil (Cremophor® EL, RH40, or RH60), and esters of edible fatty acids and various alcohols (e.g., polyglyceryl oleate (Plurol™ Oleique CC497), propylene glycol monocaprylate (Capryol™ 90), propylene glycol monolaurate (Lauroglycol™ 90), poly(ethylene glycol) (PEG)-8 stearate and PEG-40 stearate (Mirj® 45 and Mirj® 52), sorbitan monooleate and sorbitan monolaurate (Span® 80 and Span® 20), polyoxyethylene-20 sorbitan monooleate (polysorbate 80; Tween® 80), and polyoxyethylene-20 sorbitan monolaurate (polysorbate 20; Tween® 20)), poloxamers, lecithin), and cosurfactants (low molecular weight PEGs, ethanol, propylene glycol, glycerin, diethyleneglycol monoethylether).

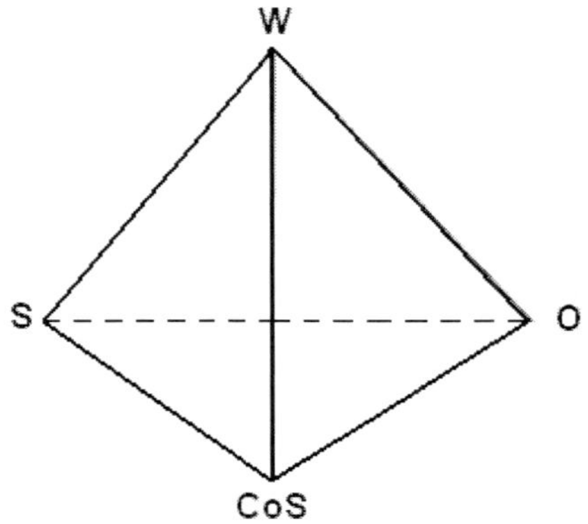

Figure 2. Tetrahedron type of the phase diagram for the four component system surfactant (S)/cosurfactant (CoS)/oil (O)/water (W).

In spite of a substantial amount of investigations of various microemulsion systems as potential drug delivery vehicles, there are no general conclusions nor guidelines to find the optimal microemulsion composition for a desired microemulsion type and structure, which subsequently affects its drug delivery potential. Pharmaceutically applicable microemulsions usually consist of four or more components, including the drug. In water–oil–tensides systems, beside microemulsions, a diverse range of other colloidal systems and coarse dispersions can be obtained (e.g. emulsions, microemulsions, micelles, lyotropic liquid crystals), depending on temperature and physico-chemical properties and composition ratios of constituents. Thus the classical trial-error approach for realizing the desired properties is time-consuming and does not guarantee success. Range of water–oil–surfactant–cosurfactant compositions, which can form microemulsions at given temperature, as well as the effect of various formulation variables on a region of existence of microemulsions, usually determines from phase behaviour investigations and represents in phase diagrams (Kahlweit, 1999). For example, the appropriate type of phase diagram for a full geometrical representation of a four component mixture at constant temperature is a tetrahedron (Figure 2) in which each corner represents 100% of one component of the system and each point inside the tetrahedron represents the one mixture of components at given percentages.

Complete differentiation of quaternary mixtures which form microemulsion from the others, would require a large number of experiments. Every 'slice' within a tetrahedron is in fact a pseudo-ternary phase triangle with two corners corresponding to 100% of two components and the third corner represents 100% of a binary mixture of two components at constant mass ratio (e.g. surfactant+cosurfactant or oil+water) (Figure 3).

Although phase diagrams represent detailed compositional maps which are of great interest to the formulators, it should be noted that the construction of complete phase diagrams requires complex and very time consuming experimental work. On the other hand, the extremely complex interactions between the components at the molecular level hinder the development of mathematical functions relating the physico-chemical properties and concentrations of the constituents with the formation of microemulsions and their structural and drug delivery characteristics.

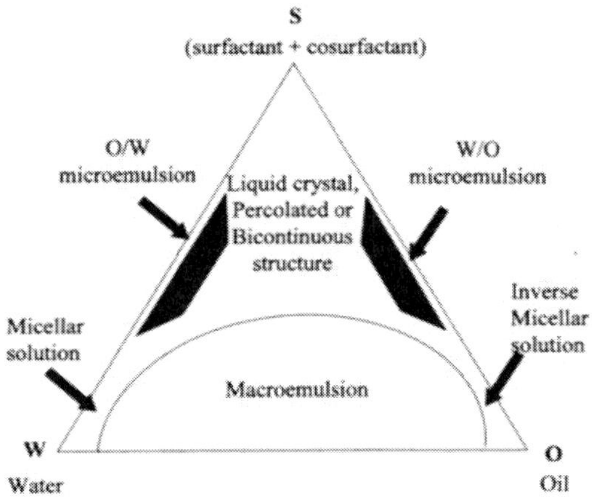

Figure 3. Hypotetical phase regions of microemulsion systems of oil (O), water (W), and surfactant+cosurfactant (S) (Bagwe et al., 2001).

A nonlinear mathematical approach, such as artificial neural networks, represents a novel strategy for *in silico* development of formulation procedures of the pharmaceutically acceptable microemulsion systems (Agatonovic-Kustrin and Alany, 2001; Agatonovic-Kustrin et al., 2003; Alany et al., 1999; Djekic et al., 2008; Mendyk and Jachowicz, 2007; Richardson et al., 1997). In the pioneering studies in this particular area the different types of ANN models were introduced for accurate differentiation and prediction of microemulsion area from the qualitative and quantitative composition of the microemulsion-forming system (Agatonovic-Kustrin et al., 2003; Alany et al., 1999, Djekic et al., 2008); ANNs were also proposed for prediction of the most favourable physico-chemical properties of the cosurfactant (Richardson et al., 1996; Richardson et al., 1997) or surfactant/cosurfactant (Agatonovic-Kustrin and Alany, 2001) molecules regarding the formation and/or drug delivery potential of microemulsions; ANN modeling was demonstrated to be effective in minimizing the experimental efforts in characterization of complex structural features of microemulsions (Podlogar et al., 2008).

4. Application of ANNs in the Development of Microemulsion Drug Delivery Systems

Understanding the mixed surfactants behavior in the presence of water and oil represents an important issue for development and optimal design of mixed surfactants based microemulsions. On the other hand, the preparation of microemulsions with the low surfactant concentrations and infinitely dilutable with water represents an important practical and theoretical interest (Fanun, 2009). ANNs provided a useful tool for the characterisation of phase behaviour of four-component microemulsion forming systems (Alany et al., 1999). Novel investigation of the phase behaviour of more complex mixtures of surfactants, cosurfactants, oil and water, by application of ANN modeling, was reported by Djekic et al., 2008. In such cases, indicators of a surfactant phase behaviour and suitability such as hydrophyle–lipophile balance (HLB) (Griffin, 1949) or critical packing parameter (CPP)

(Israelachvilli et al., 1976) are empirical and most widely used for surfactant selection. It is important to note that compositional variables (oil, presence of other amphiphiles, hydrophilic molecules (i.e. glycerol, sorbitol) or electrolytes) as well as temperature may have an influence on hydrophilic and hydrophobic properties and the geometry of the surfactant molecule and the efficiency of a surfactant to generate microemulsion (Kahlweit, 1999; Lawrence and Rees, 2000; Sjöblom et al., 1996). Additional aspect associated with the rational selection of amphiphiles is based on the fact that most commercially available surfactants are mixtures of homologous substances with different lipophilic chain length and with different degree of polymerization in hydrophilic part of a molecule. For this reason the relationship between physico-chemical characteristics of nonionic surfactants and their phase behaviour in ternary (oil/water/surfactant), pseudo-ternary (oil/water/surfactant/cosurfactant) or even more complex systems, such as microemulsion-based drug delivery systems, is still unclear.

The search for more appropriate compounds for formulation of pharmaceutically acceptable microemulsions may be facilitated by means of ANN strategy. For example, in numerous research to date the cosurfactants investigated have been entirely unsuitable for pharmaceutical formulations, and the search for more appropriate compounds has been hindered by the lack of any reliable means to predict those types of molecules that would be suitable and those that would not. The reported studies (Richardson et al., 1997, Agatonovic-Kustrin and Alany, 2001) demonstrated the predictive power of ANN strategy in assessment of the relevance of the particular physico-chemical properties of the cosurfactant molecules for the formation of microemulsions.

4.1. Prediction of Phase Behaviour

The paper of Alany et al., (Alany et al., 1999) reports the use of ANNs with back-propagation training algorithm to minimize experimental effort in characterization of phase behaviour of four component system consisting of ethyl oleate (Crodamol EO) (oil), sorbitan mono laurate (Crill 1, HLB=8.6) (primary surfactant), polyoxyethylene 20 sorbitan monooleate (Crillet 4 super, HLB=15) (secondary surfactant) and deionised water. Artificial neural networks training and testing data were extracted from several pseudo-ternary triangles which represented the cuts though the phase tetrahedron. Around 15% of the tetrahedron space was sampled: 8 samples within a range of HLB values at fixed oil-to-surfactant mass ratio (1:1) and 40 samples from the phase triangle at HLB 12.4 (at a fixed mass ratio of the two surfactants of 4:6) were used as training data providing 128 inputoutput pairs; a further 15 samples were randomly selected from the cuts as testing data. The inputs were percentage of oil, percentage of water, and HLB of the surfactant blend, and the outputs were the corresponding regions (oil-in-water emulsion (o/w EM), water-in-oil emulsion (w/o EM), microemulsion (ME), and liquid crystals (LC)). The regions were differentiated by applying phase contrast microscopy, polarized light microscopy and electrical conductivity measurements. The percentage occupied by each region was determined by a cut and weigh method (Kale and Allen, 1989). The calculations were performed by using MS-Windows based ANNs simulator software (NNMODEL Version 1.404, Neural Fusion). The ANN was trained using a different number of hidden neurons (5–25) and training cycles (0–6000). The number and size of the weights for neuron interconnections were optimized and the lowest error was obtained with 15 hidden neurons and after 4500 training cycles. The generalization ability of the optimal ANNs model were evaluated by using additional 45 sets of data selected from each of the four pseudo-ternary phase diagrams at HLB values 9, 11.5, 13, and 14.7 (Figure 4).

The trained ANN was tested on validation data and an accuracy of 85.2–92.9% was estimated, depending on the output critical values used for the classification (0, ±0.25, ±0.50, and ±0.75). Analyzing 180 validation points yielded an average of 90.5% correct answers, 3.4% unclassified and only 6.1% incorrect predictions. Narrowing the criterion of classification had little influence on the number of the wrongly classified data but increased the percentage of unclassified data (Figure 5). Although the influence of the sampling approach (i.e. fraction of the tetrahedron space sampled and the distribution of samples) and the critical values used for classification was not elucidated, the low error rate demonstrated the success in employment of ANNs in predicting the phase behavior of quaternary systems and subsequent reducing experimental effort.

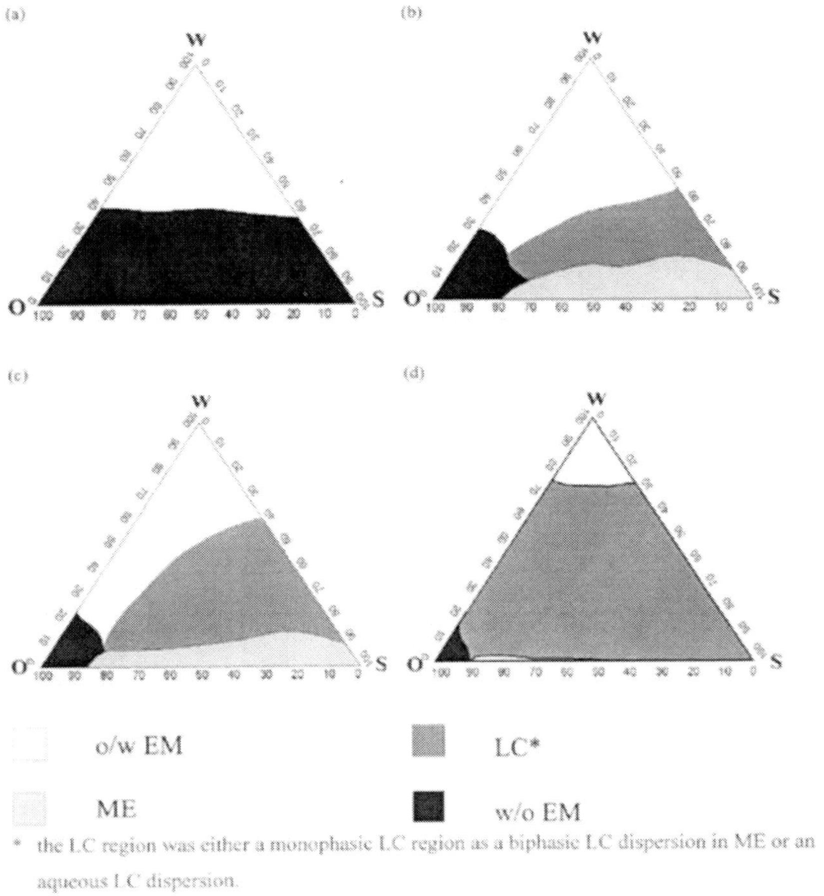

Figure 4. Predicted phase triangles at HLB (a) 9.0; (b) 11.5; (c) 13.0; (d) 14.7 (Alany et al., 1999).

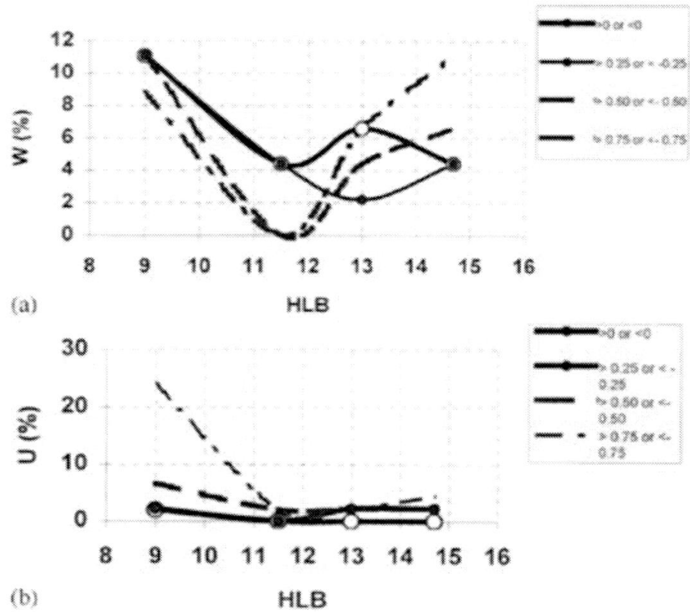

Figure 5. Accuracy of ANNs predictions for various critical values, (a) percentage wrong predictions at different HLB values; (b) percentage unclassified predictions at different HLB values (Alany et al., 1999).

ANN data modeling was applied successfully in the phase behaviour study during the development of a colloidal dosage form for the combined oral delivery of rifampicin and isoniazid (Agatonovic-Kustrin et al., 2003). The components of the investigated systems were: caprylic/capric acid triglycerides (Miglyol® 812) (oil), polyoxyl 40 hydrogenated castor oil (Cremophor® RH40), sorbitol, glyceryl monostearate (Imwitor® 308) and glycerol mono- and dicaprylate/caprate (Imwitor® 742), polyoxyethylene (10) oleyl ether (Brij® 97), and polyoxyethylene(20)sorbitane monostearate (Crillet® 3), double-distilled deionized water. The focus of this work was to identify surfactant combinations that would produce a thermodynamically stable microemulsion formulation suitable for the incorporation of the drugs (rifampicin and isoniazid) with different solubility in water (1.82 mg/ml and 128.8 mg/ml, respectively) and potential for mutual chemical reaction. Data from the 20 pseudoternary phase triangles (Figure 6) prepared by titration method at surfactant/cosurfactant (i.e. Imwitor® 308/Crille®t 3, Imwitor® 742/Crillet® 3, Cremophor® RH40/sorbitol or Brij® 97/sorbitol) ratios of 9:1, 7:3, 5:5, 3:7, and 1:9, were used to train, test, and validate the ANN model. The phases formed at increasing water fraction (in 2.5% w/w portions) were visually assessed and the observed phases were classified as isotropic (microemulsions and mixed micellar solution regions) (labeled as ME), liquid crystalline (LC), or coarse emulsion (EM). The HLB number, percentage of oil, water, and surfactants or surfactant/cosurfactant blend (the inputs for the ANN) were matched with the three different outputs (ME, LC, and EM), providing 4680 input-output data sets for the ANN. MS-Windows–based ANNs simulator software, Statistica Neural Networks 0.0F (StatSoft Inc., Tulsa, OK, USA) was used to develop a predictive model. The most successful model in the prediction of microemulsion region as well as the coarse emulsion was the radial basis function network (RBF) (Moody and C. J. Darkin, 1989) with a hidden layer of 100 neurons, thus forming 4-100-3 network architectures. However, it failed to predict LC phase. Furthermore, the composition of the final o/w microemulsion was defined (water (21.06%), Miglyol® 812 (23.68%), Imwitor® 308 (27.63%), and Crillet® 3 (27.63%)) including rifampicin (150 mg/ml) and isoniazid (100 mg/ml). The incorporation of rifampicin into the internal phase of the microemulsion vehicle improved the drugs stability from oxidative degradation and decreased its contact with isoniazid and drug–drug interaction. The formulation maintained homogeneity and integrity on excess dilution with water. The development of the ANN models in this study focuses on the evaluation and optimization of surfactant/cosurfactants for stabilization of colloidal formulations using a reduced experimental effort. Furthermore, the more general achievement of this study is the demonstration of the potential of ANNs methodology in better understanding of the process of microemulsion formation and stability within ternary and pseudoternary diagrams.

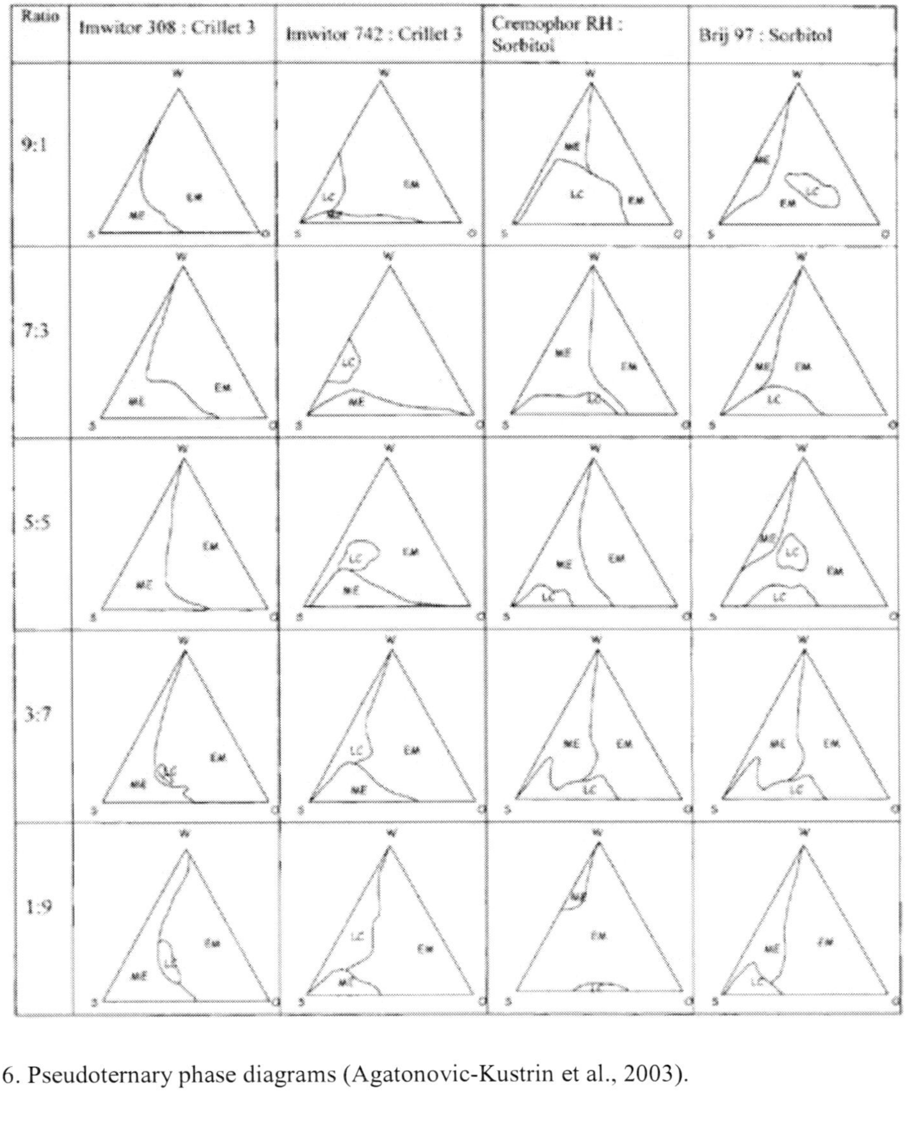

Figure 6. Pseudoternary phase diagrams (Agatonovic-Kustrin et al., 2003).

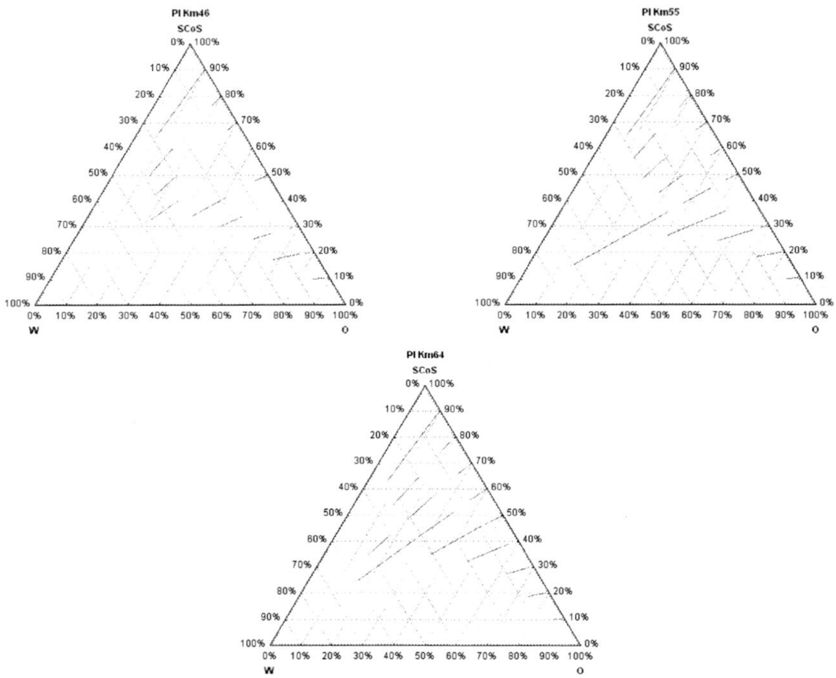

b)

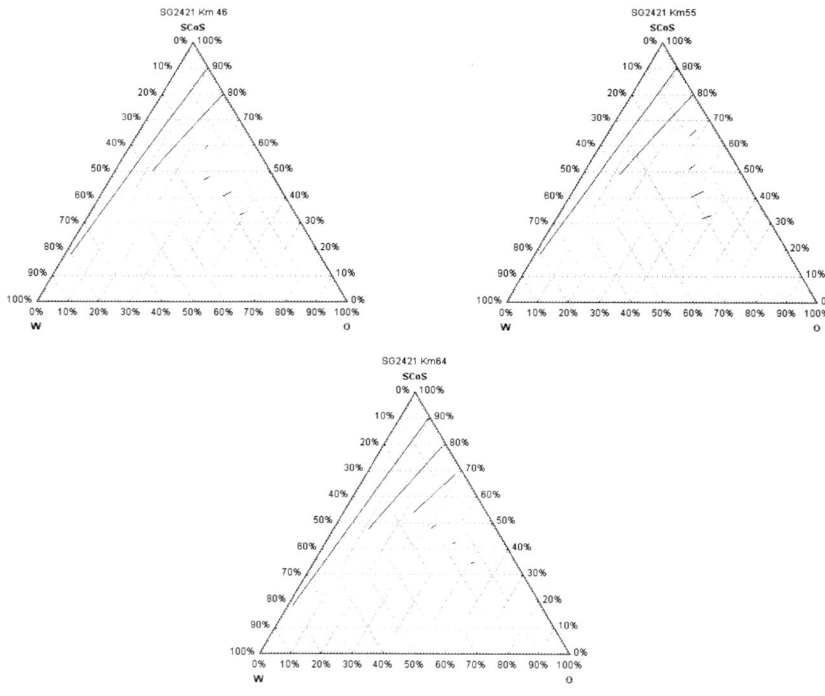

Figure 7. (Continued).

c)

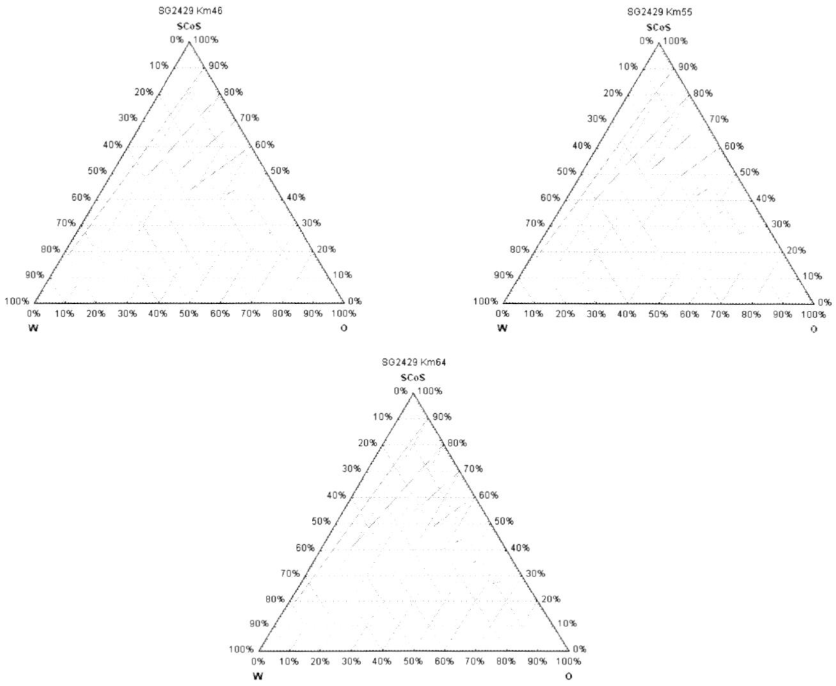

d)

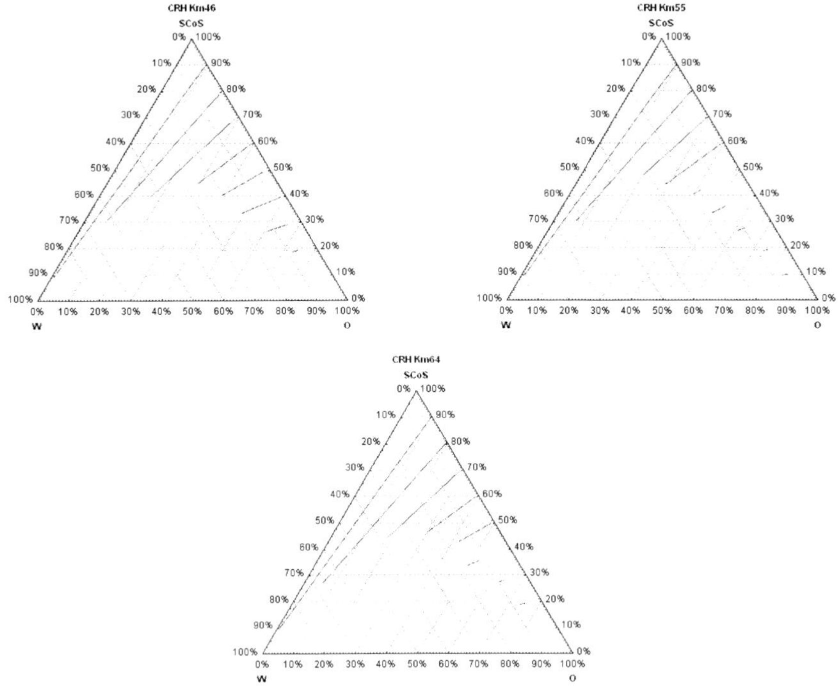

Figure 7. Microemulsion area in the pseudo-ternary phase diagram of the system Labrasol®/cosurfactant/isopropyl myristate/water at Km 4:6, 5:5 and 6:4 and at O/SCoS ratio varying from 1:9 to 9:1 using as a cosurfactant: a) polyglyceryl-6 isostearate, b) PEG-40 hydrogenated castor oil, c) Solubilisant gamma® 2421 or d) Solubilisant gamma® 2429.

A recent study has demonstrated the simplified experimental approach for investigation of phase behaviour of quaternary systems PEG-8 caprylic/capric glycerides (Labrasol®)/cosurfactant/isopropyl myristate/water using the titration method for the constuction of pseudo-ternary phase diagrams and, additionally, developing ANN model to understand the effect of formulation and compositional variables on the size and the position of microemulsion region (Djekic et al., 2008). The main goal in this study was rapid screening of microemulsion area in the system applying a reduced number of experiments. There has been investigated the effect of the cosurfactant type, the relative content of the cosurfactant (expressed as a surfactant-to-cosurfactant mass ratio (Km)) and the oil phase content (expressed as an oil-to-surfactant/cosurfactant mixture mass ratio (O/SCoS)) on the water solubilisation capacity (W_{max}, %, w/w). Pseudo-ternary phase diagrams at constant Km values (Km 4:6, Km 5:5 and Km 6:4) were constructed using titration method (Djekic and Primorac, 2008) at room temperature. The applied titration method was useful to diminish the effort to collect data requested for ANN determination, in contrast to alternative construction of phase diagrams by preparation of individual tensides/oil/water mixtures where the determination of all combinations of components which produce microemulsions is time consuming and requires huge number of individual experiments. The microemulsion domains were determined by titrating the isopropyl myristate/Labrasol®/cosurfactant mixtures with water, to the water solubilization limit (W_{max}, %, w/w), which was detected as the transition from the isotropic single phase system to a two phase system (sample became turbid), upon addition of small amount of excess of water. The microemulsion phase area along the titration lines is mapped onto pseudo–ternary phase diagrams (Figure 7) (previously unpublished pseudo-ternary diagrams)). The initial investigations were conducted on the quaternary systems employing polyglyceryl-6 isostearate or PEG-40 hydrogenated castor oil as a cosurfactant (Djekic et al., 2008). Furthermore, the investigations were expand using the novel commercial mixtures of nonionic tensides Solubilisant gamma® 2421 ((Octoxynol-12 (and) Polysorbate 20)) and Solubilisant gamma® 2429 (Octoxynol-12 (and) Polysorbate 20 (and) PEG-40 Hydrogenated castor oil) as a cosurfactants. The construction of the diagrams in Figure 7 a–d was based on a set of data from 27 independent titrations of oil/tensides mixtures with water (9 titrations at three Km values) for each of the four cosurfactants. These experiments were used to generate the inputs and output for artificial neural networks training. The inputs were Km values expressed as the surfactant concentration in surfactant/cosurfactant mixture (S, %, w/w) and O/SCoS values expressed as the oil concentration in the mixture with tensides (O, %, w/w). The output was the water solubilization limit (W_{max}, %, w/w), which represents a microemulsion systems boundary for a given quaternary mixture.

Commercially available Statistica Neural Networks (StatSoft, Inc., Tulsa, OK, USA) was used throughout the study. A Generalized Regression Neural Network (GRNN), a feed-forward network comprised of four layers, was used for modeling and optimization of boundary of the microemulsion region. The main advantage of GRNNs is that they involve a single-pass learning algorithm and are therefore much faster to train than the well-known back-propagation paradigm (Specht, 1990). Futhermore, they differ from classic neural networks in that every weight is replaced by a distribution of weights. This enables a large number of combinations of weights to be explored, and the exploration is less likely to end in a local minimum (Bruneau, 2001). Therefore, no test and verification sets are necessary and, in principle, all available data can be used for the network training. In a GRNN model, it is possible to select the number of units (nodes) in the second radial layer, the smoothing factor (which controls the deviation of the Gaussian kernel function located at the radial centres), and the clustering algorithm (e.g. subsampling, *K*-means or Kohonen). To select the optimal GRNN model, the observed versus predicted responses were shown in the regression plots drawn for the test samples, which were excluded from training data set. The GRNN model that yielded a regression plot with a slope and squared correlation coefficient (*r*2) that was closest to 1.0 was selected as the optimal GRNN model. A sum-squared error function was used in the network training. Learned GRNN was used for modelling, simulation and optimization of the microemulsion boundary region in the following ways: testing experimental points in experimental fields; searching for the optimal solutions; presenting response surfaces (or contour plots). Several training sessions were conducted using different numbers of units in the hidden layer in order to determine the optimal GRNN structure. The learning period was completed when the minimum value of the root mean square (RMS) was reached:

$$\text{RMS} = \left[\frac{\sum (y_i^p - y_i^m)^2}{n} \right]^{1/2}$$

a)

where y_i^p is the experimental (observed) response and y_i^m is the calculated (predicted) response and n is the number of experiments. The selected ANN structure had four layers: the first layer had two input units, the second layer had 27 hidden units (with negative exponential activation and radial postsynaptic function), the third layer had two units, and the fourth layer had one output unit (Figure 8). Twenty-seven units in a hidden layer were needed to obtain an excellent prediction of the response variable. Input values for test data were presented to the GRNN when network training was completed. RMS reached after the training was 0.9%, which is an acceptable value.

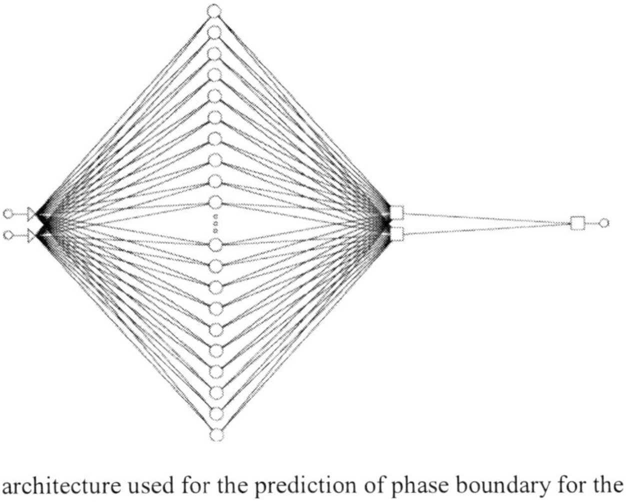

Figure 8. The GRNN architecture used for the prediction of phase boundary for the investigated Labrasol® - based microemulsions.

Figure 9 presents response surface generated by ANN presenting influence of the surfactant concentration in surfactant/cosurfactant mixture (*S*, %, w/w) and oil concentration in the mixture with tensides (*O*, %, w/ w) on the water solubilization limit (*W*max, %, w/w): a) polyglyceryl-6 isostearate, b) PEG-40 hydrogenated castor oil, c) Solubilisant gamma® 2421 or d) Solubilisant gamma® 2429.

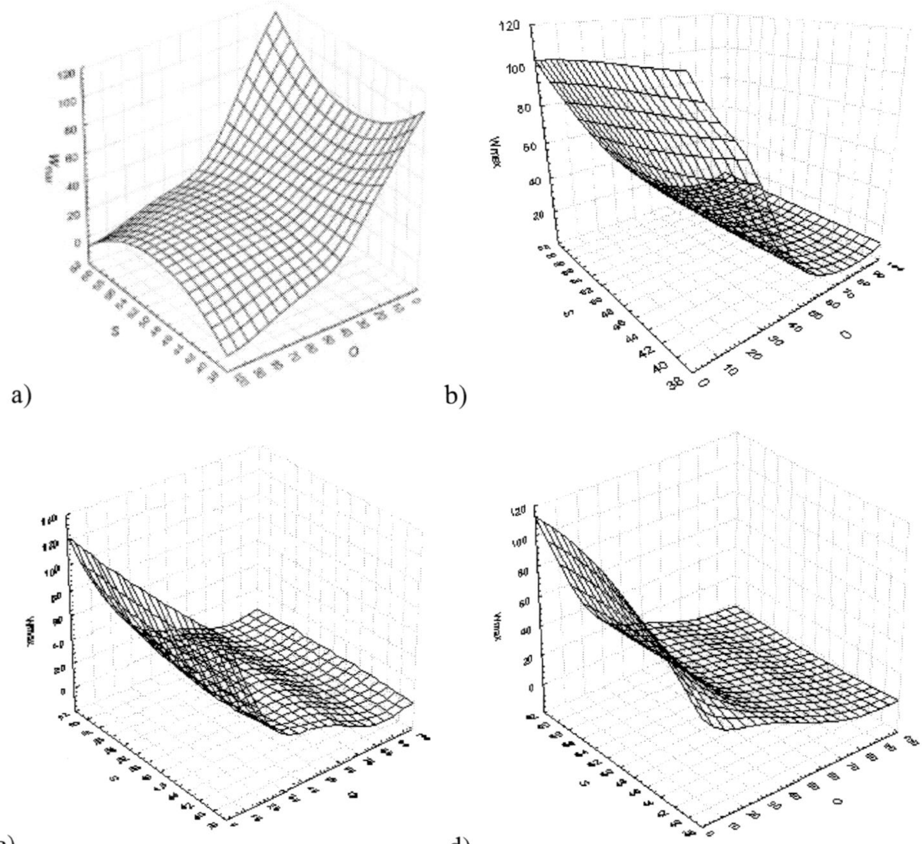

Figure 9. Response surface presenting the influence of the surfactant concentration in surfactant/cosurfactant mixture (*S*, %, w/w) and the oil concentration in the mixture with tensides (*O*, %, w/w) on the water solubilization limit (*W*max, %, w/w) which corresponds to a microemulsion region boundary for Labrasol®/cosurfactant/isopropyl myristate/water system: a) polyglyceryl-6 isostearate, b) PEG-40 hydrogenated castor oil, c) Solubilisant gamma® 2421 or d) Solubilisant gamma® 2429.

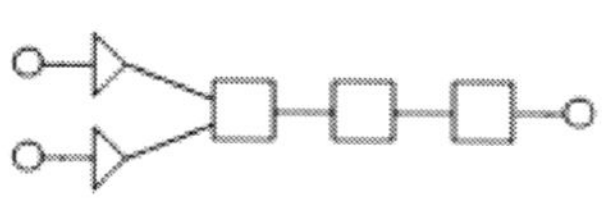

Figure 10. The MLP architecture used for the predicition of the phase boundary for the mixtures consisting of Labrasol®/ PEG-40 hydrogenated castor oil /isopropyl myristate/water.

In this study have been successfully estimated combined influences of Km and O/SCoS within a predictive mathematical model which gives accurate predictions of microemulsion formation in Labrasol®/cosurfactant/isopropyl myristate/water. The GRNN model provided deeper understanding and predicting of water solubilization limit for any combination of surfactant concentration and oil concentration in their mixture, within the investigated range.

4.1.1. The influence of ANNs type/architecture

Appropriate selection of network architecture is the milestone in utilization of ANNs. When PEG-40 hydrogenated castor oil was used as cosurfactant in sample mixtures, data from 27 independent titrations of oil/tensides mixtures with water were presented to the software. Multi-layer perceptron (MLP) network with four layers was generated (Figure 10).

Training of such network was conducted through 800.000 epochs, using learning rate 0.6 and momentum rate 0.3. Cross verification was used during training. Values of RMS error were: 0.08 for training data set, 0.07 for validation data set and 0.08 for test data set.

Experimentally observed and MLP predicted results for the test mixtures are presented in Table 1.

In the case of systems with Solubilisant gamma® 2421 and Solubilisant gamma® 2429 as cosurfactants, application of MLP as well as GRNN networks didn't get satisfied results. Therefore, third type of network was applied – Radial basis function network (RBF) (Figure 11).

RBFs have an input layer, a hidden layer of radial units and an output layer of linear units. Introduced by Broomhead and Lowe (1988) and Moody and Darkin (1989), they are described in most neural network text books (e.g. Bishop, 1995; Haykin, 1994). The radial layer has exponential activation functions; the output layer linear activation functions. RBF networks are trained in three stages:

Table 1. Experimental and predicted values of W_{max} for the test mixtures

Cosurfactant	S (%, w/w)	O (%, w/w)	W (%, w/w)	
			Experimental values	Predicted values
Polyglyceryl-6 isostearate	45.00	35.00	30.26	33.24
	43.00	65.00	30.36	32.25
	55.00	25.00	32.98	34.47
	48.00	75.00	13.04	10.10
	45.00	40.00	31.32	32.55
	43.00	40.00	34.21	34.47
	55.00	35.00	25.93	26.78
Solubilisant gamma® 2421	55.00	25.00	23.88	26.03
	55.00	35.00	15.25	15.57
PEG-40 hydrogenated castor oil	55.00	25.00	45.95	51.33
	55.00	35.00	25.93	26.78
Solubilisant gamma® 2429	45.00	35.00	20.00	23.37
	55.00	25.00	25.92	30.36
	55.00	35.00	15.25	13.02

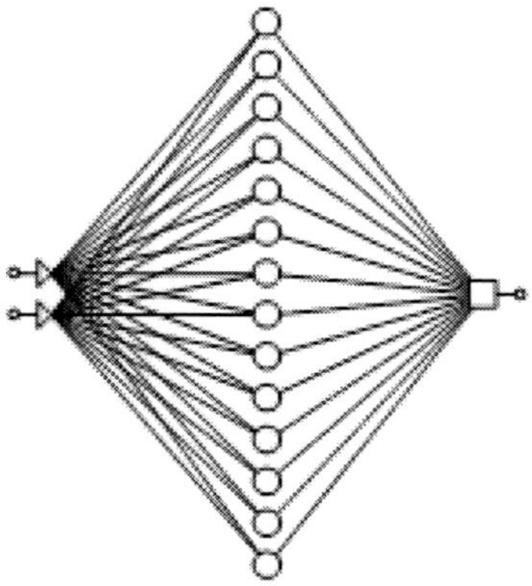

Figure 11. Radial Basis Funcion Network architecture used for the predicition of the phase boundary for the mixtures consisting of Labrasol®/ Solubilisant gamma® 2421 /isopropyl myristate/water as well as for Labrasol®/ Solubilisant gamma® 2429 /isopropylmyristate/water mixtures.

Center-assignment - the centers stored in the radial hidden layer are optimized first, using unsupervised training techniques. Centers can be assigned by a number of algorithms; by sampling, K-means or Kohonen training. These algorithms place centers to reflect clustering;

Deviation assignmen - the spread of the data is reflected in the radial deviations (stored in the threshold). Deviations are assigned by isotropic algorithm;

Linear optimization - Once centers have been assigned, the linear output layer is usually optimized using pseudo-inverse technique, minimizing the error.

RBF networks train relatively quickly and do not extrapolate too far from known data; however, they tend to be larger then MLPs and therefore execute more slowly. Using RBF network in the case of cosurfactant Solubilisant gamma® 242, values for experimentally observed and ANN predicted results for the test mixtures were very close, indicationg good prediction ability of the RBF network (Table 1).

The obtained results show that phase behaviour investigations based on titration method in combination with an optimized artificial neural network can provide useful tools which may limit the experimental effort for the formulation of pharmaceutically acceptable microemulsion vehicles.

4.2. Screening of the Microemulsion Constituents

The critical step in the formulation development is to select the proper pharmaceutically applicable components, which are able to form microemulsions. Richardson et al. (Richardson et al., 1996; Richardson et al., 1997) reported the use of ANNs employing back propagation feed-forward algorithm of learning to predict the pseudo-ternary phase diagrams for four component pharmaceutically acceptable microemulsion systems using only four computed physicochemical properties for the cosurfactants employed as well as to determine their most favourable values regarding the formation of microemulsion area. The components of the microemulsion-forming systems were lecithin (Epikuron® 200), isopropyl myristate, triple distilled water, and different types of cosurfactants including short- and medium-chain alcohols, amines, acids and ethylene glycol monoalkyl ethers. The data required for ANN training and testing were extracted from the pseudo-ternary diagrams presented by Aboofazeli et al. (Aboofazeli et al., 1994) together with the additional data from four pseudo-ternary phase diagrams constructed according to the methods described by Aboofazeli and Lawrence (Aboofazeli and Lawrence, 1993), at a fixed weight ratio of surfactant- to-cosurfactant 1:1, using the cosurfactants 2-methyl-2-butanol, 2-methyl-2-propanol, 2-methyl-1-butanol and isopropanol. In the preparation of the ANN input data, each phase diagram was overlied with a hexagonal lattice with a grid spacing of 5% (w/w) along each of the component axes. This provided a set of 171 input-target pairs labeled according to the proportions of surfactant and cosurfactant (s) and oil (o) in the mixture (input) matched with the nature of the phase structure found for this composition (ϕ) (target) (any type of microemulsion formed (i.e. L1, L2, or bicontinuous phases) or any other phase structure). The input data were then supplemented using four parameters summarizing the key properties of the different cosurfactants used in the experimental systems. These included the cosurfactant molecular volume (v), areas for its head group (a_ψ) and hydrophobe (a_ϕ), and computed octanol/water logP value. The construction and training of the ANN were carried out using the in-house software YANNI (Richardson and Barlow, 1996). A simple feed-forward back-propagation network was used with the final architecture involving 6 input neurons, a single hidden layer of 14 neurons, and 1 output neuron, as shown in Figure 12. Training was carried out in a random, on-line manner. To improve the speed and quality of learning, the time-invariant noise algorithm (TINA) of Burton and Mpitsos (Burton and Mpitsos, 1992) was employed. The trained ANNs were shown to be highly successful in predicting phase behavior for the investigetd systems given only the computed values of v, a_ψ, a_ϕ, and logP for the relvant cosurfactants, achieving mean success rates of 96.7 and 91.6% for training and test data, respectively. The established ANN can be only used to predict the phase diagrams for the investigated four-component systems at Km 1:1 (Figures 13-14), however, the obtained results give an idea of a more general network, trained with data on systems involving other oils, other surfactants, and other Km values. Within the study was assesed the potential of ANNs for evaluation of novel cosurfactants for lecithin suitable for formulation of pharmaceutically acceptable microemulsion systems. From the phase diagrams predicited for systems involving a series of fictive cosurfactants whose properties lay within the vector space of the training set cosurfactant properties, was observed that the microemulsion area increases with decreasing head group area (a_ψ), increasing hydrophobe area (a_ϕ), increasing molecular volume (v), and decreasing logP. Therefore, the most useful combiantion of cosurfactant properties would be small head group area, high molecular volume, large hydrophobe area and low logP, so that the ideal cosurfactant have log P just less than zero, a_ψ, ≤ 40 Å^2 (a small, uncharged head group such as a diol or possibly a triol moiety), a_ϕ of the order of 120 Å^2 (a hydrophobe of about hexyl size preferably with a branched-chain structure), and v around 300 Å^3. Although the properties of the cosurfactants molecule preffered for microemulsion stabilization were already recognized, the novelty brought by this study was the demonstration of the significant potential of the trained ANN to screen out cosurfactants considering simultaneously all of the features of their molecules relevant for the formation of pharmaceutically acceptable drug delivery systems.

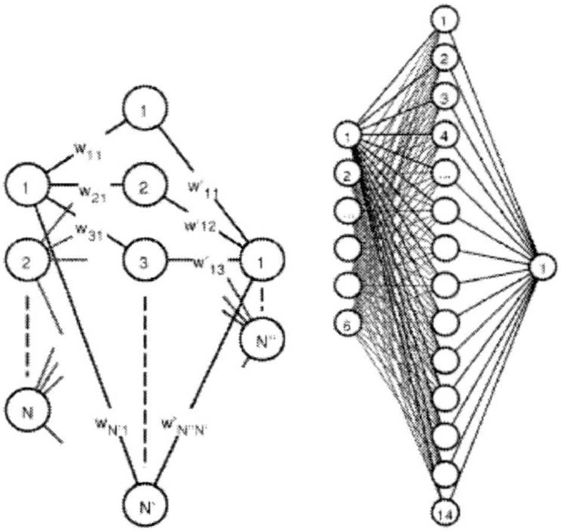

Figure 12. Structure of feed-forward back-propagation networks (Richardson et al., 1997).

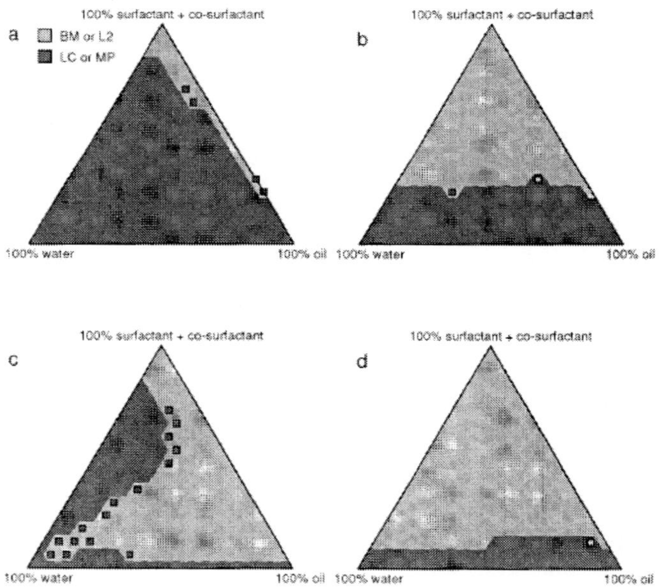

Figure 13. Network-predicted phase diagrams for four surfactant/cosurfactant systems taken from the training set: (a) *n*-pentanoic acid; (b) 1,2-hexanediol; (c) diethylene glycol monopentyl ether; (d) 3-aminopentane. Light gray areas represent microemulsion (i.e., L1, L2, or bicontinuous) regions; dark areas represent nonmicroemulsion (i.e., LC or multiphase) regions. Squares indicate prediction errors (Richardson et al., 1997).

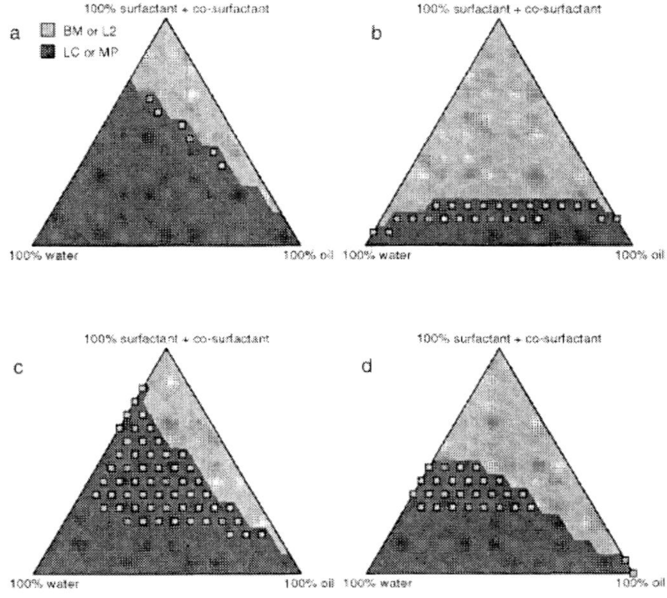

Figure 14. Network-predicted phase diagrams for four surfactant/cosurfactant systems taken from the test set: (a) *n*-pentanol; (b) 2-aminopentane; (c) diethylene glycol monobutyl ether; (d) 2-methyl-2-butanol. Light gray areas represent microemulsion (i.e., L1, L2, or bicontinuous) regions; dark areas represent nonmicroemulsion (i.e., LC or multiphase) regions. Squares indicate prediction errors (Richardson et al., 1997).

In a related study of Agatonovic-Kustrin and Alany (Agatonovic-Kustrin and Alany, 2001) a genetic neural network (GNN) model was developed to predict the phase behavior of five component systems (ethyl oleate/a mixture of sorbitan monolaurate (Crill 1, HLB = 8.6), and polyoxyethylene 20 sorbitan monooleate (Crillet 4 super, HLB = 15)/ deionised water/a cosurfactant) evaluating the influence of the cosurfactant nature (n-alcohols (1-propanol, 1-butanol, 1-hexanol, and 1-octanol) and 1,2-alkanediols (1,2-propandiol, 1,2-pentanediol, 1,2-hexanediol, and 1,2-octanediol). A nonlinear ANN model was used to correlate phase behavior of the investigated systems with cosurfactant descriptors that were preselected by a genetic algorithm (GA) input selection. A supervised network with a back-propagation learning rule and multilayer perceptron (MLP) architecture was used (8–10). In this model the inputs are fully connected to the hidden layer, and hidden layer neurons are fully connected to the outputs. The presence of a hidden layer is a crucial feature that allows the network to make generalizations from the training data. Phase behavior of microemulsion (ME), lamellar liquid crystal (LC), and coarse emulsion forming systems (w/o EM and o/w EM) was detected by carried out phase contrast and polarized light microscopy. MS-Windows®-based artificial neural network simulator software Neural Networks™ (StatSoft Inc, Tulsa, OK, USA) was used through the study. For calculating drug properties from molecular structure, Pallas 2.1 (Compu Drug Int., San Francisco, CA) and ChemSketch 3.5 freeware (ACD Inc., Toronto, Canada) were used. Eight pseudoternary phase triangles were constructed and used for training, testing, and validation purposes. A total of 21 molecular descriptors were calculated for each cosurfactant. A total of 18 descriptors including chemical composition descriptors and calculated physicochemical descriptors for each of the cosurfactants was used for the initial ANN model, using GA for selection and following a unit penalty factor of 0.000–0.004, the number of inputs was reduced from 18 to 9. A genetic algorithm was used to select important molecular descriptors, and a supervised artificial neural network with two hidden layers was used to correlate selected descriptors and the weight ratio of components in the system with the observed phase behavior. The results proved the dominant role of the chemical composition (%C, %H, and %O), HLB, number of carbon atoms, length of hydrocarbon chain, molecular volume, and hydrocarbon volume of the cosurfactant in prediction. of cosurfactant. Input selection has reduced the size and complexity of the network and focused the training on the most important data. The training and testing data set consisted of the original data from phase diagrams containing 1-butanol, 1-hexanol, 1,2-propanediol, and 1,2-hexanediol as cosurfactants (Figure 15). The total number of data points consisted of 684 input/output sets and was split randomly into 548 training sets and 136 test sets. The results of the five runs were averaged. The training set was used to train the network, and the testing set was used to determine the level of generalization produced by the training set and to monitor overtraining the network, each with corresponding root mean squared (RMS) error. The best GNN model, with 14 inputs and two hidden layers with 14 and 9 neurons, predicted the phase behavior for a new set of cosurfactants with 82.2% accuracy for ME, 87.5% for LC, 83.3% for the O/W EM, and 91.5% for the W/O EM region. The results suggest that a small number of chemically meaningful descriptors will provide the most predictive model.

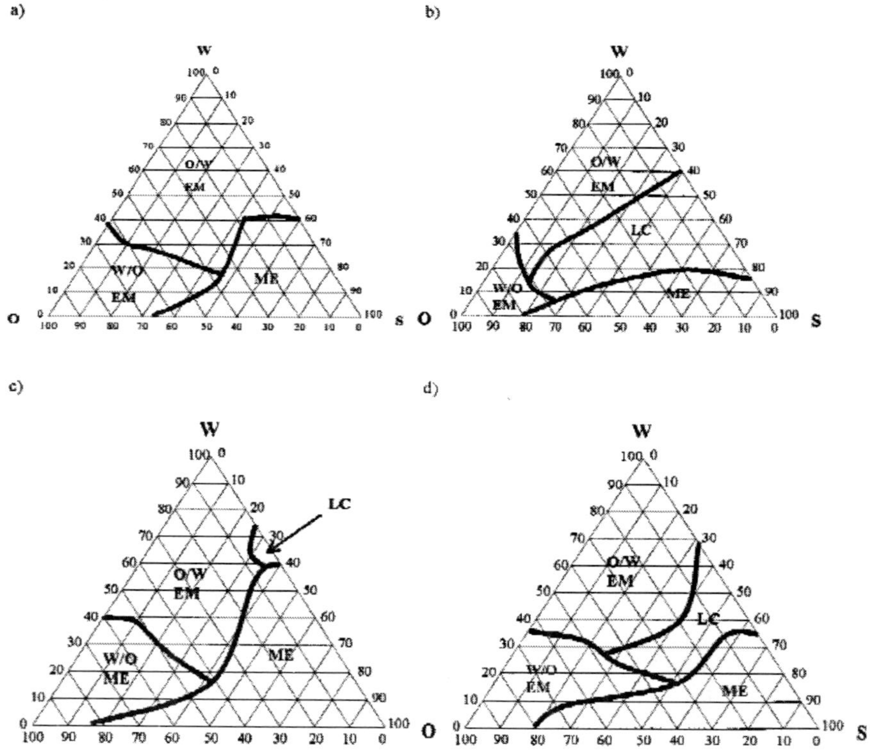

Figure 15. GNN-predicted phase triangles for four surfactant/cosurfactant systems from the validation set with (a) n-propanol, (b) n-octanol, (c) 1,2-pentandiol, and (d) 1,2-octandiol as cosurfactants (Agatonovic-Kustrin and Alany, 2001).

4.3. Prediction of Structural Features of Microemulsions

In a very recent study of Podlogar et al. (Podlogar et al., 2008) two evolutionary ANNs (Yao, 1991) has been constructed by introducing a a genetic algorithm to the feed forward ANN, one being able to predict the type of microemulsion from its composition and the second to predict the type of microemulsion from the differential scanning calorimetry (DSC) curve. The components of the quaternary microemulsion system were pharmaceutically acceptable excipients: isopropyl myristate (lipophilic phase), polyoxyethylene (20) sorbitan monopalmitate (Tween® 40) (surfactant), glyceryl caprylate (Imwitor® 308) (cosurfactant), and twice distilled water (hydrophilic phase). The type of microemulsion microstructure (i.e. O/W microemulsion, bicontinuous microemulsion, W/O microemulsion) was differentiated by measuring the freezing peak of the water in DSC thermograms. Data pool used to train both ANNs included the composition of the 170 microemulsion samples selected from three pseudo-ternary phase diagrams (constructed at Km 2:1, 1:1 and 1:2) and DSC curves. For determination of the type of microemulsion from its composition there was programmed a feed-forward network, with the final architecture involving 4 input neurons (corresponded to the weight % of the four components used to produce the microemulsion) a single hidden layer of 12 neurons, and 5 output neurons (each represented one possible structure – o/w microemulsion, bicontinuous microemulsion, w/o microemulsion, o/w emulsion or w/o emulsion). For the activation function, a sigmoid function ranging from 0 to 1 was used. A supervised form of learning, which was discontinued after the classification error dropped below 1 % was applied. For determination of the type of microemulsion from its DSC curve, a second feed-forward ANN with 1 hidden layer was constructed containing 100 input neurons (i.e. the input data of the DSC curve), a single layer of 5 hidden neurons and 5 output neurons, and trained using a genetic algorithm. A genetic algorithm (GA) was used to determine the weight (genes) values. Each weight is represented as a gene in the chromosome (solution). The initial population consisted of 50 different chromosomes where each represents a certain weight combination. The algorithms and the structures for each ANN were constructed and programmed in C++ computer language. When the ANNs were trained, the first network was instructed to predict the structures for all possible composition combinations at the surfactant to cosurfactant ratios of 1:1, 2:1, 1:2 and 1.5:1. another ANN. Also, additional microemulsion samples (previously not tested) were selected and analyzed by DSC and the results were compared with the ANN prediction. For the training set we used several DSC curves of the samples with surfactant to cosurfactant ratio of 1:1. After completing the network learning cycles, we selected several curves, not involved previously in the learning process of the ANN that related to several different types of microemulsion, in order to test the accuracy of network prediction. Both ANNs showed an accuracy of 90% in prediction of the type of microemulsion from the previously untested compositions. However, the described ANNs can be used only to predict accurately the construction of the phase diagram for four component microemulsions and within the range of the selected surfactant-to-cosurfactant ratios. Nevertheless, constructing this kind of ANN, combined with a genetic algorithm provide the tool for reducing research time and development cost for construction of pseudoternary diagrams that could facilitate the selection of the microemulsion composition as well as for characterization of the properties of the potential microemulsion drug carrier.

5. Conclusion

The overall findings raised from the efforts to apply ANN models in development of microemulsions provide the basis for reducing research time and cost for formulation and characterization of the properties of such complex colloidal systems. ANN strategy is helpful in predicting the phase behaviour of the four-component systems but also of the more complex systems. The upcoming step would be application of ANN methodology as a complementary tool in the characterization of microemulsion structure. It would then be possible to correlate data regarding type and structure to drug release or permeation and to predict biopharmaceutical properties of the systems with the desired drug depending on the composition. This will minimize the time and cost of microemulsion characterization and subsequently stimulate the further development of microemulsion carrier systems in order to exploit their advantages and avoid their disadvantages.

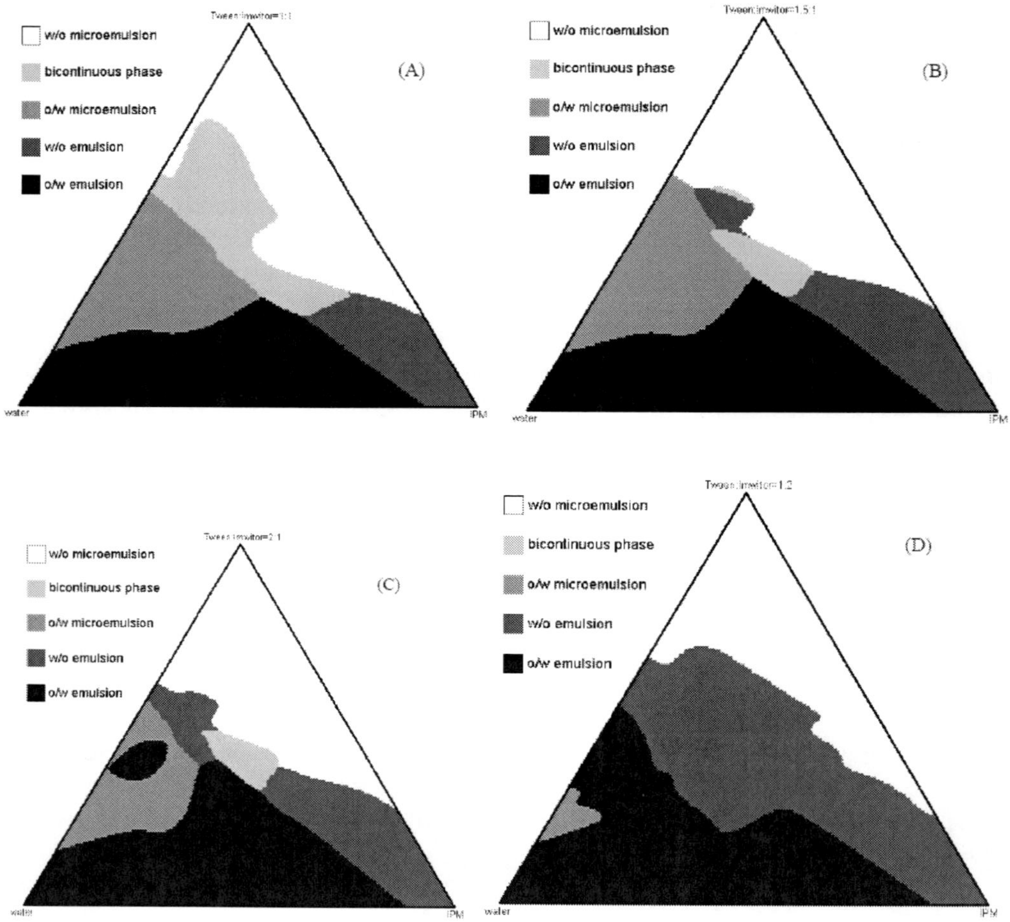

Figure 16. Solutions, predicted by ANN. The respective surfactant to cosurfactant ratios are: a) 1 : 1, b) 1,5 : 1, c) 2 : 1, d) 1 : 2, respectively. There were no training points on the diagram with 1,5 : 1 ratio (Podlogar et al., 2008).

Symbols and Terminologies

ANNs	Artificial Neural Networks
CPP	Critical packing parameter
DSC	Differential Scanning Calorimetry
GA	Genetic Algorithms
GNN	Genetic Neural Network
GRNN	Generalized Regression Neural Network
HLB	Hydrophyle–lipophile balance
Km	surfactant-to-cosurfactant mass ratio
L1	water-in-oil microemulsion
L2	oil-in-water microemulsion
LC	liquid crystals
ME	microemulsion
MLP	Multilayer Perceptron
NN	Neural Network
O/SCoS	oil-to-surfactant/cosurfactant mixture mass ratio
o/w EM	oil-in-water emulsion
RBF	Radial Basis Function Network
w/o EM	water in oil emulsion

References

[1] Aboofazeli, R., Lawrence, C. B., Wicks, S. R. & Lawrence, M. J. (1994). Investigations into the formation and characterization of phospholipid microemulsions. III. Pseudo-ternary phase diagrams of systems containing water-lecithin-isopropyl myristate and either an alkanoic acid, amine, alkanediol, polyethylene glycol alkyl ether or alcohol as cosurfactant. *Int. J. Pharm.*, *111*, 63-72.

[2] Aboofazeli, R. & Lawrence, M. J. (1993). Investigations into the formation and characterization of phospholipid microemulsions. I. Pseudo-ternary phase diagrams of systems containing water-lecithin-alcohol-isopropyl myristate. *Int. J. Pharm.*, *93*, 161-175.

[3] Agatonovic-Kustrin, S. & Alany, R. G. (2001). Role of genetic algorithms and artificial neural networks in predicting the phase behavior of colloidal delivery systems. *Pharm Res.*, *18*, 1049-1055.

[4] Agatonovic-Kustrin, S., Glass, B. D., Wisch, M. H. & Alany, R. G. (2003). Prediction of a Stable Microemulsion Formulation for the Oral Delivery of a Combination of Antitubercular Drugs Using ANN Methodology. *Pharm Res.*, *20(11)*, 1760-1765.

[5] Alany, R. G., Agatonovic-Kustrin, S., Rades, T. & Tucker, I. G. (1999). Use of artificial neural networks to predict quaternery phase systems from limited experimental data. *J. Pharm. Biomed. Anal.*, *19*, 443-452.

[6] Bagwe, R. P., Kanicky, J. R., Palla, B. J., Patanjali, P. K. & Shah, D. O. (2001). Improved drug delivery using microemulsions: rationale, recent progress, and new horizons. Crit. Rev. Ther. Drug Carrier Syst. 18, 77–140.

[7] Bishop, C.M. (1995) *Neural Networks for Pattern Recognition,* Oxford: Oxford University Press.

[8] [a] Bourquin, J., Schmidli, H., van Hoogevest P. & Leuenberger, H. (1998). Comparison of artificial neural networks (ANN) with classical modeling techniques using different experimental designs and data from a galenical study on a solid dosage form. Eur. *J. Pharm. Sci.*, 6, 287–300.

[9] [b] Bourquin, J., Schmidli, H., van Hoogevest P. & Leuenberger, H. (1998). Advantages of Artificial Neural Networks (ANNs) as alternative modeling technique for data sets showing non-linear relationships using data from a galenical study on a solid dosage form. *Eur. J. Pharm. Sci.*, 7, 5–16.

[10] [c] Bourquin, J., Schmidli, H., van Hoogevest P. & Leuenberger, H. (1998). Pitfalls of artificial neural networks (ANN) modeling technique for data sets containing outlier measurements using a study on mixture properties of a direct compressed dosage form. *Eur. J. Pharm. Sci.*, 7, 17–28.

[11] Broomhead, D. S. & Lowe, D. (1988). Multivariable functional interpolation and adaptive networks. *Complex Systems*, *2*, 321-355.

[12] Bruneau, P. (2001). Search for a predictive generic model of aqueous solubility using Bayesian neural nets. *J. Chem. Inf. Comput. Sci.*, *41*, 1605–1616.

[13] Burton, R. M. & Mpitsos, G. J. (1992). Event-dependent control of noise enhances learning in neural networks. *Neural Networks*, *5*, 627-637.

[14] Chen, Y., McCall, T. W., Baichwal A. R. & Meyer, M. C. (1999). The application of an artificial neural network and pharmacokinetic simulations in the design of controlled-release dosage forms. *J. Control. Release,* 59, 33–41.

[15] Djekic, L. & Primorac, M. (2008). The influence of cosurfactants and oils on the formation of pharmaceutical microemulsions based on PEG-8 caprylic/capric glycerides. *Int. J. Pharm.*, *352*, 231–239.

[16] Fanun, M. (2009). *Microemulsions Properties and Applications*, CRC Press, Taylor & Francis Group, Boca Raton, FL.

[17] Griffin, W. C. (1949). Classification of surface-active agents by HLB. *J. Soc. Cosmet. Chem.*, *1*, 311–326.

[18] Gupta, S. & Moulik, S. P. (2008). Biocompatible Microemulsions and Their Prospective Uses in Drug Delivery. *J. Pharm. Sci.*, *97*, 22-45.

[19] Haykin, S. (1994). Neural Networks: A Comprehensive Foundation. New York: Macmillan Publishing.

[20] Hussain, A. S., Yu, X. Q. & Johnson, R. D. (1991). Application of neural computing in pharmaceutical product development. *Pharm. Res.*, 8, 1248–1252.

[21] Ibric, S., Jovanovic, M., Djuric, Z., Parojcic J. & Solomun, L. (2002). The application of generalized regression neural network in the modeling and optimization of aspirin extended release tablets with Eudragit® RS PO as matrix substance. *J. Control. Release,* 82, 213–222.

[22] Israelachvilli, J. N., Mitchell, D. J. & Ninham, B. W. (1976). Theory of self assembly of hydrocarbon amphiphiles into micelles and bilayers. *J. Chem. Soc. Faraday Trans.*, II 72, 1525-1567.

[23] Kahlweit, M. (1999). *Microemulsions Ann. Rep. Prog. Chem. Sect.*, C 95, 89-115.

[24] Koza, J. R. (1992). *Genetic Programming – On the Programming of Computers by Means of Natural Selection.* MIT Press, Cambridge, Massachusetts, USA.

[25] Lawrence, M. J. & Rees, G. D. (2000). Microemulsion-based media as novel drug delivery systems, *Adv. Drug Deliv. Rev.*, *45*, 89-121.
[26] Malmstein, M. (1999). Microemulsion in pharmaceuticals. In: Kumar P, Mittal KL (Eds.) *Handbook of Microemulsion: Science and Technology* Marcel Dekker, New York, Basel, 755-772.
[27] Mendyk, A. & Jachowicz, R. (2007). Unified methodology of neural analysis in decision support systems built for pharmaceutical technology. *Expert Systems with Applications.*, *32*, 1124–1131.
[28] Moody J. & Darkin. C. J. (1989). Fast learning in networks of locally tuned processing units. *Neural Comput.*, *1(2)*, 281-294.
[29] Murtoniemi, E., Yliruusi, J., Kinnunen, P., Merkku P. & Leiviskä, K. (1994). The advantages by the use of neural networks in modeling the fluidized bed granulation process. *Int. J. Pharm.*, 108, 155-164.
[30] Kale, N. J. & Allen, L. V. (1989). Studies on microemulsions using Brij 96 as surfactant and glycerin, ethylene glycol and propylene glycol as cosurfactants. *Int. J. Pharm.*, *57*, 87-93.
[31] Peh, K. K., Lim, C. P., Quek, S. S. & Khoh, K. H. (2000). Use of artificial neural networks to predict drug dissolution profiles and evaluation of network performance using similarity factor. *Pharm. Res.*, 17, 1384–1388.
[32] Podlogar, F., Šibanc, R. & Gašperlin, M. (2008). Evolutionary Artificial Neural Networks as Tools for Predicting the Internal Structure of Microemulsions. *J. Pharm. Pharmaceut. Sci.*, (www. cspsCanada.org) *11(1)*, 67-76.
[33] Reis, M. A. A., Sinisterra, R. D. & Belchior, J. C. (2004). An alternative approach based on artificial neural networks to study controlled drug release, *J. Pharm. Sci.*, 93, 418-430.
[34] Richardson, C. J., Mbanefo, A., Aboofazeli, R., Lawrence, M. J. & Barlow, D. J. (1996). Neural network prediction of microemulsion phase behaviour. *Eur. J. Pharm. Sci.*, *4*, (S1), S139.
[35] Richardson, C. J., Mbanefo, A., Aboofazeli, R., Lawrence, M. J. & Barlow, D. J. (1997). Prediction of Phase Behavior in Microemulsion Systems Using Artificial Neural Networks. *J. Colloid Interface Sci.*, *187*, 296–303.
[36] Richardson, C. J. & Barlow, D. J. (1996). Neural network computer simulation of medical aerosols. *J. Pharm. Pharmacol.*, *48(6)*, 581-591.
[37] Rowe, R. C. & Colbourn, E. A. (1996). Modelling and optimization of a tablet formulation using neural networks and genetic algorithms. *Pharm. Tech. Eur.*, *9*, 46-55.
[38] Rowe, R. C. & Roberts, R. J. (1998). Artificial intelligence in pharmaceutical product formulation: neural computing and emerging technologies. *PSTT* 1, 200–205.
[39] Sjöblom, J., Lindbergh, R. & Friberg, S. E. (1996). Microemulsions – phase equilibria characterization, structures, applications and chemical reactions. *Adv. Colloid Interf. Sci.*, *65*, 125–287.
[40] Specht, D. F. (1990). Probabilistic neural networks. *Neural Networks, 3*, 109–118.
[41] Spernath, A. & Aserin, A. (2006). Microemulsions as carriers for drugs and nutraceuticals. *Adv. Colloid Interface Sci.*, 128-130, 47-64.
[42] Sun, Y., Peng, Y., Chen, Y. & Shukla, A. (2003). Application of artificial neural networks in the design of controlled release drug delivery systems. *Adv. Drug Deliv. Rev.*, 55, 1201–1215.

[43] Turkoglu, M., Aydin, I., Murray, M. & Sakr, A. (1999). Modeling of a roller-compaction process using neural networks and genetic algorithms. *Eur. J. Pharm. Biopharm.*, 48, 239–245.

[44] Yao, X. (1991). Evolution of connectionist networks," in *Preprints Int. Symp. AI, Reasoning and Creativity,* T. Dartnall, Ed., Queensland, Australia. Griffith Univ., 49-52.

[45] Zupancic Bozic, D., Vrecer, F. & Kozjek, F. (1997). Optimization of diclofenac sodium dissolution from sustained release formulations using an artificial neural network. *Eur. J. Pharm. Sci.*, 5, 163-169.

In: Focus on Artificial Neural Networks
Editor: John A. Flores, pp. 29-53
ISBN: 978-1-61324-285-8

Chapter 2

INVESTGATIONS OF APPLICATION OF ARTIFICIAL NEURAL NETWORK FOR FLOW SHOP SCHEDULING PROBLEMS

T. Radha Ramanan
National Institute of Technology Calicut, Calicut, Kerala, India

ABSTRACT

The objective of this chapter is to present the research findings, of the author, that primarily use Artificial Neural Network (ANN) as a tool to find an improved solution for the performance measure(s) taken under consideration. The following studies are undertaken to investigate the applicability of ANN:

A bicriterian approach considering makespan and total flow time as performance measures to flow shop scheduling problem applying ANN with competitive network structure is made as a first attempt. With this objective, the architecture is constructed in two stages, viz. initial learning stage and implementation stage. In the initial learning stage the nodes of the network learns the scheduling incrementally and implements the same in the implementation stage. A number of problems are solved for different combinations of jobs and machines by varying jobs from 5 to 30 in steps of 5 and by varying machines from 5 to 30 in steps of 5. A total of 180 problems are solved by taking 5 problems in each set. The work is then extended to seek solutions for multicriteria flow shop scheduling considering makespan, earliness and lateness as performance measures. The result of the ANN is discussed in comparison with particle swarm optimization (PSO).

The next part of the study is modeled with the back propagation network of ANN and tested for seeking solutions to makespan as a performance measure. The results of ANN is sought to be further improved with improvement heuristics, Genetic algorithm (GA) and Simulated Annealing (SA). The problems are also tested against Taillard's benchmark problems (1993). The work aims at obtaining improved solutions by initializing SA and GA with a good starting solution provided by ANN. El-Bouri et al. (2005) show that neural sequences exhibit the potential to lead neighborhood search methods to lower local optima. This aspect is investigated in the study by making a comparison of the performance of the perturbation search and a non-perturbation search when starting from ANN initial solutions. The results show that neural sequences when made a perturbation, exhibit the potential to lead neighborhood search methods to lower local optima.

1.0 INTRODUCTION

Artificial neural network is used by the researchers in diverse fields to determine their characteristics of interest or performance measure. Ranging from Medical applications to Industrial applications; financial management to human resources management; data mining to sports prediction ANN is used.

Artificial Neural Network is used by the researchers for its many desirable characteristics, such as massive parallelism (Cristea and Okamoto [1]) distribution representation and computation (Elman [2]), generalization ability (Wu and Liu[3]), adaptivity (Davoian, and Lippe [4]).

ANNs are used for solving a variety of problems, such as pattern recognition (Jeson[5]; El-Midany et al. [6]), financial applications (Wong and Selvi [7]) such as bankruptcy prediction (Zhang et al., [8], Pendharkar [9], Tsai and Wu [10]), stock market prediction (Kim and Han[11]; Cao et al.[12]), forecasting (Zhang et al. [13]) optimization (Shen and Li [14]; Ghaziri and Osman [15]; Song and Zhang, [16])

When compared with the huge volume of literatures that are available for the applications of ANN; the usage of ANN as an optimization tool in scheduling of the shops is very much limited. ANN has been applied in job shop applications and also a few applications in flexible manufacturing systems. The applicability of ANN in a flow shop environment is not a much researched area. This can be also understood by the review work of Akyol and Bayhan [17]. Hence, the applicability of the ANN to the flow shop environment is explored by the author.

1.1. Flow Shop Scheduling

Elsayed and Boucher [18] state that the job sequencing could be stated as follows: Given 'n' jobs to be processed, each has a setup time, processing time, and a due date. To be completed, each job is to be processed at several machines. It is required to sequence these jobs on the machines to optimize a certain performance criterion. Most of the research in the flow-shop sequencing problem has concentrated on the development of a permutation flow shop schedule. The machines in a flow shop are dedicated to processing at most one job, and each job can be processed on at most one machine at any time. Preemption of individual jobs is not allowed. The jobs must be processed in the same sequence by each of the 'm' machines, given the processing times of each job on each machine. The objective of the sequencing problem is usually to decide the sequence of jobs, which minimizes the makespan.

1.2. Methodologies used In Flow shop Scheduling

The complexity of the flow shop scheduling problem renders exact solution methods impractical for instances of more than a few jobs and/or machines. This is the main reason for the various heuristic methods proposed in the literature. Many constructive heuristic approaches have been proposed in the research [19]. Johnson's algorithm [20] is the earliest known heuristic for the permutation flow shop problems. Constructive Heuristics (Palmer's,[21] Campbell Dudek and Smith(CDS)[22], Nawaz, Encore and Ham(NEH)[23], Gupta[24]) Improvement heuristics (Koulamas[25], Rajendran[26], Suliman[27]), Branch and bound methods have been applied to determine a solution for the NP-hard problems of flow shop scheduling. Tabu search (Nowicki and Smutnicki [28] , Moccellin[29], Ben-Daya and Al-Fawzan[30], Widmer and Hertz[31]) Genetic Algorithm (Reeves[32], Murata et al.[33], Sridhar and Rajendran[34], Tsutsui and Miki [35])and Simulated Annealing (Osman and Potts[36], Ogbu and Smith[37], Ishibuchi et al.,[38] Peng Tia et al.[39]) have been used to determine solutions for the performance of interest.

2.0. ANN Approach for Scheduling a Bicriterion Flow Shop

The applicability of ANN to the flow shop scheduling is done in two parts. In the first part, competitive network architecture is proposed for multicriterian optimization. The second part proposes a back propagation network along with hybrid approaches.

The first part of the research has sought to extend the work of Lee and Shaw [40] to include bigger size problems (up to 30 machines and 30 jobs) and seek solution for performance measures makespan (MS) and total flow time (TFT). The seed sequence for improvement is taken using Campbell, Dudek and Smith (CDS) heuristics. Artificial neural network with a competitive network and winner-take-all strategy is used for sequencing the jobs

The objectives of this part of the research are twin fold:

(i) To develop an ANN approach for bicriterian flow shops to give a solution to the sequencing problems of the shop.
(ii) To develop an optimal sequence considering both TFT and MS to reduce the total time.

The results of the ANN are compared with a traditional heuristic (CDS) and an improvement heuristics (C Rajendran's heuristic(CR))

2.1. Problem Description

In a flow shop scheduling problem, there is a set of n jobs, tasks or items (1... n) to be processed on a set of m machines or processors (1 . . . m) in the same order, i.e. first on machine 1, then on machine 2, and so on until machine m. The objective is to find a sequence for the processing of the jobs on the machines so that the total completion time or makespan and total flow time of the schedule is minimized. The processing times needed for the jobs on the machines are denoted as p_{ij}, where i = 1...n and j = 1...m; these times are fixed, known in advance and non-negative. There are several assumptions that are made regarding this problem:

- Jobs arrive in various combinations of batches
- A job has a fixed machining time on each of the machine
- The shop is capable of producing only n jobs
- The jobs pass through all m machines and preemption of jobs is not possible.
- Machines have unlimited buffer

2.2. Architecture of the Proposed System

With the objective of optimizing the MS and TFT, the architecture is constructed in two stages, viz. learning stage and implementation stage. In the initial learning stage the nodes of the network learns the scheduling incrementally and implements the same in the implementation stage. Figure 1 shows the architecture of the proposed system.

2.2.1. Initial learning stage

The optimization module: In this module the batches of job are first generated randomly. Processing time for each job is also generated randomly. To determine the optimum sequences to be given as input to the training module, the traditional heuristic is used. The traditional heuristic used in this experiment is CDS heuristic. First, using this heuristic the least MS for a sequence is identified for the in-coming jobs. Next taking this MS, a pair wise interchange of job is made to find out if performance measure could be further optimized. After finding the optimum MS, during pair wise interchange of jobs a lesser TFT is found. Thus an optimal sequence with both minimum TFT and minimum MS is identified. Since the machining times are assumed to be constant, whenever the jobs arrive in that same combination the sequence for machining will also remain constant. This optimal sequence is the output of the optimization module. Figure 2 shows the block diagram depicting the input and output of optimization module

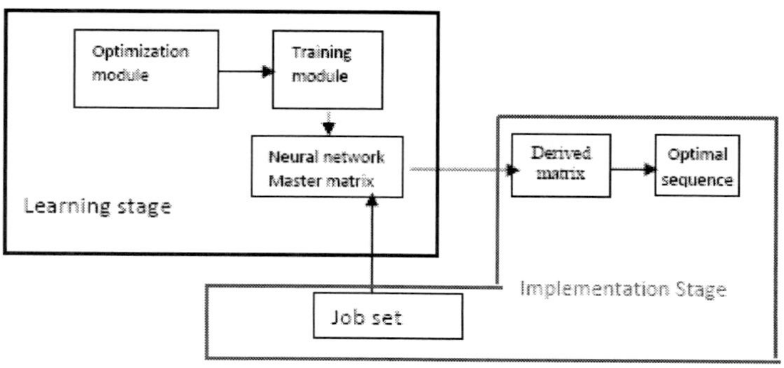

Figure 1. Architecture of the ANN system.

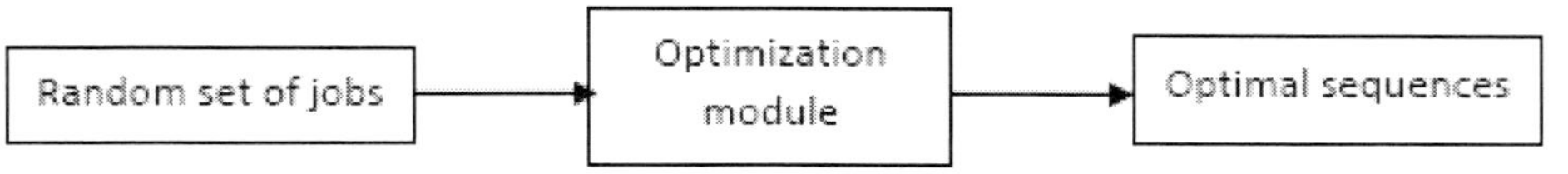

Figure 2. Block diagram of optimization module

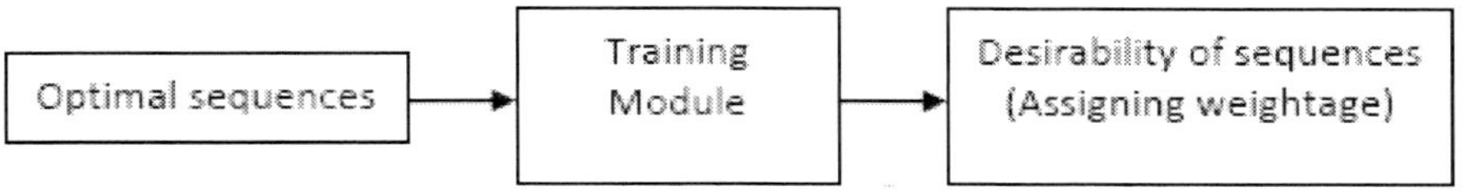

Figure 3. Block diagram of training module.

The Training module: The optimal sequence obtained from the optimization module is given as input to the training module in the form of vector pair. Each vector pair consists of a predecessor and a successor. Assume that the flow shop can machine a set of 15 different jobs and jobs arrive at the flow shop in various batches at random for processing. Machining time for these jobs, which the shop is capable of handling, will be constant. Assume that a set of jobs say, {15 18 21 24 25} arrive at the shop, the optimal sequence determined by using CDS heuristic is found to be {18 21 15 25 24}. From the five sets of job that has arrived, 4 vector pairs that represent adjacent jobs are identified. The vector pairs are (18 21) (21 15) (15 25) (25 24). These vector pair shows the desirability of job sequences. Thus the sequences are identified and are assigned a weightage of 1 for each vector pair. The training module assigns a weightage of 1 for each of the vector pair of the sequence and is aggregated every time the same predecessor and successor is repeated for the further generated arrival of jobs. The assigned weightage indicates the desirability of the sequence. Figure 3 shows the block diagram depicting the input and output of the training module.

Neural network master matrix (NNMM): The aggregated weights are the acquired knowledge given through the training module and are stored in the form of Master matrix. The magnitude of weights is the indicator of the desirability between jobs and NNMM is constructed before the implementation stage. The NNMM is formed after a certain number of training to the nodes.

2.2.2. Implementation stage

The weights of NNMM are considered as neurons. Successors and predecessors are considered as two layers of the network. These layers are fully connected two-layer network. ANN consists of two layers with an equivalent number of processing elements and each processing element is connected to all processing element in the other layer.

Job set: During the implementation stage, when a combination of jobs arrives for processing, it is given as input to the NNMM. The job set is initialized to the NNMM to take the desirability of the sequences.

Derived matrix (DM): From the NNMM the desirability of the sequences is taken and the weightages of the relevant nodes are derived in the form of derived matrix for further processing (sequencing) for the job set initialized.

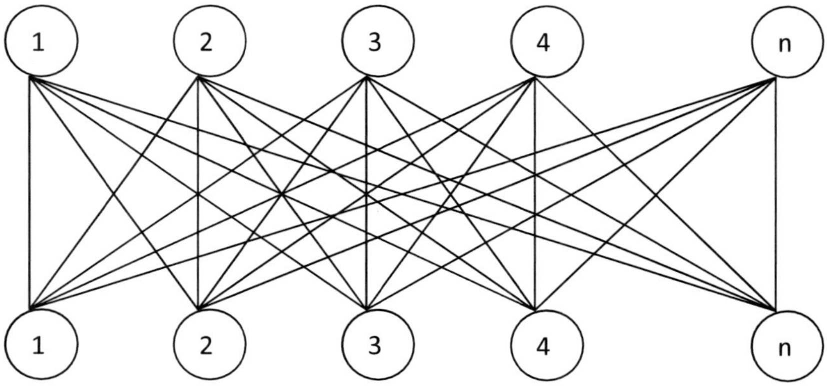

Figure 4. Bidirectional network structure.

Optimal sequence: Each job in the job set is considered as the starting job and the sequence is obtained from the DM. For these sequences the MS and TFT is calculated and the optimal sequence for the arrived job set is found. To generate an optimal and feasible sequence of different jobs, each job is not allowed to choose more than one other job as its successor, and each job is not allowed to choose more than one other job as its predecessor. Once the job is sequenced its weightage is made to zero to avoid obtaining infeasible solution.

2.3. Bidirectional Neural Network Structure

The network consists of two layers X and Y. The network is initialized with a job set. For each element in X layer, a connection with the highest weight from X-layer to Y-layer is chosen and vice-versa. The figure 4 shows the bi-directional nature of the network. This is referred to as winner-take-all strategy (Zurada [41]. Repeat the process until all the jobs are sequenced and a feasible solution is obtained.

At any step of the procedure at which elements in x-layer and y-layer must make a choice between two or more connections with the same weights, the value that is first read is assigned. The procedure must stop since there are only a finite number of jobs and no connection between x-layer and y- layer is activated more than once. The final outcome generated by the neural-net approach is complete and feasible sequence of jobs, since each job is linked at any step to exactly another job.

2.4. An Illustration

The neural network is designed to find the sequences for n jobs and m machines. Suppose that the flow shop has the capability to process 15 different jobs the training module constructs a master neural matrix of 15×15. The NNMM constructed in Figure 5 is obtained after giving 950 training exemplars to the network. Assuming that all the 15 jobs have arrived, in the implementation stage the DM takes its desirable sequences and derives the matrix from NNMM and the matrix in this case will be the same as NNMM. From the DM

(Which is NNMM in this case) it can be seen that in the 1st row (layer X) the 11th column (layer Y) has the highest value. This indicates that 11th job is the next desirable job for sequencing after 1st job. The network then goes to the 11th row and finds the highest weight, which is found in the 7th column, which stands for job number 7. Thus job 7 is sequenced after 11th job.

	1	2	3	4	5	6	7	8	9	10	11	12	13	14	15
1	0	0	0	0	64	0	105	0	0	0	449	0	7	0	0
2	0	0	0	943	0	0	0	0	0	0	0	0	0	7	0
3	0	0	0	0	0	0	0	29	893	0	7	0	21	0	0
4	0	0	0	0	0	0	0	140	0	0	0	781	0	29	0
5	265	0	7	0	0	0	302	0	0	0	207	0	0	0	0
6	0	0	0	0	0	0	0	0	0	615	0	0	0	328	7
7	255	0	0	0	577	0	0	0	0	0	118	0	0	0	0
8	0	0	774	0	7	0	0	0	0	0	0	169	0	0	0
9	29	0	0	7	0	0	0	0	0	0	21	0	893	0	0
10	0	0	0	0	0	0	0	0	7	0	0	0	29	140	774
11	401	0	0	0	0	0	71	0	29	0	0	0	0	0	0
12	0	0	140	0	0	0	0	781	0	0	29	0	0	0	0
13	0	0	0	0	302	0	472	0	21	0	119	0	0	0	29
14	0	586	29	0	0	0	0	0	0	335	0	0	0	0	0
15	0	364	0	0	0	140	0	0	0	0	0	0	0	446	0

Figure 5. Neural Network Master matrix.

The seq is ...														
1	11	7	5	3	9	13	15	14	2	4	12	8	6	10

Figure 6. The output sequence.

The sequences are:														
13	7	5	1	11	9	4	12	8	3	6	10	15	14	2
3	9	13	7	5	1	11	6	10	15	14	2	4	12	8
8	3	9	13	7	5	1	11	6	10	15	14	2	4	12
2	4	12	8	3	9	13	7	5	1	11	6	10	15	14
10	15	14	2	4	12	8	3	9	13	7	5	1	11	6
7	5	1	11	9	13	15	14	2	4	12	8	3	6	10
15	14	2	4	12	8	3	9	13	7	5	1	11	6	10
12	8	3	9	13	7	5	1	11	6	10	15	14	2	4
4	12	8	3	9	13	7	5	1	11	6	10	15	14	2
11	1	7	5	3	9	13	15	14	2	4	12	8	6	10
14	2	4	12	8	3	9	13	7	5	1	11	6	10	15
9	13	7	5	1	11	6	10	15	14	2	4	12	8	3
1	11	7	5	3	9	13	15	14	2	4	12	8	6	10
6	10	15	14	2	4	12	8	3	9	13	7	5	1	11
5	7	1	11	9	13	15	14	2	4	12	8	3	6	10
The makespans are:														
200	195	196	184	181	203	190	194	191	198	186	196	198	171	203
The total flow times are:														
2121	2044	2091	2003	1934	2104	1958	2068	2082	2057	2010	2021	2038	1856	2112

Figure 7. Possible sequences with corresponding MS and TFT.

Table 1. Effect of training on Network

ANN		CDS	
MS	TFT	MS	TFT
Training exemplars 300			
110	780	108	784
106	793	106	807
102	764	102	809
108	831	108	831
116	896	114	920
Training exemplars 600			
110	831	110	789
108	755	110	765
119	887	117	905
120	829	116	846
116	868	116	898
Training exemplars 900			
104	797	106	800
107	801	110	818
104	736	101	753
118	879	118	879
112	796	112	816

Since the job number 1 is already assigned, the weightage value is suitably reduced to get a feasible solution (i.e. the network takes care that it is not scheduled). Thus the sequence is identified until all the jobs are sequenced. The output of the sequence is given in figure 6. Figure 7 gives all the possible sequences and its corresponding optimum MS and TFT.

It is observed as shown in Table 1 that the ANN incrementally learns the sequencing. That is, higher the number of training given to the network, the results obtained are better.

2.5. Results and Discussions

The source code of the program for ANN, CDS and CR was written in C language on a Pentium III processor machine. A number of problems were solved for different combinations of jobs and machines by varying jobs from 5 to30 in steps of 5 and by varying machines from 5 to 30 in steps of 5. A total of 180 problems were solved by taking 5 problems in each set. The figure 8 shows the comprehensive results of the performance of ANN along with the performances of CDS and CR heuristics. The results obtained show that more than one heuristic give optimal results. Thus, it can be seen that ANN approach yields better results than the constructive or improvement heuristics.

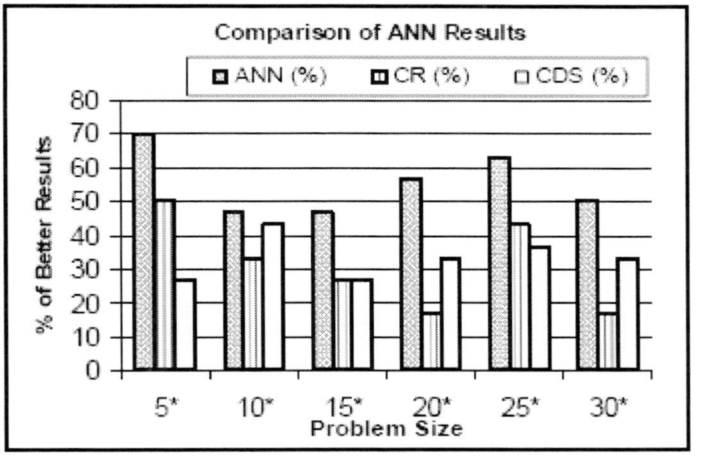

Figure 8. Comparison of ANN, CR and CDS heuristics results.

3.0. ANN Approach for Scheduling a Multi Criterion Flow Shop

The objective of this second part of the study is to optimize the makespan and total flow time and then the earliness and the tardiness of the flow shop with 'm' machines and 'n' jobs using Artificial Neural Network and compare the results with Particle Swarm Optimization (PSO). The architecture, network and strategy of the previous work are used for the current study also. Hence, they are not elaborated here.

The objective function for the current study is the one given by Baker [42]

Minimize Z = $\sum_{i=1}^{n} \alpha E_i + \beta T_i$

E_i – earliness of job i, T_i – tardiness of job i, α - earliness penalty, β - tardiness penalty

The earliness penalty is assumed to be 8 monetary units and lateness penalty is assumed to be 10 monetary units. In this illustration 10 particles (population size) are considered. Dimension of particle is considered as 5. The population size is the shop capacity, which is assumed to be 10, and dimension of particle is the orders on hand, which is assumed to be 5. Initial solution is generated randomly for all 10 particles.

3.1. Illustration

The neural network is designed to find the sequences for n jobs and m machines. Suppose that the flow shop has the capability to process 10 different jobs the training module constructs a master neural matrix of 10 x 10. The NNMM constructed in Figure 9 is obtained after giving 400 training exemplars to the network.

	1	2	3	4	5	6	7	8	9	10
1	0	18	22	7	22	21	10	19	25	15
2	14	0	12	12	23	20	27	22	20	20
3	13	17	0	11	25	16	18	26	22	11
4	20	20	17	0	29	12	14	15	21	11
5	17	21	14	20	0	21	15	25	25	9
6	19	18	11	27	9	0	17	17	15	18
7	8	13	28	15	21	19	0	26	23	17
8	12	15	14	19	14	20	15	0	25	9
9	19	19	23	10	12	21	18	18	0	15
10	5	7	43	19	27	11	16	20	23	0

Figure 9. Neural Network Master Matrix.

	8	1	7	3	6
8	0	14	15	14	25
1	17	0	11	24	18
7	26	12	0	28	14
3	23	11	19	01	18
6	18	20	20	14	0

Figure 10. Derived Matrix.

Sequences					Penalty
8	6	1	3	7	2816
1	3	8	6	7	2488
7	3	8	6	1	2608
3	8	6	1	7	2864
6	1	3	8	7	2528

Figure 11. Output with sequences and their penalties.

The batch of jobs that has randomly been generated is {8 1 7 3 6}. The Derived Matrix represented in the figure 10 is as follows.

From the derived matrix, it can be seen that 5th column has the highest value. The fifth column stands for job 6. Hence as per the weightage assigned the sequence will be after the 8th job the 6th job is sequenced. The network then goes to the 5th row and finds the highest weight, which is found in the second column. The second column corresponds to job number 1.

Thus the sequence is given as {8 6 1 3 7} and the penalty is 2816. The Figure 11 gives all the possible sequences and its corresponding penalty.

3.2. Results and Discussions

The source code of the program for ANN and PSO was written in C language on a Pentium IV processor machine. A number of problems were solved for different combinations of jobs and machines by varying jobs from 5 to 25 and by varying machines from 5 to 25 in different steps. The table 2 gives the comparison of results of the penalty obtained through PSO and ANN when the number of machines is kept constant as five. The results show that the ANN performs better than PSO in all the instances. The early convergence of PSO enables only in finding the local optima rather than global optima.

The table 3 shows the results of the penalty obtained through PSO and ANN when the number of machines through which the jobs have to be processed is fixed at 10.

4.0. A Hybrid Neural Network-Meta Heuristic Approach for Permutation Flow Shop Scheduling

4.1. Introduction

This work deals with the solution of permutation flow shop scheduling problem with the objective of minimizing makespan. The sequence obtained using ANN approach is further improved with an improvement heuristic suggested by Suliman [27]. The results of the ANN-Suliman heuristics are found to be not better than NEH heuristics (Refer Tables 10 and 11). Hence, improvement is made by hybrid approaches of ANN with Genetic Algorithm (GA) and Simulated Annealing (SA). For the purpose of experimentation, the benchmark problems provided by Taillard [136] are used.

These problems are as follows.

Set - I: Ten instances of 20 jobs, 10 machines problem.
Set – II: Ten instances of 50 jobs 10 machines problem.

These problems are used for comparing the performance of all the approaches.

The solution for the problem is obtained through two stages.

In the first stage, an initial solution or sequence of jobs is obtained by using Artificial Neural Network approach.

Table 2. Comparison of results of penalty of PSO and ANN (machine fixed as 5)

Number of jobs	PSO	ANN
5	592	476
7	768	644
10	3936	2896
13	4012	3584
15	9336	6712
17	10756	8564
19	12788	9654
20	15786	10248
23	22234	14652
25	24864	18542

Table 3. Comparison of penalty of PSO and ANN (machine fixed as 10)

Number of jobs	PSO	ANN
5	678	546
7	894	788
10	4016	3564
13	7564	6548
15	10876	9746
17	14788	12456
19	21278	18746
20	24866	21486
23	26684	24568
25	28462	26548

The second stage which is improvement stage consists of improving the solution obtained in the first stage by using improvement heuristics. The improvement heuristics used in present study are: Suliman's Two Phase Heuristic, GA and SA.

4.2. Architecture of the ANN

In the present study the feed-forward back propagation neural network (BPN) is used to solve the problem. This architecture consists of one input layer, one output layer and at least one hidden layer of neurons. The BPN is trained by presenting patterns of data at the input layer together with the corresponding desired output patterns at the output layer. The weights placed on connections between the nodes in the network's processing layers are updated by using an algorithm called Back-Propagation (BP) algorithm. The BP algorithm minimizes the error between network's output and desired output. A series of input-output pairs are presented to the neural network to train the network. When mean square error between desired output and actual output is minimized, the neural network is deemed trained. A trained neural network is expected to produce an output, based on the relationship it has learnt, whenever a new pattern (one not seen in the training set) is introduced at the input layer.

The proposed neural network has two hidden layers containing 30 and 20 neurons respectively. This particular design, in which the size of the input layer depends on the number of machines, means that separate and individually trained networks are needed for flowshops having different numbers of machines. The number of hidden layers and number of neurons in that layer are determined empirically by trying different combinations. To improve the performance of the Back propagation algorithm the input values should be normalized i.e. the values for the input patterns must be between zero and one. The minimum and maximum processing times in the flowshops considered in this study are 1 and 99 minutes. Therefore, the divisor of 100 in Equation (1) guarantees that processing times in the range (1, 99) are covered in the normalization.

4.3. Methodology

Each job is represented by a vector that contains a number of elements that describe the processing time of the job individually and relative to the machine workloads and other jobs to be scheduled. The size of this vector is three times the number of machines (3m). The input layer of the proposed neural network has 3m nodes allocated as follows:

- The first m nodes contain a job's processing times on each of the m machines.
- The middle m nodes contain the average processing times on each of the m machines.
- The last m nodes contain the standard deviation of the processing times on each of the m machines.

Table 4. Data for the test problem

JOBS	Machines							
	M1	M2	M3	M4	M5	M6	M7	M8
1	86	83	56	18	65	17	38	54
2	97	85	94	29	82	33	31	35
3	66	38	89	29	28	93	77	79
4	28	41	72	22	42	97	84	27
5	95	90	29	84	11	95	56	14
6	46	13	64	99	72	29	47	65
7	32	59	29	19	60	12	76	23
8	92	20	89	19	77	41	18	47
9	25	49	43	96	59	71	74	14
10	32	20	23	85	25	70	38	19
11	76	89	41	33	41	58	15	41
12	10	18	64	90	26	45	13	30
13	36	100	57	87	99	81	33	59
14	81	54	25	76	27	81	63	28
15	78	39	57	17	21	45	98	65
16	22	50	95	91	60	74	21	13
17	56	32	18	83	21	76	70	64
18	61	47	14	34	16	46	51	57
19	86	41	42	14	37	59	95	11
20	57	12	81	47	49	100	89	43
21	94	14	36	90	53	75	18	16
22	59	62	93	22	63	63	94	54
23	61	97	49	92	45	54	99	14
24	21	23	81	72	38	36	52	89
25	62	24	18	78	78	38	45	42
26	34	93	97	44	54	13	45	75
27	24	65	17	45	18	59	99	84
28	74	10	89	49	69	51	24	73
29	94	36	90	14	81	64	13	19
30	43	75	69	28	87	12	59	52

The output layer has only one node, regardless of m. The output node assumes values between 0.1 and 0.9.

First this approach is tested by the randomly generated problem of size 30 jobs and 8 machines. Table 4 shows the processing times of this problem. The values in the table 4 are processing time of jobs on respective machines. Processing times are taken as uniformly distributed random numbers between one and ninety nine. From the 30 job types considered, combination of three jobs each is selected and by complete enumeration, the optimal sequence for these three jobs is determined. In a similar manner, $^{30}C_3$ combinations are possible. 1000 combinations are selected randomly and optimum solution for those problems is determined.

Table 5. Makespan comparison with optimal sequences for test problem

Problem no.	Makespan for neural network sequence	Optimal Makespan	% Error
1	739	727	2
2	742	711	4
3	723	711	2
4	813	721	13
5	720	664	8
6	753	732	3
7	722	701	3
8	755	744	1
9	675	675	0
10	647	647	0
11	673	641	5
12	722	708	2
13	726	683	6
14	822	734	12
15	780	759	3
16	785	749	5
17	673	648	4
18	743	711	5
19	704	675	4
20	715	676	6
21	682	654	4
22	715	705	1
23	858	756	13
24	763	698	9
25	739	720	3
26	785	738	6
27	808	730	11
28	705	668	6
29	750	726	3
30	763	725	5

Each problem will give three input-output pairs; in all 3000 pairs are formed. The network is trained with these 3000 input-output pairs. The performance criterion is Mean Square Error between neural networks output and desired output. The network is trained until there is no improvement possible is this mean square error. After sufficient training, the network is presented with 30 problems of size 5 jobs, derived from the same problem in Table 4 and the sequence is obtained by arranging the jobs in ascending values of neural network's output. The makespan obtained for the sequences generated by the neural network for these problems are compared with the makespan for optimal sequences. The results are shown in Table 5.

It is found that for twenty out of thirty problems, the makespan values are only 5% more than that for optimal sequence, for six problems it is in between 5 to 10 % while for four problems; it is between 10 to 15%. Thus, near-optimal sequence can be obtained by using this approach. Hence this approach is further used for benchmark problems, taken from literature of size 20 jobs and 50 jobs with 10 machine flow shop.

The first step is to train the neural network. The training is tried with 5 jobs problems and also by varying the problem size from five to seven, derived from the given problem of 20 and 50 jobs. These problems are solved by complete enumeration to get the optimal solution. In case of multiple optima, the sequence having lesser total flow time is selected as an optimal one, to bring the consistency. Each five job problem will give five input-output pairs called exemplar. Ten thousand input output exemplars i.e. 2000 problems of size 5 are enumerated for training the network. In case of varying jobs size training exemplars i.e. from five to seven job size total 10005 exemplars are generated. In these exemplars, half of the population i.e. 5005 exemplars represent the seven job problem while 3000 are from six job problems and remaining 2000 exemplars are from five jobs problem.

The values assumed by the nodes of the input layer when a job i is presented to the network, are computed by using following equations (Elbouri et al.[43]):

$$\text{Node q} = \begin{cases} \dfrac{t_{iq}}{100} & q = 1\ldots m \\ \dfrac{\bar{t}_{(q-m)}}{100} & q = m+1, \ldots, 2m \\ \sqrt{\dfrac{X_{(q-2m)} - n\bar{t}^{2}_{(q-2m)}}{(n-1)\times 10^{4}}} & q = 2m+1 \ldots 3m \end{cases} \tag{1}$$

where,

$$\bar{t}_k = \frac{1}{n}\sum_{i=1}^{n} t_{i,k} \tag{2}$$

and

$$X_{(k)} = \sum_{i=1}^{n} t^2_{i,k} \tag{3}$$

The target output (O_i) for the ith job in the optimal sequence is determined by

$$o_i = 0.1 + \frac{0.8(i-1)}{(n-1)} \tag{4}$$

Equation (4) distributes the targets for the three jobs in equal interval between 0.1 for the first job to 0.9 for the last job. The network is trained with these 10005 exemplars by using *Neural Network Toolbox of Matlab 7.0.*

After sufficient training, the network is presented with the 20 and 50 jobs problem and the sequence is obtained by arranging the jobs in non-decreasing order of values of neural network's output. The first step in applying a trained network to produce a sequence for a new problem is to use Equation (1) to create n input patterns representing the n jobs. These patterns are applied one at a time, to the input layer of the trained network. An output value is generated in response to each input pattern. The outputs obtained after all input patterns have been processed and then sorted in non-decreasing order, and the job associated with each of these outputs is noted. The resulting sequence of these jobs represents the neural network's recommended sequence.

To demonstrate how the neural network is employed to construct a sequence from a set of jobs, the example of 5 job 10 machine problem from the data set is considered. Suppose jobs1, 2, 3, 4 and 5 from the example under study are to be sequenced. Their processing times on each machine are shown in Table 6.

Each of the five jobs is first expressed as a thirty element vector (vector consisting of thirty rows) of input nodes by using Equations (1) to (3). The results of this step are displayed in Table 7. Next, these vectors are introduced one at a time, at the input layer of the neural network that has been trained for the 10-machine flowshop.

The corresponding output from the network in each instance is given in the last row of Table 7. By sorting the five jobs in non-decreasing order of their neural outputs, the neural sequence is obtained, which is shown in Table 8.

Table 6. Processing Times for the Five Jobs

Jobs	Processing Times on Machines									
	M1	M2	M3	M4	M5	M6	M7	M8	M9	M10
1	46	61	3	51	37	79	83	22	27	24
2	52	87	1	24	16	93	87	29	92	47
3	79	51	58	21	42	68	38	99	75	39
4	45	25	85	57	47	75	38	25	94	66
5	97	73	33	69	94	37	86	98	18	41

Table 7. Input Layer Vectors and Outputs for the Five Jobs Ten Machine Problem

Jobs	1	2	3	4	5
1	0.4600	0.5200	0.7900	0.4500	0.9700
2	0.6100	0.8700	0.5100	0.2500	0.7300
3	0.0300	0.0100	0.5800	0.8500	0.3300
4	0.5100	0.2400	0.2100	0.5700	0.6900
5	0.3700	0.1600	0.4200	0.4700	0.9400
6	0.7900	0.9300	0.6800	0.7500	0.3700
7	0.8300	0.8700	0.3800	0.3800	0.8600
8	0.2200	0.2900	0.9900	0.2500	0.9800
9	0.2700	0.9200	0.7500	0.9400	0.1800
10	0.2400	0.4700	0.3900	0.6600	0.4100
11	0.0300	0.0300	0.0300	0.0300	0.0300
12	0.6380	0.6380	0.6380	0.6380	0.6380
13	0.5940	0.5940	0.5940	0.5940	0.5940
14	0.3600	0.3600	0.3600	0.3600	0.3600
15	0.4440	0.4440	0.4440	0.4440	0.4440
16	0.4720	0.4720	0.4720	0.4720	0.4720
17	0.7040	0.7040	0.7040	0.7040	0.7040
18	0.6640	0.6640	0.6640	0.6640	0.6640
19	0.5460	0.5460	0.5460	0.5460	0.5460
20	0.6120	0.6120	0.6120	0.6120	0.6120
21	0.0158	0.0158	0.0158	0.0158	0.0158
22	0.2315	0.2315	0.2315	0.2315	0.2315
23	0.2347	0.2347	0.2347	0.2347	0.2347
24	0.3608	0.3608	0.3608	0.3608	0.3608
25	0.2104	0.2104	0.2104	0.2104	0.2104
26	0.2870	0.2870	0.2870	0.2870	0.2870
27	0.2078	0.2078	0.2078	0.2078	0.2078
28	0.2597	0.2597	0.2597	0.2597	0.2597
29	0.4015	0.4015	0.4015	0.4015	0.4015
30	0.3623	0.3623	0.3623	0.3623	0.3623
Output	0.99997	0.41032	0.33002	0.10090	0.99331

Table 8. Sequence Obtained after Sorting the Neural Output

Sorted Output	Sequence
0.10090	4
0.33002	3
0.41032	2
0.99331	5
0.99997	1

Table 9. Comparison of Makespan and CPU Time for ANN-SA-I and ANN-SA

Seed No.	ANN-SA-I		ANN-SA	
	Makespan	Time	Makespan	Time
1	1582	156.17	1586	92.906
2	1672	144.36	1669	85.76
3	1508	150.77	1507	91.609
4	1379	155.87	1385	93.469
5	1425	153.47	1419	93.313
6	1401	150	1404	87.297
7	1486	145.92	1490	84.484
8	1550	144.09	1556	86.984
9	1608	143.61	1600	85.125
10	1600	149.59	1613	86.875

4.4. Results and Discussion

The objective of this study is to find a sequence of jobs for the permutation flow shop to minimize makespan. A feed forward back propagation neural network is employed as described earlier. The sequence obtained using neural network is used to generate initial sequence for the second phase of Suliman's heuristic, SA, GA initialized with random population and GA using Random Insertion Perturbation Scheme (RIPS). Makespan of the sequence obtained by all these approaches are presented in the tables 10 and 11. It is found that the ANN-GA-RIPS approach performs better than ANN-GA starting with random population. ANN-SA approach performs better than all other approaches.

4.4.1. Suliman's heuristic

Suliman heuristic consists of two phases as follows:

(1) Phase I: constructing initial sequence using CDS heuristic
(2) Phase II: Improving the initial sequence using pairwise exchange controlled by directionality constraint.

In the present study, ANN approach is used to construct the initial sequence for Phase I.

In the first phase of Suliman heuristic, the sequence is generated by using neural network (ANN-Suliman approach). The sequence is also generated by using CDS heuristic. These two sequences are given to the phase II of the Suliman heuristic to obtain the final sequences. Makespan of both the approaches are tabulated in the tables 10 and 11. Ten instances each of 20 and 50 jobs problems are solved. Also, the makespan values thus obtained are tabulated alongside to that of NEH.

4.4.2. Genetic algorithm

The second approach is improvement of ANN solution by Genetic algorithm. Based on generation of initial population from ANN sequence, two types of Genetic Algorithms are used as described earlier. In the first approach (ANN-GA), initial population contains one sequence from ANN and others are generated randomly. In the second approach (ANN-GA-RIPS), initial population contains one sequence from ANN and others are generated by perturbing ANN sequence using RIPS strategy, explained earlier.

In this study GA with roulette wheel selection procedure, partially matched crossover operator and shift mutation operator is used. The premature convergence problem is one that has been noted by several researchers in applying GA. In this study, the following method is used to come out of the premature convergence. If for a specified number of generations there is no improvement of the solution then, the GA again restarts with the available best solution.

The values of parameters are as follows:

Population size = 60.
Probability of crossover = 0.8
Probability of mutation = 0.2
Maximum number of generations = 50
Number of generation to restart the GA if solution is not improved = 10
No of iterations =50

Generation of initial population

Since GA deals with a population of solutions and not with a single solution, and ANN gives only one sequence as solution, there is need to initialize GA from this ANN solution. Two methods are adapted; in the first method called ANN-GA, the population is initialized with one sequence of ANN and the remaining M-1 sequences are generated randomly, where M is population size. The choice of neighbourhood technique greatly influences algorithm performance. Choosing a rich neighbourhood containing a large number of candidate solutions will increase the likelihood of finding good solutions. In the second method technique proposed by Parthasarathy and Rajendran [44] and known as RIPS (Random Insertion Perturbation Scheme). In the second method known as ANN-GA-RIPS, the population contains one solution given by ANN and the other M-1 sequences are generated from the ANN sequence using the Random Insertion Perturbation Scheme (RIPS). The results for both the approaches are shown in Tables 10 and 11 for both the Set – I and Set – II problems

4.4.3. Simulated annealing

This approach adopts improvement by simulated annealing algorithm (ANN-SA). It is found that the ANN-SA results are within 5% from the respective upper bounds. The initial temperature (T) was set at 475. The iteration is carried for every 0.9T. The stopping criterion is when the temperature reaches 20.

Simulated Annealing algorithm given by Parthasarathy and Rajendran [16], for sequence dependent set up time flow shop environment, is used. In the original algorithm for coming out of the specific temperature stage, the conditions given are either total moves made are greater than that of two times of number of job or number of accepted moves reaches beyond n/2 where n is number of jobs. This condition is slightly modified, in the present work, while using this algorithm. As the simulated annealing is started with the good sequence given by Neural Network, more number of initial moves, for inferior solutions, at the higher temperature are not required hence these moves are restricted up to n/4 with the other criterion i.e. total number of moves made kept the same (2n). It is observed that by using this method, the solution quality remains the same as when applied the previous SA but the time required to terminate the algorithm is reduced. Table 9 shows the comparison of makespan and time for both the approaches. In table 9 ANN-SA-I is the original Simulated Annealing applied to ANN solution while ANN-SA is modified simulated annealing applied to ANN solution.

Time is CPU time required for solving the problem 30 times. Makespan values are the minimum among these thirty iterations.

4.5. Results and Discussions

Ten instances of 20 jobs and 50 jobs problems are solved and performance is compared with respect to makespan values. For the twenty job problems, ANN-Suliman approach gives better solutions than CDS-Suliman and NEH. This is because, in the initial sequence which is given by ANN, the jobs are placed as close as possible with respective to their positions in optimal sequences. Hence, the second phase of Suliman Heuristic is able to improve the solution with the constrained improvement mechanism. But as job sizes are increased (i.e. in this case 50 jobs, refer table 11), more numbers of iterations are required to improve the solution quality. But, because of directionality constraint imposed by Suliman Heuristic, it is not being able to improve the solution further. For the problem size 50, NEH is found to give better solution than ANN-Suliman heuristic, because of directionality constraint imposed by the Suliman Heuristic. Hence, there is scope to use other methods of improving solution such as Genetic Algorithm, Simulated Annealing, etc.

Table 10. Makespan for a 20 job problem for different heuristics

	MAKESPAN								
Seed No	CDS	ANN	CDS-Suliman	ANN-Suliman	NEH	ANN-GA	ANN-GA-RIPS	ANN-SA	Upper Bound *
1	1774	1770	1697	1667	1680	1624	1604	1586	1582
2	1873	1817	1803	1732	1786	1691	1691	1669	1659
3	1658	1644	1626	1542	1555	1525	1517	1507	1496
4	1548	1514	1498	1465	1450	1405	1392	1385	1377
5	1606	1649	1553	1504	1502	1453	1434	1419	1419
6	1530	1626	1459	1456	1453	1444	1421	1404	1397
7	1610	1665	1539	1550	1562	1500	1487	1490	1484
8	1813	1768	1728	1658	1609	1574	1551	1556	1538
9	1796	1830	1685	1661	1692	1642	1612	1600	1593
10	1766	1839	1711	1703	1653	1644	1625	1613	1591

* Upper bounds of Taillard's Benchmarks problems.

Table 11. Makespan for a 50 job problem for different heuristics

Seed No	MAKESPAN								
	CDS	ANN	CDS-Suliman	ANN-Suliman	NEH	ANN-GA	ANN-GA-RIPS	ANN-SA	Upper Bound*
1	3464	3288	3241	3171	3168	3175	3126	3037	3037
2	3231	3366	3095	3122	3150	3030	2976	2911	2911
3	3369	3288	3219	3187	3198	3010	2975	2943	2871
4	3436	3408	3280	3235	3183	3224	3101	3071	3067
5	3362	3427	3182	3198	3128	3126	3079	3021	3011
6	3392	3374	3300	3176	3135	3143	3104	3037	3021
7	3473	3521	3411	3316	3289	3259	3243	3195	3124
8	3597	3439	3397	3252	3200	3211	3123	3072	3048
9	3346	3304	3108	3126	3075	3062	3025	2957	2910
10	3362	3466	3182	3215	3253	3274	3195	3140	3100

* Upper bounds of Taillard's Benchmarks problems.

The tables 10 and 11 give the makespan obtained by ANN-GA-RIPS, ANN-GA and ANN-SA for a 20 job and 50 job benchmark problem set respectively. It can be observed from the table that the former approach gives better solution than the latter in the entire problem sets considered. ANN-GA-RIPS approach performs better than ANN-GA because in the ANN-GA-RIPS approach, the initial population is obtained by perturbing the ANN sequence which provides a better starting population as compared to ANN-GA. Because of this perturbation, it is also observed the ANN-GA-RIPS is able to converge faster that ANN-GA. ANN-SA is found to give better results than ANN-GA-RIPS. This is due to the inherent robustness and the higher initial temperature set for the search process. SA is able to jump out of the local optima even better than ANN-GA-RIPS approach.

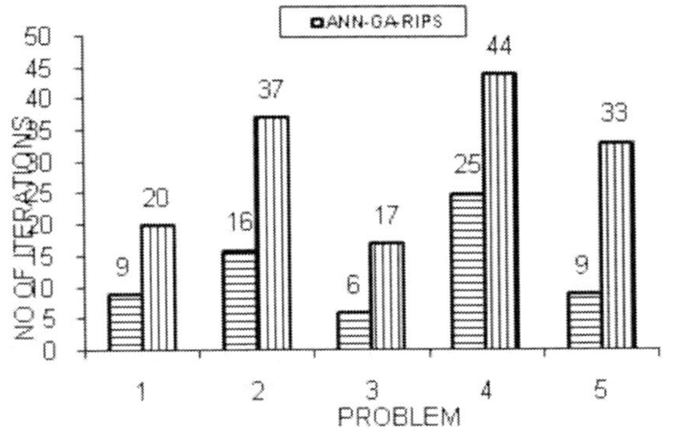

Figure 9. No of Iterations Required to Reach Final Solution for 20 Jobs Problem.

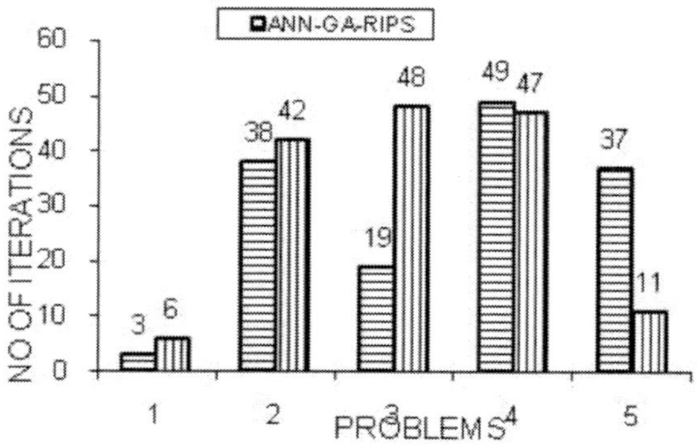

Figure 10. No of Iteration Require to Reach Final Solution for 50 Jobs Problem.

It is found that the makespan values obtained using ANN-GA-RIPS and ANN-SA approach are within 5% from the Upper Bounds stated by Taillard [45] for the respective problems.

The figures 9 and 10 show the number of iterations required for ANN-GA-RIPS and ANN-GA to reach the final solution. For the set-I problems, the ANN-GA-RIPS has converged faster. However, for the set-II problems, in 2 out of 5 instances, the convergence of ANN-GA is faster, though the quality of solution is better in the former approach.

4.6. Inferences

The inferences from the study are follows:

(i) The ANN incrementally improves the solution quality with the increase in numbers of training exemplars.
(ii) The ANN achieves a solution quality better to that of traditional heuristics or at least comparable to it.
(iii) The time required for ANN to process the inputs is considerably lesser. That is, the processing time of ANN is considerably lesser than other heuristics. Once the NNMM is constructed the results are extracted immediately, as the job set arrives doesn't require any further process.
(iv) The competitive network provides encouraging results. However, the network requires further research before arriving at the adoptability of the network. The author is currently working in this direction.
(v) ANN understands the inputs in terms of patterns. Hence the inputs are normalized and given in terms of the average and standard deviation of the processing times
(vi) ANN is able to provide better results when the inputs provided are of better quality. Hence completely enumerated results of 5, 6 and 7 machines are used for training.

5.0. Conclusions and Future Directions

Two different networks, viz. competitive network and back propagation network are adopted in the study to find the suitability of the ANN for bigger size permutation flow shop scheduling problems. The performance measures taken for optimization include makespan, total flow time, earliness and tardiness. The study investigated ANN based heuristics and improvement heuristics such as Suliman's, GA and SA approaches for obtaining the solution of permutation flow shop scheduling problem. The investigation aims at obtaining improved solutions by initializing genetic algorithm with a good starting solution provided by ANN. The investigations show that neural sequences exhibit the potential to lead neighborhood search methods to lower local optima when a local perturbation is made to the solution. The results of ANN-SA approach are better than ANN-GA-RIPS approach, especially for bigger size problems. Based on the results, it can be inferred that ANN-SA approach outperforms all the other approaches. The ANN approach with the architecture adopted in the present work provides better performance measure when combined with GA and SA.

In spite of the hybrid approaches, the results obtained are not optimum. Hence, the research requires further changes in the architecture of the neural network. Further comparison of the performance of competitive network and back propagation network could be done. Different training methodology can be adopted to ensure that the input of the network is nearer to the optimal values of bigger problems.

References

Alexandra Cristea, Toshio Okamoto. *"ANN Parallelization on a Token-Based Simulated Parallel System,"* ICCIMA, pp.24, Third International Conference on Computational Intelligence and Multimedia Applications (ICCIMA'99), 1999(DOI Bookmark: http://doi.ieeecomputersociety.org/10.1109/ICCIMA.1999.798495).

Baker, K. R. (1974). "Introduction to sequencing and scheduling" *Wiley*, New York.

Ben-Daya, M. & Al-Fawzan, M. (1998). "A tabu search approach for the flow shop scheduling problem", *European Journal of Operational Research, 109*, 88–95.

Bo, K. Wong & Yakup Selvi (1998) "Neural Network applications in finance: A review and analysis of literature (1990-1996)" *Information & Management*, *Vol.34, Issue 3*, Pages 129-139.

Campbell, H. R. & Smith, D. M. (1970). "A heuristic algorithm for n-jobs m-machines sequencing problem". *Management Science*, 16B, 630-637

Changyu Shen, Lixia Wang & Qian Li, (2007). "Optimization of injection molding process parameters using combination of artificial neural network and genetic algorithm method" *Journal of Materials Processing Technology*, *Volume 183, Issues 2-3*, Pages 412-418.

Chih-Fong Tsai & Jhen-Wei, Wu. (2008). "Using neural network ensembles for bankruptcy prediction and credit scoring" *Expert Systems with Applications*, *Vol., 34, Issue 4*, Pages 2639-2649.

Derya Eren Akyol & G. Mirac Bayhan, (2007). "A review on evolution of production scheduling with neural networks" *Computers & Industrial Engineering*, *Vol.53, No. 1*, Pages 95-122.

El-bouri, B. Subramaniam & Popplewell N, (2005) "A neural network to enhance local search in the permutation flowshop", *Computers & Industrial Engineering*, *Vol. 49*, 182-196.

Elman, J. L. (1991). Distributed representations, simple recurrent networks, and grammatical structure. *Machine Learning, vol. 7*, no. 2/3, 195-226.

El-Midany, T. T., El-Baz, M. A. & Abd-Elwahed, M. S. (2010). "A proposed framework for control chart pattern recognition in multivariate process using artificial neural networks" *Expert Systems with Applications*, *Volume 37, Issue 2*, *Pages 1035*-1042.

Elsayed, A. Elsayed & Boucher, Thomas, O. (1985) *"Analysis and Control of Production Systems"* Prentice Hall Inc. Upper saddle River, NJ.

Guoqiang Zhang, Michael, Hu, Y., Eddy Patuwo, B. & Indro, Daniel, C. (1999). "Artificial neural networks in bankruptcy prediction: General framework and cross-validation analysis" *European Journal of Operational Research*, *Vol.116, No.1*, Pages 16-32.

Guoqiang Zhang, Patuwo, Eddy B. & Hu, Michael, Y. (1998). " Forecasting with artificial neural networks:: the state of the art" *International Journal of Forecasting*, *Vol. 14,* Issue 1, Pages 35-62.

Gupta, J. (1971). "A functional heuristic algorithm for the flow shop scheduling problem". *Operational Research*, *22*, 27–39

Hassan Ghaziri & Osman Ibrahim, H. (2003). "A neural network algorithm for the travelling salesman problems with backhauls" *Computers & Industrial Engineering*, *Vol. 44,No. 2*, Pages 267-281.

Ishibuchi, H., Misaki, S. & Tanaka, H. (1995). "Modified Simulated Annealing Algorithms for the Flow Shop Sequencing Problem", *European Journal of Operational Research*, *Vol. 81*, 388-398.

Jacek, M. & Zurada, (1992). "Introduction to Artificial Neural Systems", *Jaico Publishing House.*

Jiansheng Wu and Mingzhe Liu (2005). Improving generalization performance of artificial neural networks with genetic algorithms. *Proceedings of the IEEE International Conference on Granular Computing.* P.288-291.

John Peter Jeson. (2004). "The Neural Approach to Pattern Recognition" *Ubiquity, Volume 5, Issue* 7, April 14 – 20.

Johnson, S. (1954). "Optimal two- and three-stage production schedules with setup times included". Naval Research Logistics Quarterly 1, vol. 61.

Koulamas, C. (1998). "A new constructive heuristic for the flowshop scheduling problem". *European Journal of Operational Research Society, Vol. 105*, 66-71.

Kristina Davoian & Wolfram-M. Lippe (2006). *A New Self-Adaptive EP Approach for ANN Weights Training.* World Academy of Science, Engineering and Technology 15 2006.

Kyoung-jae Kim & Ingoo Han (2000). "Genetic Algorithms approach to feature discretization in artificiant neural networks for the prediction of stock price index" *Expert Systems with Applications*, *Vol.19,* Issue 2, Pages 125-132.

Lee, I. & Shaw, M. J. (2000). "A neural-net approach to real time flow- shop sequencing" *Computers and Industrial Engineering*, *38*, 125-147.

Moccellin, J. A. V. (1995). "A new heuristic method for the permutation flow shop scheduling problem", *Journal of the Operational Research Society*, *46*, 883–886.

Murata, T., Ishibuchi, H. & Tanaka, H. (1996). "Genetic Algorithms for Flowshop Scheduling Problems", *Computers and Industrial Engineering*, *Vol. 30*, No. 4, 1061–1071.

Nawaz, M., Enscore, J. & Ham, I. (1983). "A Heuristic algorithm for the m-machine, n-job flowshop sequencing problem". OMEGA: *International Journal of Management Science*, *Vol. 11*, No. 1,91-95

Nowicki, E. & Smutnicki, C. (1996). "A fast tabu search algorithm for the permutation flow-shop problem", *European Journal of Operational Research*, *91*, 160–175.

Ogbu, F. A. & Smith, D. K. (1990). "The Application Of The Simulated Annealing Algorithm To The Solution Of The N/M/Cmax Flow Shop Problem", *Computer and Operation Research*, *Vol. 17*, 243–253.

Osman, I. H. & Potts, C. N. (1989). "Simulated Annealing for Permutation flow shop scheduling" *Omega*, *Vol. 17(6)*, 551-557.

Palmer, D. (1965). "Sequencing jobs through a multi-stage. Process in the minimum total time – a quick method of obtaining a near optimum". *Operations Research*, *16*, 45−61.

Parthasarathy, S. & Rajendran, C. (1998). "An experimental evaluation of heuristics for scheduling in a real-life flowshop with sequence-dependent setup times of jobs", *International Journal of Production Economics*, *Vol. 49*, 255-263.

Pendharkar, Parag, C. (2005). "A threshold-varying artificial neural network approach for classification and its application to bankruptcy prediction problem" *Computers & Operations Research*, *Vol. 32, Issue 10, Pages,* 2561-2582.

Peng Tia, Jian Ma & Dong-Mo Zhang, (1999). "Application of the simulated annealing algorithm to the combinatorial optimization problem with permutation property: An investigation of generation mechanism" *European Journal of Operational Research*, *118*, 81-94.

Qing Cao, Leggio, Karyl, B. & Schniederjans, Marc, J. (2005) "A comparison between Fama and French's mode and artificial neural network in predicting the Chinese stock market" *Computers & Operations Research*, *Vol. 32,* Issue 10, Pages 2499-2512.

Rajendran, C. (1995). "Theory and methodology heuristics for scheduling in flow shop with multiple objectives". *European Journal of Operations Research*, *82*, 540-555.

Reeves, C. A. (1995). "Genetic Algorithm For Flow Shop Sequencing", *Computer & Operations Research*, *Vol. 22*, 5-13.

Saghafian, Hejazhi, S. (2005). "Flowshop scheduling problems with makespan criterion: a review", *International Journal for Production Research, Vol. 43,* No 14, 895-2929.

Shiegeyoshi Tsutsui, & Mitsunori Miki (2002). "Solving Flow shop scheduling problems with probabilistic Genetic Algorithms". *Proc. of the 4th Asia-Pacific Conference on Simulated Evolution And Learning (SEAL-2002)*,465-471.

Song, R. G. & Zhang, Q. Z. (2001). "Heat treatment technique optimization for 7175 aluminium alloy by an artificial neural network and a genetic algorithm" *Journal of Materials Processing Technology*, *Vol. 117,* No.1-2,2 2001, Pages 84-88.

Sridhar, J. & Rajendran, C. (1996). "Scheduling in flowshop and cellular manufacturing systems with multiple objectives–a genetic algorithmic approach", *Production Planning and Control*, *7*, 374–382.

Suliman, S. (2000). "A Two-Phase Heuristic Approach To The Permutation Flow-Shop Scheduling Problem", *International Journal of Production Economics*, *Vol. 64*, 43-152.

Taillard, E. (1990). "Some efficient heuristic methods for the flowshop sequencing problem". *European Journal of Operational Research, 47*, 67–74A.

Widmer, M. & Hertz, A. (1989). "A new heuristic method for the flow shop sequencing problem". *European Journal of Operational Research, 41*, 186-193.

In: Focus on Artificial Neural Networks
Editor: John A. Flores, pp. 55-74
ISBN: 978-1-61324-285-8

Chapter 3

ARTIFICIAL NEURAL NETWORKS IN ENVIRONMENTAL SCIENCES AND CHEMICAL ENGINEERING

F. G. Martins**, *D. J. D. Gonçalves and J. Peres
LEPAE, Departamento de Engenharia Química, Faculdade de Engenharia,
Universidade do Porto, Porto, Portugal

ABSTRACT

Artificial neural networks have been used for a long time in a wide range of fields inside Environmental Sciences and Chemical Engineering. The main reason for this extensive utilization is the ability of this technique to model easily the complexity of the systems related with these fields, keeping most of the valuable original information about each system.

The feedforward artificial neural networks are the most commonly used topology due to the inherent simple architecture, the diversity of the available training algorithms, and the good performances. Besides feedforward artificial neural networks, the self organizing maps, or also called Kohonen neural networks, have as well relevant applications.

In Environmental Sciences, the most relevant applications appear in modelling for both environmental and biological processes. In Chemical Engineering, artificial neural networks have been applied mainly in: i) modelling; ii) control; and iii) development of software sensors.

This chapter compiles several applications that have been published recently concerning the subjects referred above. A special attention is given to the relevance of the cases, the procedures/techniques, and the ability to be extrapolated to other applications.

---

* Corresponding author, e-mail: fgm@fe.up.pt

INTRODUCTION

The seeking of knowledge trying to understand relationships between variables or entities is a very difficult task. To discover these relationships, the common approximation is through models, laws, phenomenological relations, etc. Though, this approach can be very often rather complicated or unknown.

Nevertheless, a different approach can be implemented to search these relationships, called the statistical approach. Almost all processes present high levels of complexity to fully understand them. However, a large amount of data is often available. This data can be correlated to extract important information.

Several techniques are available for this purpose, each one with different levels of complexity and accuracy. In this context, the Artificial Neural Networks (ANN) are a technique that has stood out over the last decades.

During this period, the applicability of ANN has continually grown covering a wide range of areas, such as: medicine, biology, chemistry, finance, engineering, social sciences, etc. The main reasons for this extensive utilization are: i) the capacity to learn by examples and then generalise its knowledge, ii) the capability to model complex, non-linear processes without having to express a formal relationship between input and output variables, iii) the ability to keep original valuable information and relationships within the final structure, iv) the robustness of the network itself tending to be fault-tolerant and v) the accuracy and precision of the final network.

This chapter is focused specifically on the latest applications of ANN in two very important areas: Environmental Sciences and Chemical Engineering.

As well as in the other areas, research and development of ANN in the fields mentioned above have been also very wide. However, some applications can be highlighted given their importance in each field.

In Environmental Sciences, the main applications have been in modelling environmental and biological processes. In Chemical Engineering, the use of ANN has been more focused on: i) modelling, ii) control and iii) software sensors development.

BRIEF DESCRIPTION OF ANN

ANN are tools that mimics the neural structure of the human brain. The brain basically learns from experience (De et al., 2007). These structures adaptively respond to inputs according to a learning rule. The ability to learn by examples and to generalize are the principal characteristics of ANN (Ham and Kostanic, 2001).

Nonlinearity is another key feature of ANN, expressed on their vital units, the neurons, corresponding to the information processing elements. The neurons are linear combinations of the weighted inputs and bias values that perform nonlinear evaluations through the activation functions. The election of the activation functions depend on the nature of the problem, but are usually the sigmoid function, the hyperbolic tangent function or the linear function.

There are different ANN configurations, although Feedforward Artificial Neural Networks (FANN) are the most common architecture, where neurons are grouped into successive layers and the information goes from the input layer subsequently to the hidden layers and finally to the output layer (Sousa et al., 2006). An example of this configuration with one hidden layer is presented in Figure .

The learning process is carried out by an optimization process where, based on the training data or examples, weights and biases are updated looking for minimizing the error between the predicted variable and the real one. At the same time of the training phase, the validation phase must be conducted to evaluate the generalization capacity of the final network configuration. This evaluation is done through a cross-validation test (usually the data set is divided into two subsets, one for training and another for validation). Even though, it is important to remark that the optimization problem is not convex, and for this reason the optimization process is not a simple task. On the other hand, the generalization or prediction power of the network will be better when the training data correspond to a wide range of problem situations.

Besides the qualities mentioned above (adaptability and nonlinearity), the ANN have another important one, their robustness and resistance tending to be fault-tolerant. This means that even if a neuron or a connection is damaged, it will only decrease a little the overall performance of the network, because the information is distributed in total neural network. These reasons have contributed to transform ANN as powerful tools used in different interdisciplinary areas (Ham and Kostanic, 2001).

Another class of artificial neural networks often used are the Self-organizing Maps (SOM) (Kohonen, 1990) or also called Kohonen's networks.

According to Kohonen (1990), SOM create "spatially organized internal representations of various features of input signals and their abstractions".

SOM consist in high-dimensional input layer and output layer (also called competitive layers) which are assembled in a two or three dimensional grid (Worner and Gevrey, 2006; Kadlec et al., 2009). Inside this grid, each cell or neuron become specifically tuned to diverse input signal patterns or classes of patterns through an unsupervised learning process. The final result is a grid where the location of a cell corresponds to a particular domain of the input signal patterns (Kohonen, 1990). In other words, the final network is a low dimensional representation of the data preserving the high dimensional topology (Kadlec et al., 2009).

The training method begins with a set of initial values in the nodes. After that an unsupervised method is initiated based on the 'winner' node, more similar to the input vector. This comparison is usually done with a Euclidean metric. Then the winner node is modified

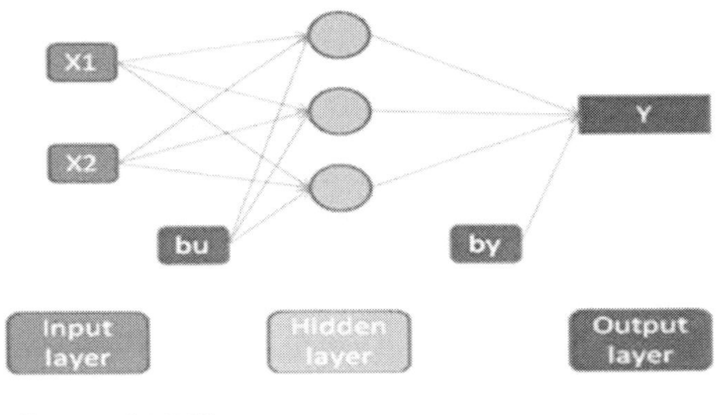

Figure 1. Schematic diagram of FANN structure.

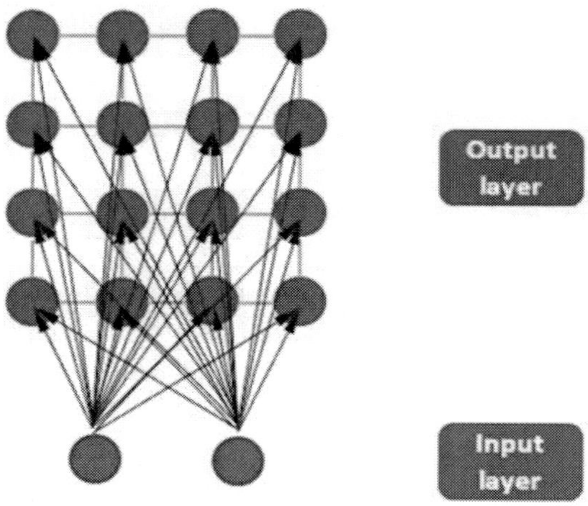

Figure 2. Schematic diagram of SOM structure.

with a specific learning rate. Its neighbors are also updated through a neighborhood function which is often a Gaussian. The method goes on while the learning rate decreases monotonically to zero (Kohonen, 1998). A representation of a grid of 4x4 and two inputs can be seen in Figure 2.

These characteristics originate the main application of SOM: visualization of high dimensional data (Kangas and Kohonen, 1996; Kadlec et al., 2009).

It is also important to highlight that even SOM are normally trained with an unsupervised routine, if they are used for pattern recognition, their accuracy can be improved with a fine-tune supervised learning algorithm (Kohonen, 1990).

LITERATURE REVIEW

In the following sections, a detailed analysis is addressed, covering the relevant works published in the last years concerning the topics mentioned above. A particular attention is given to the importance of the cases, the procedures/techniques, and the ability to be extrapolated to other applications.

The selected articles are those published since 2006 with more citations (according to the list represented in SCOPUS during March 2010).

Table 1. Publications in the areas of interest

Area	Articles	Citations
Environmental Sciences		
Modelling	13	324
Chemical Engineering		
Modelling	15	258
Control	10	113
Software Sensors	10	168

Table 1 shows the number of selected articles in each area, along with the amount of citations.

Table 2. Publications in the areas of interest

Area	Articles	Citations
Environmental Sciences		
Modelling	13	324
Chemical Engineering		
Modelling	15	258
Control	10	113
Software Sensors	10	168

Environmental Sciences

Getting to know the processes that occur in nature represents a milestone in the way for beginning to understand their relationships. Describing those processes symbolizes a very difficult and complex task. The variables involved in those processes can be so many that establishing deterministic associations is almost impossible.

This is the main reason why statistical approaches have been rising as a good alternative trying to explain these systems. Among these techniques, ANN have been found as a valuable technique to strive these situations.

As mentioned before, the main applications of ANN in environmental sciences is to describe, in other words, modelling. Among the several applications that can be found in the literature, this section will be focused on a few of them which have presented an intense activity over the last years. Table shows these applications, the techniques employed and the authors of the works.

Water is a highly valuable resource, essential to all kind of lives on earth. Even though water is found in more than 2/3 of the earth, only a very small fraction is suitable for consumption and utilization. With the world population growing every year with a rapid pace, it is important to set strategies to manage and optimize the available resources.

Rainfall-runoff process represents a key step in the hydrologic cycle. Runoff forecast models are useful in many applications such as: flood control, optimal reservoir operation (irrigation, hydropower generation, water supply, etc.) and design of hydraulic structures (dams, bridges, etc). Normally these models are developed using hydrologic and climatic data (Srinivasulu and Jain, 2006).

As mentioned before, creating a deterministic model of this process is a very complex task. Rainfall-runoff transformation is a very complex hydrological phenomena with many factors involved, such as: i) rainfall patterns, ii) watershed physical phenomena (evapotranspiration, infiltration, etc), iii) geomorphological characteristics of the watershed and iv) climatic characteristics (Chen and Adams, 2006).

Srinivasulu and Jain (2006) studied the creation of a rainfall-runoff model applied to a complete catchment using a conjunction of techniques. First, applying self organizing maps, it was possible to separate the input data into several categories with similar dynamic flows. This fragmentation applied to a large database improves the performance of the feedforward neural network model, describing the rainfall-runoff process for each section. This FANN had another technique involved, a real coded genetic algorithm, handling the optimization process in the training phase of the network.

Table 3. Environmental applications

Application	Techniques	Authors
Hydrological Modelling	FANN, Self organizing maps, Fuzzy Logic, Time series, Genetic algorithms	Chang and Chang, 2006 Srinivasulu and Jain, 2006 Dawson et al., 2006 Jain and Kumar, 2007 Han et al., 2007 Kisi, 2008
Air modelling	FANN, Principal components	Agirre-Basurko et al., 2006 Nagendra and Khare, 2006 Sousa et al., 2006 Sousa et al., 2007
Mapping biodiversity information	FANN, Self organizing maps,	Harrison et al., 2006 Worner and Gevrey, 2006 Foody and Cutler, 2006

The combination of the two techniques proved to be more successful and efficient when compared with a single FANN handling all the dataset.

In the same line, Jain and Kumar (2007) demonstrated how FANN performed streamflow forecasting more accurate than time series techniques. The time series forecasting only deliver acceptable performance when the linearity and stationary conditions are present. However, the authors concluded that coupling time series techniques for data pretreatment (de-trending and de-seasonalizing) and FANN improves substantially the previous results.

Han et al. (2007) also modelled the runoff process addressing their efforts to the common problem in ANN modelling (uncertainties and meaning of the model). To overcome the problem of uncertainties, a bootstrapping technique was used to divide the global dataset, assuring similar characteristics in the training and validation datasets. In addition, the training dataset is kept stored (instead of discarded) and used to compare it with the new data entered in the model. These two procedures aim to reduce the uncertainties of the model: i) reducing the differences between the datasets; and ii) comparing data (after training) with the data used in the training stage. These procedures can give an idea of how reliable the obtained prediction can be.

The meaning problem is related to the black box nature of ANN. The work of Han et al. (2007) proposed exciting ANN model with some standard input signals to reveal the non-linear behaviour under different circumstances, which should be checked against the idealized response cases. This approach revealed to be non-conclusive in the results but very promising in the opportunities that this procedure could lead to solve this issue.

Expanding the ANN applications, Dawson et al. (2006) covered the information of several catchments across UK (850) demonstrating their potential predicting flood events, especially in those engaged sites. Though, it realized the necessity to perform a previous separation of the dataset in clusters with similarities to improve its forecasting capability.

As referred before, managing water resources efficiently is crucial. In order to do that, reservoirs are the most important and effective water storage facilities. They not only provide water, hydroelectric energy and irrigation, but also smooth out extreme inflows to mitigate floods or droughts. To make the best use of the available water, the optimal operation of reservoirs in a system is undoubtedly very important. Reservoir operation requires a series of decisions that determine the accumulation and release of water over time. Even with the natural uncertainty of this process, forecasting the future reservoir inflow can be very helpful in making efficient operating decisions (Chang and Chang, 2006).

In this case, Chang and Chang (2006) presented a model using two techniques; first, mapping the input with fuzzy logic and then the outputs were fed to a FANN. A key parameter was added to improve its accuracy and reliability: the human operating decisions. The result was a model capable to make water level forecasting with a superior performance when compared to the model without the incorporation of the human factor.

Another interesting application looking for managing water resources is the determination of the evapotranspiration coefficient (ET). This is an important term in the water balance of an irrigated area, helping the calculation of the water consumed by the crops. According to Kisi (2008), correlating variables like solar radiation, air temperature, relative humidity and wind speed, it is possible to estimate ET through a FANN with similar accuracy to those obtained with complex deterministic expressions.

Air modelling has been another special area of interest in environmental sciences. With the rising of pollution in the atmosphere, it is vital to perform forecasting of these key components. The developing of mathematical tools capable to predict the concentrations is very important to provide early warnings to the population and to reduce the number of measuring sites (Sousa et al., 2007).

In this area, the lack of sufficient data and the difficulty to model the different interactions between pollutants are the main causes of the complexity of this modelling (Borrego et al., 2000).

Agirre-Basurko et al. (2006) developed an ANN to forecast hourly (up to 8 hours) the ozone and nitrous dioxide based on meteorological data, air pollutant concentrations and traffic variables. The addition of the latter improved the performance of the model.

In the same line of research, Sousa et al. (2006) studied the next day daily mean of ozone. The performances of FANN, multiple linear regression (MLR) and time series techniques (TS) were compared. The FANN outperformed the previous techniques, TS was incapable of deliver good results (as mentioned earlier, this techniques have a strong connection with linearity and stationary of the processes) and MLR was not good enough in describing the non-linear behaviour of the presented relationship.

Later on, Sousa et al. (2007) developed an improved model, coupling with the FANN model a principal component pre-treatment. This conjunction allowed: i) to handle more easily all the predictors into a few principal components, ii) to evaluate the influence of these variables on ozone formation, and iii) to predict the next day hourly ozone concentrations keeping or improving its performance but decreasing its complexity.

Nagendra and Khare (2006) presented an ANN to model the dispersion of the vehicular exhaust emissions. The model took into account not also meteorological variables but also traffic variables. Finally, the model was able also to predict the nitrous dioxide concentration. This study presented acceptable performances, especially when both types of variables were used.

Over the years, with the increasing human pressure over nature, expanding over large portions of the earth is affecting not only the ground, water, but also the clime. The necessity of study this influence becomes a major concern around the world.

Two main questions that arise now are: i) how do these changes affect the other living creatures on earth? and ii) how can the man act to reduce or control this influence?

Obviously, to perform field studies about this impact is almost impossible and it is crucial to develop new techniques to map the biodiversity and to perform it on large scales. With the development of imaging techniques, the capability of getting images of large areas with a high level of details has become reality. Along with them, methods to analyze those images and extract useful and valuable information have been developed too.

Foody and Cutler (2006) presented a study to determine the species richness, evenness and composition over a region. Two neural network models were used to derive biodiversity information from the remotely sensed imagery. First, standard feedforward neural networks were used to estimate the species richness and evenness of the sample plots from the remotely sensed response. Second, self organizing maps were implemented for mapping species composition. The results presented were very promising when compared to traditional approaches; even though in this case it was considered only one specie, the procedure showed potentialities to expand it to cover more species.

Biodiversity research does not necessarily focus only in benign or in danger species. Worner and Gevrey (2006) studied the global insect pest species assemblages with the objective to determine the risk of invasion. Data comprising the presence and absence of 844 insect pest species recorded over 459 geographical regions world-wide were analyzed using a self-organizing map. The SOM analysis classified the high dimensional data into two-dimensional space such that geographical areas that had similar pest species assemblages were organized as neighbours on a map or grid. It allowed ranking the species in terms of their risks of invasion in each area based on the strength of the association with the assemblage that was characteristic for each geographical region. This risk assessment is valuable, giving information that can be reinforced with local studies that can lead to important decisions in terms of biosecurity.

Finally, Harrison et al. (2006) worked on the impact of climatic changes on species' distribution across Europe. The study focused on 47 species over 10 specific places in Europe, correlating their behaviours with bioclimatic data. The results were satisfactory showing different patterns for each species, allowing simulations of the climate changes and the effects over the species.

CHEMICAL ENGINEERING

Artificial Neural Networks have been successfully applied in many areas of engineering. Their ability to describe complex and nonlinear relationships within their structures has been valuable. In chemical engineering, the applications can be found in almost all areas. In this document three major topics are highlighted: Modelling, Control and Software Sensors.

Modelling

Modelling is always difficult and challenging. To describe the interactions taking place into one specific process, in a rigorous and deterministic way, is often not possible. Models too complex affect the quality of the modelling in cases where there is no processing capacity available. This finding is increased when there is lack of knowledge.

As mentioned before, the nonlinear nature of the ANN has been very useful for developing models. The variety of modelling applications in chemical engineering is huge, but here only a few are covered (see Table).

Membrane filtration is a separation process used often across the chemical industry. Its performance is governed by the hydraulic and physicochemical conditions of the system (Bowen and Jenner, 1995).

Table 4. Chemical engineering/Modelling applications

Application	Technique	Authors
Membrane separation	FANN, Genetic algorithms	Sahoo and Ray, 2006 Al-Zoubi et al., 2007
Wastewater treatment processes	FANN, Principal components	Tomenko et al., 2007 Aleboyeh et al., 2008 Moral et al., 2008
Emissions of fuel engines	FANN	Canakci et al., 2006 Sayin et al., 2007
Kinetics	FANN, Principal components	Durán et al., 2006 Ni andWang, 2007
Biochemical processes	FANN	Bas and Boyaci, 2007
Photovoltaic systems	FANN	Mellit et al., 2007
Impact sensitivity of explosives	FANN	Keshavarz and Jaafari, 2006
Food drying processes	FANN	Erenturk and Erenturk, 2007
Air properties in cooling processes	FANN	Yigit and Ertunc, 2006

Desalination is one process where this technique can be applied. Al-Zoubi et al. (2007) studied the treatment of highly concentrated salty solutions with nanofiltration membranes. Through the development of FANN models it was possible to predict with a high accuracy the behaviour of rejection versus pressure and flux. This study demonstrates the potential of ANN models in desalination processes.

Sahoo and Ray (2006) worked on the prediction of flux decline in crossflow membranes. The research was focused on the optimization of the neural network topology. Applying genetic algorithms instead of using the traditional trial and error methodology made possible to establish the best ANN configuration for a given problem. It demonstrates the effect of network topology on the prediction performance efficiency of ANN models and it presents a method to determine the optimal design network using genetic algorithms.

Environmental regulations, high disposal costs, recycling policies, among others, have made wastewater treatment a key process not just to industry, but to society in general. Constructed treatments wetlands represent a cost/efficiency alternative to deal with wastewater treatments.

In Tomenko et al. (2007), a comparison of multiple regression analysis and ANN models to predict the biochemical oxygen demand (BOD) is presented. In addition to these techniques, a pre-treatment with principal components was applied to improve their performances. The ANN models outperformed the regression model.

Moral et al. (2008) worked on predicting the performance of a wastewater treatment plant, creating a routine to study the influence of the ANN parameters to search the optimal configuration. In addition to the topology characteristics mentioned before, the program included also several training methods and different neural functions. The obtained network was able to predict successfully the behaviour of the plant.

Even though, in wastewater treatment plants the biological processes are predominant in the facilities, sometimes those processes are not enough. That is why it is important to carry out certain pre-treatments to handle those substances that are non-biodegradable. This is the case presented by Aleboyeh et al. (2008), where the photochemical decolourisation (with hydrogen peroxide and UV light) of a non-biodegradable dye (C.I. Acid Orange 7) was modelled with good results.

Several studies have been carried out trying to reduce the global pollution generated by transportation. The objectives of these studies were not only to study the efficiency of the engines, but also the influence of the fuel selected and the subsequent emissions. In these ideas Canakci et al. (2006) and Sayin et al. (2007) based their works. They developed ANN which successfully predicted the performance and emissions of engines fuelled with biodiesel and gasoline respectively. In both cases the models obtained presented a great accuracy allowing them to study the systems in other conditions.

Kinetics can be very troublesome when facing complex, consecutive and multicomponents reactions. The ANN can be used to obtain empirical expressions or even the kinectic parameters. Ni and Wang (2007) proposed a simple method based on chemometrics. In this special case, the determination of iodate and periodate use data from a spectrophometer that was correlated with regression techniques (multiple linear regression, principal components, partial least squares (PLS) and ANN). Reliable results were obtained with PLS and with principal components associated with ANN.

Similar approaches are found in biochemical processes. In this case, ANN are used to describe the physicochemical behaviour present in processes with living organisms. One of these situations corresponds to the enzymatic reactions. Bas and Boyaci (2007) worked with these reactions highlighting the power of the ANN to adapt and explain these systems. In comparison of a traditional method used in biochemical processes, such as response surface methodology (RSM), the ANN used had better performance.

Seeking for renewable and cleaner sources of energy, the photovoltaic systems are emerging as alternatives forms of energy. Mellit et al. (2007) developed an ANN model to simulate a standalone photovoltaic system. The model, not only described the solar cell, but also included a model for every component of the whole system.

The final result was a model capable to predict the photovoltaic system performance based on meteorological data. Even more, the models developed for each piece of the system allowed to know the intermediate signals, therefore improving the sizing process of the equipment and the possibility of coupling a control system.

Developing and testing explosives materials always represent big risk and security measures. Keshavarz and Jaafari (2006) presented an ANN model capable of predict the impact sensibility (is the height that a given mass must be dropped onto the sample to produce an explosion of 50% of the test trial) based on the molecular structure of the explosive compound. The model showed superior performance than the theoretical model based on quantum mechanical, besides its generality and simplicity.

Drying food is a common activity looking forward to obtain dehydrated goods. Erenturk and Erenturk (2007) presented a study where dynamic modelling of drying carrots was performed using diverse models such as: a deterministic model, data driven models using genetic algorithms and ANN. The drying kinetics were expressed in terms of the moisture content. The ANN showed a superior performance facing the others techniques. The accuracy of the results subject to diverse drying conditions demonstrates the high potential of this technique even when used for online monitoring.

Getting to know the outlet conditions in a heat exchanger is vital for a performance assessment. Yigit and Ertunc (2006) developed an ANN model capable of predicting the outlet temperature and humidity of a wire-o-tube type heat exchanger. This model is valuable, not only because of its accuracy (less than 2% error) but also because it allows the manufacturer to expose virtually the system to any operation conditions without experimenting.

Control

The potentialities of ANN are not only present on the modelling activities. The inner qualities of this technique result very appealing in process control. In this context, ANN offer diverse alternatives.

The power showed by ANN in modelling is closely related to their application to model predictive control (MPC). Besides this, ANN can perform other functions as proportional-integral-derivative (PID) autotuning, addressing control issues like robustness and fault-tolerance, etc. Table summarizes some of these applications in chemical engineering.

After the development of process models, it is possible to use these expressions for control purposes. Having a complete description of the process allows to engineers performing optimization procedures and determining a control vector trajectory to achieve certain objectives over the future. These predicted values are obtained from the process models (Nagy, 2007). This is the rationale behind model predictive control (MPC).

Lu and Tsai (2007) developed a system coupling fuzzy logic and recurrent neural networks (RNN). These RNN have a similar topology as FANN, but also take into consideration the past events to predict the future ones. The fuzzy logic is used to handle the uncertainties that are always present in industrial operations. The system showed a good performance in terms of setpoint tracking and disturbance rejection.

Table 5. Chemical engineering/Control applications

Nu	Technique	Authors
Generalized predictive control for industrial processes	RNN, Fuzzy logic	Lu and Tsai, 2007
Control of membrane separation processes	FANN	S. Curcio et al., 2006
Yeast fermentation control	FANN	Nagy, 2007
Temperature control in long ducts	FANN	Aggelogiannaki et al., 2007
Control of distributed parameter systems (DSP)	FANN, Fuzzy logic	Aggelogiannaki and Sarimveis, 2008
Predictive control of an evaporator system	RNN	Al Seyab and Cao, 2008
Batch reactor control	FANN	Mujtaba et al., 2006 Zhang, 2008
Robust fault-tolerant control	FANN	Wang et al., 2007
PID autotuning	FANN, Principal components	D'Emilia et al., 2007

In a specific case, Curcio et al. (2006) presented the application of model predictive control in membrane separation processes. Through the development of a model that describes the complex behaviour of this process, it was possible to determine the feed flow rate that maximizes the permeate flux across the membrane. It demonstrated that is possible to operate these systems with a variable feed (instead of the traditional fixed feeds) improving the separation process.

Nagy (2007) made a comparison between linear model predictive control (LMPC), non-linear predictive control (developed with FANN) and simple PID control applied to the yeast fermentation process. The performance presented by the neural application was superior to the linear and PID devices. It was also presented an algorithm, called Optimal Brain Surgeon (OBS), which improves the performance of a FANN by pruning the network. Through this methodology, those nodes that added no significant information to the final result were excluded, resulting in smaller networks with the same performance as the full network topology.

Aggelogiannaki et al. (2007) developed a MPC to control the temperature distribution in a one dimensional long duct. The model related this distribution with the flow rate changes. This technique outperformed the traditional control approaches showing an accelerate response and an efficient handling of the system delays. Later, these ideas where expanded to any distribute parameter system (e.g. packed bed reactors, crystallizers, etc, systems which may vary in space and time; Aggelogiannaki and Sarimveis, 2008). This work highlights the importance of developing those models with historical data serving as predictors of the controlled variables.

Al Seyab and Cao (2008) reported satisfactory results controlling an evaporator system. In this case, the model used inside the MPC routine was developed with RNN.

A comparison of diverse control methods using neural networks is found in the work presented by Mujtaba et al. (2006). Three control strategies were studied: generic model control (GMC), direct inverse model control (DIC) and internal model control (IMC). All approaches were applied to a batch reactor case. In GMC, an FANN estimator is coupled with an inverse model of the process based on its fundamental equations. In the DIC, an inverse neural network model works as a controller, without feedback, supplying the control signal according to the desired set point. In the final case, IMC, an inverse model acts as the controller taking into account the error signal corrected with the possible model mismatches existing between the plant and the model itself. The results showed a superior performance of the GMC. However, it is important to highlight the potential of the neural-based controllers when they are trained with data covering almost all possible situations.

Continuing with batch operations, Zhang (2008) presented a methodology to address the lack of data in batch processes. The work showed that due to the successive runs it is possible to improve the next batch operation with the information of the previous runs. It is also presented a structure of three parallel networks to perform a bootstrap procedure, obtaining a better final representation than using only a single network.

Other application of the neural network is fault-tolerance. Tracking control of delicate systems (e.g. non-minimum phase process) is a difficult task, especially when the system can present mismatches. These faults can compromise the system performance and even its stability. Wang et al. (2007) demonstrated the utilization of a FANN to approximate and estimate online when a fault happens.

Finally, D'Emilia et al. (2007) showed a methodology to execute autotuning of PID controllers based on FANN. This method is capable to determine the parameters without waiting until the steady-state. The performances of the final controllers were comparable with the traditional methods, but the tuning time was significantly reduced.

Software Sensors

Nowadays in the industrial world, granting quality standards and raising productivity have become a lemma in the daily activities. To achieve these goals, it is necessary to monitor the process behaviour; which is commonly represented by some crucial variables (e.g. purity, physical or chemical properties, etc), frequently called primary variables. However, measuring online these variables often represents a difficult task due to some reasons, such as: i) technological (no equipment available for the required measurement) and ii) economical (the necessary equipment is too expensive; Dai et al., 2006; Kadlec et al., 2009).

Nevertheless, there are other process variables that can be easily measured online (secondary variables) and through these it is possible to build a relationship with the primary variables and 'infer' their conditions (Feyo de Azevedo et al., 1993). This is the rationale behind a soft sensor, it holds a relationship between primary and secondary variables generating, or inferring, a virtual measurement to replace a real sensor measurement (Yan et al., 2004; Desai et al., 2006; Lin et al., 2007).

There are mainly two types of software sensors: model-driven sensors and data-driven sensors. The first are based on phenomenological knowledge of the process; they should be preferred as far as it is available (Lin et al., 2007). Unfortunately, they present some drawbacks: i) models are not always available for this purpose (Lin et al., 2007; Kadlec et al., 2009) or ii) models are computationally intensive for real-time applications (Lin et al., 2007).

Table 6. Chemical engineering/Software sensor applications

Application	Objective	Variables	Technique	Authors
Food industry	Discrimination of yogurt varieties	P: Type of yogurt B: NIRS of yogurt	PCA + FANN	He et al., 2006
Food industry	Determination of the color of red grapes.	P: Anthocyanin concentration. B: Visible-NIRS.	PLS+FANN PCA+FANN	Janik et al., 2007
Membrane separation	Hydrogen determination	P: Permeate hydrogen concentration; permeate gas flux and residue hydrogen concentration. B: Operating conditions (temperature, pressures and flux)	FANN	Wang et al., 2006
Oil industry	Gasoline properties	P: density and boiling points. B: NIRS of gasoline.	FANN	Balabin et al., 2007
Oil industry	Measure ionic liquids in water in the working range (0 and 5 ppm),	P:Concentrations of toluene and EMISE[1] B: UV absorbance of a water solution	FANN	Torrecilla et al., 2007
Bioprocesses	Determination of invertase from *Saccharomyces carlsbergensis.*	P: Invertase concentration B: Glucose concentration, ethanol concentration, bioreactor volume, biomass concentration	FANN SVR	Desai et al., 2006
Bioprocesses	Predicting the values of the active biomass and recombinant protein concentrations in the streptokinase process	P: Active biomass and streptokinase. B: reactor volume, biomass concentration and substrate.	FANN SVR	Desai et al., 2006
Wastewater treatments	Determination of the effluent quality in a wastewater treatment plants.	P: Suspended solids and chemical oxygen demand. B:pH, temperature, Suspended solids and chemical oxygen demand.	Genetic algorithms + FANN	Pai et al., 2007
Wastewater treatments	Estimation of nutrient concentrations.	P: Concentrations of NH_4^+, PO_4^{3-}, NO_3^-. B:pH, Oxidation-Reduction Potential and Dissolved Oxygen	FANN	Hong et al., 2007
Energy processes	Evaluate the biomass boiler fouling.	P: Combustion gases compositions, heat transferred in each section and boiler fouling indexes. B: initial mass flows, initial compositions and boiler temperatures	FANN	Romeo and Gareta, 2006
Polymers production	Monitoring industrial polymerization processes	P: Polyethylene terephthalate (PET) viscosity B: Operating conditions (Pressures, temperatures and mass flow)	FANN	Gonzaga et al., 2009

[1] EMISE: 1-ethyl-3-methylimidazolium ethylsulphate ionic liquid.

The latter models (data-driven models) aren't based on any knowledge about the process dynamics but only on historically process databases. To correlate these databases, several methods can be used, such as ANN and multivariate statistical methods like MLR, principal component analysis, PLS (Lin et al., 2007) and support vector regression (Lin and Liu, 2005) among others.

In this particular case, a special attention will be focused in the implementation of ANN to build software sensors. As referred earlier, the capacity of ANN to approximate non-linear behaviours, simple operation after the training phase, their robustness or fault tolerance and the possibility to couple ANN with other techniques to improve their performance, have reinforced their utilization in this field (Dai et al., 2006; Romeo and Gareta, 2006).

Table 6 shows some applications of ANN in software sensors development. It shows not only the general application, but also the main objective in each application and the correlated variables, predicted (P) and base (B).

CONCLUSIONS

In the course of this chapter was demonstrated, through many examples, the relevance that the artificial neural networks have in the engineering world. In this specific case applied in Environmental Sciences and Chemical Engineering.

This vast field of applications is a result of the powerful of artificial neural networks' characteristics, such as: i) a non-linear structure able to describe complex non-linear relationships, ii) the capacity to learn by examples (i.e. data driven technique) requiring no previous knowledge about the process and its variables, iii) the robustness of their structures, and iv) the accuracy and precision of the final network.

Also it was shown that even though feedforward artificial neural networks were the most used topology in the referred works, there were others used for more specific applications. Self-organizing maps revealed a big impact when visualization or clustering analysis was required. Recurrent neural networks had a special attention in some control situations due to their ability to use data from the past to predict the future.

An important general feature of artificial neural networks was the possibility to couple them with other techniques, such as: i) principal components, ii) fuzzy logic, iii) genetic algorithms, iv) partial least squares, v) time series methods, among others. This coupling in all cases resulted in an important improvement of the final result.

Into the specific areas, it could be seen that modelling was the main activity among the applications in environmental sciences. On the other hand, in chemical engineering, were highlighted three main topics (modelling, control and software sensors). In each and every one of the publications, the artificial neural networks performed in an efficient way.

However, it is also important to mention the drawbacks present by artificial neural networks:

- Data driven technique. It is crucial that the training and validation datasets are as broad as possible, related to almost every likely situation in the case study. Doing this, the final network will be more robust and will be prepared for any situation without compromising its performance. It is also important to keep in mind (when the datasets are too large and diverse) the possibility to perform any clustering pretreatment and developing modular neural models.

- Black box. As a consequence of the previous point, artificial neural networks' are frequently considered as black box models, which bring some issues when trying to give a meaning to the model or getting intermediate responses. In addition, this black box consideration is the cause of a certain resistance in the scientific community towards the application of these models.
- Lack of standardization. There are neither explicit rules nor established consensus towards the implementation of artificial neural networks and the definition of their parameters (topology, training methods, splitting dataset, transfer functions, etc). Nowadays trial and error continues to be the preferred methodology to obtain an 'optimal' network.

Finally, and being aware not only of the potential benefits, but also of the possible drawbacks of this technique, it can be stated that artificial neural networks are very powerful for future applications in several fields.

ACKNOWLEDGMENTS

This work was supported by *Fundação para a Ciência e a Tecnologia* (FCT). D.J.D. Gonçalves also thanks the FCT for the fellowship SFRH/BD/33644/2009.

REFERENCES

Aggelogiannaki, E., Sarimveis, H. & Koubogiannis, D. (2007). Model predictive temperature control in long ducts by means of a neural network approximation tool, *Applied Thermal Engineering*, *27(14-15)*, 2363-2369.

Aggelogiannaki, E. & Sarimveis, H. (2008). Nonlinear model predictive control for distributed parameter systems using data driven artificial neural network models, *Computers & Chemical Engineering*, *32(6)*, 1225-1237.

Agirre-Basurko, E., Ibarra-Berastegi, G. & Madariaga, I. (2006). Regression and multilayer perceptron-based models to forecast hourly O3 and NO2 levels in the Bilbao area, *Environmental Modelling & Software*, *21(4)*, 430-446.

Al-Zoubi, H., Hilal, N., Darwish, N. A. & Mohammad, A. W. (2007). Rejection and modelling of sulphate and potassium salts by nanofiltration membranes: neural network and Spiegler-Kedem model, *Desalination*, *206(1-3)*, 42-60.

Al Seyab, R. K. & Cao, Y. (2008). Nonlinear system identification for predictive control using continuous time recurrent neural networks and automatic differentiation, *Journal of Process Control*, *18(6)*, 568-581.

Aleboyeh, A., Kasiri, M. B., Olya, M. E. & Aleboyeh, H. (2008). Prediction of azo dye decolorization by UV/H2O2 using artificial neural networks, *Dyes and Pigments*, *77(2)*, 288-294.

Balabin, R. M., Safieva, R. Z. & Lomakina, E. I. (2007)., Comparison of linear and nonlinear calibration models based on near infrared (NIR) spectroscopy data for gasoline properties prediction, *Chemometrics and Intelligent Laboratory Systems*, *88(2)*, 183-188.

Bas, D., Boyaci, I. H. (2007). Modeling, and optimization II: Comparison of estimation capabilities of response surface methodology with artificial neural networks in a biochemical reaction, *Journal of Food Engineering*, *78(3)*, 846-854.

Borrego, C., Tchepel, O., Barros, N. & Miranda, A. I. (2000). Impact of road traffic emissions on air quality of the Lisbon region, *Atmospheric Environment*, *34(27)*, 4683-4690.

Bowen, W. R. & Jenner, F. (1995). Theoretical Descriptions of Membrane Filtration of Colloids and Fine Particles - an Assessment and Review, *Advances in Colloid and Interface Science*, *56*, 141-200.

Canakci, M., Erdil, A. & Arcaklioglu, E. (2006)., Performance and exhaust emissions of a biodiesel engine, *Applied Energy*, *83(6)*, 594-605.

Chang, F. J. & Chang, Y. T. (2006)., Adaptive neuro-fuzzy inference system for prediction of water level in reservoir, *Advances in Water Resources*, *29(1)*, 1-10.

Chen, J. & Adams, B. J. (2006). Integration of artificial neural networks with conceptual models in rainfall-runoff modeling, *Journal of Hydrology*, *318(1-4)*, 232-249.

Curcio, S., Calabro, V. & Iorio, G. (2006). Reduction and control of flux decline in cross-flow membrane processes modeled by artificial neural networks, *Journal of Membrane Science*, *286(1-2)*, 125-132.

D'Emilia, G., Marra, A. & Natale, E. (2007)., Use of neural networks for quick and accurate auto-tuning of PID controller, *Robotics and Computer-Integrated Manufacturing*, *23(2)*, 170-179.

Dai, X., Wang, W., Ding, Y. & Sun, Z. (2006). "Assumed inherent sensor" inversion based ANN dynamic soft-sensing method and its application in erythromycin fermentation process, *Computers & Chemical Engineering*, *30(8)*, 1203-1225.

Dawson, C. W., Abrahart, R. J., Shamseldin, A. Y. & Wilby, R. L. (2006). Flood estimation at ungauged sites using artificial neural networks, *Journal of Hydrology*, *319(1-4)*, 391-409.

De, S., Kaiadi, M., Fast, M. & Assadi, M. (2007). Development of an artificial neural network model for the steam process of a coal biomass cofired combined heat and power (CHP) plant in Sweden, *Energy*, *32(11)*, 2099-2109.

Desai, K., Badhe, Y., Tambe, S. S. & Kulkarni, B. D. (2006). Soft-sensor development for fed-batch bioreactors using support vector regression, *Biochemical Engineering Journal*, *27(3)*, 225-239.

Erenturk, S. & Erenturk, K. (2007). Comparison of genetic algorithm and neural network approaches for the drying process of carrot, *Journal of Food Engineering*, *78(3)*, 905-912.

Feyo de Azevedo, S., Chorão, J., Gonçalves, M. J. & Bento, L. S. M. (1993). Monitoring Crystallization, Part I, *International Sugar Journal*, *95(1140)*, 483-488.

Foody, G. M. & Cutler, M. E. J. (2006). Mapping the species richness and composition of tropical forests from remotely sensed data with neural networks, *Ecological Modelling*, 195 (1-2):37-42.

Gonzaga, J. C. B., Meleiro, L. A. C., Kiang, C. & Maciel, R. (2009). ANN-based soft-sensor for real-time process monitoring and control of an industrial polymerization process, *Computers & Chemical Engineering*, *33(1)*, 43-49.

Ham, F. M. & Kostanic, I. (2001). *Principles of neurocomputing for science and engineering*. New York [etc.]: McGraw Hill.

Han, D., Kwong, T. & Li, S. (2007). Uncertainties in real-time flood forecasting with neural networks, *Hydrological Processes*, *21(2)*, 223-228.

Harrison, P. A., Berry, P. M., Butt, N. & New, M. (2006). Modelling climate change impacts on species' distributions at the European scale: implications for conservation policy, *Environmental Science & Policy*, *9(2)*, 116-128.

He, Y., Feng, S., Deng, X. & Li, X. (2006). Study on lossless discrimination of varieties of yogurt using the Visible/NIR-spectroscopy, *Food Research International*, *39(6)*, 645-650.

Hong, S. H., Lee, M. W., Lee, D. S. & Park, J. M. (2007). Monitoring of sequencing batch reactor for nitrogen and phosphorus removal using neural networks, *Biochemical Engineering Journal*, *35(3)*, 365-370.

Jain, A. & Kumar, A. M. (2007). Hybrid neural network models for hydrologic time series forecasting, *Applied Soft Computing*, *7(2)*, 585-592.

Janik, L. J., Cozzolino, D., Dambergs, R., Cynkar, W. & Gishen, M. (2007). The prediction of total anthocyanin concentration in red-grape homogenates using visible-near-infrared spectroscopy and artificial neural networks, *Analytica Chimica Acta*, *594(1)*, 107-118.

Kadlec, P., Gabrys, B. & Strandt, S. (2009). Data-driven Soft Sensors in the process industry, *Computers & Chemical Engineering*, *33(4)*, 795-814.

Kangas, J. & Kohonen, T. (1996). Developments and applications of the self-organizing map and related algorithms, *Mathematics and Computers in Simulation*, *41(1-2)*, 3-12.

Keshavarz, M. H. & Jaafari, M. (2006). Investigation of the various structure parameters for predicting impact sensitivity of energetic molecules via artificial neural network, *Propellants Explosives Pyrotechnics*, *31(3)*, 216-225.

Kisi, O. (2008). The potential of different ANN techniques in evapotranspiration modelling, *Hydrological Processes*, *22(14)*, 2449-2460.

Kohonen, T. (1990). The Self-Organizing Map, *Proceedings of the Ieee*, *78(9)*, 1464-1480.

Kohonen, T. (1998). The self-organizing map, *Neurocomputing*, *21(1-3)*, 1-6.

Lin, B., Recke, B., Knudsen, J. K. H. & Jorgensen, S. B. (2007). A systematic approach for soft sensor development, *Computers & Chemical Engineering*, *31(5-6)*, 419-425.

Lin, J. P. & Liu, J. H. (2005). A wavelet kernel for support vector machine based on frame theory, *ISTM/2005: 6th International Symposium on Test and Measurement, Vols 1-9, Conference Proceedings*:4413-4416.

Lu, C. H. & Tsai, C. C. (2007). Generalized predictive control using recurrent fuzzy neural networks for industrial processes, *Journal of Process Control*, *17(1)*, 83-92.

Mellit, A., Benghanem, M. & Kalogirou, S. A. (2007). Modeling and simulation of a stand-alone photovoltaic system using an adaptive artificial neural network: Proposition for a new sizing procedure, *Renewable Energy*, *32(2)*, 285-313.

Moral, H., Aksoy, A. & Golcay, C. F. (2008). Modeling of the activated sludge process by using artificial neural networks with automated architecture screening, *Computers & Chemical Engineering*, *32(10)*, 2471-2478.

Mujtaba, I. M., Aziz, N. & Hussain, M. A. (2006). Neural Network Based Modelling and Control in Batch Reactor, *Chemical Engineering Research and Design*, *84(8)*, 635-644.

Nagendra, S. M. S. & Khare, M. (2006). Artificial neural network approach for modelling nitrogen dioxide dispersion from vehicular exhaust emissions, *Ecological Modelling, 190(1-2)*, 99-115.

Nagy, Z. K. (2007). Model based control of a yeast fermentation bioreactor using optimally designed artificial neural networks, *Chemical Engineering Journal, 127(1-3)*, 95-109.

Ni, Y. & Wang, Y. (2007). Application of chemometric methods to the simultaneous kinetic spectrophotometric determination of iodate and periodate based on consecutive reactions, *Microchemical Journal, 86(2)*, 216-226.

Pai, T. Y., Tsai, Y. P., Lo, H. M., Tsai, C. H. & Lin, C. Y. (2007). Grey and neural network prediction of suspended solids and chemical oxygen demand in hospital wastewater treatment plant effluent, *Computers & Chemical Engineering, 31(10)*, 1272-1281.

Romeo, L. M. & Gareta, R. (2006). Neural network for evaluating boiler behaviour, *Applied Thermal Engineering, 26(14-15)*, 1530-1536.

Sahoo, G. B. & Ray, C. (2006)., Predicting flux decline in crossflow membranes using artificial neural networks and genetic algorithms, *Journal of Membrane Science*, 283 (1-2):147-157.

Sayin, C., Ertunc, H. M., Hosoz, M., Kilicaslan, I. & Canakci, M. (2007). Performance and exhaust emissions of a gasoline engine using artificial neural network, *Applied Thermal Engineering, 27(1)*, 46-54.

Sousa, S. I. V., Martins, F. G., Pereira, M. C. & Alvim-Ferraz, M. C. M. (2006). Prediction of ozone concentrations in Oporto city with statistical approaches, *Chemosphere, 64(7)*, 1141-1149.

Sousa, S. I. V., Martins, F. G., Alvim-Ferraz, M. C. M., Pereira, M. C. (2007). Multiple linear regression and artificial neural networks based on principal components to predict ozone concentrations, *Environmental Modelling & Software, 22(1)*, 97-103.

Srinivasulu, S. & Jain, A. (2006). A comparative analysis of training methods for artificial neural network rainfall-runoff models, *Applied Soft Computing, 6(3)*, 295-306.

Tomenko, V., Ahmed, S. & Popov, V. (2007). Modelling constructed wetland treatment system performance, *Ecological Modelling, 205(3-4)*, 355-364.

Torrecilla, J. S., Fernandez, A., Garcia, J. & Rodriguez, F. (2007). Determination of 1-ethyl-3-methylimidazolium ethylsulfate ionic liquid and toluene concentration in aqueous solutions by artificial neural network/UV spectroscopy, *Industrial & Engineering Chemistry Research, 46(11)*, 3787-3793.

Wang, L., Shao, C., Wang, H. & Wu, H. (2006). Radial Basis Function Neural Networks-Based Modeling of the Membrane Separation Process: Hydrogen Recovery from Refinery Gases, *Journal of Natural Gas Chemistry, 15(3)*, 230-234.

Wang, Y. Q., Zhou, D. H. & Gao, F. R. (2007). Robust fault-tolerant control of a class of non-minimum phase nonlinear processes, *Journal of Process Control, 17(6)*, 523-537.

Worner, S. P. & Gevrey, M. (2006). Modelling global insect pest species assemblages to determine risk of invasion, *Journal of Applied Ecology, 43(5)*, 858-867.

Yan, W. W., Shao, H. H. & Wang, X. F. (2004). Soft sensing modeling based on support vector machine and Bayesian model selection, *Computers & Chemical Engineering, 28(8)*, 1489-1498.

Yigit, K. S. & Ertunc, H. M. (2006). Prediction of the air temperature and humidity at the outlet of a cooling coil using neural networks, *International Communications in Heat and Mass Transfer*, *33(7)*, 898-907.

Zhang, J. (2008). Batch-to-batch optimal control of a batch polymerisation process based on stacked neural network models, *Chemical Engineering Science*, *63(5)*, 1273-1281.

In: Focus on Artificial Neural Networks
Editor: John A. Flores, pp. 75-95
ISBN: 978-1-61324-285-8

Chapter 4

ESTABLISHING PRODUCTIVITY INDICES FOR WHEAT IN THE ARGENTINE PAMPAS BY AN ARTIFICIAL NEURAL NETWORK APPROACH

R. Alvarez *** and J. De Paepe***
Facultad de Agronomía, Universidad de Buenos Aires-CONICET,
Buenos Aires, Argentina

ABSTRACT

The Pampas of Argentina is a vast fertile plain that covers approximately 60 Mha and is considered as one of the most suitable regions for grain production worldwide. Wheat production represents a main national agricultural activity in this region. Usually, regression techniques have been used in order to generate wheat yield models, at regional and subregional scales. In a whole regional analysis, using these techniques, climate and soil properties explained 64% of the spatial and interannual variability of wheat yield. Recently, an artificial neural network (ANN) approach was developed for wheat yield estimation in the region. In this chapter we compared the performance of multiple regression methods with the ANN approach as wheat yield estimation tools and propose developing productivity indexes by the latter technique. The ANN approach was able to generate a better explicative model than regression, with a lower RMSE. It could explain 76% of the interannual wheat yield variability with positive effects of harvest year, soil available water holding capacity, soil organic carbon, photothermal quotient and the ratio rainfall/crop potential evapotranspiration. Considering that the input variables required to run the ANN can be available 40-60 days before crop harvest, the model has a yield forecasting utility. The results of the ANN model can be used for estimating climate and soil productivity. A climate productivity index developed assessed the effect of the climate scenario and its changes on crop yield. A soil productivity index was also elaborated which represents the capacity to produce a certain amount of harvest grain per hectare, depending on soil characteristics. These indices are tools for characterizing climatic regions and for identifying productivity capabilities of soils at regional scale.

* Corresponding author: Email: ralvarez@agro.uba.ar

The methodology developed can be applied in other cropping areas of the World and for different crops.

Keywords: wheat yield, productivity indices, Argentine Pampas.

Environmental Factors Controlling Wheat Yield in the Pampas

The Argentinean Pampas (located between 28 and 40°S and 68 and 57°W) covers approximately 60 Mha (Alvarez and Lavado, 1998), and is considered as one of the most suitable areas for grain production worldwide (Satorre and Slafer, 1999). The region is a vast plain with a relief that is flat or slightly rolling. Grasslands represent its natural vegetation and graminaceous plants species dominate. Mean annual temperature ranges from 14°C in the south to 23°C in the north and mean annual rainfall varies from 200 mm to 1200 mm from west to east. Cropping is a regular feature of land use in the humid and semiarid portions of the region on well-drained soils, mainly Mollisols formed on loess like materials, and areas with hydromorphic soils are devoted to pastures (Hall et al., 1992). At present, nearly 50% of the area is under agricultural use, and wheat is one of the main crops widespread over the whole region (Hall et al., 1992) with an annual sown area of around 5 Mha (MinAgri, 2010). The wheat growing cycle starts in July and ends the last weeks of November, and the fallow period usually runs from April to June; although the exact dates of both periods can vary through the differing pampean subregions.

The effect of climate on crop yield has been extensively studied in the Argentinean Pampas. For field experiments widespread along the region, under water and nutrient non limiting scenarios, the phototermal quotient (ratio between incident radiation to temperature during the critical period of one month prior to anthesis) accounted for nearly 52 % of the interannual wheat yield variability (Magrin et al., 1993). In a whole regional analysis of the effect of climate factors on above ground net primary productivity of wheat, combining county yield statistics and the scarce information available on harvest index, it was demonstrated that rainfall and temperature accounted for 27 % of the variance (Veron et al., 2002). Similarly, it has been showed (Verón et al., 2004) that 34 % of wheat yield variability is explained by the photothermal quotient when using also yield information at county level. Yield is also lower in areas with drainage problems (Verón et al., 2002).

In order to perform researches about the effect of climate on wheat yield in some pampean subregions, down scaling and disaggregating data over space is necessary. At this smaller scale, the variability of climate factors is smaller. In the southernmost pampean area, a humid plain with fine textured soils of high organic matter content, it has been assessed that water deficit from 30 days before to 10 days after the flowering period, and mean temperature during grain filling, accounted for more than 50% of on-farm yield variance (Calviño and Sadras, 2002). In this subregion some researches have been performed on the effect of soil properties on wheat yield. Also, it has been demonstrated that yields are higher in deep soils (100-120 cm free rotting depth) than in shallower ones (Sadras and Calviño, 2001). The western subregion of the Pampas has a semiarid climate and soils of coarse texture with a medium to low organic matter content. Wheat yield is correlated to soil organic matter following a linear-plateau trend (r^2= 0.48) with a critical level at 72 t SOM ha^{-1} in the upper 20 cm of the soil profile (Diaz-Zorita et al., 1999). In the central Pampas, a humid subregion with fine deep soils of medium organic carbon levels, 50-70 % of wheat yield variability was accounted for by rainfall and nutrient availability (Alvarez and Grigera, 2005, Sain and Jauregui, 1993).

Attempts for Predicting Wheat Yield in the Pampas Using Regression Techniques

A study was performed in the Pampas, employing commonly used regression methodologies, in order to generate models capable of predicting wheat yield at the entire regional scale using both climate and soil factors as independent variables (Alvarez, 2009). The region was subdivided into ten geographic units accounting for geomorphologic properties and soil classification considerations previously defined (INTA, 1980, 1989) (Figure 1), in which rainfall and temperature patterns were distributed homogenously within. Total surveyed area that was integrated in this analysis was about 26 Mha and it included around 60% of the surface devoted to wheat production of the Pampean Region.

Wheat yield data from 10 growing seasons were used (1995-2004) from statistics at county level and integrated to the geomorphologic level of the geographic units applying weighted averages per county, influenced by the corresponding surfaces. Wheat yield variability was very high among geographic units and years, ranging from 950 to 4130 kg ha^{-1}, with an average of 2500 kg ha^{-1}.

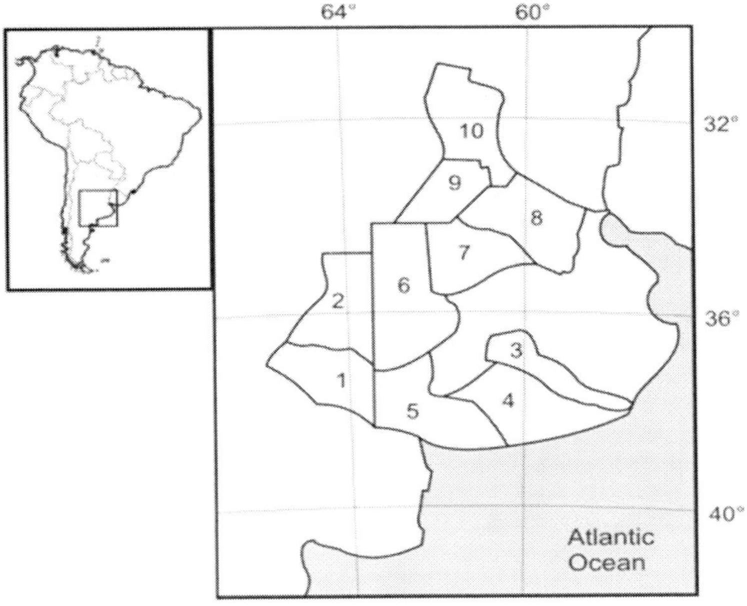

Figure 1. Map of the Pampean Region and the ten geographic units defined.

A climate dataset was generated using records provided by the National Meteorological Service (SMN, 2010). Crop potential evapotranspiration was calculated by a modification of the Penman formula (Linacre, 1977), with k_c coefficients adjusted locally (Doorenbos and Pruitt 1977, Totis and Perez, 1994). No k_c´s were available for the fallow period, therefore a value equivalent to the one corresponding to the sowing period (0.5) was assumed. At the top of the atmosphere, radiation was calculated using algorithms included in the RadEst 3.00 (Donatelli et al, 2003). The incoming solar radiation was estimated using a locally developed modification of the Hunt et al. (1998) method to calculate the atmospheric transmittance which allows a closer agreement between estimated radiation vs. radiometric measurements in the Pampean Region (Alonso et al., 2002). For the fallow period, and for different crop growing cycle periods, the ratio rainfall/crop potential evapotranspiration was calculated. Anthesis dates varied according to latitude in the Pampas from September 30 in the north to November 10 in the south. At different latitudes the dates were taken from experiments published in Magrin et al. (1993) and for the intermediate latitudes, the estimations were performed using unpublished data (F. Menéndez, personal communication). The photothermal quotient was calculated for the wheat critical period, the period of one month before anthesis, using the estimated incoming radiation and mean daily temperature above a base temperature of 4.5 °C (Magrin et al. 1993).

In the region there was a four-fold difference in incoming solar radiation during the crop critical period, and this variation resulted in a photothermal quotient range of 1.09 to 2.22 MJ m^{-2} d^{-1} $°C^{-1}$ (Table 1). Similarly, a five-fold difference was observed of the sum of rainfall during fallow and the entire crop growing period which resulted in a rainfall/crop potential evapotranspiration range of 0.30 to 2.0. Rainfall during fallow and crop growing periods were positively and significantly associated. Conversely, a negative correlation was observed between the photothermal quotient and rainfall during the wheat reproductive stage.

Table 1. Variability of climate variables in the ten geographic units. [1] during the crop growing cycle. [2] for the critical period of one month before anthesis

Geographic unit	Temperarure[1] (°C)	Radiation[2] (MJ m^{-2} d^{-1})	Photothermal quotient (MJ m^{-2} d^{-1} $°C^{-1}$)	Rainfall		
				Fallow period (mm)	Vegetative period (mm)	Reproductive period (mm)
1	14,8	21,9	1,71	126	100	171
2	15,9	19,5	1,73	144	87,0	192
3	14,1	20,2	1,93	201	173	213
4	14,5	20,0	1,82	194	171	195
5	14,4	21,3	1,72	123	121	176
6	15,8	15,5	1,76	182	104	233
7	16,5	16,8	1,59	242	117	245
8	16,9	18,0	1,52	227	133	230
9	17,2	18,1	1,58	177	68,3	191
10	18,5	18,3	1,68	218	96,6	264

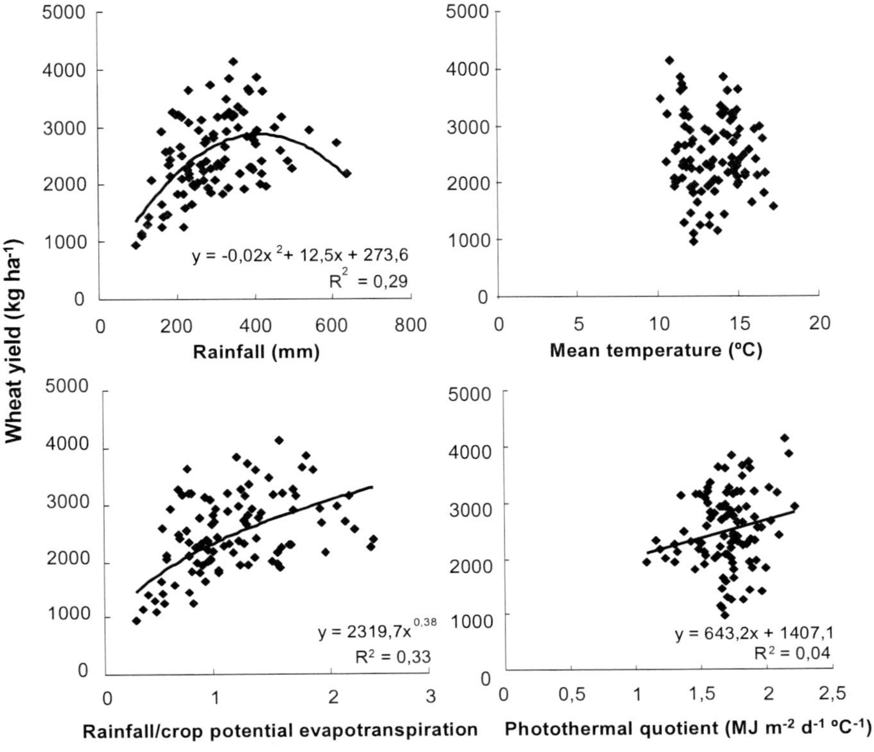

Figure 2. Simple regressions of wheat yield vs. climate variables. The variable rainfall and the ratio rainfall/crop potential evapotranspiration were measured during the fallow and crop vegetative growing periods. Temperature was the average of the temperature from the vegetative and the reproductive stages. The photothermal quotient was calculated for the crop critical period.

Yield was significantly correlated to rainfall during fallow and crop vegetative periods. Nevertheless, no significant association was detected between yield and rainfall during the reproductive stage. A quadratic model accounted for 33 % of yield variability when regressing yield against the ratio rainfall/crop potential evapotranspiration of fallow and vegetative periods summed (Figure 2). The fitted model could not be improved including data from rainfall during the reproductive stage. Yield was not significantly correlated to mean temperature along the growing cycle. This result can be attributed to the importance of soil water availability at sowing on crop yield as it has been reported in other agricultural regions of the World (Baier and Robertson, 1968). Considering that soil water content data at sowing are usually not available in the Pampas, this information was taken into account indirectly including rainfall during the fallow period in the ratio rainfall/crop potential evapotranspiration. Reduction of above-ground crop growth and yield are strongly influenced by early drought during the vegetative growth of wheat in the Pampas and it is mainly related to the plasticity (the ability to adapt their growth pattern to limited water supply using strategies like enhancement of effective rooting) of wheat cultivars (Brisson et al., 2001). In the southern portion of the Pampas, water deficit near the critical flowering period decreased wheat yield (Calviño and Sadras, 2002), but deficits during the vegetative period or during grain filling may also affect negatively crop yield (González Montaner et al., 1997). For the southern (Travasso and Delécolle, 1995) and the northern Pampas (Savin et al., 1995), wheat yield estimation, using CERES-wheat model, reached the maximum obtainable yield when rainfall during the crop growing period ranged from 450 to 500 mm. Even though the correlation was low, yield and the photothermal quotient were positively correlated; with an increase of 1 MJ m^{-2} d^{-1} $°C^{-1}$ in the photothermal quotient, wheat yield increased in average 640 kg ha^{-1} (Figure 2).

A soil dataset was collected from soil surveys (INTA 1980, 1981, 1983 and 1989). Based on the description of soil profile characteristics and its corresponding area, the weighted average values of the following independent variables were obtained: soil organic carbon, clay, silt, and sand contents, which were calculated using previously described techniques (Alvarez and Lavado, 1998) for three different soil layers: 0-20 cm, 20-50 cm and 50-100 cm. The obtained value of each variable was the mean soil organic carbon, clay, silt or sand content of the geographic unit at one of the three layers. Texture and organic matter contents were used to estimate soil bulk density (Rawls, 1983) and soil organic carbon contents were estimated to a depth of 50 cm expressed on an areal basis. Using the method of Rawls et al. (1982), soil available water holding capacity was estimated to a depth of 100 cm. In part of the southern and western Pampean subregions, the main soil-related constraint to cropping is a strong petrocalcic layer within the upper 100 cm of the profile that completely restricts root growth. During soil data integration, this constraining factor was accounted for when calculating soil water holding capacity within the free rotting depth.

Five out of the 10 geographical units had no impedance constraints within the first 100 cm of the soil profile, meanwhile in the other five, the average depth to the presence of a petrocalcic horizon varied from 77 to 88 cm (Table 2). Soil texture varied from sandy loam to silty clay loam. As a consequence, soil available water holding capacity, a soil variable that results from the combination of texture and free rooting depth, varied from 79 to 187 mm. Soil fertility was evaluated indirectly using the soil organic carbon content, and the resulting fertility rating was very different between units. Organic carbon content ranged from 41 to 126 t C ha^{-1} in the 0-50 cm soil layer.

Table 2. Variability of soil properties in the geographic units. [1] 0-100 cm depth or up to the petrocalcic horizon, [2] 0-50 cm depth

Geographic unit	Clay[1] (%)	Silt[1] (%)	Sand[1] (%)	Available water holding capacity[1] (mm)	Organic carbon[2] (%)	Average integrated depth (cm)
1	10,0	21,4	68,6	78,7	1,22	36,5
2	11,0	24,9	64,1	86,1	1,37	51,3
3	27,2	33,7	39,1	114	3,98	92,2
4	31,6	32,9	35,5	115	3,41	84,0
5	26,4	31,2	42,3	106	3,30	88,3
6	17,3	21,9	60,9	94,7	1,63	74,9
7	21,6	24,0	54,4	105	2,22	81,7
8	30,3	54,9	14,7	172	2,10	135
9	25,2	43,6	31,2	147	2,18	108
10	31,0	62,8	6,23	187	2,12	134

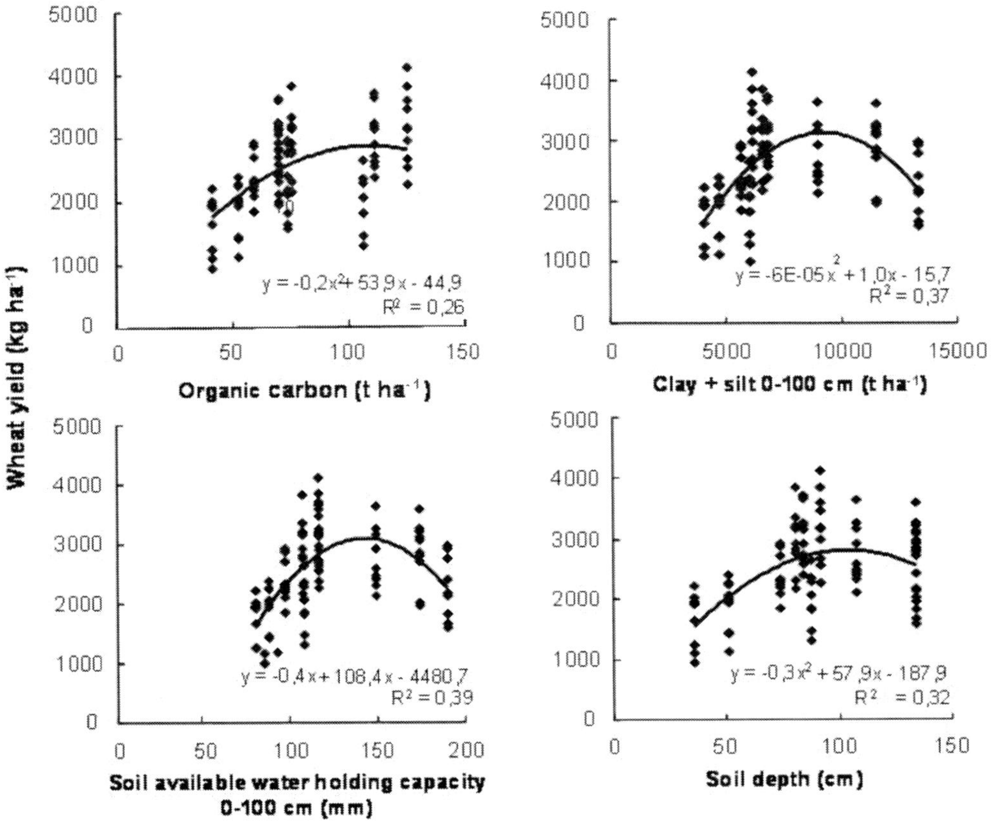

Figure 3. Simple regressions of wheat yield vs. soil variables.

The association between climate and soil variables was low. Positive correlations were observed between soil organic carbon content and rainfall. However, this latter variable was not significantly correlated to soil available water holding capacity and this soil property was mainly dependent on, and significantly determined by, soil clay and silt contents (Soil available water holding capacity (mm) = 31 + 0.012 clay + silt (t ha^{-1}), r^2= 0.98, P= 0.01).

Between yield and soil available water holding capacity a curvilinear relationship was observed, with the maximum yield attained in soils that can store up to 150 mm of available water in the upper 100 cm of the soil profile (Figure 3). The main soil factor controlling wheat yield appeared to be texture through its indirect effect on soil water properties. Soil water holding capacity is mainly derived from particle composition and free rotting depth and yields are higher in deeper soils with greater capacity to store water (Quiroga et al. 2001, Wong and Asseng 2006). As soil available water holding capacity was closely correlated to the sum of clay and silt contents of pampean soils, using the quadratic model, 37 % of yield variability (P = 0.01) was explained by the sum of clay and silt masses in the first 100 cm of the profile.

In other parts of the World, studies performed at different scales about the effects of soil properties on crop productivity have demonstrated the significant influence of texture or free rotting depth. At sub-field scale, plant available water storage capacity of soils regulated wheat productivity in Southern Australia (Wong and Asseng 2006) and soil texture was highly correlated to soybean yield in Mississippi soils (Cox et al., 2003). When collecting data from field experiments, significant relationships were demonstrated between texture and cotton yield in Central Greece (Kalivas and Kollias, 2001) and free rooting depth and crop yield in Spain (De la Rosa et al., 1981). In other words, the determination of the soil water holding capacity appeared to be necessary to develop yield prediction models under different scales of analysis and different soil-climate situations.

Soil organic carbon content was significantly correlated to yield, increasing until 90 t C ha^{-1}, stabilizing afterwards (Figure 3). The observed relationship seems to be more based on the impact of organic matter as a source of nutrients than on its influence on soil available water holding capacity. In the Pampas many soils in the southern portion of the region had high soil organic carbon contents but shallow depths which contrasted with soils in the northern portion, of medium soil organic carbon contents but with high fine particle contents and very deep profiles (Quiroga et al. 2001, Wong and Asseng 2006). At the regional level in the Pampas, soil organic carbon was not correlated to soil available water holding content capacity to 100 cm depth. In this regional assessment, soil water holding capacity was estimated using the Rawls et al. (1982). Organic matter contents of pampean soils only accounted for 1 % of soil available water holding content capacity variability estimated by this method.

Moreover, differing results have been obtained in studies on the effects of soil organic carbon on crop yields worldwide. Positive correlations have been found between soil organic carbon and crop yield as related to soil fertility in studies performed in other cropping areas (Catching et al. 2002; García-Paredes et al., 2000), while in other studies, no significant association was detected between both variables (Alvarez and Grigera, 2005, Jiang and Thelen, 2004). In the semiarid portion of the Pampean Region, soils present a wide spectrum of organic carbon contents and textures and free rotting depths. An on-farm survey of wheat yield related to soil properties carried out in this subregion demonstrated that soil organic carbon is correlated to wheat yield, independently of soil texture and depth (Bono and Alvarez, 2006). As a consequence, the inclusion of this soil property in yield prediction models can be useful in some situations, especially when the variability range is large; that is, including low soil organic carbon contents that restrict crop yield (Diaz-Zorita et al., 1999).

Multiple regression techniques were also tested for yield forecasting in the research. A polynomial surface response model was developed of the form:

$$\text{Yield} = a_0 + a_1 v_1 - a_2 v_1^2 + a_3 v_2 - a_4 v_2^2 + a_5 v_1 v_2 + \ldots + a_{n-2} v_x - a_{n-1} v_x^2 + a_n v_x v_{x-1}$$

Where:
a_0 to a_n: regression coefficients
v_1 to v_x: independent variables

In this model, linear and quadratic terms are incorporated as they assess linear and curvilinear effects, and the interaction terms between independent variables are also tested. This method has been of common use in agronomic experiment evaluation, with expected positive linear effects and negative quadratic effects (Colwell, 1994). In order to obtain the simplest model and the one with the highest r^2, a combination of forward, backward and stepwise regression adjustments were used. The final regression model was selected by a $P = 0.01$ by the F test and it included only statistically significant terms at $P = 0.05$. The VIF value was used to check the autocolinearity of independent variables (Neter et al., 1999). For assessing the generalization ability to other possible datasets of the selected regression model, a ten-fold cross validation technique was used. A hierarchical approach was also implemented to combine variables for calculation of other variables with the purpose of including the effects of the variables in the first level and allowing the simplification of the selected models (Schaap et al., 1998). From the regression of predicted vs. observed yield, slopes and intercepts were compared by the t test using IRENE (Fila et al., 2003).

The surface regression response model accounted for 64 % of the interannual wheat yield variance (Figure 4) and it included the following independent variables: harvest year, soil available water holding capacity, ratio rainfall/crop potential evapotranspiration and the photothermal quotient. The regression of observed against estimated values had an intercept of 0 and a slope of 1 ($P = 0.05$). Wheat yield was positively affected by harvest year and the photothermal quotient, whereas soil available water holding capacity and rainfall/crop potential evapotranspiration presented linear positive effects and curvilinear negative terms. The generalization ability of the regression model was not high ($R^2 = 0.53$) and this result was established by calculating the average determination coefficient of the ten-fold cross validation. Consequently, the model may be use for wheat yield forecasting in the Pampas, but it only can explain around 50 % of the variability.

Use of Artificial Neural Networks to Predict Wheat Yield

Artificial neural networks (ANN) have become a popular technique in biological sciences because of their predictive quality and due to the fact that they are simpler than process based models (Jorgensen and Bendoricchio, 2001, Özesmi et al. 2006). ANN are based on neural structures and processing of the brain and are adaptive analytical methodologies capable of learning relationships in information patterns (Jorgensen and Bendoricchio, 2001). Compared to empirical modeling techniques, ANN have the advantage of not assuming an a priory structure for the data, they are well suited for fitting non-linear relationships and complex interactions, and can expose hidden relationships among input variables (Batchelor et al., 1997).

The typical structure of an ANN is with three neural layers: an input layer in which the number of neurons corresponds to the number of input variables, a hidden layer with a complexity determined empirically during the ANN development, and an output layer with a neuron for each output variable (Figure 5). The information flow starts at the input layer, ending in the output layer, and this happens through the hidden layer. The learning process consists in adjusting weights associated to the transfer functions between neurons of the different layers and comparing the ANN outputs with observed data by an iterative procedure (Jorgensen and Bendoricchio, 2001). Usually, it is the back propagation algorithm that fits the weights during the learning process starting at the output layer and through the input layer (Kaul et al., 2005). A sigmoidal transfer function is commonly used between the hidden layer and the output layer, and a lineal function passes information from the input layer to the hidden layer (Kaul et al., 2005). The results of a neural network cannot be extrapolated outside the range of the input data, a common feature of empirical models. Some examples of the agronomic uses of ANN (Park and Vlek 2002) are: soil organic carbon content prediction

(Somaratne et al., 2005), fertilization recommendations (Broner and Comstock 1997), estimation of soil hydraulic properties (Nemes et al., 2003), crop development prediction (Elizondo et al., 1994), evaluation of epidemic severity (Batchelor et al., 1997), and yield prediction (Kaul et al., 2005).

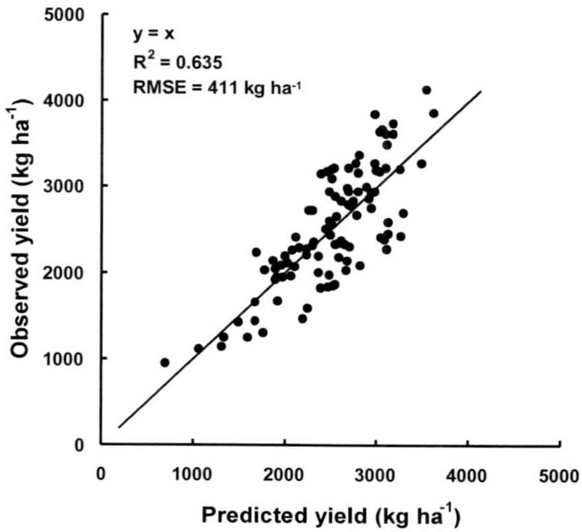

Figure 4. Observed vs. estimated wheat yield generated with a lineal multiple regression model.

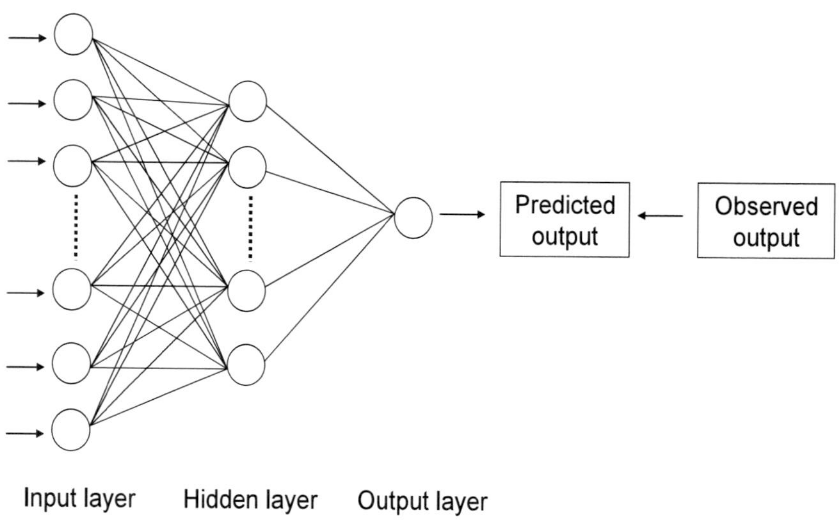

Figure 5. Representation of a feed-forward artificial neuronal network showing layers and connections.

An artificial neural network approach was tested in the Argentinean Pampas to estimate wheat yield using the same yield, climate and soil dataset previously described Alvarez (2009). It has been demonstrated that multilayer preceptors are well suited for managing datasets of similar size than this one for different agronomic purposes (Kaul et al., 2005,

Starrett et al. 1997). From the input layer to the hidden layer and from the output layer to the network output lineal transfer functions were used (Lee et al., 2003); while sigmoidal functions connected the hidden layer to the output layer. The minimax procedure was applied to scale the input variables between 0 and 1 to create uniform variation ranges and make data suitable for the sigmoid function (Park and VleK, 2002). Network outputs were de-scaled to original units. In the development of the ANN, a back propagation algorithm was used, in a supervised learning procedure, for weight fittings (Rogers and Dowla, 1994). For model simplification during the selection of input variables, a hierarchical approach was implemented in which the preferred variables were those that resulted from the integration of variables used for the network construction (Park and VleK, 2002). For input selection during the ANN testing the stepwise methodology was applied (Gevrey et al., 2003). The size of weight change made by the back propagation algorithm is controlled by the learning rate (Kaul et al. 2005). If the learning rate is large, it can result in a faster convergence but also in a local minimum (Lee et al. 2003). Accordingly, a low learning rate (0.1) was used during ANN development. The number of epochs (iterations) for which the algorithm will run is represented by the epoch size. At each epoch, the entire training set is fed through the network and it is used to adjust the network weights (Somaratne et al., 2005). In some situations, around 50 epochs are adequate for convergence (Schaap and Bouten, 1996, Schaap et al. 1998), in this study an epoch size of 100 was used. A selected model can fit better to the training dataset as the number of neurons in the hidden layer increases. However, the overlearning possibility increases too (Özesmi et al. 2006). As a consequence, a balance between ANN prediction ability and complexity must be reached. Using the methods described by Somaratne et al. (2005) the maximum initial number of neurons in the hidden layer was set and using the r^2 as a decision criterion, the neurons were deleted one at a time until model simplification reduced the model ability to fit the data. To avoid overlearning, cross validation is recommended (Özesmi et al. 2006). If the weight adjustments stop early, then the deviation from the verification dataset becomes larger than that from the training dataset (Park and VleK, 2002). The dataset was randomly partitioned into 2 sets: 70 % for training and 30 % for verification. Iteration and the network construction was stopped when the r^2 of the verification set tended to be lower compared to the r^2 of the training set. A modification of the procedure outlined by Schaap and Bouten (1996) was applied in order to test the generalization capacity of the models. The entire dataset was partitioned ten times into a 70:30 subsets of data, for training and verification respectively, and the best generated models with the first 70:30 partition were run against the remaining 70:30 groups. Making comparisons between the r^2 of the groups showed which model was able to predict wheat yield independently of the partitioned dataset and consequently could be apt for generalizing.

The best ANN accounted for 76 % of the wheat yield variance of the dataset (Figure 6). The network was structured with five neurons in the hidden layer and the inputs were: harvest year, soil available water holding capacity, soil organic carbon ratio rainfall/crop potential evapotranspiration during the fallow and vegetative growing period and photothermal quotient. The regression of observed against estimated values had an intercept non different from 0 and a slope of 1 (P = 0.05). The average determination coefficient, after the ten-fold 70:30 partitioning of the dataset for training and verification respectively, ranged from 0.76 to 0.80; which implies a good generalization ability of this method.

The ANN estimated a positive effect of harvest year on wheat yield, increasing along the studied period in average 52 kg ha^{-1} y^{-1}. This high increase of yield with time may be explained by the fact that the surveyed area included much of the more suitable regions for wheat production of Argentina, where genetic improvement (Calderini et al. 1995) and better management practices (Satorre and Slafer, 1999) are commonly implemented. The ANN model predicted a yield decrease for high ratio rainfall/crop potential evapotranspiration values that can be attributed to leaf diseases, which are a very serious and demonstrated constraint for wheat yield in the Pampean Region (Annone 2001). The use of this ratio, which integrated variables related to water available for crop development at different stages, allowed a better explanation of yield variance than the use of the simple variables in the construction of the ANN model (results not presented). Therefore, in spite of its simplicity it can be highlighted that this index explained 33 % of yield variance.

Wheat yield estimations performed by the multiple regression model and the ANN were contrasted. In order to compare the performance of both methodologies the mean square error (RMSE) (Kobayashi and Salam, 2000) was calculated and the differences between them tested by an F test (Xiong and Meullenet, 2006). The RMSE of the ANN model was significantly lower than the one obtained using the surface regression response technique.

When using both soil and climate variables for yield prediction through an ANN, it was demonstrated that it is a better tool than regression techniques. In this regional-scale dataset, the correlation between independent variables was generally low and only the variables significantly not correlated were included in the ANN in order to discard confounding effects. In this kind of studies, confounding effects generated by aotocolineality between independent variables is a potential problem and it may be partially eliminated by experimentation by fixing all conditions except the one tested (Bakker et al. 2005). Integration of information at regional scales allowed improving the fit of models by averaging outliers, and as the surface of the assessed region increases, the results also improve (Bakker et al. 2005).

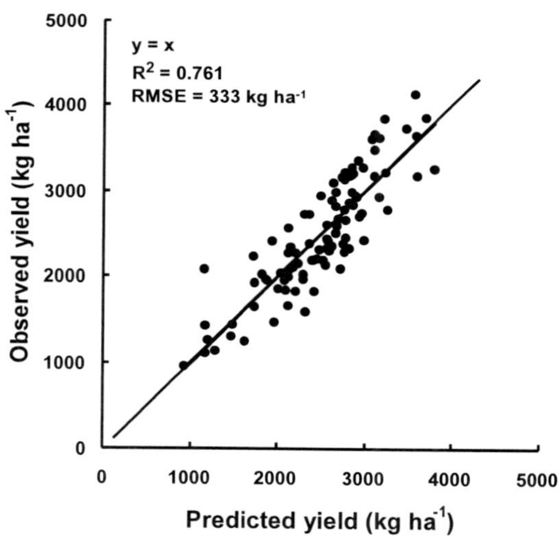

Figure 6. Relationship between observed and predicted wheat yield using an ANN.

In-season regional wheat yield prediction is possible using the results of an ANN as all variables needed to run the model are available 40-60 days before wheat harvest. In other agricultural regions different methodologies have been tested for in-season wheat yield prediction, for example using the NDVI (Freeman et al. 2003), or the application of agro-climatic models for sorghum (Potgieter et al., 2005); but these techniques are not available at present in the Pampean Region. The hierarchical ANN approach, meaning the use of combined independent variables, resulted in a simple model with good predictive capacity (higher R^2 and lower RMSE compared to the response surface regression model).

Establishing Productivity Indices by an Artificial Neural Network Approach

Characterization of the climate and/or soil situation, according to its influence on crop yield, can be useful to determine optimal management practices and adequate soil uses. Productivity indices can be elaborated inductively through environmental characterization, applying models based on previous theoretical knowledge or, on the contrary, deductively (empirical methods) with direct validation (Huddleston, 1984). When the calculation of productivity indices is empirical, it is based on the results of explicative yield models (for example multiple regressions, classification or regression trees, artificial neural networks, etc.) and validated against environmental variables of the evaluated region (García-Paredes et al., 2000).

As an option for improved and timely monitoring of crop production, climate indices have been generated related to yield in different cropping areas worldwide. Allen and Bakayama (2003) developed a crop water-stress index that is related to physiological indicators of the plant water status. They performed field experiments with guayule under well-watered and water-stressed conditions and concluded that there was no significant difference in net photosynthesis in both situations. Zhang et al., (2005) defined a climate-variability impact index as the monthly contribution to anomalies in annual growth, quantifying the percentage of the crop productivity either gained of lost due to climatic variability during a given month. The index uses remotely sensed data, specifically MODIS information, is based on the crop LAI and can provide both fine-scale and aggregated information on vegetation productivity of various crop types. The results show that 60 % of the variance in crop production is explained by variations in this index. Finally, by determining the estimated production as a function of the growing-season period it is possible to determine when, in the crop cycle, the predictive value of the index plateau is attained and which months provide the greatest forecasting capacity. Another climate-variability impact index was calculated by Irmak et al., (2000), that represents a potential tool for irrigation scheduling and yield estimation. The research was based on three irrigation treatments for corn grown under Mediterranean semiarid cropping conditions. The value of this water-stress index is determined by the relationship between canopy temperature minus air temperature and vapour pressure deficit of summer grown corn.

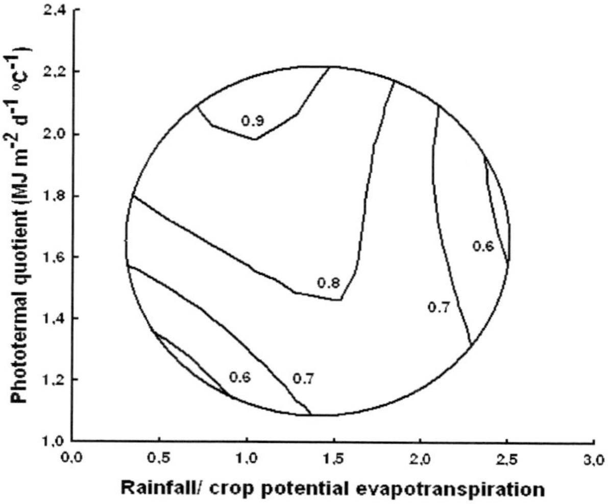

Figure 7. Climate productivity index under the average soil conditions estimated by an ANN approach. The circle indicates the dataset range with which the ANN was developed. Numbers near the curves represent the wheat productivity index.

In order to generate climate productivity indices for wheat in the Argentine Pampas, results obtained with the ANN model described in the previous section have been used (Alvarez, 2008). Maximum wheat yield was estimated by the ANN, which received a value of 1, and all the other yield results were expressed as relative terms under varying climate scenarios. The climate productivity index was calculated for an average soil condition. Climate impact on productivity, as predicted by the ANN, was characterized by a positive impact of the photothermal quotient on yield and an optimum ratio rainfall/crop potential evapotranspiration around the value of 1 (Figure 7). Wheat yield decreased as the ratio rainfall/crop potential evapotranspiration became lower or higher than 1, indicating possible drought negative effects or stress resulting from water excess. The index was only calculated within the range of the input variables.

Soil quality represents the combination of soil physical, chemical and biological properties that allows a soil to function within the limits of an ecosystem, maintain biologic cycling and environmental quality, and promote vegetal and animal health (Doran y Parkin, 1994; Arshad y Martin, 2002). As it is related to the soil function, this concept reflects an appreciation of soil's fitness for use and the capacity of soils to resist and recover from contamination and degradation (Wander et al., 2002). The concept considers not only the productive capacity of soil but also it's potential as a filter of toxic substances (Wander et al., 2002).

Soil productivity, a component of soil quality, can be defined as the capacity of a soil to produce plant biomass or crop seed (Yang et al., 2003, Sauerborn, 2002). The degradation of soil productivity, by human actions and/or natural processes, has increased the need to develop methods for the quantification of soil properties in the context of productivity (Kim et al., 2000; Yang et al., 2003). A soil productivity index represents the capacity to produce a certain amount of harvest per hectare and per year, expressed as a percentage related to the optimal productivity that an ideal soil would have in its first year of cropping. In general, productivity indices are multiplicative and related to soil properties and they are used as an evaluation method related to crop yield (FAO, 2007; Laya et al., 1998). Usually, to elaborate soil productivity indices, topographic factors (Iqbal et al., 2005), depth of the A horizon (Yang et al., 2003), factors related to water storage capacity of soil profiles (Martín et al., 2006), or factors associated to soil chemical fertility (Udawatta and Henderson, 2003) are used. Many times, to quantify soil fertility, the organic matter content is used (Bauer y Black, 1994; Stenberg, 1998). Nevertheless, it has not been possible until present to define thresholds of organic matter below which yield is limited (Loveland y Webb, 2003). In the Semiarid Pampean Subregion, for instance, a lineal-plateau tendency, with a critical level of 72 t ha^{-1} of organic matter in the first 20 cm, was found for wheat yield (Díaz-Zorita, 1999). Nevertheless, this threshold can be the result of a confounding effect between organic matter and soil available water holding capacity, because it was demonstrated that soil organic carbon is higher in this pampean subregion as soil capacity of retention of available water increases (Quiroga anf Funaro, 2004).

The soil available water storage capacity and its fertility, represented by the organic carbon content, have been used in the Argentine Pampas to characterize its soil productivity (Alvarez, 2008) using the ANN model. Maximum wheat yield obtained by the ANN received a value of 1 and all the other wheat yield results were expressed as relative terms of this maximum, dependent on soil properties. A productivity index for a soil condition was developed by characterizing soils under an average climate scenario. Since the ANN predicted effects of soil available water holding capacity and organic carbon on wheat yield, the soil productivity index was developed using these variables. The calculated index increased as both these soil properties increased (Figure 8). Soil productivity increases when the soil available water holding capacity increases along with the organic carbon content, meaning that there exists a positive interaction between these two variables. When soil available water holding capacity was low, yield increments related to greater soil organic carbon levels were low or even inexistent, but in soils with medium to high soil available water holding capacity, wheat yield increased as soil organic carbon rose. In the same way, the effect of soil available water holding capacity on yield was more pronounced in soils with high soil organic carbon levels.

In other parts of the World soil productivity indices have been generated employing soil variables as those used here. When interannual climate variance was eliminated, by using ten-year climate averages, up to 50% of the yield variance of corn and soybean in Illinois soils was explained using texture, rooting depth, organic matter and other properties to generate specific soil productivity indices (García-Paredes et al., 2000). Soil productivity has been also determined for soybean using the water holding capacity of Mexican soils (Yang et al., 2005), and the suitability of soils to produce sorghum in Australia was mainly established using indices accounting for soil water holding capacity and water balance (Potgieter et al., 2005).

Climate and soil productivity indices developed in the Pampas have only been calculated within the range of the data set, and so far no extrapolations have been performed. The effects of different combinations of climate and soil scenarios on wheat yield may be modeled with the ANN for productivity characterization in specific areas.

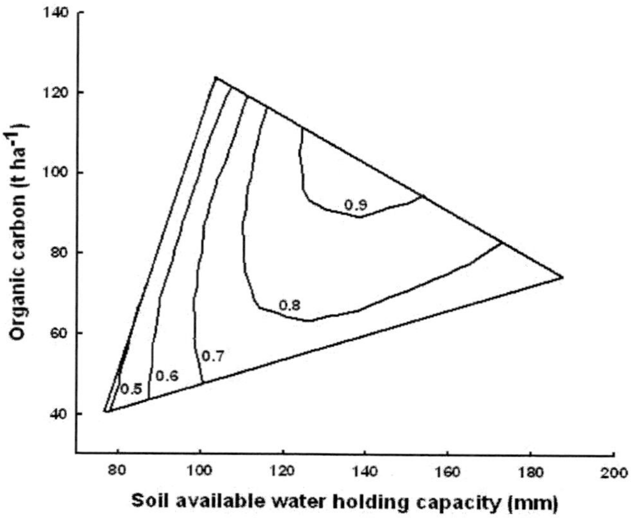

Figure 8. Soil productivity index under average climatic conditions estimated by an ANN approach. The triangle indicates the range of data from which the ANN was developed. Numbers near the curves represent the wheat productivity index.

Concluding Remarks

ANN has demonstrated to be a stronger statistical method than regression techniques to generate predictive wheat yield models at the regional scale of the Argentine Pampas. Compared to these other techniques, ANN explained a higher percentage of the wheat yield variance, with a smaller RMSE. The information needed in order to run the ANN yield model, can be available 40-60 days before crop harvest. Subsequently, this tool can predict wheat yield in-season at the whole pampean scale and the methodology could be applied in other cropping areas of the World and for different crops.

A useful application of the wheat yield model developed by an ANN approach was to generate climate and soil productivity indices. The climate productivity index, that can allow a climate characterization of the pampean subregions, and the soil productivity index, which can characterize soil productivities, are useful tools that may be used by farmers and decision-makers when assessing site potential for wheat production.

References

Allen, S. G. & Nakayama, F. S. (1988). Relationship between crop water stress index and other physiological plant water status indicators in guayule. *Field Crops Research*, *18(4)*, 287-296.

Alonso, M. R., Rodriguez, R. O., Gomez, S. G. & Giagnoni, R. E. (2002). Un método para estimar la radiación global con la amplitud térmica y la precipitación diarias. *Rev.* Fac. Agron. UBA *22*, 51-56.

Alvarez, R. (2008). Predicción del rendimiento y la producción de trigo en la Región Pampeana usando una red neuronal artificial. Congreso Nacional de Trigo. INTA. Santa Rosa, La Pampa, Argentina, 5.

Alvarez, R. (2009). Predicting average regional yield and production of wheat in de Argentine Pampas by an artificial neural netword approach. *Europ. J. Agronomy*, *30*, 70-77.

Alvarez, R. & Grigera, S. (2005). Analysis of soil fertility and fertilizer effects on wheat and corn yield in the Rolling Pampa of Argentina. *J. Agron. Crop. Sci.*, *191*, 321-329.

Alvarez, R. & Lavado, R. S. (1998). Climate, organic matter and clay content relationships in the Pampa and Chaco soils, *Argentina. Geoderma, 83*, 127-141.

Annone, J. G. (2001). Criterios empleados para la toma de decisiones en el uso de fungicidas en trigo. *Revista de Tecnología Agropecuaria*, *6*, 16-20.

Arshad, M. A. & Martin, S. (2002). Identifying critical limits for soil quality indicators in agro-ecosystems. *Agriculture, Ecosystems & Environment*, 8802), 153-160.

Baier W. & Robertson G.W. (1961). The performance of soil moisture estimates as compared with the direct use of climatological data for estimating crop yields. *Geoderma 5*. 17-31.

Batchelor, W. D., Yang, X. B. & Tschanz, A. T. (1997). Development of a neural network for soybean rust epidemics. *Trans. ASAE*, *40*, 247-252.

Bauer, A. & Black, A. (1994). Quantification of the effect of soil organic matter content on soil productivity. *Soil Sci Soc Am J, 58*, 185-193.

Bono, A. & Alvarez, R. (2006). Rendimiento de trigo en la Región Semiárida y Subhumeda Pampeana: un modelo predictivo de la respuesta a la fertilización nitrogenada. XX Congreso Argentino de la Ciencia del Suelo, *Proceedings* on CD, 5.

Bralmor, A. K. & Vlek, P. L. G. (2006). Soil quality and other factors influencing maize yield in northern Ghana. *Soil Use Managem.*, *22*, 165-171.

Brisson, N., Guevara, E., Meira, S., Maturano, M. & Coca, G. (2001). Response of five wheat cultivars to early drought in the Pampas. *Agronomie*, *21*, 483-495.

Broner, I. & Comstock, C. R. (1997). Combining expert systems and neural networks for learning site-specific conditions. *Comp. Elec. Agric.*, *19*, 37-53.

Calderini, D. F. & Slafer, G. A. (1998). Changes in yield and yield stability in wheat during the 20th century. *Field Crops Res.*, *57*, 335-347.

Calderini, D. F., Dreccer, M. F. & Slafer, G. A. (1995). Genetic improvement in wheat yield and associated traits. A re-examination of previous results and the latest trends. *Plant Breeding, 114*, 108-112.

Calviño, P. & Sadras, V. (2002). On-farm assessment of constraints to wheat yield in the southeastern Pampas. *Field Crop Res.*, *74*, 1-11.

Catching, W. E., Hawkins, K., Sparrow, L. A., McCorkell, B. E. & Rowley, W. (2002). Crop yields and soil properties on eroded slopes of red ferrosols in north-west *Tasmania. Austr. J. Soil Res.*, *40*, 625-642.

Colwell, J. D. (1994). *Estimating Fertilizer Requirements. A Quantitative Approach.* CAB International, UK, 259.

Cox, M. S., Gerard, P. D., Wardlaw, M. C. & Abshire, M. J. (2003). Variability of selected soil properties and their relationship with soybean yield. *Soil Sci. Soc. Am. J.*, *67*, 1296-1302.

De la Rosa, D., Cardona, F. & Almorza, J. (1981). Crop yield predictions based on properties of soil in Sevilla, Spain. *Geoderma*, *25*, 267-274.

Díaz-Zorita, M., Buschiazzo, D. E. & Peinemann, N. (1999). Soil organic matter and wheat peoductivity in the Semiarid Argentine Pampas. *Agron. J.*, *91,* 276-279.

Donatelli, M., Bellocchi, G. & Fontana, F. (2003). RasEst3.00: software to estimate daily radiation data from commonly meteorological variables. *Eur. J. agron.*, *18*, 363-367.

Doorenbos, J. & Pruitt, W. O. (1977). *Crop water requirements*. FAO. Irrigation ad drainage Paper Nº 24, Rome, Italy, 193.

Doran, J. & Parkin, T. (1994). Defining and assessing soil quality. En: *Defining soil quality for sustainable environment.* S. S. P. *35*, 3-21.

Elizondo, D. A., McClendon, R. W. & Hoogenboom, G. (1994). Neural network models for predicting flowering and physiological maturity of soybean. *Trans ASAE*, *37*, 981-988.

FAO. (2007). Land Evaluation: towards a revised framework. *Land & water discussion paper 6;* Rome: FAO 6.

Fila, G., Bellocchi, G., Acutis, M. & Donatelli, M. (2003). IRENE: a software to evaluate model performance. *Eur. J. Agron.*, *18,* 369-372.

Freeman, K. W., Raun, W. R., Jonson, G. V., Mullen, R. W., Stone, M. L. & Solie, J. B. (2003). Late-season prediction of wheat yield and grain protein. *Commun. Soil Sci. Plant Anal.*, *34*, 1837-1852.

García-Paredes, J. D., Olson, K. R. & Lang, J. M. (2000). Predicting corn and soybean productivity for Illinois soils. *Agric. Sys.*, *64,* 151-170.

Gevrey, M., Dimopoulos, I. & Lek, S. (2003). Review and comparison of methods to study the contribution of variables in artificial neural network models. *Ecol. Mod.*, *160*, 249-264.

González Montaner, J. H., Maddonni, G. A. & DiNapoli, M. R. (1997). Modeling grain yield and grain yield response to nitrogen in spring wheat crops in the Argentinean Pampa. *Field Crop Res.*, *51*, 241-252.

Hall, A. J., Rebella, C. M., Ghersa, C. M. & Culot, J. P. (1992). Field crop systems of the Pampas. In *Field crop ecosystems of the World 18.* C.J. Pearson Ed. Elsevier, Amsterdam. 413-450.

Huddleston, J. H. (1984). Development and use of soil productivity ratings in the United States. *Geoderma*, 3204, 297-317.

Hunt, L. A., Kuchar, L. & Swanton, C. J. (1998). Estimation of solar radiation for use in crop modeling. *Agric. Forest Meteorol.*, *91*, 293-300.

INTA, MAGPSF., (1981). *Mapa de suelos de la Provincia de Santa Fe. Parte I.*, 245.

INTA, MAGPSF., (1983). *Mapa de suelos de la Provincia de Santa Fe. Parte II.* 216.

INTA, MEPLP, FALP., (1980). Inventario de los recursos naturales de la Provincia de la Pampa. 493.

INTA., (1989). *Mapa de suelos de la Provincia de Buenos Aires*. 525.

Iqbal, J., Read, J., Thomasson, A. & Jenkins, J. (2005). Relationships between soil-landscape and dryland cotton lint yield. Soil Sci. Soc. *Am. J*, *69*, 872-882.

Jiang, P. & Thelen, K. D. (2004). Effect of soil and topographic properties on crop yield in a north-central corn-soybean cropping system. *Agron. J.*, *96*, 252-258.

Irmak, S., Haman, D. Z. & Bastug, R. (2000). Determination of crop water stress index for Irrigation timing and yield estimation of corn. *Agronomy Journal*, *92*, 1221-1227.

Jorgensen, S. E. & Bendoricchio, G. (2001). *Fundamentals of Ecological Modelling,* Third edition. Elsevier, Oxford, UK, 530.

Kalivas, D. P. & Kollias, V. J. (2001). Effects of soil, climate and cultivation techniques on cotton yield in Central Greece, using different statistical methods. *Agronomie*, *21*, 73-89.

Kaul, M., Hill, R. L. & Walthall, C. (2005). Artificial neural networks for corn and soybean yield prediction. *Agric. Sys.*, *85*, 1-18.

Kim, K., B. L. Barham, & Coxhead, I. (2000). Recovering soil productivity attributes from experimental data: a statistical method and an application to soil productivity dynamics. *Geoderma* 9603, 239-259.

Kobayashi, K. & Salam, M. U. (2000). Comparing simulated and measured values using mean square deviation and its components. *Agron. J.*, *92*, 345-352.

Laya, D., Van Ranst, E. & Debaveye, J. (1998). A modified parametric index to estimate yield potentials for irrigated alfalfa on soils with gypsum in Quinto0Aragon, Spain). *Geoderma*, *87*, 111-122.

Lee, J. H. W., Huang, Y., Dickman, M. & Jayawardena, A. W. (2003). Neural network modeling of coastal algal blooms. *Ecol. Mod.*, *159*, 179-201.

Linacre, E. T. (1977). A simple formula for estimating evapotranspiration rates in various climates, using temperature data alone. *Agric. Meteorol.*, *18*, 409-424.

Loveland, P. & Webb, J. (2003). Is there a critical level of organic matter in the agricultural soils of temperate regions: a review. *Soil and Tillage Research,* 7001, 1-18.

Magrin, G. O., Hall, A. J., Baldy, C. & Grondona, M. O. (1993). Spatial and interannual variations in the phototermal quotient: implications for the potential kernel number of wheat crops in Argentina. Agric. *Forest. Meteorol.*, *67*, 29-41.

Martín, N., Bollero, G., Kitchen, N., Kravchenko, A., Sudduth, K., Wiebold, W. & Bullock. D. (2006). Two classification methods for developing and interpreting productivity zones using site properties. *Plant Soil*, *288*, 357-371.

Nemes A., Schaap M. G. & Wösten, J. H. M. (2003). Functional evaluation of pedotransfer functions derived from different scales of data collection. *Soil Sc. Soc. Am. J.*, *67*, 1093-1102.

Neter, J., Wasserman, W. & Kutner, M. H. (1990). Applied linear statistical models. *Irwin inc. Eds*., Illinois, USA. 1172.

Özesmi, S. L., Tan, C. O. & Özesmi, U. (2006). Methodological issues in building, training, and testing artificial neural networks in ecological applications. *Ecol. Mod.*, *195*, 83-93.

Park, S. J. & Vlek, P. L. G. (2002). Environmental correlation of three-dimensional soil spatial variability: a comparison of three adaptive techniques. *Geoderma*, *109*, 117-140.

Potgieter, A. B., Hammer, G. L., Doherty, A. & Voil, P. (2005). A simple regional-scale model for forecasting sorghum yield across North-Eastern Australia. *Agric. Forest Meteorol., 132,* 143-153.

Quiring, S. M. & Papykryiakou, T. N. (2003). An evaluation of agricultural drought indices for the Canadian prairies. *Agric. Forest. Meteorol.*, *118*, 49-62.

Quiroga, A. R., Dias-Zorita, M. & Buschiazzo, D. E. (2001). Safflower productivity as related to soil water storage and management practices in semiarid regions. *Commun. Soil Sci. Plant Anal.*, *32*, 2851-2862.

Quiroga, A. R. & Funaro, D. (2004). Materia orgáncia, factores que condicionana su utilización como indicador de calidad en Molisoles de las Regiones Semiárida y Subhúmeda Pampeana. XIX Congreso Argentino de la Ciencia del Suelo. *Procceding*, *6*,.

Rawls W. J. (1983). Estimating soil bulk density from particle size analysis and organic matter content. *Soil Sci.*, *135*, 123-125.

Rawls, W. J., Brakensiek, D. L. & Saxton, K. E. (1982). Estimation of soil water properties. *Trans. ASAE*, *25*, 1316-1328.

Rogers, L. L. & Dowla, F. U. (1994). Optimization of groundwater remediation using artificial neural networks with parallel solute transport modeling. *Water Res. Res.*, *30*, 457-481.

Sadras, V. O. & Calviño, P. O. (2001). Quantification of grain response to soil depth in soybean, maize, sunflower, and wheat. *Agron. J.*, *93*, 577-583.

MinAgri. Ministerio de Agricultura, Ganadería y Pesca, (2010). *Estadísticas de producción agrícola*. http://www.minagri.gob.ar/.

Sain, G. E. & Jauregui, M. A. (1993). Deriving fertilizer recommendation with a flexible functional form. *Agron. J.*, *85*, 934-937.

Satorre, E. H. & Slafer, G. A. (1999). *Wheat Production systems of the Pampas. In Wheat. Ecolocy and physiology of yield determination*. E.M. Satorre and G.A. Slafer Eds. The Haworth Press, Inc. New York. 333-348.

Sauerborn, J. (2002). Site productivity, the Key to crop productivity. *J. Agronomy & Crop Science*, *188*, 363-367.

Savin, R., Satorre, E. H., Hall, A. J. & Slafer, G. A. (1995). Assessing strategies for wheat cropping in the monsoonal climate of the Pampas using the CERES-wheat simulation model. *Field Crop Res.*, *42*, 81-91.

Schaap, M. G. & Bouten, W. (1996). Modeling water retention curves of sandy soils using neural networks. *Water Res. Res.*, *32*, 3033-3040.

Schaap, M. G., Leij, F. J. & van Genuchten, M. T. H. (1998). Neural networks analysis for hierarchical prediction of soil hydraulic properties. *Soil Sci. Soc. Am. J.*, *62*, 847-855.

Somaratgne, S., Seneviratne, G. & Coomaraswamy, U. (2005). Prediction of soil organic carbon across different land-use patterns: a neural network approach. *Soil Sci. Soc. Am. J. 69*, 1580-1589.

Starrett, S. K., Starrett, S. K. & Adams, G. L. (1997). Using artificial neural networks and regression to predict percentage of applied nitrogen leached under turfgrass. Commun. *Soil Sci. Plant Anal.* , *28*, 497-507.

Stenberg, B. (1998). Soil attributes as predictors of crop production under standardized conditions. *Biol. Fert. Soils*, *27*, 104-112.

Totis, L. & Perez, O. (1994). Relaciones entre el consumo de agua máximo de la secuencia de cultivo trigo/soja y la evapotranspiración potencial para el cálculo de la dosis de riego. INTA *Pergamino-Carpeta de Producción Vegetal, 12*, 1-4.

Travasso, M. I. & Delécolle, R. (1995). Adaptation of the CERES-wheat model for large area yield estimation in Argentina. *Eur. J. Agron.*, *4*, 347-353.

Udawatta, R. P. & Henderson, G. S. (2003). Root Distribution Relationships to Soil Properties in Missouri Oak Stands: A Productivity Index Approach. *Soil Sci Soc Am J, 6706*, 1869-1878.

Veron, S. V., Paruelo, J. M. & Slafer, G. A. (2004). Interannual variability of wheat yield in the Argentine Pampas during the 20th century. *Agric. Ecosys. Environm.*, *103*, 177-190.

Verón, S. R., Paruelo, J. M., Sala, O. E., Laurenroth, Y. W. K. (2002). Environmental Controls of Primary Production in Agricultural Systems of the *Argentine Pampas. Ecosystems*, *5*, 625–635

Wander, M., Walter, G., Nissen, T., Bollero, G., Andrewss, S. & Cavanaugh-Grant, D. (2002). Soil quality: science and process. *Agron. J.*, *94*, 23-32.

Wong, M. T. F. & Asseng, S. (2006). Determining the causes of spatial and temporal variability of wheat yields at sub-field scale using a new method of upscalling a crop model. *Plant and Soil, 283*, 203-215.

Xiong, R. & Meullenet, J. F. (2006). A PLS dummy variable approach to asses the impact of jar attributes on linking. *Food Qual. Preferen.*, *17*, 188-198.

Yang, J., Hammer, R. D., Thompson, A. L. & Planchar, R. W. (2005). Predicting soybean yield in a dry and wet year using a spoil productivity index. *Plant and Soil*, *250*, 175-182.

Zhang, P., Anderson, B., Tan, B., Huang, D. & Myneni, R. (2005). Potential monitoring of crop production using a satellite-based Climate-Variability Impact Index. agricultural and *Forest Meteorology*, *132(3-4)*, 344-358.

In: Focus on Artificial Neural Networks
Editor: John A. Flores, pp. 97-127
ISBN: 978-1-61324-285-8

Chapter 5

DESIGN OF ARTIFICIAL NEURAL NETWORK PREDICTORS IN MECHANICAL SYSTEMS PROBLEMS

İkbal Eski, Eyüp Sabri Topal and Şahin Yildirim
Mechatronics Engineering Department, Engineering Faculty,
Erciyes University, Kayseri, Turkey

ABSTRACT

Due to nonlinearity of the mechanical systems, it is necessary to use adaptive predictors for analysing system parameters. Neural networks could be used as an alternative to overcome such problems. In this chapter, two approaches of mechanical systems are presented for CAD-CAM systems and vehicle suspension systems. In the first approach, surface roughness prediction studies on end milling operations are usually based on three main parameters composed of cutting speed, feed rate and depth of cut. The step-over ratio is usually neglected without investigating it. The aim of this study is to discover the role of the step-over ratio in surface roughness prediction studies in flat end milling operations. In realising this, machining experiments are performed under various cutting conditions by using sample specimens. The surface roughnesses of these specimens are measured. Two Artificial neural networks (ANN) structures were constructed. First of them was arranged with considering, and the second without considering the step-over ratio. ANN structures were trained and tested by using the measured data for predicting surface roughness. Average RMS error of the ANN model with considering step-over ratio is 0.04 and without considering stepover ratio is 0.26. The first model proved capable of prediction of average surface roughness (Ra) with a good accuracy and the second model revealed remarkable deviations from the experimental values.

Other approach is consisted of analyzes effects of vibrations on comfort and road holding capability of vehicles as observed in the variations of suspension springs, road roughness etc. Also, design of non-linear experimental car suspension system for ride qualities using neural networks is presented. Proposed active suspension system has been found more effective in vibration isolation of car body than linear active suspension system. Proposed neural network predictor could be used in vehicle's suspension vibration analysis.

The results of both approaches improved that ANN structure has superior performance at adapting large disturbances of mechanical systems.

Keywords: CAD-CAM, surface roughness, end milling, Quarter car test-rig, Vehicle vibrations, Neural network, Active suspension system.

1. Introduction

Surface roughness is a criterion of the product quality of machined parts and a factor that greatly influences tribological characteristics of a part. Several factors will influence the final surface roughness in a CNC end milling operation such as cutting speed, depth of cut, feed rate, stepover ratio etc.. Developing surface roughness prediction models aimed to determine the optimum cutting conditions for minimum surface roughness on account of time and money saving.

A number of studies have been made in recent years on estimation of surface roughness in end milling using different approaches such as statistical, analytical, mathematical, neuro fuzzy and neural network modelling [1-3]. Artificial neural network (ANN) modelling has become more widely used in prediction surface roughness and optimisation of machining conditions. Sağlam and Ünüvar [4] used an artificial neural network model for future selection in order to estimate flank wear of tool and surface roughness during face milling depending on cutting speed, feed rate, and depth of cut, feed force and vertical force. Topal et al. [5] proposed an ANN model for predicting surface roughness from machining parameters such as cutting speed, feed rate, and depth of cut in milling of AISI 1040 steel.

Özcelik et al. [6] investigated optimum machining parameters of Inconel 718 Al alloy to obtain minimum surface roughness by employing an ANN model and a genetic algorithm. Balic and Korosec [7] estimated average surface roughness (Ra) of free surfaces using ANN. Çolak et al. [8] predicted surface roughness of milling surface with related to cutting parameters by using genetic expression programming method. They considered cutting speed, feed rate and depth of cut of end milling operations for predicting surface roughness and predicted a linear equation for surface roughness related to experimental study. Lou and Chen [9] also considered spindle speed, feed rate and depth of cut in their study on the surface roughness of end milling processes. They used a neural fuzzy network and in-process surface roughness recognition (ISRR) system to predict the surface roughness. Alauddin et al. [10] predicted the surface roughness of 190 BHN steel after end milling using a mathematical model depending on cutting speed, feed rate and depth of cut. They used the response surface methodology (RSM) to explore the effect of these parameters on surface roughness. Luo et al. [11] investigated the effects of machining variables and tooling characteristics on surface generation through simulations. They also evaluated and validated their approach and simulations by practical trials. Liu and Cheng [12] presented a practical method for modelling and predicting the machining dynamics and surface roughness/waviness in peripheral milling. Various neuro fuzzy inference systems have also been used to determine operation conditions in machining. Lo [13] used an adaptive network-based fuzzy inference system to predict the surface roughness in and milling. Dweiri et al. [14] modelled the down milling machining process of Alumic-79 using an adaptive neuro fuzzy inference system to predict the effect of machining variables such as spindle speed, feed rate, depth of cut, and number of flutes of tool on the surface finish. Lou and Chen [15] used an in-process surface recognition system that used the fuzzy-nets and a sensor-testing system to measure surface roughness in end milling operations.

In the literature, it has been shown that the main machining parameters usually considered in surface roughness modelling studies are cutting/spindle speed, feed rate and depth of cut, and additional parameters are tool wear, vibration, tool errors etc., while the stepover is usually neglected without investigating. In fact, the stepover ratio must be studied because it determines the fact that how many times the tool being passed and scraped again on a finished surface and may influence the final surface roughness by this way. The present work proposes two ANN surface roughness prediction models, consisting of cases of with and without considering stepover on account of comparing performance of these models and discovering the role of the stepover in surface roughness prediction studies in flat end milling operations.

Vehicle suspension system isolates vehicle body from vertical accelerations that are generated by variations in road surface and provides a more comfortable ride for passengers inside the vehicle. Guglielmino&Edge [16] used a servo-driven dry-friction damper in a car suspension application; as a potential alternative to a traditional viscous damper. Modular adaptive robust control (MARC) technique, which was applied to design force loop controller of an electro-hydraulic active suspension system, has this modular design approach lies in the fact that a key advantage of the adaptation algorithm that can be designed for explicit estimation convergence [17]. Inherited challenge and possible remedies of servo-loop control design for active suspension systems have been presented [18]. Methods and algorithms have been developed to identify, control and diagnose faults in case of suspension systems proposing a mechatronic vehicle suspension design concept for active and semi-active suspensions [19]. Generic control structure was derived based on physical insight in car and semi active suspension dynamics without explicitly using a model [20]. Choi&Han [21] presented vibration control performance of a semi-active electrorheological seat suspension system using a robust sliding mode controller. A functional approximation based adaptive sliding controller with fuzzy compensation has been developed and successfully employed to control a quarter-car hydraulic active suspension system [22, 23]. Gao et al. [24] presented a load-dependent controller design approach to solve the problem of multi-objective control for vehicle active suspension systems by using linear matrix inequalities. For a quarter-car model, vehicle roll and pitch motions are ignored and only degrees of freedom included are the vertical motions of sprung mass and un-sprung mass [25].

The chapter is organized as following; section 1 describes broad review of mechanical applications with neural network predictors. Broad description of artificial neural networks and algorithms are given in Section 2. Mathematical description of mechanical systems are out-lined in Section 3. Simulation and experimental results for both approaches are discussed in section 4. Paper is concluded with the section of 5.

2. ARTIFICIAL NEURAL NETWORKS (ANNS)

Traditionally, the term neural network had been to refer to a network or circuit of biological neurons. The modern usage of the term often refers to artificial neural networks, which are composed of artificial neurons or nodes. Thus the term has two distinct usages:

1. Biological neural networks are made up of real biological neurons that are connected or functionally related in the peripheral nervous system or the central nervous system. In the field of neuroscience, they are often identified as groups of neurons that perform a specific physiological function in laboratory analysis.
2. Artificial neural networks are made up of interconnecting artificial neurons (programming constructs that mimic the properties of biological neurons). Artificial neural networks may either be used to gain an understanding of biological neural networks, or for solving artificial intelligence problems without necessarily creating a model of a real biological system. The real, biological nervous system is highly complex and includes some features that may seem superfluous based on an understanding of artificial neural networks.

Artificial Neural Network (ANN) are made up of simple, highly interconnected processing units called neurons each of which performs two functions: aggregation of its inputs from other neurons or the external environment and generation of an output from aggregated inputs. ANN can be classified into two main categories based on their connection structures: feedforward and recurrent networks. Feedforward networks are the most commonly used type, mainly because of difficulty of training recurrent networks, although the last mentioned are more suitable for representing dynamical systems [26].

The application of ANN typically comprises two phases: a learning phase and a testing phase. Learning is the process through which the parameters and structure of the network are adjusted to reflect the knowledge contained within the distributed network structure. A trained network subsequently represents a static knowledge base which can be recalled during its operation phase. There are three general learning schemes in neural networks:

- Supervised learning, for example, error back propagation which requires the correct output signal for each input vector to be specified.
- Unsupervised, competitive learning or self organizing, in which the network self-adjusts its parameters and structure to capture the regularities of input vectors, without receiving explicit information from the external environment.
- Reinforcement or graded learning in which the network receives implicit scalar evaluations of previous inputs.

ANN can be classified as being feedforward and recurrent neural networks. Feedforward Neural Networks are straight forward networks allowing signals to travel only in one way, i.e., the perceptions are arranged in layers with the first layer taking in an input and the last layer producing an output, thus information is constantly “fed forward” from one layer to the next. There is no sense of time or memory of previous layers.On the contrary, Recurrent neural networks contain feedback connections neurons in the hidden layers. The ANN approaches are described in the following subsection.

2.1. Feedforward Neural Networks

Feedforward neural networks are the most popular and most widely used in various mechanical systems application. Feedforward Neural Networks are made up of one or more hidden layers between the input and output layers, as illustrated in Figure 1. The functionality of the network is determined by specifying the strengths of the connection paths called weights and the neuron activation function. The input layer distributes inputs to the first hidden layer. The inputs propagate forward through the network and each neuron computes its output according to;

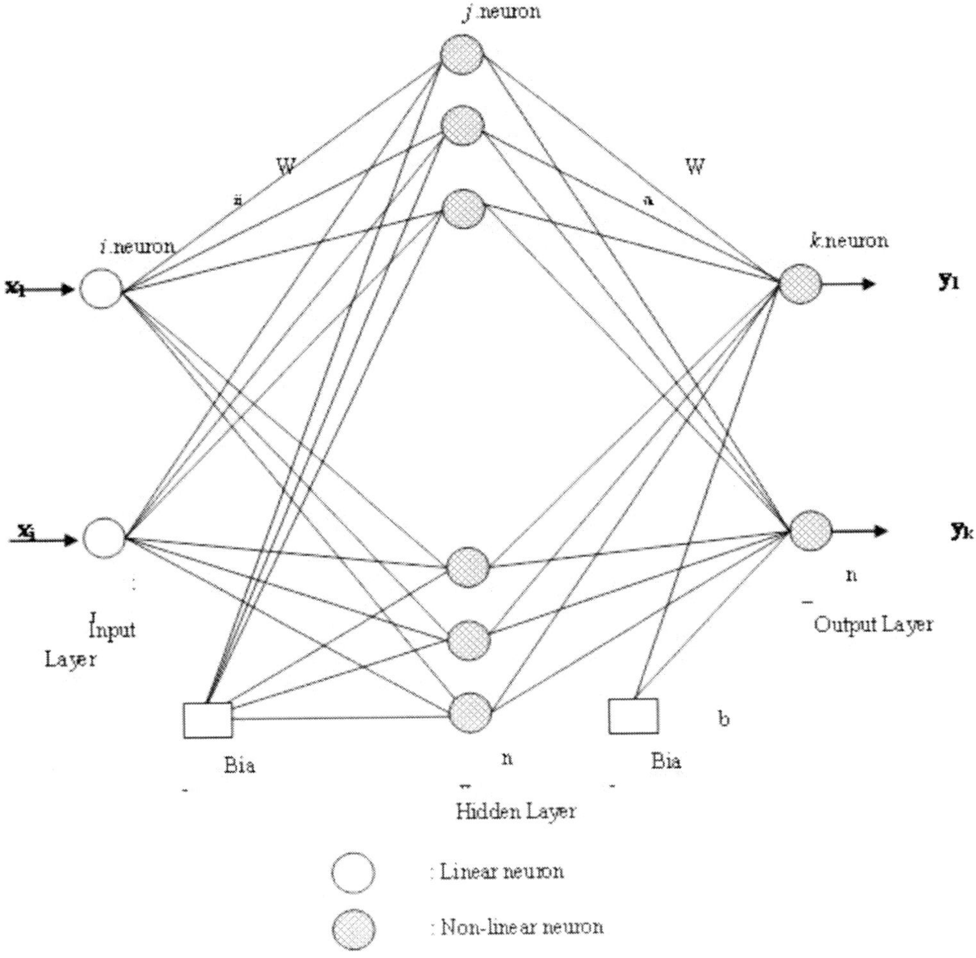

Figure 1. Schematic representation a feedforward neural network structure.

$$z_j(t) = f\left(\sum_{i=1}^{n} \sum_{j=1}^{m} W_{ij} x_i(t) + b_j \right) \tag{1}$$

where z_j is the output of the jth neuron in the hidden layer, W_{ij} is the weight of the connection between the input layer neurons and the hidden layer neurons, b_j is the bias of the jth neuron in the hidden layer. b_j can be regarded as the weight of the connection between a fixed input of unit value and neuron j in the hidden layer. The function f(.) is called the activation function of the hidden layer. The output signal of the neural network can be expressed in the following form;

$$y_k(t) = g\left(\sum_{j=1}^{m} \sum_{k=1}^{r} W_{jk} z_j(t) + b_k \right) \tag{2}$$

where W_{jk} are the weights between j^{th} neurons the hidden layer and k^{th} neurons output layer and b_k are the bias of the kth neurons in the output layer and g(.) is the activation function of the output layer.

2.2. Recurrent Neural Networks

Recurrent neural networks have been an important focus of research. Recurrent neural network structures have been applied to a wide variety of applications. Recurrent networks are more powerful than nonrecurrent networks, particularly for uses in comlex applications.

The architectures range from fully interconnected to partially connected networks, including multilayer feedforward networks with distinct input and output layers. Fully connected networks do not have distinct input layers of neurons, and each neuron has input from all other neurons. Feedback to the neuron itself is possible. Although some neurons are part of a feedforward structure, Figure 2. [27]. A feedback connection is used to pass output of a neuron in a certain layer to the previous layers [28]. The output recurrent neural network of the hidden layer;

$$z_j(t) = f\left(\sum_{i=1}^{n} \sum_{j=1}^{m} W_{ij} x_i(t) + \sum_{i=1}^{n} \sum_{h=1}^{m} W_{jh} z_h(t-1) + b_j \right) \tag{3}$$

where W_{jh} is the additional recurrent weight. The output signal of the recurrent neural network can be expressed in the following form;

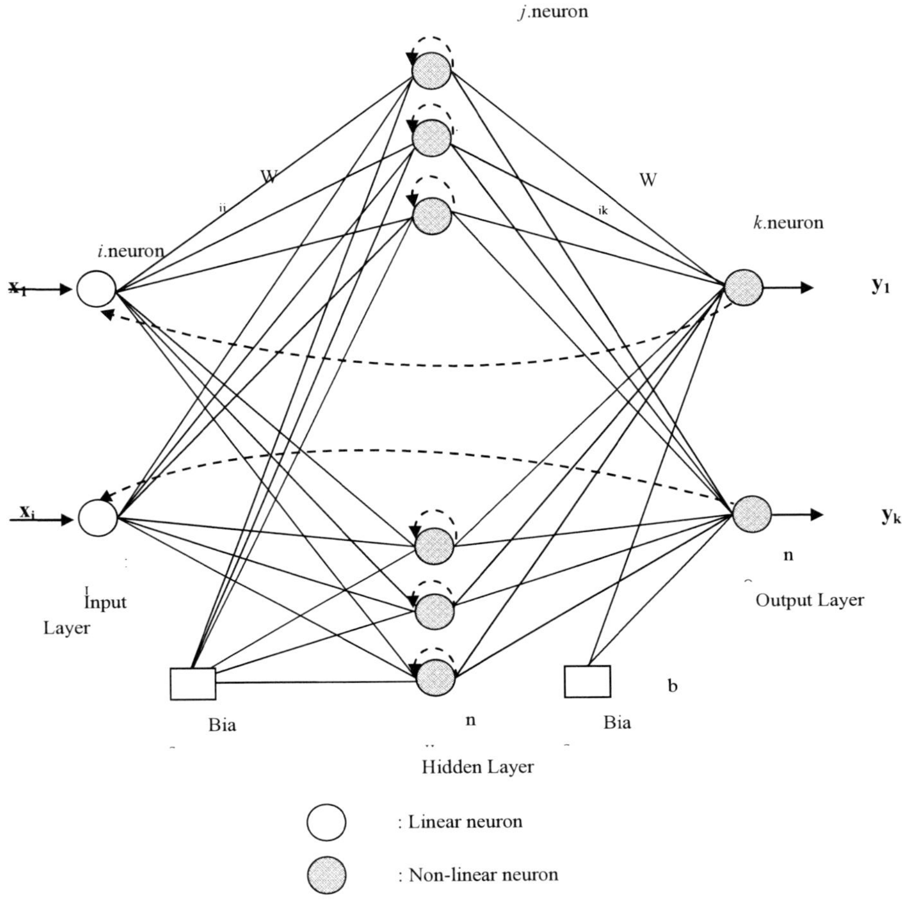

Figure 2. Schematic representation a recurrent neural network structure.

$$y_k(t) = g\left(\sum_{j=1}^{m} \sum_{k=1}^{r} W_{jk} z_j(t) + b_k \right) \tag{4}$$

Some supervised learning methods which are used to prediction and analyze in a mechanical systems, briefly described in the following subsections.

2.1.1. Back Propagation neural network (BPNN)

The BPNN is a method of supervised neural network learning. During training, the network is presented with a large number of input patterns. The experimental outputs are then compared to the neural network output nodes. The error between the experimental and neural network response is used to update the weights of the network inter-connections. This update is performed after each pattern presentation. One run through the entire pattern set is termed an epoch. The training process continues for multiple epochs, until a satisfactorily small error is produced. The test phase uses a different set if input patterns. The neural network outputs are again compared to a set of experimental outputs. This error is used to evaluate the networks ability to generalize. Usually, the training set and/or architecture needs to be assessed at this point. The BPNN is the most commonly used to update the weights of the neural networks. The weights between input layer and the hidden layer are updated as follows;

$$\Delta W_{ij}(t) = -\eta \frac{\partial E_2(t)}{\partial W_{ij}(t)} + \alpha \Delta W_{ij}(t-1) \tag{5}$$

The weights between the hidden layer and the output layer are updated in the following equation;

$$\Delta W_{jk}(t) = -\eta \frac{\partial E_1(t)}{\partial W_{jk}(t)} + \alpha \Delta W_{jk}(t-1) \tag{6}$$

where η is the learning rate, and α is the momentum term. E_2 *(t)* is the propagation error between hidden layer and output layer. E_1 *(t)* is the error between experimental and neural network output signals.

2.1.2. General regression neural network (GRNN)

GRNN are paradigms of the Probabilistic and Radial Basis Function used in functional approximation. To apply GRNN to analyze, a vector f_j and f_k are formed. The output y is the weighted average of the target values t_k of training cases f_k close to a given case f_j, as given by;

$$y_k = \frac{\sum_{j=1}^{m} \sum_{i=1}^{n} z_j W_{jk}}{\sum_{j=1}^{m} \sum_{i=1}^{n} W_{jk}} \tag{7}$$

where,

$$y_k = \frac{\sum_{j=1}^{m} \sum_{i=1}^{n} z_j W_{jk}}{\sum_{j=1}^{m} \sum_{i=1}^{n} W_{jk}} \tag{8}$$

The only weights that need to be learned are the smoothing parameters, h of the RBF units, which are set using a simple grid search method. The distance between the computed value k and each value in the set of target values T is given by;

$$T = \{1,2\} \tag{9}$$

The values 1 and 2 correspond to the training class and all other classes respectively. The class corresponding to the target value with the minimum distance is chosen. As, GRNN exhibits a strong bias towards the target value nearest to the mean value μ of T so we used target values 1 and 2 because both have the same absolute distance from μ [29].

2.1.3. Modular neural network (MNN)

MNN refer to the adaptive mixtures of local experts. The most attractive point of MNN architecture is that different experts handle different parts of the data. The problem of pattern interference can be alleviated by this type of architecture. Each local network receives the same input vector applied to the network. A gating network determines the contribution of each local network to the total output as well as what range of the input space each network should learn. The back propagation algorithm is used to train the gating and local networks. The outputs of the local networks are combined to give the network output such that:

$$\mathbf{y_k} = \sum_{j=1}^{m} \sum_{k=1}^{r} \mathbf{g_j y_j} \tag{10}$$

where y_j are the output of the jth local network and y_j are the normalized output vector elements of the gating network given by:

$$\mathbf{g_j} = \frac{\mathbf{e^{u_i}}}{\sum_{j=1}^{m} \sum_{n=1}^{n} \mathbf{e^{u_j}}} \tag{11}$$

where u_j are the weighted input received by the jth output unit of the gating network.

2.1.4. Radial basis neural network (RBNN)

Traditionally, RBNN which model functions y(x) mapping $x \in R^n$ to $y \in R$ have a single hidden layer so that the model,

$$\mathbf{f(x)} = \sum_{j=1}^{m} \mathbf{w_j h_j(x)} \tag{12}$$

is linear in the hidden layer to output weight $(w_j)_{j=1}^{m}$. The characteristic feature of RBFNN is the radial nature of the hidden unit transfer functions, $(h_j)_{j=1}^{m}$, which depend only on the distance between the input x and the centre c_j of each hidden unit, scaled by a metric R_j,

$$h_j(x) = \varphi\left[(x - c_j)^T R_j^{-1} (x - c_j)\right] \quad (13)$$

where φ is some function which is monotonic for non negative numbers. Gaussian basis function so that the transfer functions can be written

$$h_j(x) = \exp\left(-\sum_{k=1}^{r} \frac{(x_k - c_{jk})^2}{r_{jk}^2}\right) \quad (14)$$

Using the Gaussian approximation, the output network is approximately;

$$f(x) = \sum_{j=1}^{m} w_j \exp\left(-\sum_{k=1}^{r} \frac{(x_k - c_{jk})^2}{r_{jk}^2}\right) \quad (15)$$

where r_j is the radius vector of the j_{th} hidden unit. A direct approach to the model complexity issue is to select a subset of centres from a larger set which, if used in its entirety, would over fit the data.

2.1.5. Learning vector quantization neural network (LVQNN)

LVQNN is supervised neural network, which was developed by Kohonen and is based on the Self-Organizing Map (SOM) or Kohonen feature map, LVQNN methods are simple and effective adaptive learning techniques. They rely on the nearest neighbor classification model and are strongly related to the condensing methods, where only a reduced number of prototypes are kept from the whole set of samples. This condensed set of prototypes is then used to classify unknown samples using the nearest neighbor rule. LVQNN has a competitive and linear layer in the first and second layer, respectively. The competitive layer learns to classify the input vectors and the linear layer transforms the competitive layer's classes into the target classes defined by the user. In the learning process, the weights of LVQNN are updated by the following Kohonen learning rule if the input vector belongs to the same category.

$$\Delta W_I(i, j) = \eta a_I(i)(p(j) - W_I(i, j)) \quad (16)$$

where η is the learning rate and $a_I(i)$ is the output of competitive layer.

2.1.6. Self organizing map neural network (SOMNN)

SOMNN consists of a regular, usually two-dimensional, grid of map units. Each unit is represented by a prototype vector, where is input vector dimension. The units are connected to adjacent ones by a neighborhood relation. The number of map units, which typically varies from a few dozen up to several thousand, determines the accuracy and generalization capability of the SOMNN. Data points lying near each other in the input space are mapped onto nearby map units. Thus, the SOMNN can be interpreted as a topology preserving mapping from input space onto the 2-D grid of map units. The SOMNN is trained iteratively. At each training step, a sample vector z is randomly chosen from the input data set. Distances between z and all the prototype vectors are computed. The best matching unit, which is denoted here by b, is the map unit with prototype closest to z;

$$\|\mathbf{z}-\mathbf{m_b}\| = \min_{\mathbf{i}}\{\|\mathbf{z}-\mathbf{m_i}\|\} \tag{17}$$

The update rule for the prototype vector of unit i is

$$\mathbf{m_i}(\mathbf{t}+1) = \mathbf{m_i}(\mathbf{t}) + \beta(\mathbf{t})\mathbf{h_{bi}}(\mathbf{t})[\mathbf{z}-\mathbf{m_i}(\mathbf{t})] \tag{18}$$

where $\beta(t)$ is the adaptation coefficient, $h_{bi}(t)$ is the neighborhood kernel centered on the winner unit.

$$\mathbf{h_{bi}}(\mathbf{t}) = \exp\left(-\frac{\|\mathbf{r_b}-\mathbf{r_i}\|^2}{2\sigma^2(\mathbf{t})}\right) \tag{19}$$

where r_b and r_i are positions of neurons b and i on the SOMNN grid. Both $\beta(t)$ and $\sigma(t)$ decrease monotonically with time. In the case of a discrete data set and fixed neighborhood kernel, the errors function of SOMN as follows;

$$\mathbf{E} = \sum_{\mathbf{i}=1}^{\mathbf{N}}\sum_{\mathbf{j}=1}^{\mathbf{M}} \mathbf{h_{bj}}\|\mathbf{z_i}-\mathbf{m_j}\|^2 \tag{20}$$

where N is number of training samples and M is the number of map units. Neighborhood kernel h_{bj} is centered at unit b, which is the best matching unit of vector z_i, and evaluated for unit j [30].

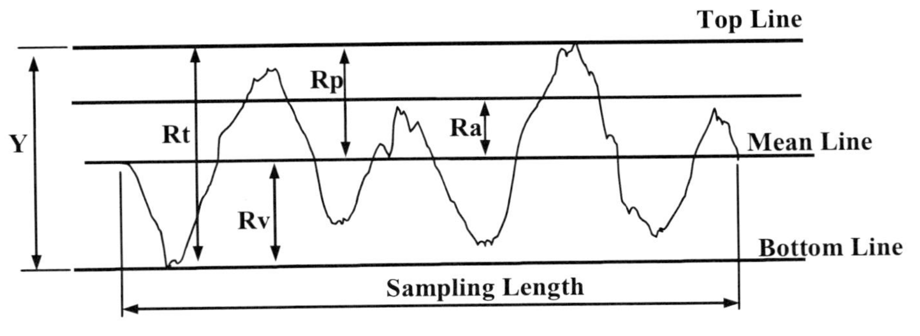

Figure 3. Surface roughness parameters.

3. DESCRIPTION OF THE MECHANICAL SYSTEMS

The following subsection presents two analysis approaches for such as surface roughness in end milling and vehicle suspension systems.

3.1. Surface Roughness

Surface roughness can be described as the residual irregularities on machined workpiece produced by machining process and can be specified with many different parameters. Due to the need for different parameters in a wide variety of machining operations, a large number of surface roughness parameters were developed. In this study, the average roughness (Ra) was preferred as the parameter of surface finish specification since Ra is the most used international parameter of roughness. This parameter is also known as the arithmetic mean roughness value, AA (arithmetic average) or CLA (centre line average). It can be expressed by the following relationship [31],

$$Ra = \frac{1}{L}\int_0^L |\,Y(x)\,|\,dx \tag{21}$$

where *Ra* is the arithmetic average deviation from the mean line, *L* is the sampling length and *Y* is the ordinate of the profile. It is the arithmetic mean of the departure of the roughness profile from the mean line (Figure 3).

3.1.1. The Term of stepover

Stepover is a milling parameter that defines the distance between two neighbouring passes over the workpiece. It is usually given as a percentage (ratio) of the tool diameter and usually called as stepover ratio. The term of stepover is illustrated in Figure 4. In finishing passes of flat end milling operations, stepover ratio may affect the final surface roughness by determining the fact that how many times the tool being passed and scraped again on a finished surface band. Various experiments are realised in this study under conditions of constant stepover (100% of tool diameter) and varying stepover (varying from 10% to 100% of tool diameter).

3.1.2. Influence of stepover ratio on surface roughness

In the earlier literature, surface roughness is usually considered as a result of three main machining parameters such as cutting speed, feed rate and depth of cut. Afterwards, some additional parameters are considered such as tool geometry, machine tool errors, vibrations etc. But the stepover ratio (SR) is usually neglected. Figure 5 represents the effects of main machining parameters on surface roughness including SR during finishing passes of flat end milling. As the figure, SR has a considerable effect on surface roughness.

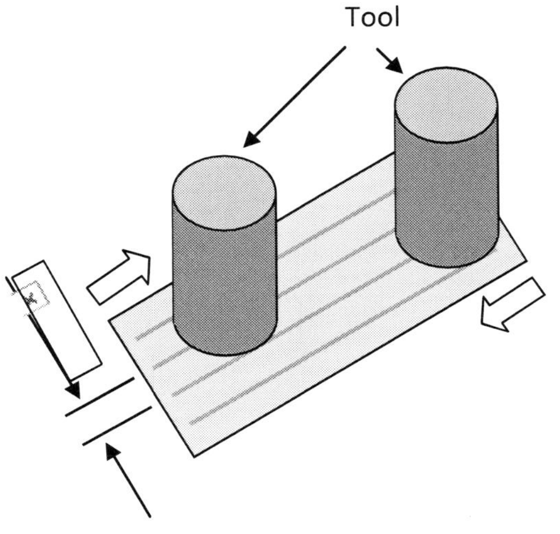

Figure 4. The term of stepover. 50% ratio of tool diameter is illustrated on the example.

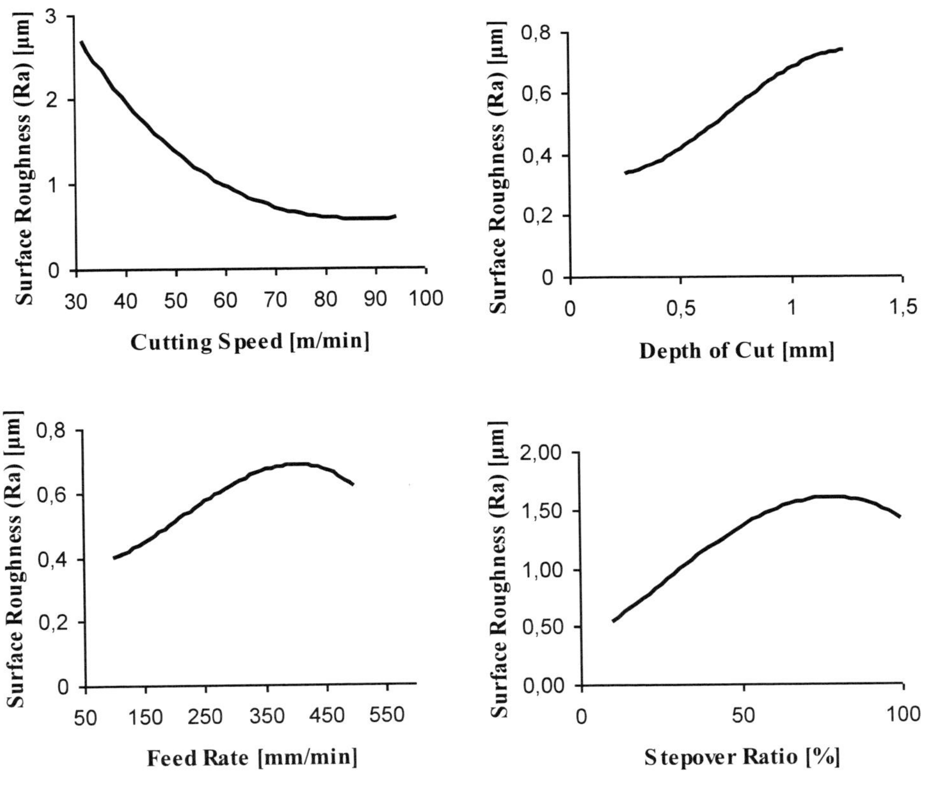

Figure 5. Effects of main machining parameters including stepover ratio on surface roughness in conditions used in the experiments of the study.

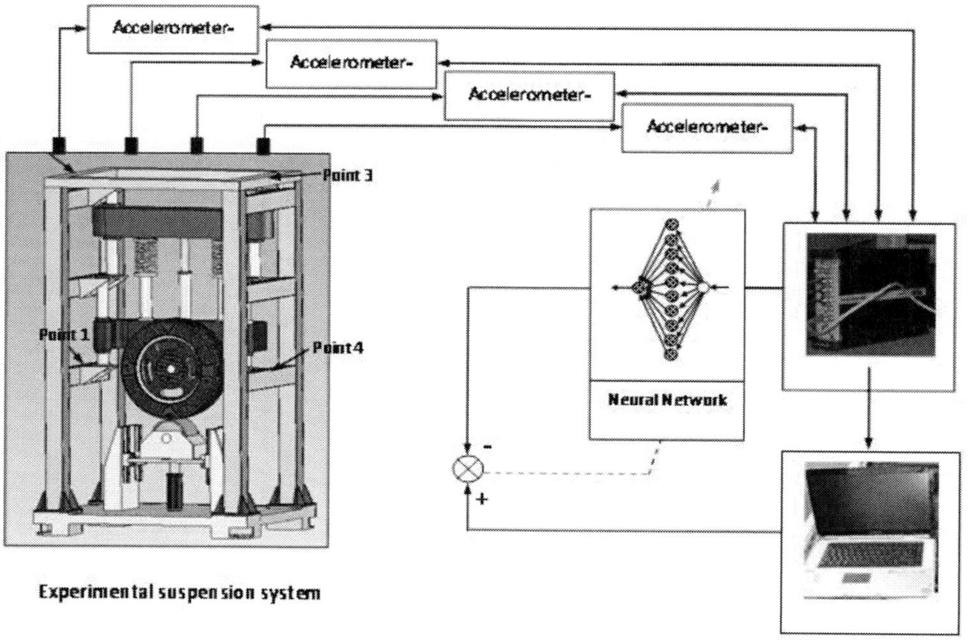

Figure 6. Block diagram of experimental and simulation approaches for proposed ASS of a car.

3.2. Proposed Active Suspension System (ASS)

Important components of proposed active suspension system (Figure 6) are as follows;

i) 200 kg metal plate that represents the weight of the ¼ car model (Sprung mass);
ii) Tachometer is used to measure wheel speeds;
iii) Linear Variable Differential Transformer (LVDT) is used to measure road displacement, wheel displacement, and vehicle body displacement, respectively;
iv) A pneumatic actuator is used to actuate road excitation;
v) Two springs, active damper, a wheel and tyre assembly (Unsprung mass);
vi) A compressor is used to supply air pressure for pneumatic actuator;
vii) A computer is used for recording experimental and simulations data;
viii) Different type valves are used to control pressure of pneumatic actuator; and
ix) Programmable Logic Controller (PLC) and Data Acquisition card are used to control speed of wheel and hydraulic cylinder. ASS is described as

$$m_s\ddot{z}_s(t)+c_s[\dot{z}_s(t)-\dot{z}_{us}(t)]+k_{s1}[z_s(t)-z_{us}(t)]+k_{s2}[z_s(t)-z_{us}(t)]=F_a \quad (22)$$

$$k_{s1}+k_{s2}=k_s \quad (23)$$

$$m_{us}\ddot{z}_{us}(t)+c_s[\dot{z}_{us}(t)-\dot{z}_s(t)]+k_s[z_{us}(t)-z_s(t)]+k_t[z_{us}(t)-z_s(t)]+c_t[\dot{z}_{us}(t)-\dot{z}_r(t)]=-F_a \quad (24)$$

where m_s is sprung mass, which represents vehicle chassis; m_{us} is unsprung mass, which represents wheel assembly; z_s and z_{us} are displacements of sprung and unsprung masses, respectively; c_s, k_s are damping and stiffness of suspension system, respectively; c_t and k_t are damping and stiffness of pneumatic tyre; z_r is road displacement; F_a represents active control force of suspension system. Layout of ASS test rig with its control loop is given in Figure 7. State variables are defined as [22]

$$x_1(t) = z_s(t) - z_{us}(t) \tag{24}$$

$$x_2(t) = z_{us}(t) - z_r(t) \tag{25}$$

$$x_3(t) = \dot{z}_s(t) \tag{26}$$

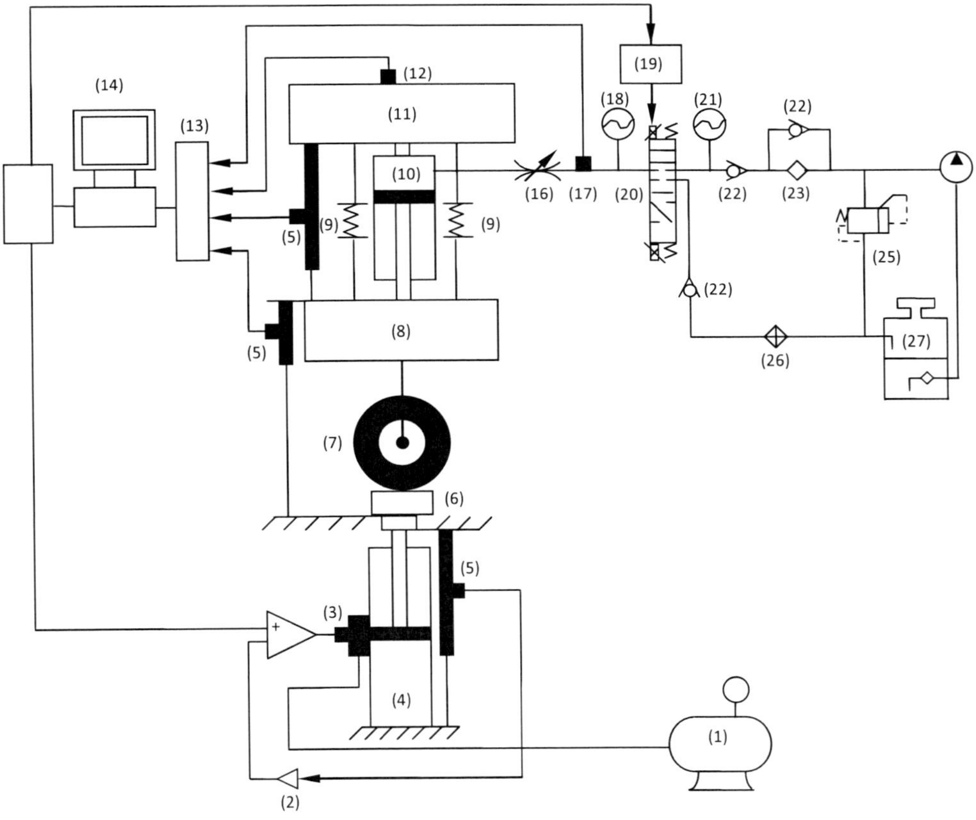

Figure 7. Layout for ASS test rig
(1)Compressor and pressure meter (2)Transducer amplifier (3)Servo valve (4)Pneumatic cylinder (5)Displacement transducer (6)Roller (7)Tyre (8)Unsprung mass (9)Spring (10)Hydraulic cylinder (11)Sprung mass (12)Accelerometer (13)Analogue to digital converter (14)Computer (15) Digital to analogue converter (16)Throttle valve (17)Pressure transducer (18)Gas spring (19)Proportional amplifier (20)Proportional control valve (21) Main accumulator (22)Check valve (23)Hydraulic filter (24) Hydraulic pump (25)By-pass valve (26) Cooler (27)Oil tank.

$$x_4(t) = \dot{z}_{us}(t) \tag{27}$$

and

$$\mathbf{x(t)} = \begin{bmatrix} x_1(t) \\ x_2(t) \\ x_3(t) \\ x_4(t) \end{bmatrix} \tag{28}$$

where $x_1(t)$ represents suspension deflection, $x_2(t)$ represents tyre deflection, $x_3(t)$) represents sprung mass speed, and $x_4(t)$ represents unsprung mass speed. Eqs of system model (22) and (24) can be rewritten as

$$\dot{\mathbf{x}}(\mathbf{t}) = {}_A\,\mathbf{x(t)}\,{}_{+B}\,\mathbf{z(t)}\,{}_{+E}\,\mathbf{u(t)} \tag{29}$$

where

$$A = \begin{bmatrix} 0 & 0 & 1 & -1 \\ 0 & 0 & 0 & 1 \\ -\frac{k_s}{m_s} & 0 & -\frac{c_s}{m_s} & \frac{c_s}{m_s} \\ \frac{k_s}{m_{us}} & -\frac{k_t}{m_s} & \frac{c_s}{m_s} & -\frac{(c_s + c_t)}{m_s} \end{bmatrix},\ B = \begin{bmatrix} 0 \\ -1 \\ 0 \\ \frac{c_t}{m_{us}} \end{bmatrix}$$

$$E = \begin{bmatrix} 0 \\ 0 \\ \frac{1}{m_s} \\ -\frac{1}{m_{us}} \end{bmatrix},\ \mathbf{z(t)} = \dot{\mathbf{z}}_{\mathbf{r}}(\mathbf{t}),\ \mathbf{u(t)} = \mathbf{F_a(t)}$$

$$\dot{\mathbf{x}}(\mathbf{t}) = \begin{bmatrix} 0 & 0 & 1 & -1 \\ 0 & 0 & 0 & 1 \\ -\frac{k_s}{m_s} & 0 & -\frac{c_s}{m_s} & \frac{c_s}{m_s} \\ \frac{k_s}{m_{us}} & -\frac{k_t}{m_s} & \frac{c_s}{m_s} & -\frac{(c_s + c_t)}{m_s} \end{bmatrix} \mathbf{x(t)} + \begin{bmatrix} \mathbf{0} \\ \mathbf{-1} \\ \mathbf{0} \\ \frac{\mathbf{c_t}}{\mathbf{m_{us}}} \end{bmatrix} \dot{\mathbf{z}}_{\mathbf{r}}(\mathbf{t}) + \begin{bmatrix} 0 \\ 0 \\ \frac{1}{m_s} \\ -\frac{1}{m_{us}} \end{bmatrix} \mathbf{F_a(t)} \tag{30}$$

4. Simulation Experimental Results

In this section, the analysis results from surface roughness in end milling and vehicle suspension system using different neural networks approaches are presented [32, 33].

In first analysis, 3-layered Feed Forward Multi Layer Perceptron network type has been used with Back Propagation (BP) training technique [34, 35]. Two Feed Forward Neural Network (FNN) structures were employed. First of them was arranged with considering, and the second without considering the stepover ratio (Figure 8). The first ANN structure was configured with four neurons in the input layer corresponding to cutting speed, depth of cut, feed rate and stepover ratio, and one neuron corresponding to the error in the output layer. The second was configured with three neurons in the input layer (excluding stepover ratio). The hidden layers of the both of ANN structures have 10 neurons.

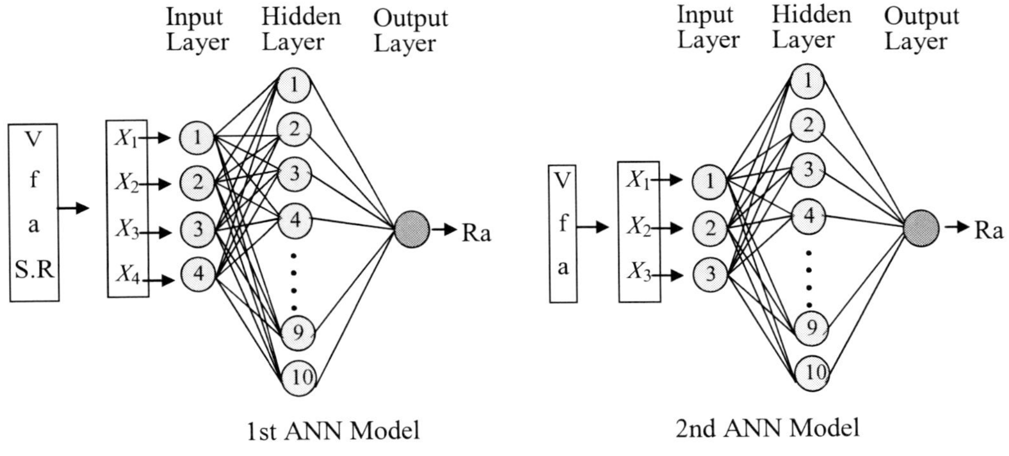

Figure 8. The ANN structures used in surface roughness.

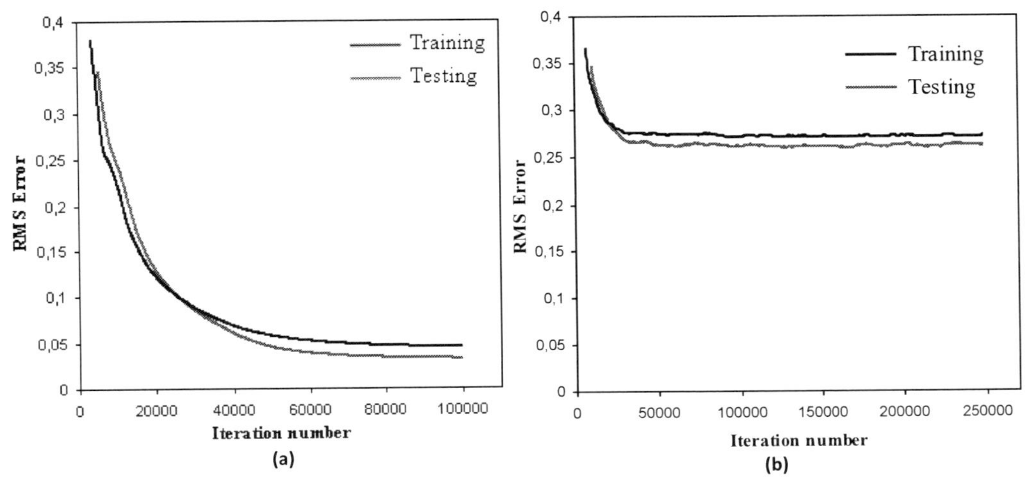

Figure 9. RMS error for training and testing models, (a) for the first ANN model with considering stepover and (b) for the second ANN model without considering stepover.

Table 1. Structural and training parameters of the ANN approach for surface roughness in end milling

ANN Model	Learning Algorithm	η	μ	n_I	n_H	n_O	N	AF	RMSE
1st ANN Model	BPNN	0.1	0.001	4	10	1	10^5	Logistic	0.04
2nd ANN Model	BPNN	0.1	0.001	3	10	1	$25*10^4$	Logistic	0.26

Around 100000 iterations were required to train the first network (ANN Model with considering stepover) satisfactorily. Structural and training parameters of this network are summarized in Table 1. The training was continued until no further significant decrease in the convergence error of the first ANN model was observed. The final average RMS error of the first ANN model was 0.04 after mentioned iterations. The error convergence of training process is illustrated in Figure 9(a). As the Figure, the error reduces to small values after 20000 iterations and stays constant subsequent to 65000. The training dataset of the first ANN model was given in Appendix A.

The training of the second ANN model was realised similar to first ANN model excepting the experimental data concerning to stepover ratio from the training dataset of the first ANN model listed in the Appendix A. Structural and training parameters of the second network are also summarized in Table 1. The average RMS error of training process of the second ANN model is illustrated in Figure 9(b). Accordingly, the second ANN model manifested a poorer prediction performance (average RMS error of 0.26) despite to train with less data and with more iteration than the first ANN model.

After a satisfactory training, both of networks were tested for their accuracy of prediction by a set of experimental data. The testing dataset of the first ANN model was given in Appendix B. The testing dataset of the second ANN model was obtained by excepting the experimental data concerning to stepover ratio from the testing dataset of the first ANN model. A few stray samples are eliminated for more satisfactorily training. The trained and tested ANN models are used to predict the surface roughness likely to be occurred on the workpiece. The predicted ANN output data's were compared with each other and also with the actual (experimental) data.

Experiments were realised by up milling cutting method with compressed air cooling using a TAKSAN TMC 500 CNC Vertical Machining Centre. Machining parameters (cutting speed (V), depth of cut (a), feed rate (f) and Stepover ratio S.R.) used in the experiments can be shown in the table of training datasets in Appendix A. Sample workpieces were cut from extruded AISI 1040 steel flat bar with dimensions 20x50x100 [mm]. A pre-machining with mild cutting conditions was done on the specimens on account of obtaining a uniform surface at initial. A TiAlN coated solid carbide flat end mill with 10 mm diameter, 45° helix angle and 4-flutes was used. The machining setup is represented in Figure 10. The average surface roughness (R_a) was measured in micrometer [μm] by a Mitutoyo profilometer and was listed in the Appendix A.

The results depicted in Figures 11-14 showed the generalization capabilities of prediction of the proposed ANN models comparatively in cases of with and without considering stepover ratio. Tests conducted for varying depth of cut and constant cutting speed and feed rate showed that increasing depth of cut has a different effect on surface roughness under V=31.41 m/min and V=62.83 m/min. The ANN model predicted the variation of surface roughness in case of with stepover almost exactly while producing a limited deviation in case of without stepover in Figure 11(a) and a remarkable deviation in Figure 11(b).

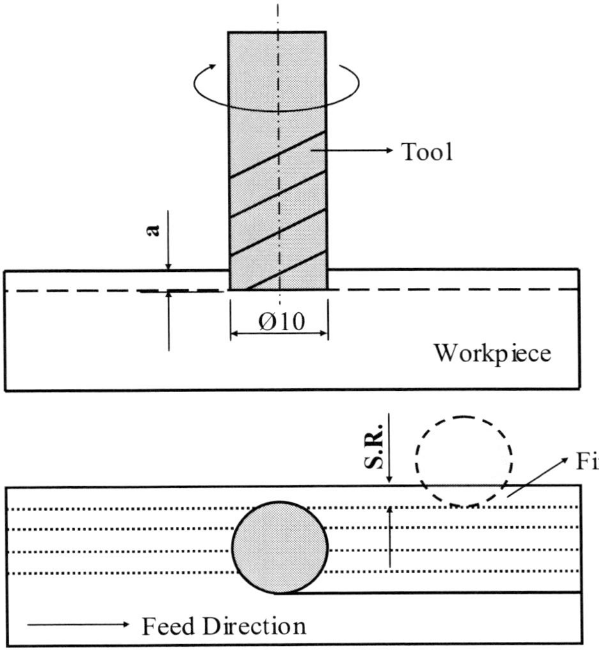

Figure 10. Schematic illustration of flat end milling setup (a: Depth of cut, S.R.: Stepover ratio).

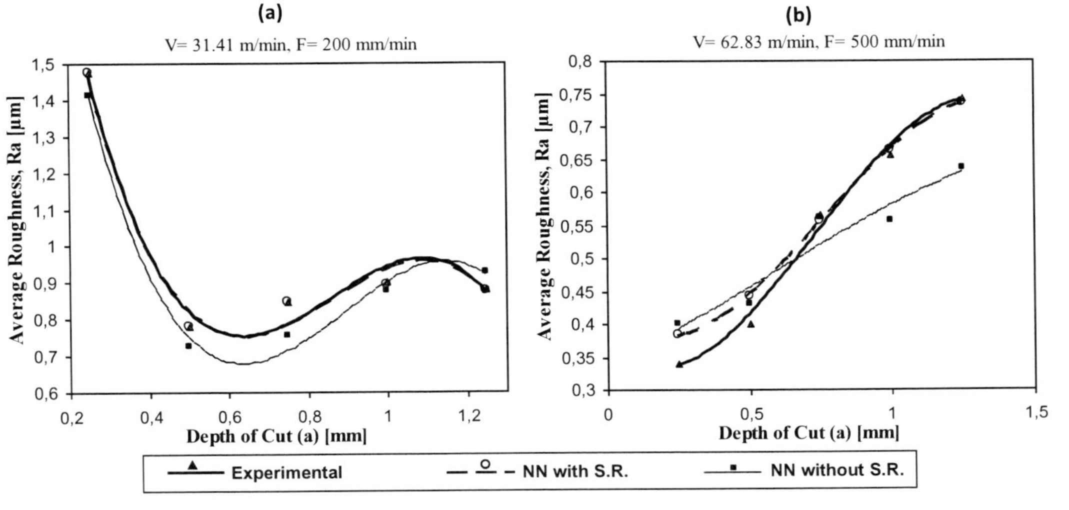

Figure 11. Prediction performance of ANN models under varying depth of cut (Comparatively between experimental, ANN model with stepover and ANN model without stepover).

Figure 12 illustrates the performance of ANN model under varying feed rate. For constant values of V=47.12 m/min and a=0.25 mm (Figure 12(a)), a negligible deviation is noticed in predicting performance of the model with stepover. But the deviation of the case of without stepover is remarkable. For a comparatively high cutting speed of V=94.24 m/min (Figure 12(b)), deviation of the case of with stepover is the maximum value of all tests. This situation may be due to sharply varying trend of experimental surface roughness values from decreasing to increasing in these test conditions instead of all other tests. On the other hand, performance of prediction model without stepover is poorer than the model with stepover.

Tests implemented for varying cutting speed showed that increasing cutting speed have no unfavourable effect on prediction performance of the ANN model with stepover. Although a small error in low cutting speed and low dept of cut values (Figure 13(a)), the best fit of curves of experimental and ANN model with stepover is observed in a=0.75 mm and F= 400 mm/min (Figure 13(b)), such that he curves are completely coincident. Besides, similar to other tests, the performance of the model without stepover is still not satisfied in both of Figure 13(a) and (b).

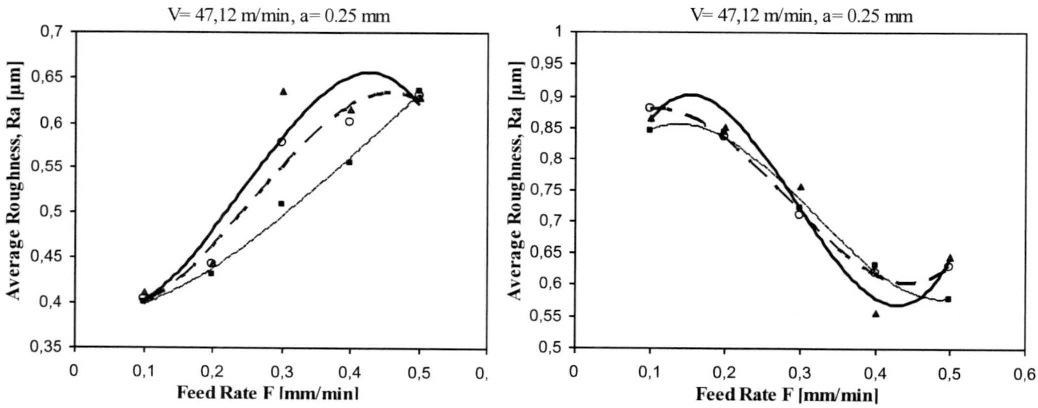

Figure 12. Prediction performance of ANN models under varying feed rate.

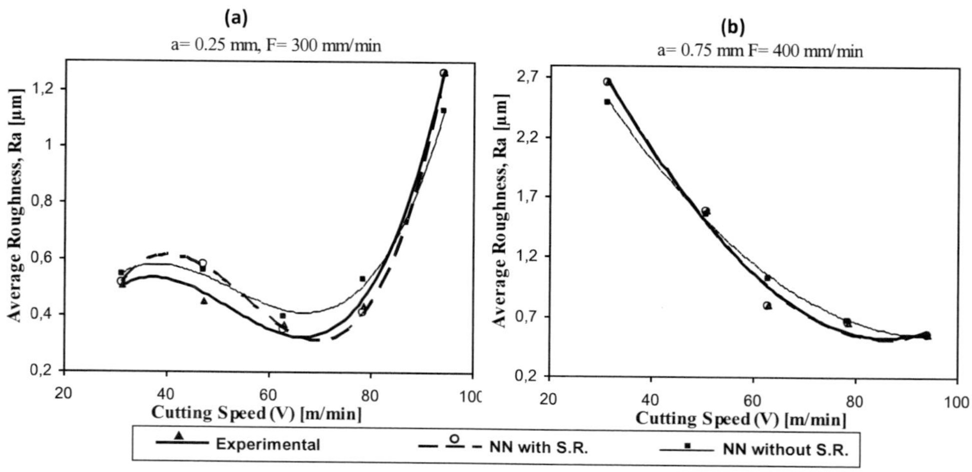

Figure 13. Prediction performance of ANN models under varying cutting speed.

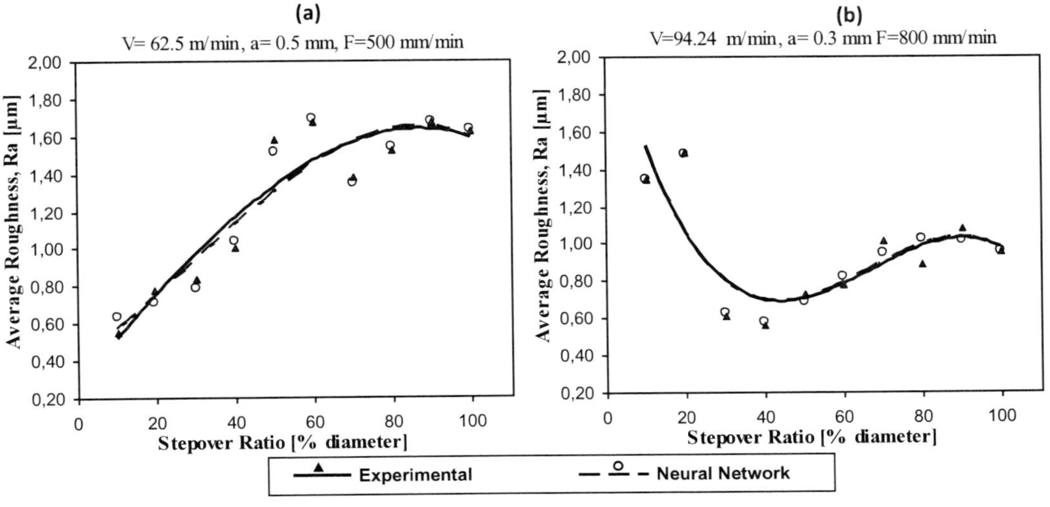

Figure 14. Prediction performance of the ANN model with considering stepover under varying stepover ratio.

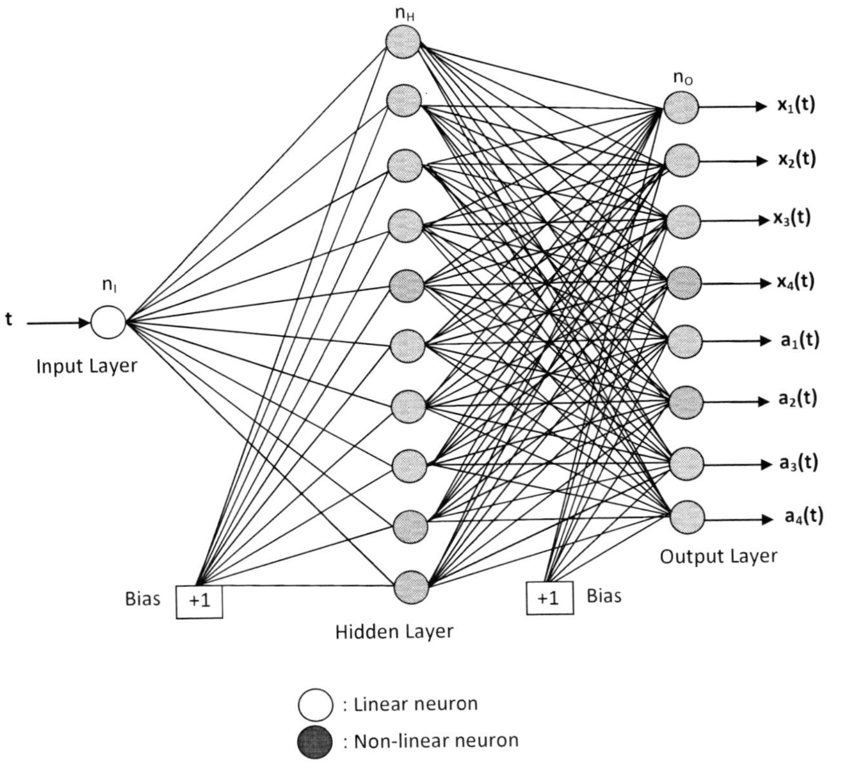

Figure 15. The ANN structure used in the vibration analysis of vehicle suspension system.

In Figure 14, the curves of the ANN model without stepover are not existent due to these graphs are already dependent to varying of stepover ratio. These graphs showed that the ANN model with stepover fit to the experimental data almost exactly under varying stepover ratio.

The Figure also indicated that the stepover ratio has a remarkable effect on surface roughness while the other parameters are kept constant.

In second analysis, the vibration analysis results of vehicle suspension system using two neural network approaches are presented. The ANN structure used in the vibration analysis of vehicle suspension system is shown in Figure 15. Ride comfort is examined by running test on a smooth surface road providing ground at 90 km/h speed of system tyre. Performance of BPNN and RBNN structures are compared for finding robust NN predictor. Two accelerometers are mounted on wheel axes that measure wheel acceleration and displacement. Two other accelerometers were used to measure vehicle body acceleration and displacement.

BPNN and experimental results of displacements obtained from four measuring points on vehicle test rig are given in Figure 16 show that BPNN approach gives poor performance. RBNN proved to be effective in analyzing vehicle suspension vibration and achieve better performance than BPNN approach (Figure 17).

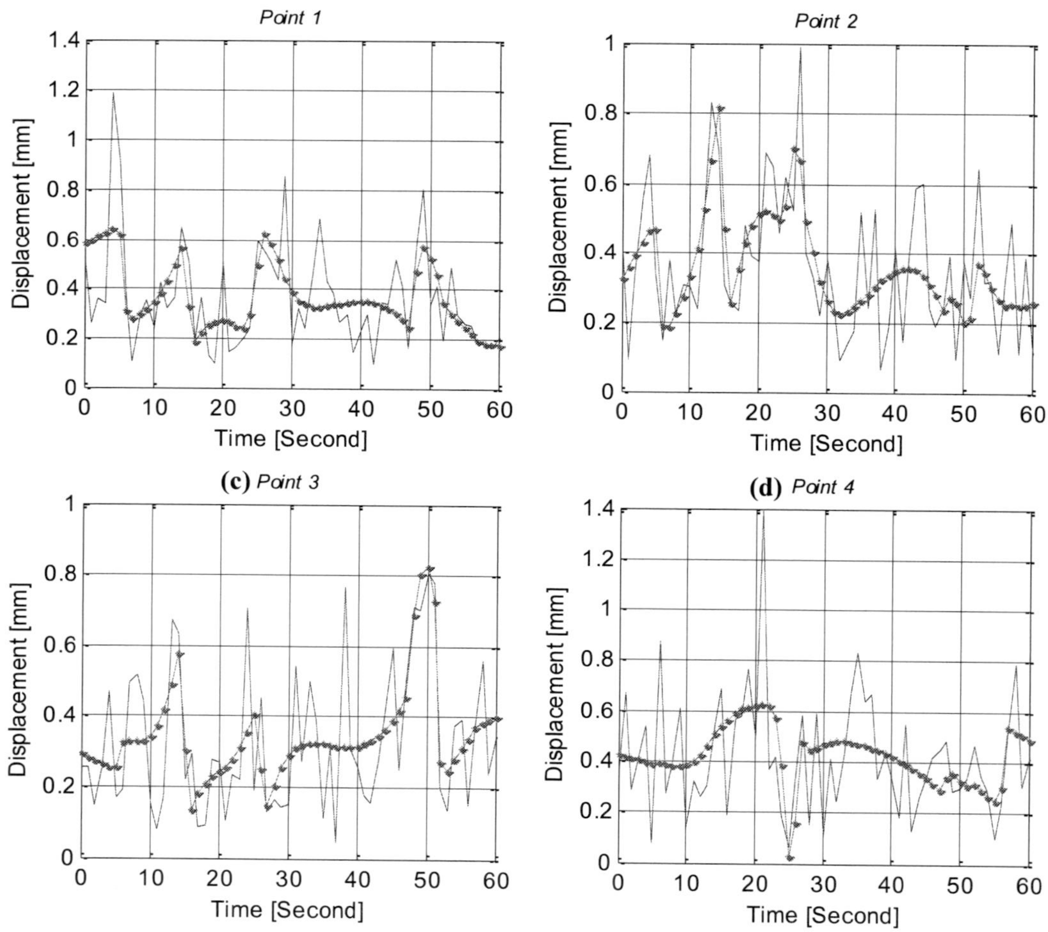

Figure 16. Displacement results of measure points on experimental suspension system using BPNN.

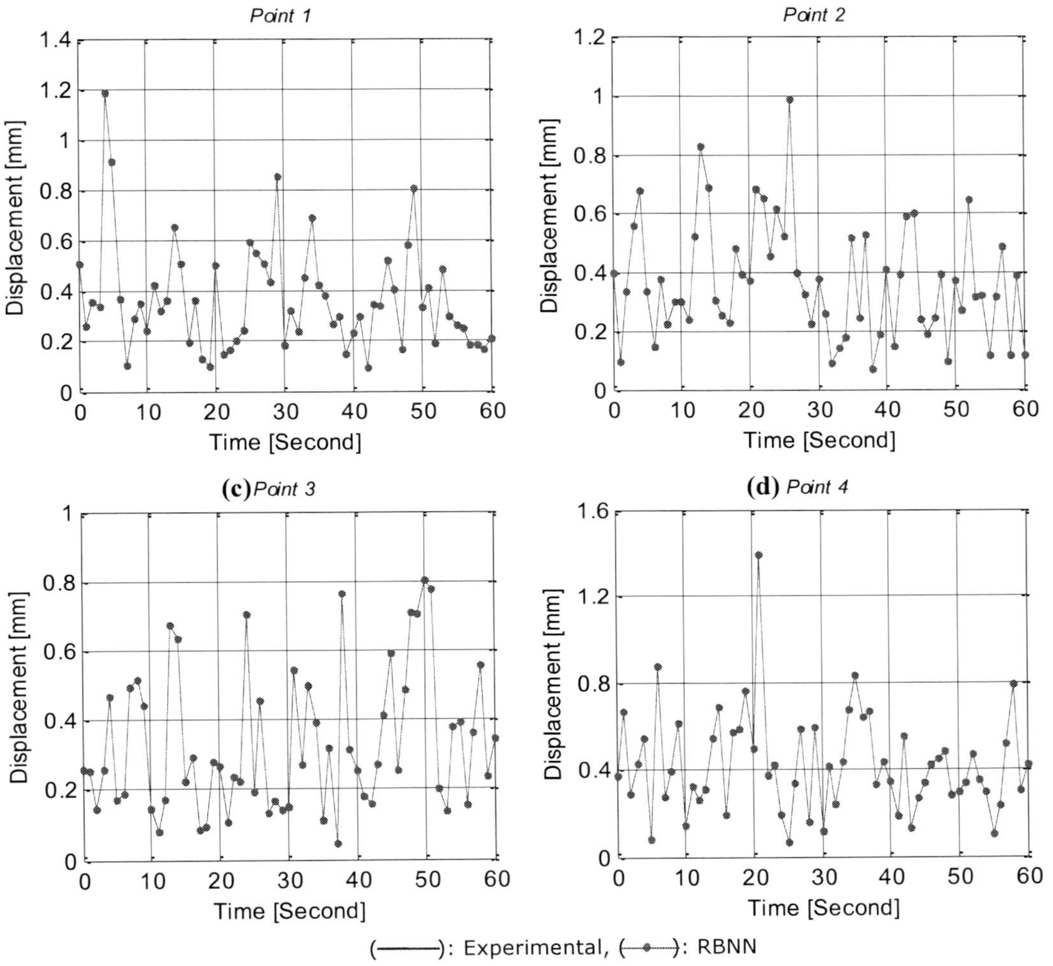

Figure 17. Displacements results of measure points on experimental suspension system using RBNN.

Acceleration results of four points using BPNN (Figure 18) is not acceptable because of NN structure. Root Mean Square Errors (RMSEs) variations after training are given in Table 2. Simulation and experimental studies showed that RBNN analyzer achieved superior performance because of radial basis function (Figure 19).

Table 2. Structural and training parameters of the ANN approach for vehicle suspension system

Results	Learning Algorithm	η	μ	n_I	n_H	n_O	N	AF	RMSE
Displacement	BPNN	0.3	0.4	1	10	8	$5*10^7$	Sigmoid	0.0935
	RBNN	0.3	0.4	1	10	8	$5*10^7$	Sigmoid	0.0001
Acceleration	BPNN	0.3	0.4	1	10	8	$5*10^7$	Sigmoid	0.1059
	RBNN	0.3	0.4	1	10	8	$5*10^7$	Sigmoid	0.0001

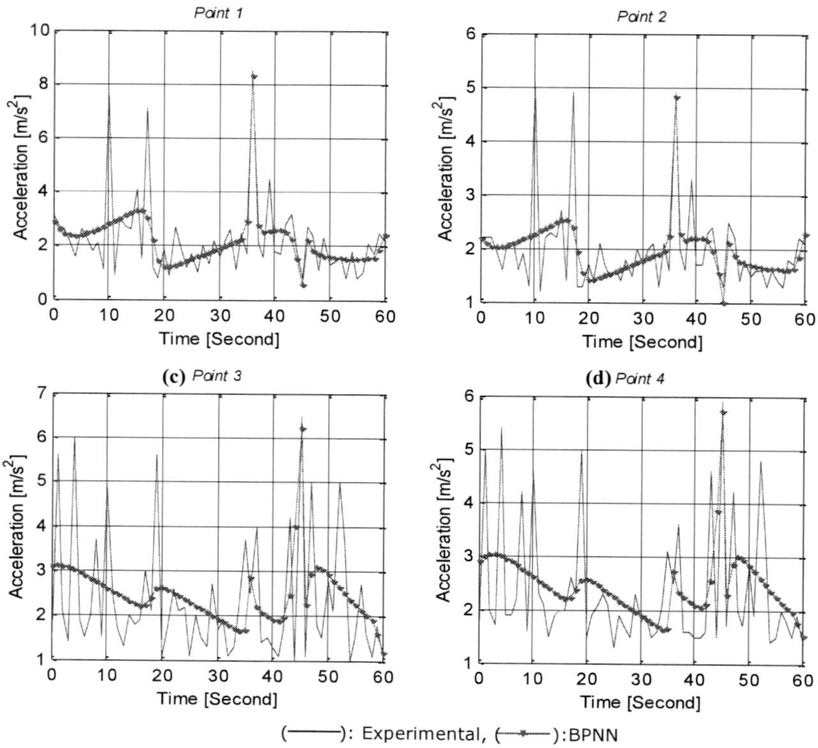

Figure 18. Acceleration results of measure points on experimental suspension system using BPNN.

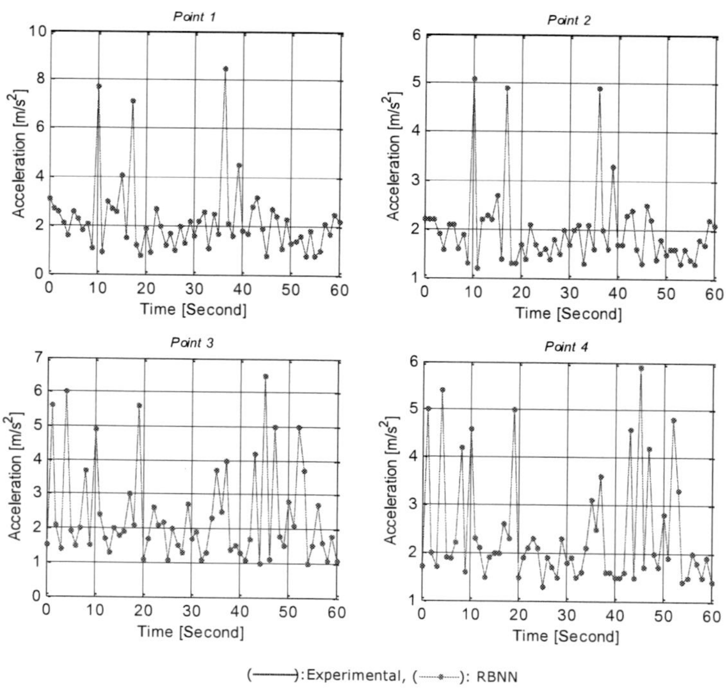

Figure 19. Acceleration results of measure points on experimental suspension system using RBNN.

5. Conclusion

This chapter presents different ANN approaches for the surface roughness prediction and the vehicle suspension system's analyses. Two ANN approaches is used for surface roughness prediction models. The prediction models consisted of cases of with and without considering stepover on account of comparing performance of the models and discovering the role of stepover in surface roughness prediction studies. The ANN models were trained with a Back Propagation Algorithm. RMS error of the model with considering stepover decreased to 0.04 after 100000 iterations, while error of the other model (without considering stepover) stayed at 0.26 after 250.000 iterations. Simulation results (graphs in Figs. 11-14) also indicated that the model with stepover proved capable of prediction of surface roughness. But the model without stepover produced remarkable deviations from experimental values. It can be yielded from general examination of results that surface roughness values are remarkably influenced by stepover, and it is not possible to predict the surface roughness accurately without considering stepover.

Another approach, the problem of the design of a non-linear hybrid car suspension system for ride quality using neural network predictors has been presented. Quarter car suspension test-rig model was considered a non-linear two degrees of freedom system subject to excitation from random road profiles. Performance of RBNN structure is found better than that of BPNN structure. RBNN is used for its advantages of rapid training, generality and simplicity over feed-forward neural network. Thus, RBNN structure could be employed for analyzing such systems in vehicle systems design.

The structure of recurrent neural network is unusual in that constitution of its hidden layer and feedback is entirely different from that of its output units. With radial-basis functions providing the foundation for the design of the hidden units, the theory of radial basis function networks is linked closely with that of radial-basis functions, which is one of the main fields of study in numerical analysis. Another interesting point is that with linear weights of the output layer providing a set of adjustable parameters, much can be gained ploughing through the extensive literature on linear adaptive filters.

Acknowledgments

Authors would like to their express to Scientific & Technological Research Council of Turkey (TUBITAK) 105M22 and Technological Research Centre of Erciyes University for supporting this study.

Nomenclature

AA	Arithmetic average
ANN	Artificial Neural Network
$a_1(t)$	Acceleration of suspension deflection
$a_2(t)$	Acceleration of tyre deflection
$a_3(t)$	Acceleration of sprung mass

$a_4(t)$	Acceleration of unsprung mass
$a_I(i)$	Output of competitive layer
ASS	Active Suspension System
b_j	Bias of the jth neuron in the hidden layer
b_k	Bias of the kth neurons in the output layer
BPNN	Back Propagation Neural Network
c_s	Damping of suspension system
c_t	Damping of pneumatic tyre
CLA	Centre line average
E_1 (t)	Error between experimental and neural network output signals
E_2 (t)	Propagation error between hidden layer and output layer
f(.)	Activation function of the hidden layer
F_a	Active control force of suspension system
FNN	Feed Forward Neural Network
g(.)	Activation function of the output layer
GRNN	General Regression Neural Network
$h_{bi}(t)$	Neighborhood kernel centered on the winner unit
k_s	Stiffness of suspension system
k_t	Stiffness of pneumatic tyre
L	Sampling length
LVQNN	Learning Vector Quantization Neural Network
LVDT	Linear Variable Differential Transformer
M	Number of map units
MNN	Modular Neural Network
m_s	Sprung mass
m_{us}	Unsprung mass (vehicle chassis)
N	Number of training samples
PLC	Programmable Logic Controller
Ra	Average roughness
r_b	Position of neurons b on the SOMNN grid
r_j	Radius vector of the j_{th} hidden unit
r_i	Position of neurons i on the SOMNN grid
RMSE	Root Mean Square Error
SR	Stepover ratio
T	Target value
$\mathbf{u}_j$	Weights input received by the jth output unit of the gating network
W_{ij}	Weight of the connection between the input layer neurons and the hidden layer neurons
W_{jh}	Additional recurrent weight
W_{jk}	Weights between j[th] neurons the hidden layer and k[th] neurons output layer
$x_1(t)$	Suspension deflection
$x_2(t)$	Tyre deflection
$x_3(t)$	Sprung mass speed
$x_4(t)$	Unsprung mass speed
Y	Ordinate of the profile

y_j	Output of the jth local network
$\mathbf{z_j(t)}$	Output of the jth neuron in the hidden layer
z_s	Displacements of sprung
z_{us}	Displacements of unsprung masses
z_r	Road displacement
$\beta(t)$	Adaptation coefficient
α	Momentum term
η	Learning rate
φ	Some function which is monotonic for non negative numbers

REFERENCES

[1] Lee, W. B. & Cheung, C. F. (2001). A dynamic surface topography model for the prediction of nano-surface generation in ultra-precision machining. *Int Jnl of Mech*, *43*, 961-991.

[2] Benardos, P. G. & Vosniakos, G. C. (2003). Predicting surface roughness in machining: a review. *Int J Mach Tools Manuf*, *43*, 833-844.

[3] Ozcelik, B. & Bayramoglu, M. (2006). The statistical modeling of surface roughness in high-speed flat end milling. *Int J Mach Tools Manuf*, *46*, 1395–1402.

[4] Sağlam, H. & Ünüvar, A. (2003). Tool condition monitoring in milling based on cutting forces by a neural network. *Int J Prod Res*, *41*, 1519–1532.

[5] Topal, E. S., Sinanoglu, C., Gercekcioglu, E. & Yildizli, K. (2007). Neural network prediction of surface roughness in milling of AISI 1040 steel. *J Balkan Trib Assoc*, *13*, 18-23.

[6] Özcelik, B., Öktem, H. & Kurtaran, H. (2005). Optimum surface roughness in end milling Inconel 718 by coupling neural network model and genetic algorithm. *Int J Adv Manuf Technol*, *27*, 234–241.

[7] Balic, J. & Korosec, M. (2002). Intelligent tool path generation for milling of free surfaces using neural networks. *Int J Mach Tools Manuf*, *42*, 1171–1179.

[8] Çolak, O., Kurbanoğlu, C. & Kayacan, M. C. (2007). Milling surface roughness prediction using evolutionary programming methods. *Materials&Design*, *28*, 657–666.

[9] Lou, S. J. & Chen, J. C. (1999). In-process surface roughness recognition (ISRR) system in end-milling operation. *Int J Adv Manuf Technol*, *15*, 200–209.

[10] Alauddin, M., El Baradie, M. A. & Hashmi, M. S. J. (1995). Computer-aided analysis of a surface-roughness model for end milling. *J Mater Process Technol*, *55*, 123–127.

[11] Luo, X., Cheng, K. & Ward, R. (2005). The effects of machining process variables and tooling characterisation on the surface generation. *Int J Adv Manuf Technol*, *25*, 1089-1097.

[12] Liu, X. & Cheng, K. (2005). Modelling the machining dynamics of peripheral milling, *Int J Mach Tools Manuf*, *45*, 1301-1320.

[13] Lo, S. P. (2003). An adaptive-network based fuzzy inference system for prediction of workpiece surface roughness in end milling. *J Mater Process Technol*, *142*, 665–675.

[14] Dweiri, F., Al-Jarrah, M. & Al-Wedyan, H. (2003). Fuzzy surface roughness modeling of CNC down milling of Alumin-79. *J Mater Process Technol*, *133*, 266–275.

[15] Lou, S. J. & Chen, J. C. (1997). In-process surface recognition of a CNC milling machine using the fuzzy nets method, *Comput in Ind Eng*, *33*, 401–404.

[16] Guglielmino, E. & Edge, K. A. (2004). A controlled friction damper for vehicle applications, *Contr Engin Prac*, *12*, 431–443.

[17] Chantranuwathana, S. & Peng, H. (2004). Adaptive robust force control for vehicle active suspensions, *Int J Adapt Contr Sign Process*, *18*, 83–102.

[18] Shen, X. & Peng, H. (2003). Analysis of active suspension systems with hydraulic actuators, *Proceed of the*, IAVSD Conf, Japan.

[19] Fischer, D. & Isermann, R. (2004). Mechatronic semi-active and active vehicle suspensions, *Contr Engin Pract*, *12*, 1353–1367.

[20] Swevers, J., Lauwerys, C., Vandersmissen, B., Maes, M., Reybrouck, K. & Sas, P. (2007). A model-free control structure for the on-line tuning of the semi-active suspension of a passenger car, *Mech Syst Sign Process*, *21*, 1422–1436.

[21] Choi, S. B. & Han, Y. M. (2007). Vibration control of electrorheological seat suspension with human-body model using sliding mode control, *Sound Vibrat*, *303*, 391–404.

[22] Du, H. & Zhang, N. (2007). H∞ control of active vehicle suspensions with actuator time delay, *Sound Vibrat*, *301*, 236–252.

[23] Huang, S. & Chen, H. Y. (2006). Adaptive sliding controller with self-tuning fuzzy compensation for vehicle suspension control, *Mechatronics*, *16*, 607–622.

[24] Gao, H., Lam, J. & Wang, C. (2006). Multi-objective control of vehicle active suspension systems via load-dependent controllers, *Sound Vibrat*, *290*, 654–675.

[25] Yıldırım, Ş. & Eski, İ. (2006). Design of robust model based neural controller for controlling vibration of active suspension system, *Sci Indus Res*, *65*, 646-654.

[26] Yıldırım, Ş., Eski, İ. Noise analysis of robot manipulator using neural networks, *Robot Cim-Int Manuf*, (in press).

[27] Jain, L. C. & Medsker, L. (2010). *Recurrent neural networks: design and applications*, Taylor and Francis Group.

[28] Saliza, İ. & Ahmad, A. M. B. (2004). Recurrent neural network with backpropagation through time algorithm for Arabic recognition, *Proceedings 18th European Simulation Multiconference*, Graham Horton (C) Scs Europe.

[29] Yıldırım, Ş., Erkaya, S., Eski, İ. & Uzmay, İ. (2009). Noise and vibration analysis of car engines using proposed neural network, *J Vib Contr.*, *15*, 133-146.

[30] Vesanto, J. & Alhoniemi, E. (2000). Clustering of the Self-Organizing Map, *IEEE Transactions on Neural Networks*, *11*, 586-600.

[31] Dagnall, H. M. A. (1986). *Exploring surface texture*. Rank Taylor Hobson, Leicester.

[32] Topal, E. S. The role of stepover ratio in prediction of surface roughness in flat end milling. *Int. J. Mech. Sci.* (in press).

[33] Yıldırım, Ş. & Eski, İ. (2009). Vibration analysis of an experimental suspension system using artificial neural networks, *J. Sci. Ind. Res.*, *68*, 496-505.

[34] Lippmann, R. P. (1987). An introduction to computing with neural nets. IEEE *ASSP Magazine*, *2*, 4-22.

[35] Yıldırım, Ş. & Eski, İ. (2006). A QP Artificial neural network inverse kinematic solution for accurate robot path control. *KSME Int. J. 20*, 917-928.

APPENDIX A

The training dataset (**V**:Cutting speed[m/min], **a**:Depth of cut[mm], **f**:Feed rate[mm/min], **S.R**: Stepover ratio[%], **Ra**: Average surface roughness [μm])

No	V	a	f	S.R.	Ra
1	47.12	0.25	100	100	0.41
2	47.12	0.25	200	100	0.44
3	47.12	0.25	300	100	0.72
4	47.12	0.25	400	100	0.62
5	47.12	0.25	500	100	0.63
6	94.24	0.5	100	100	0.87
7	94.24	0.5	200	100	0.85
8	94.24	0.5	300	100	0.76
9	94.24	0.5	400	100	0.56
10	94.24	0.5	500	100	0.65
11	31.41	0.25	300	100	0.51
12	47.12	0.25	300	100	0.45
13	62.83	0.25	300	100	0.37
14	78.53	0.25	300	100	0.43
15	94.24	0.25	300	100	1.27
16	31.41	0.75	400	100	2.66
17	47.12	0.75	400	100	1.59
18	62.83	0.75	400	100	0.80
19	78.53	0.75	400	100	0.66
20	94.24	0.75	400	100	0.57
21	62.83	0.25	500	100	0.34
22	62.83	0.5	500	100	0.40
23	62.83	0.75	500	100	0.57
24	62.83	1	500	100	0.66
25	62.83	1.25	500	100	0.74
26	31.41	0.25	200	100	1.48
27	31.41	0.5	200	100	0.78
28	31.41	0.75	200	100	0.85
29	31.41	1	200	100	0.90
30	31.41	1.25	200	100	0.88
31	62.83	0.5	500	10	0.55
32	62.83	0.5	500	20	0.80
33	62.83	0.5	500	30	0.84
34	62.83	0.5	500	40	0.92
35	62.83	0.5	500	50	1.78
36	62.83	0.5	500	60	1.92
37	62.83	0.5	500	70	1.28

Appendix A Continued

38	62.83	0.5	500	80	1.42
39	62.83	0.5	500	90	1.61
40	62.83	0.5	500	100	1.42
41	94.24	0.3	800	10	1.45
42	94.24	0.3	800	20	1.59
43	94.24	0.3	800	30	0.61
44	94.24	0.3	800	40	0.56
45	94.24	0.3	800	50	0.73
46	94.24	0.3	800	60	0.78
47	94.24	0.3	800	70	1.01
48	94.24	0.3	800	80	0.89
49	94.24	0.3	800	90	1.08
50	94.24	0.3	800	100	0.95

APPENDIX B

The testing dataset (**V**:Cutting speed[m/min], **a**:Depth of cut[mm], **f**:Feed rate[mm/min], **S.R**: Stepover ratio[%], **Ra**: Average surface roughness [μm])

No	V	a	f	S.R.	Ra
1	47.12	0.25	100	100	0.41
2	47.12	0.25	200	100	0.44
3	47.12	0.25	400	100	0.62
4	47.12	0.25	500	100	0.63
5	94.24	0.5	200	100	0.85
6	94.24	0.5	300	100	0.76
7	94.24	0.5	500	100	0.65
8	31.41	0.25	300	100	0.51
9	62.83	0.25	300	100	0.37
10	78.53	0.25	300	100	0.43
11	31.41	0.75	400	100	2.66
12	47.12	0.75	400	100	1.59
13	78.53	0.75	400	100	0.66
14	94.24	0.75	400	100	0.57
15	62.83	0.5	500	100	0.40
16	62.83	0.75	500	100	0.57
17	62.83	1.25	500	100	0.74
18	31.41	0.25	200	100	1.48
19	31.41	0.75	200	100	0.85
20	31.41	1	200	100	0.90
21	62.83	0.5	500	10	0.55
22	62.83	0.5	500	20	0.80

23	62.83	0.5	500	40	0.92
24	62.83	0.5	500	50	1.78
25	62.83	0.5	500	70	1.28
26	62.83	0.5	500	80	1.42
27	62.83	0.5	500	100	1.42
28	94.24	0.3	800	10	1.45
29	94.24	0.3	800	30	0.61
30	94.24	0.3	800	40	0.56
31	94.24	0.3	800	60	0.78
32	94.24	0.3	800	70	1.01
33	94.24	0.3	800	90	1.08
34	94.24	0.3	800	100	0.95

In: Focus on Artificial Neural Networks
Editor: John A. Flores, pp. 129-150

ISBN: 978-1-61324-285-8

Chapter 6

MASSIVE-TRAINING ARTIFICIAL NEURAL NETWORKS FOR SUPERVISED ENHANCEMENT/SUPPRESSION OF LESIONS/PATTERNS IN MEDICAL IMAGES

Kenji Suzuki [*]
Department of Radiology, Division of the Biological Sciences,
The University of Chicago, Chicago, Illinois, USA

ABSTRACT

Medical imaging is an indispensable tool for patients' healthcare in modern medicine. Machine learning plays an important role in the medical imaging field, including medical image processing, medical image analysis, computer-aided diagnosis, organ/lesion segmentation, lesion classification, functional brain mapping, and image-guided therapy, because objects in medical images such as lesions, structures, and anatomy often cannot be modeled accurately by simple equations; thus, tasks in medical imaging require some form of "learning from examples." Pattern enhancement (or suppression: enhancement of specific patterns means suppression of other patterns) is one of the fundamental tasks in medical image processing and analysis. When a doctor diagnoses lesions in medical images, his/her tasks are detection, extraction, segmentation, classification, and measurement of lesions. If we can enhance a specific pattern such as a lesion of interest in a medical image accurately, those tasks are almost complete. What is left to do is merely thresholding of the enhanced lesion. For the tasks of detection and measurement, calculation of the centroid of and the area in the thresholded region may be needed. Thus, enhancement (or suppression) of patterns is one of the fundamental tasks. In this chapter, the basic principles and applications of supervised enhancement/suppression filters based on machine learning, called massive-training artificial neural networks (MTANN), for medical image processing/analysis are presented.

[*] Corresponding author: Department of Radiology, Division of the Biological Sciences, The University of Chicago, 5841 South Maryland Avenue, MC 2026, Chicago, IL 60637, USA, Phone: (773) 834-5098, Fax: (773) 702-0371, E-mail: suzuki@uchicago.edu

1. INTRODUCTION

Medical imaging is an indispensable tool for patients' healthcare in modern medicine. Machine learning plays an important role in the medical imaging field, including medical image processing, medical image analysis, computer-aided diagnosis, organ/lesion segmentation, lesion classification, functional brain mapping, and image-guided therapy, because objects in medical images such as lesions, structures, and anatomy often cannot be modeled accurately by simple equations; thus, tasks in medical imaging require some form of "learning from examples."

What is a fundamental task in medical imaging? There are many answers to this broad question. If we limit the question to the image-processing and analysis portion in the imaging chain (i.e., image acquisition/reconstruction, image processing, image analysis, and image evaluation), we can realize that pattern enhancement (or suppression: enhancement of specific patterns means suppression of other patterns) is one of the fundamental tasks in medical image processing and analysis. When a medical doctor diagnoses lesions in medical images, his/her tasks are detection, extraction, segmentation, classification, and measurement of lesions. If we can enhance a specific pattern such as a lesion of interest in a medical image accurately, those tasks are almost complete. What is left to do is merely thresholding of the enhanced lesion. For the tasks of detection and measurement, calculation of the centroid of and the area in the thresholded region may be needed. Thus, enhancement (or suppression) of patterns is one of the fundamental tasks.

Although we now know that enhancing of patterns is one of the fundamental tasks in medical image processing and analysis, this is not so easy in the real world. Lesions, structures, and anatomy in medical images are not simple enough to be represented accurately by a simple equation in many cases. For example, a blob-enhancement filter based on the Hessian matrix can enhance sphere-like objects in medical images [1]. Actual lesions, however, often differ from a simple sphere model. A lung nodule is often modeled as a solid sphere, but there are nodules of various shapes and inhomogeneous nodules such as nodules with spiculation and ground-glass nodules. Thus, conventional filters often fail to enhance actual lesions.

In recent years, as computational power has dramatically increased, machine-learning-based image-processing filters which learn to enhance patterns in medical images have emerged [2-6]. The machine-learning-based filters are trained with input images and corresponding "teaching" images to learn the relationship between the two images directly (as opposed to a machine-learning technique used as a classifier which learns the relationship between features and classes [7-10]). By training with different image sets, the machine-learning-based filters can acquire different functions, such as noise reduction [2, 6, 11], edge enhancement from noisy images [12], enhancement of contours of an organ such as the left ventricle traced by a medical doctor [13], enhancement of a specific anatomy such as ribs in chest radiographs [4, 5, 14, 15], and enhancement of lesions such as lung nodules [3, 16-21] and colorectal polyps [22-25]. Because of the versatility of the machine-learning-based filters, there are a wide variety of applications in medical imaging, such as improvement of image quality, quantitative measurement of anatomic structures and lesions, anatomic structure separation, computer-aided detection and diagnosis, function measurement, and distinction between benign and malignant lesions. Other image-based machine-learning models including shift-invariant artificial neural networks (ANNs) [26, 27] and convolution ANNs [28-30] have been investigated for classification of patterns, i.e., classification of objects into nominal classes such as normal or abnormal, but they do not enhance patterns (i.e., as a form of images).

In this chapter, the basic principles and applications of supervised enhancement/suppression filters based on machine learning for medical image processing/ analysis are presented.

2. Convolution Filters and Their Network Representations

2.1. Averaging Filter for Noise Reduction

Convolution filters are widely used for image processing such as suppression (smoothing) of noise, enhancement of edges, and enhancement or suppression of a specific frequency range. The convolution is defined as the integral of the product of two functions after one of them is reversed and shifted. In a discrete case, the integral becomes a summation. A convolution filter for image processing involves the convolution of an original image $f(x,y)$ and a filter kernel $K(i,j)$, represented by

$$g(x,y) = \sum_{(i.j)\in R} f(x,y) \cdot K(i-x, j-y), \tag{1}$$

where R is a kernel region. It is often written as

$$g = f * K, \tag{2}$$

where * denotes the convolution operation. By changing a kernel function, it becomes a filter with a different characteristic. For example, when all elements (filter coefficients) of a kernel function have the same signed values, the filter becomes a smoothing filter. An averaging filter would be the simplest one and is often used for noise reduction. The kernel of an averaging filter is represented by

$$K(i,j) = \frac{1}{N}, \tag{3}$$

where N is the number of pixels in the kernel region, R. A simple example of a 3 by 3 averaging filter is shown in Figure 1(a). All filter coefficients of the kernel function have 1. This averaging filter is represented by

$$g(x,y) = \sum_{(i.j)\in R} f(x-i, y-j) \cdot \frac{1}{9}. \tag{4}$$

To normalize the output, input pixels are multiplied by 1/9. This has the effect of smoothing out (suppressing) noise and forms a low-pass filter. The Fourier transform of a kernel function provides the transfer function of a convolution filter (which is a linear system). In the case of averaging, the Fourier transform of the kernel function, which is a two-dimensional rectangular function, is represented by a sinc function in the Fourier domain, which provides a low-pass filtering characteristic. High-frequency components in the Fourier domain, which represent noise and edges in an image, are cut by the sinc function. Although simple averaging reduces noise in an image, important details such as edges of an object tend to be smoothed out as well.

2.2. Laplacian Filter for Edge Enhancement

When filter coefficients in a kernel function have different signed values, the filter becomes an enhancement filter. The Laplacian filter is such a filter based on a second derivative, and it is often used for enhancement of edges. The digital Laplacian filter (which is a discrete approximation of the Laplacian operation) is illustrated in Figure 1(b). Note that the sign of filter coefficients is reversed so that the gray scale of the output image is not reversed. The Laplacian filter is represented by

$$g(x,y) = 8 \cdot f(x,y) - \sum_{(i,j) \neq (x,y) \in R} f(x-i, y-j). \tag{5}$$

It has the effect of enhancing edges and forms a high-pass filter. The drawback of the Laplacian filter is that it enhances not only edges, but noise as well. A more complex characteristic which mixes smoothing and edge-enhancement effects can be designed and obtained by use of a more complex kernel function.

It is very interesting to think of a convolution filter in a graphic way. A convolution filter can be represented by a network. The network representations of the smoothing filter and the Laplacian filter are shown in Figures 1(a) and (b). Filter coefficients in the kernel function correspond to weights in the network. By changing of weights in the network, the network can have various characteristics.

3. Supervised Neural-Network Filter

3.1. Architecture

As described above, a convolution filter can be represented by a network. What if we use an ANN [31, 32] as the kernel function of a convolution filter, as shown in Figure 1(c)? Filter coefficients of a supervised neural-network (NN) filter can be determined by training with input images and corresponding "teaching" images. The universal approximation property of an ANN guarantees diverse capabilities of a supervised NN filter, because it has been proved theoretically that a three-layered ANN can approximate any continuous mapping with an arbitrary precision [33, 34]; thus, it is expected that the supervised NN filter can be a universal convolution filter for image processing. In other words, conventional convolution filters are special cases of the supervised NN filters. The high degree of nonlinearity of an ANN would overcome the limitations of the linear or quasi-linear performance of conventional convolution filters. The supervised NN filter can realize, through training, many image-processing functions, including high-pass, low-pass, and band-pass filtering, noise reduction, edge enhancement, object enhancement, segmentation, and classification. For example, the supervised NN filter can act as an averaging operation, gradient operation, Laplacian operation, templates, a part of a sinusoidal function, etc.

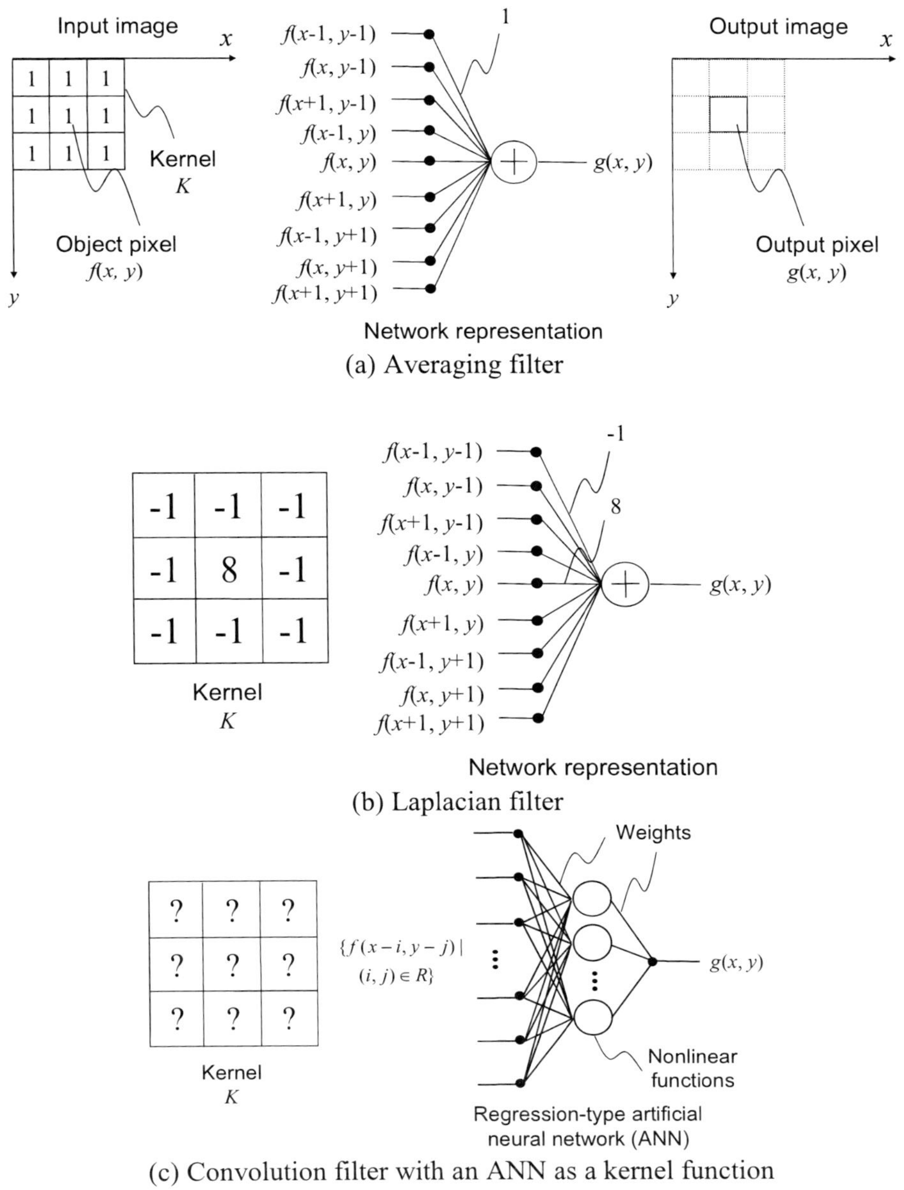

Figure 1. Convolution filters and their network representations.

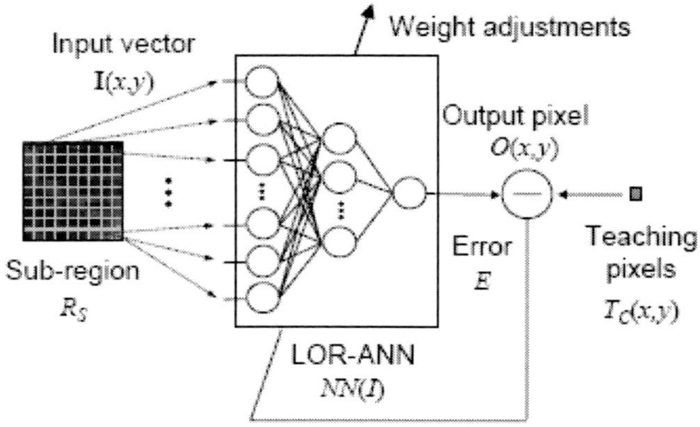

Figure 2. Architecture and training of a supervised NN filter based on a linear-output regression ANN (LOR-ANN) model.

The architecture of a supervised NN filter is shown in Figure 2. A supervised NN filter consists of a linear-output regression ANN (LOR-ANN) model [12], which is a regression-type ANN capable of operating on pixel data directly. The supervised NN filter is trained with input images and the corresponding "teaching" images that are ideal (or desired) images. The input to the supervised NN filter consists of pixel values in a sub-region, R_S, extracted from an input image. The output of the supervised NN filter is a continuous scalar value, which is associated with the center pixel in the sub-region and is represented by

$$O(x, y) = LORANN\{I(x-i, y-j) \mid (i, j) \in R_S\}, \tag{6}$$

where x and y are the coordinate indices, $LORANN(\cdot)$ is the output of the LOR-ANN model, and $I(x,y)$ is a pixel value in the input image. The LOR-ANN employs a linear function,

$$f_L(u) = a \cdot u + 0.5, \tag{7}$$

instead of a sigmoid function,

$$f_S(u) = 1/\{1+\exp(-u)\}, \tag{8}$$

as the activation function of the output layer unit because the characteristics and performance of an ANN are improved significantly with a linear function when applied to the continuous mapping of values in image processing [12]. Note that the activation function in the hidden layers is still a sigmoid function. The input vector can be rewritten as

$$\mathbf{I}(x, y) = \{I_1, I_2, \cdots, I_m, \cdots, I_{N_I}\}, \tag{9}$$

where m is an input unit number and N_I is the number of input units. The output of the n-th unit in the hidden layer is represented by

$$O_n^H = f_S\left\{\sum_{m=1}^{N_I} w_{mn}^H \cdot I_m - w_{0n}^H\right\}, \tag{10}$$

where $w^H{}_{mn}$ is a weight between the m-th unit in the input layer and the n-th unit in the hidden layer, and $w^H{}_{0n}$ is an offset of the n-th unit in the hidden layer. The output of the output layer unit is represented by

$$O(x,y) = f_L\left\{\sum_{m=1}^{N_H} w_m^O \cdot O_m^H - w_0^O\right\}, \tag{11}$$

where $w^O{}_m$ is a weight between the m-th unit in the hidden layer and the unit in the output layer, and $w^O{}_0$ is an offset of the unit in the output layer. For processing of the entire image, the scanning of an input image with the supervised NN filter is performed pixel by pixel in a raster scan order, like a convolution filter does.

3.2. Training of a Supervised NN Filter

The supervised NN filter involves training with a large number of pairs of sub-regions and pixels; we call it a massive-sub-region training scheme. For enrichment of the training samples, a training region, R_T, extracted from the input image is divided pixel by pixel into a large number of sub-regions. Note that close sub-regions overlap each other. Single pixels are extracted from the corresponding teaching (desired) image as teaching values. The supervised NN filter is massively trained by use of each of a large number of input sub-regions together with each of the corresponding teaching single pixels. The error to be minimized by training of the supervised NN filter is given by

$$E = \frac{1}{P}\sum_{c}\sum_{(x,y)\in R_T}\{T_c(x,y) - O_c(x,y)\}^2 , \tag{12}$$

where c is a training case number, O_c is the output of the supervised NN filter for the c-th case, T_c is the teaching value for the supervised NN filter for the c-th case, and P is the number of total training pixels in the training region, R_T. The supervised NN filter is trained by a linear-output back-propagation algorithm where the generalized delta rule [31] is applied to the LOR-ANN architecture [12]. The correction of the weight between hidden units and output unit can be represented by

$$\Delta W^O = -\eta \frac{\partial E}{\partial W^O} = -\eta a(T - f)O^H . \tag{13}$$

where η is a learning rate. Please refer to Refs. [12, 35] for the details and the property of the linear-output BP algorithm. After training, the supervised NN filter is expected to output values similar to values in the teaching (desired) images.

4. Applications

4.1. Reduction of Quantum Noise in Medical X-Ray Images

Although conventional averaging filters can reduce noise in images, they smooth out details in the images as well. To address this issue, we developed a "neural filter" [2, 6, 11] based on the supervised NN filter for reduction of noise while preserving image details. For reduction of quantum noise (quantum noise is dominant in relatively low-radiation-dose x-ray images used in diagnosis) in diagnostic x-ray images, we need noisy input images and corresponding noiseless "teaching" images. To this end, we start from high-radiation-dose x-ray images which have little noise. We synthesized a noisy input image by addition of simulated quantum noise (which is modeled as signal-dependent noise) to a noiseless original high-radiation-dose image $f_O(x,y)$, represented by

$$f_N(x,y) = f_O(x,y) + n\{\sigma(f_O(x,y)\}, \tag{14}$$

where $n\{\sigma(f_O(x,y)\}$ is noise with the standard deviation $\sigma\{f_O(x,y)\} = k_N \sqrt{f_O(x,y)}$,

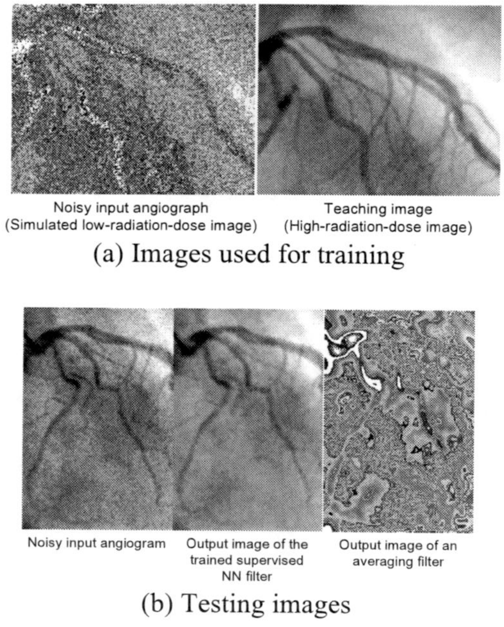

Figure 3. Reduction of quantum noise in angiograms by using a supervised NN filter called a "neural filter".

and k_N is a parameter determining the amount of noise. A synthesized noisy image and a noiseless original high-radiation-dose image shown in Figure 3(a) were used as the input image and as the teaching image, respectively, for training the neural filter. For sufficient reduction of noise, the input region of the neural filter consisted of 11x11 pixels. For efficient training of features in the entire image, 5,000 training pixels were extracted randomly from the input and teaching images. The training of the neural filter was performed for 100,000 iterations. The output image of the trained neural filter for a non-training case is shown in Figure 3(b). The noise in the input image is reduced, while image details such as the edges of arteries and peripheral vessels are maintained, whereas an averaging convolution filter reduces image details together with noise.

4.2. Enhancement of Edges from Very Noisy X-Ray Images

Although conventional edge enhancers can enhance edges in images with little noise very well, they do not work well on noisy images. To address this issue, we developed a "neural edge enhancer" [12] based on the supervised NN filter for enhancing edges from very noisy images. We started from a noiseless high-radiation-dose x-ray image. We added quantum noise to the original noiseless image to create a noisy input image. We applied a Sobel edge enhancer to the original noiseless image to create a teaching clear-edge image, as shown in Figure 4. The key here is that the Sobel edge enhancer works very well on noiseless images. We trained the neural edge enhancer with the noisy input image together with the corresponding teaching (desired) edge image. For comparison, we applied the trained neural edge enhancer and the Sobel edge enhancer to noisy non-training images. The resulting non-training edge-enhanced images are shown in Figure 5. Edges are enhanced clearly in the output image of the neural edge enhancer while noise is suppressed, whereas the Sobel edge enhancer enhances not only edges, but also noise.

4.3. Enhancement of Contours Traced by a Cardiologist

"Physical" edges enhanced by a conventional edge enhancer do not necessarily agree with edges a person determines. In critical applications such as measurement of the size of an organ or a lesion, the accuracy of edge enhancement affects the accuracy of final diagnosis. Thus, edges enhanced by an edge enhancer need to agree well with "subjective" edges determined by a medical doctor who uses clinical knowledge and experience for.

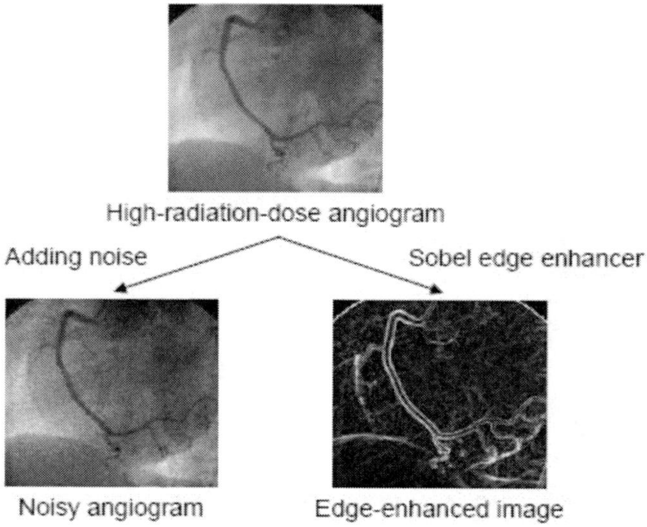

Figure 4. Creating a noisy input image and a teaching clear-edge image from a high-radiationdose image with little noise for training a supervised NN filter, called a "neural edge enhancer," for enhancing edges from very noisy images.

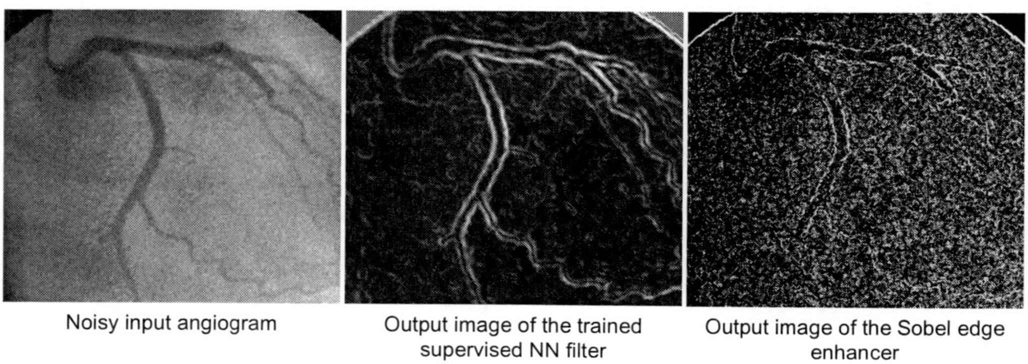

Figure 5. Comparison of enhancement of edges in a very noisy image by the trained supervised NN filter, called a "neural edge enhancer," with that by the Sobel edge enhancer.

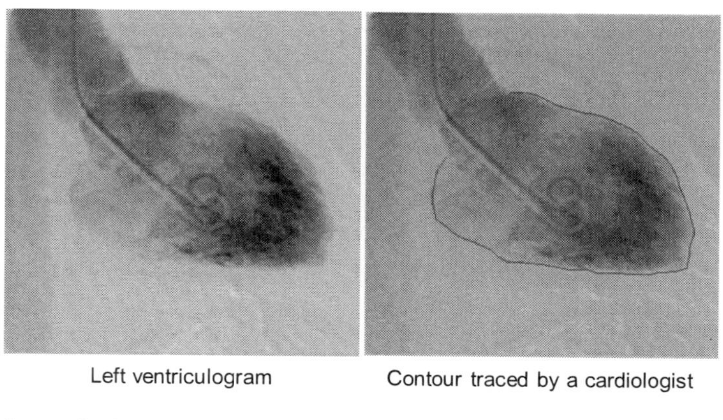

Figure 6. Input left ventriculogram and the corresponding "teaching" contour traced by a cardiologist.

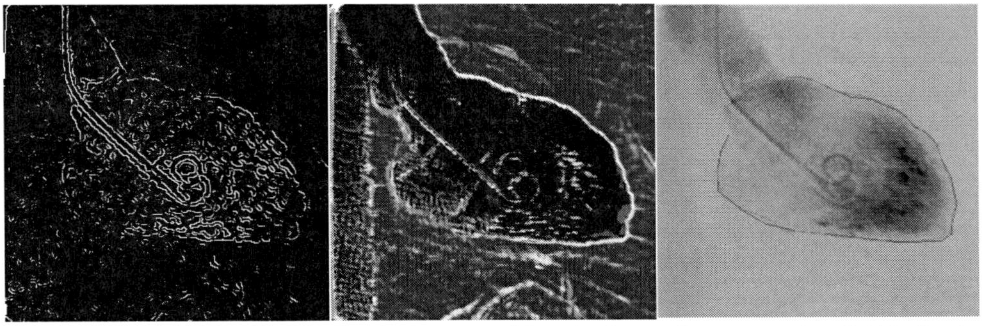

Edges enhanced by the Marr-Hildreth edge enhancer | Output image of the trained supervised NN filter | Computer contour (light line) and cardiologist contour (dark line)

Figure 7. Comparison of edge enhancement by the trained supervised NN filter, called a "neural edge enhancer," with that by the Marr-Hildreth edge enhancer, and a comparison of the contour traced by a computer based on the NN-filter-enhanced edges with the "gold-standard" contour traced by a cardiologist. Reprinted with permission from Suzuki et al. [13].

Accurate tracing of the contour of the left ventricle on ventriculograms is very difficult with an automated tracing method, because some portions of the margins of the left ventricle can be poorly defined due to the dilution of contrast media in the blood. In clinical practice, a cardiologist corrects manually the incorrect or incomplete aspects of the contour provided by the automated tracing tool. To address this issue, we developed a "neural edge enhancer" based on the supervised NN filter that can "learn" the contours traced by a cardiologist [13]. We trained the neural edge enhancer with input left ventriculograms at end-diastole in the 30-degree right anterior oblique projection and the corresponding teaching contours traced by a cardiologist, as illustrated in Figure 6. It should be noted that some parts of the contours of the left ventricle are very subtle. For comparison of the neural edge enhancer with a well-known Marr-Hildreth edge enhancer, the parameters of the Marr-Hildreth edge enhancer were optimized with the images used for training the neural edge enhancer under the minimum-mean-square error criterion. The results of edge enhancement are shown in Figure 7. In the edges enhanced by the trained neural edge enhancer, the enhanced edges are prominent, continuous, and are similar to the "gold-standard" contour, whereas the edges enhanced by the Marr-Hildreth edge enhancer are fragmented and some parts are missing. We applied a tracing method to the edges enhanced by the neural edge enhancer. The contour traced by our method agrees well with the "gold-standard" contour traced by a cardiologist, as shown in Figure 7.

4.4. Separation of Ribs from Soft Tissue in Chest Radiographs

Chest radiography is the most frequently used diagnostic imaging examination for chest diseases such as lung cancer, tuberculosis, and pneumonia. More than 9 million people worldwide die annually from chest diseases [36]. Lung cancer causes 945,000 deaths [36], and is the leading cause of cancer deaths in the world [36] and in countries [37] such as the United States, the United Kingdom, and Japan. Lung nodules (i.e., potential lung cancers) in chest radiography, however, can be overlooked by radiologists in from 12 to 90% of cases with nodules that are visible in retrospect [38, 39]. Studies showed that 82 to 95% of the missed lung cancers were partly obscured by overlying bones such as ribs and/or a clavicle [38, 39]. To address this issue, dual-energy imaging has been investigated. The dual-energy imaging uses the energy dependence of the x-ray attenuation by different materials; it can produce two tissue-selective images, i.e., a "bone" image and a "soft-tissue" image [40-42]. Major drawbacks of dual-energy imaging, however, are that (a) the radiation dose can be double, (b) specialized equipment for obtaining dual-energy x-ray exposures is required, and (c) the subtraction of two-energy images causes an increased noise level in the images.

To resolve the above drawbacks with dual-energy images, we developed an image-processing technique based on the supervised NN filter for separation of ribs from soft tissue, which we call it a massive-training ANN (MTANN) [4, 14]. The basic idea is to train the MTANN with soft-tissue and bone images acquired with a specialized radiography system with dual-energy imaging. For separation of ribs from soft tissue, we trained the MTANN with input chest radiographs and the corresponding teaching dual-energy bone images, as illustrated in Figure 8.

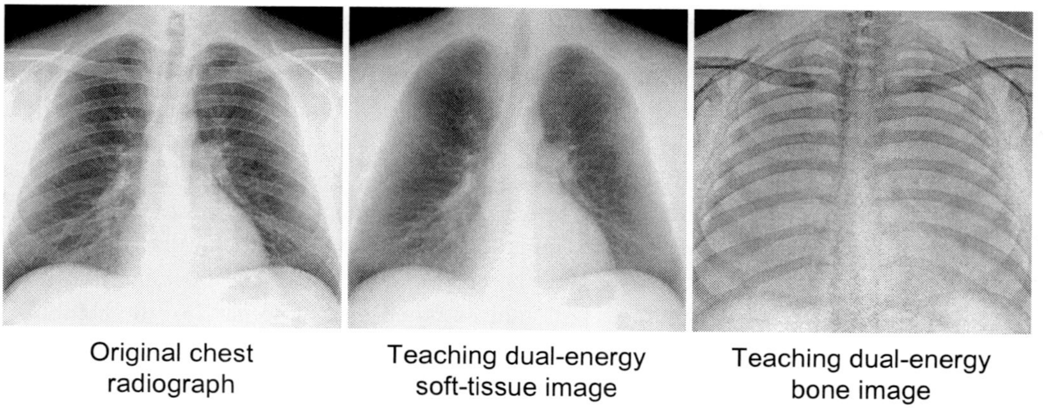

Figure 8. Images used for training the supervised NN filter called an MTANN. The soft-tissue and bone images were acquired with a dual-energy radiography system where two x-ray exposures at different energy levels were used to create those two images. Reprinted with permission from Suzuki et al. [4].

Ribs in chest radiographs include various spatial-frequency components. For a single MTANN, suppression of ribs containing such various frequencies is difficult, because the capability of a single MTANN is limited, i.e., the capability depends on the size of the input kernel of the MTANN. Because the training of the MTANN takes a substantially long time, it is difficult in practice to train the MTANN with a large subregion. In order to overcome this issue, we employed multi-resolution decomposition/composition techniques [43, 44]. The multi-resolution decomposition is a technique for decomposing an original high-resolution image into different-resolution images. First, one obtains a medium-resolution image $g_M(x,y)$ from an original high-resolution image $g_H(x,y)$ by performing down-sampling with averaging, i.e., four pixels in the original image are replaced by a pixel having the mean value for the four pixel values, represented by

$$g_M(x,y) = \frac{1}{4} \sum_{i,j \in R_{22}} g_H(2x-i, 2y-j)\,, \tag{15}$$

where R_{22} is a 2-by-2-pixel region. The medium-resolution image is enlarged by up-sampling with pixel substitution, i.e., a pixel in the medium-resolution image is replaced by four pixels with the same pixel value, as follows:

$$g_M^U(x,y) = g_M(x/2, y/2). \quad (16)$$

Then, a high-resolution difference image $d_H(x,y)$ is obtained by subtraction of the enlarged medium-resolution image from the high-resolution image, represented by

$$d_H(x,y) = g_H(x,y) - g_M^U(x,y). \quad (17)$$

These procedures are performed repeatedly, producing further lower-resolution images. Thus, multi-resolution images having various frequencies are obtained by use of the multi-resolution decomposition technique.

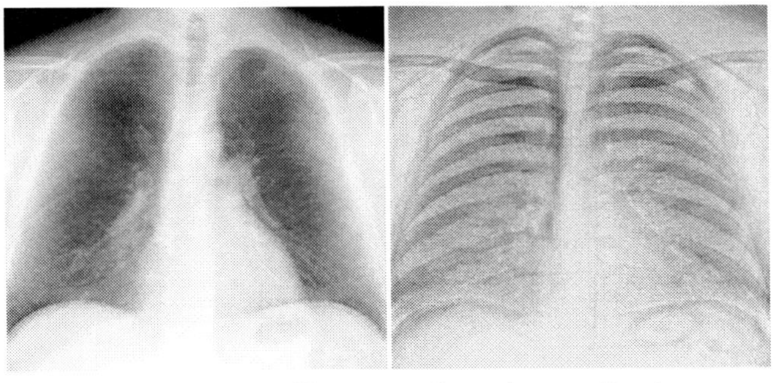

Figure 9. Soft-tissue-image-like and bone-image-like images obtained by using the trained supervised NN filter called an MTANN. Reprinted with permission from Suzuki et al. [4].

An important property of this technique is that exactly the same original-resolution image $g_H(x,y)$ can be obtained from the multi-resolution images, $d_H(x,y)$ and $g_M(x,y)$, by performing the inverse procedures, called a multi-resolution composition technique, as follows:

$$g_H(x,y) = g_M(x/2, y/2) + d_H(x,y). \quad (18)$$

Therefore, we can process multi-resolution images independently instead of processing original high-resolution images directly; i.e., with these techniques, the processed original high-resolution image can be obtained by composing of the processed multi-resolution images. An MTANN only needs to support a limited spatial frequency rage in each resolution image instead of the entire spatial frequencies in the original image.

With the multi-resolution decomposition technique, input chest radiographs and the corresponding "teaching" bone images are decomposed into sets of different-resolution images, and then these sets of images are used for training three MTANNs in the multi-resolution MTANN. Each MTANN is an expert for a certain resolution, i.e., a low-resolution MTANN is in charge of low-frequency components of ribs, a medium-resolution MTANN is for medium-frequency components, and a high-resolution MTANN for high-frequency components. Each resolution MTANN is trained independently with the corresponding resolution images. After training, the MTANNs produce different-resolution images, and then these images are composed to provide a complete high-resolution image by use of the multi-resolution composition technique. The complete high-resolution image is expected to be similar to the teaching bone image; therefore, the multi-resolution MTANN would provide a "bone-image-like" image in which ribs are separated from soft tissues.

Figure 9 shows soft-tissue-image-like and bone-image-like images obtained by using the trained MTANN. Ribs are extracted effectively in the bone-image-like image, and this image is similar to the "gold-standard" dual-energy bone image shown in Figure 8. The contrast of ribs is substantially suppressed in the soft-tissue-image-like image, whereas the visibility of soft tissue such as lung vessels is maintained. The soft-tissue-image-like image is very similar to the "gold-standard" dual-energy soft-tissue image shown in Figure 8.

4.5. Enhancement of Lesions in Medical Images

Computer-aided diagnosis (CAD) has been an active area of study in medical image analysis [45-47]. Some CAD schemes employ a filter for enhancement of lesions as a preprocessing step for improving sensitivity and specificity; and some do not employ such a filter. The filter enhances objects similar to a model employed in the filter; e.g., a blob-enhancement filter based on the Hessian matrix enhances sphere-like objects [1]. Actual lesions, however, often differ from a simple model, e.g., a lung nodule is generally modeled as a solid sphere, but there are nodules of various shapes and inhomogeneous nodules such as nodules with spiculation and ground-glass nodules. Thus, conventional filters often fail to enhance actual lesions.

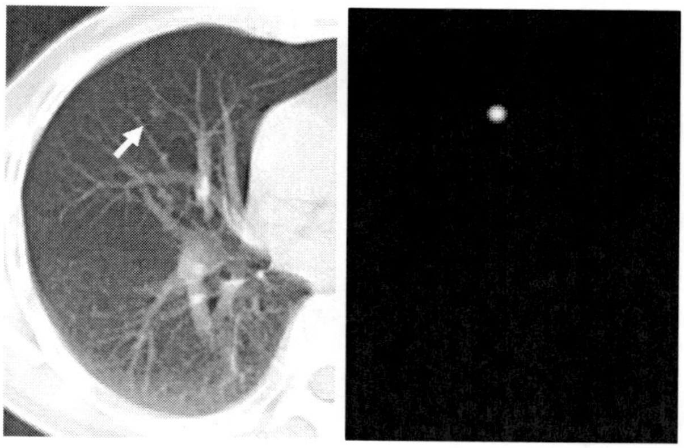

Figure 10. Lung CT image with a nodule (i.e., potential lung cancer indicated by an arrow) and the corresponding teaching image containing a map for the "likelihood of being a nodule."

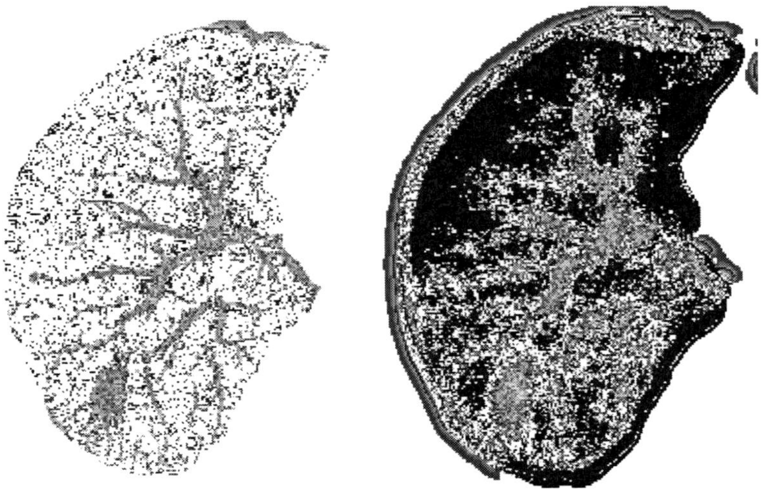

Figure 11. Non-training lung CT image with a nodule (indicated by an arrow) and the corresponding output image of the trained supervised NN filter called an MTANN. The nodule is enhanced, whereas most of the normal structures such as lung vessels are suppressed. Reprinted with permission from Suzuki [48].

To address this issue, we developed a "lesion-enhancement" technique based on the supervised NN filter called an MTANN for enhancement of actual lesions (as opposed to a lesion model) [48] in a CAD scheme for detection of lung nodules in CT [3, 16, 18]. For enhancement of lesions and suppression of non-lesions in CT images, the teaching image contains a map for the "likelihood of being lesions." The input lung CT image with a nodule and the corresponding teaching image are shown in Figure 10. We placed a 2D Gaussian distribution at the location of the nodule in the teaching image, as a model of the likelihood of being a lesion. To test the performance, we applied the trained MTANN filter to non-training lung CT images. The nodule is enhanced in the output image of the trained MTANN filter, while normal structures such as lung vessels are suppressed, as shown in Figure 11. Note that small remaining regions due to vessels can easily be separated from nodules by use of their area information which can be obtained by using connected-component labeling [49-52].

4.6. Classification of Lesions and Non-Lesions in Medical Images

A major challenge with CAD development is to reduce the number of false positives while maintaining a high sensitivity, because there are various normal structures similar to lesions in medical images. To address this issue, we developed a false-positive-reduction technique based on a supervised NN filter, called an MTANN, in a CAD scheme for lung nodule detection in CT [3, 16, 18, 19]. For enhancement of nodules (i.e., true positives) and suppression non-nodules (i.e., false positives) in CT images, the teaching image contains a 2D distribution of values that represent the "likelihood of being a nodule." We used a 2D

Gaussian distribution as the teaching image for a nodule and an image that contains all zeros (i.e., completely dark) for non-nodules, as illustrated in Figure 12.

We trained an MTANN with typical nodules and typical types of false positives (non-nodules). Figure 13 shows various types of nodules and non-nodules and the corresponding output images of the trained MTANN. Various types of nodules such as a solid nodule, a part-solid nodule, and a non-solid nodule are enhanced, whereas various types of non-nodules such as different-sized lung vessels and soft-tissue opacity are suppressed around the centers of regions-of-interest. To combine output pixels into a single score for each nodule candidate, we developed a scoring method for distinction between a nodules and a non-nodule. A score for a given nodule candidate from an MTANN is defined as

$$S = \sum_{x,y \in R_E} f_G(\sigma; x, y) \times O(x, y), \tag{19}$$

where

$$f_G(\sigma; x, y) = \frac{1}{\sqrt{2\pi}\sigma} \exp\left\{ -\frac{(x^2 + y^2)}{2\sigma^2} \right\} \tag{20}$$

is a 2D Gaussian weighting function with standard deviation σ, R_E is the region for evaluation, and $O(x,y)$ is the output of the trained MTANN. Thus, a single score for each nodule candidate is obtained by multiplying the output image by a 2D Gaussian weighting function, as illustrated in Figure 14. The use of the 2D Gaussian weighting function allows us to combine the individual pixel-based responses (outputs) of a trained MTANN as a single score. The score obtained by the above equations represents the weighted sum of the estimates for the likelihood that the image (nodule candidate) contains an actual nodule near the center. The concept of this scoring is similar to that of a matched filter. We use the same 2D Gaussian weighting function as is used in the teaching images. A higher score would indicate a nodule, and a lower score would indicate a non-nodule. Thresholding of scores is made for classification of nodule candidates into nodules or non-nodules.

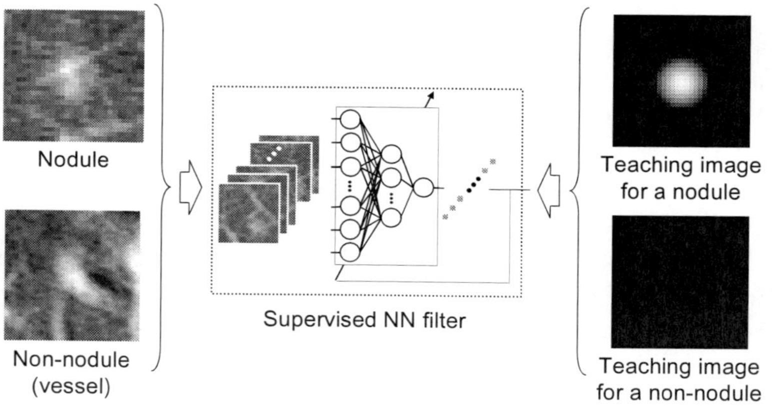

Figure 12. A supervised NN filter, called an MTANN, for enhancement of nodules and suppression of non-nodules for reduction of false positives in a CAD scheme for lung nodule detection in CT. A teaching image for a nodule contains a 2D Gaussian distribution at the center of the image, whereas that for a non-nodule contains zero (i.e., it is completely dark).

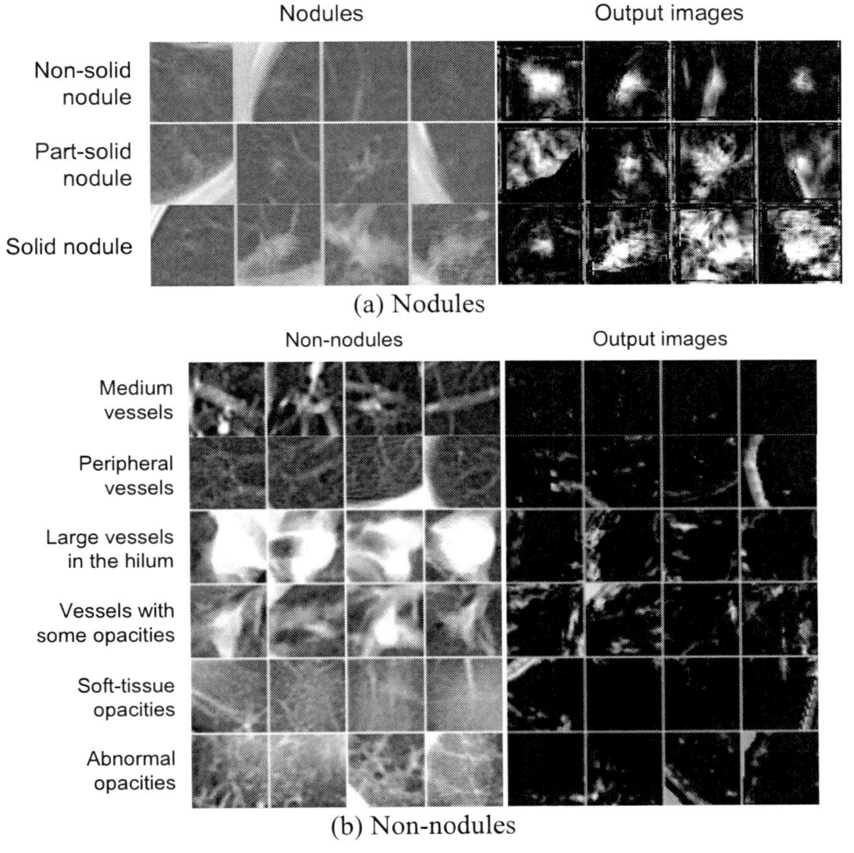

Figure 13. Input images containing various types of nodules and non-nodules and the corresponding output images of the trained supervised NN filter. Reprinted with permission from Suzuki et al. [3].

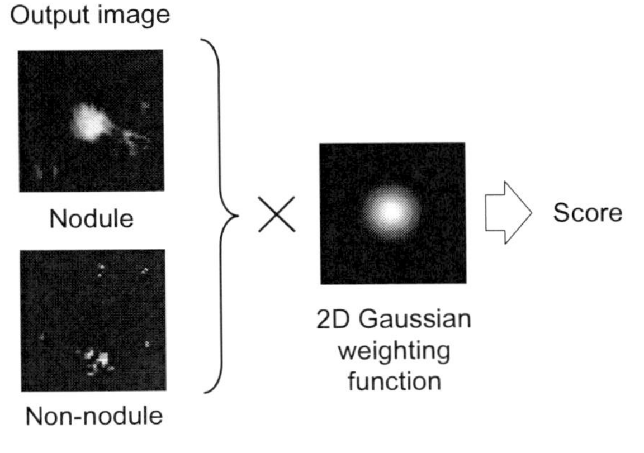

Figure 14. Scoring method for combining output pixels in the output image into a single score for distinction between a nodule and a non-nodule.

It is difficult to distinguish a small distribution for a small nodule in the output image from a small distribution due to noise; this difficulty can lower the ability of the MTANN to differentiate nodules from non-nodules. To force the MTANN to output a standard-sized (regular-sized) distribution for different-sized nodules, the same-sized Gaussian distribution is used in the teaching images. After training in this manner, the MTANN is expected to output relatively regular-sized distributions for different-sized nodules, e.g., a relatively large output distribution for a small nodule and a relatively small output distribution for a large nodule. This property of the regular-sized output distributions is expected to increase the scores for small nodules and to improve the overall performance of an MTANN.

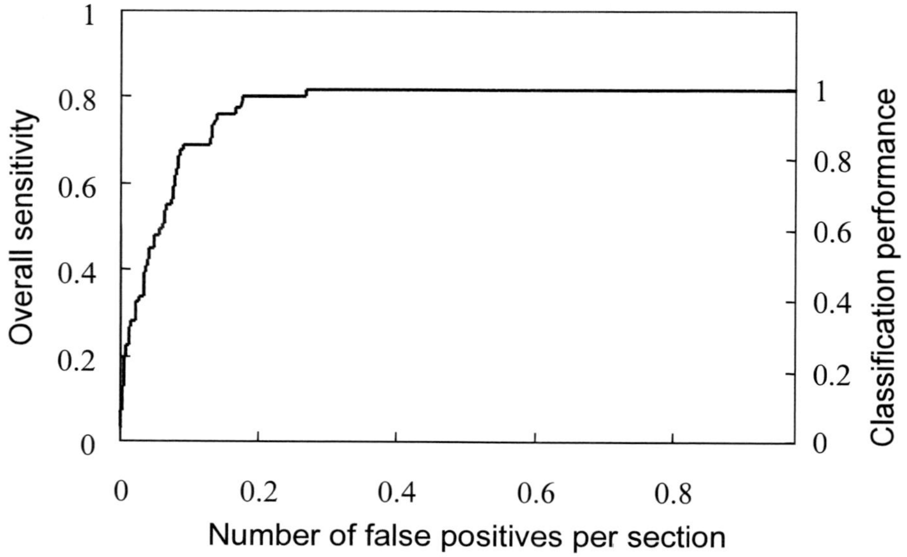

Figure 15. Performance of the supervised NN filter, called an MTANN, in false-positive reduction in a CAD scheme for lung nodule detection in CT. The FROC curve indicates that the MTANN yielded a reduction of 54% of false positives (non-nodules) without any loss of true positive.

We applied the MTANN to 57 true positives (nodules) and 1,726 false positives (non-nodules) produced by our CAD scheme [53, 54]. Free-response receiver operating characteristic (FROC) analysis [55] was performed to evaluate the performance of the trained MTANN. The FROC curve for the MTANN indicates 80.3% overall sensitivity and a reduction in the false-positive rate from 0.98 to 0.18 per section, as shown in Figure 15.

An MTANN is applicable to false-positive reduction in other CAD schemes for detection of lesions in medical images, such as lung nodules in chest radiographs [56] and polyps in CT colonography [23, 57].

5. Conclusion

The supervised NN filter is a fundamental tool for enhancing/suppressing patterns such as noise, edges, normal structures, and lesions in medical images and has a wide variety of applications in medical image processing and analysis. The supervised NN filter unifies convolution filters, neural filters, neural edge enhancers, and MTANNs. The supervised NN filter was effective for reduction of quantum noise in x-ray images, enhancement of edges from very noisy images, enhancing contours traced by a physician, suppression of ribs in chest radiographs, enhancement of lesions in chest CT images, and reduction of false positives in a CAD scheme for lung nodule detection in CT and chest radiography, and polyp detection in CT colonography.

ACKNOWLEDGMENTS

The author is grateful to all members of the Kenji Suzuki Laboratory in the Department of Radiology at the University of Chicago for their valuable suggestions and contributions to the research, and to Ms. E. F. Lanzl for improving the manuscript. The author is also grateful to Harumi Suzuki for her help with figures and graphs, and Mineru Suzuki and Juno Suzuki for cheering me up. This work was supported partially by a National Cancer Institute Grant (R01CA120549) and by NIH S10 RR021039 and P30 CA14599.

REFERENCES

[1] Frangi, AF; Niessen, WJ; Hoogeveen, RM; van Walsum, T; Viergever, MA. Model-based quantitation of 3-D magnetic resonance angiographic images. IEEE *Trans Med Imaging*, 1999, 18(10), 946-56.

[2] Suzuki, K; Horiba, I; Sugie, N. Efficient approximation of neural filters for removing quantum noise from images. IEEE *Transactions on Signal Processing*, 2002, 50(7), 1787-1799.

[3] Suzuki, K; Armato, SG; Li, F; Sone, S; Doi, K. Massive training artificial neural network (MTANN) for reduction of false positives in computerized detection of lung nodules in low-dose CT. *Medical Physics*, 2003, 30(7), 1602-1617.

[4] Suzuki, K; Abe, H; MacMahon, H; Doi, K. Image-processing technique for suppressing ribs in chest radiographs by means of massive training artificial neural network (MTANN). IEEE *Transactions on Medical Imaging*, 2006, 25(4), 406-416.

[5] Loog, M; van Ginneken, B; Schilham, AM. Filter learning: application to suppression of bony structures from chest radiographs. *Med Image Anal,* 2006, 10(6), 826-40.

[6] Suzuki, K; Horiba, I; Sugie, N; Nanki, M. Neural filter with selection of input features and its application to image quality improvement of medical image sequences. IEICE *Transactions on Information and Systems,* 2002, E85-D(10), 1710-1718.

[7] Bishop, CM. *Neural Networks for Pattern Recognition*. In. New York: Oxford University Press; 1995.

[8] Duda, RO; Hart, PE; Stork, DG. In: Pattern Recognition. 2nd ed. Hoboken; NJ: Wiley *Interscience*, 2001, 117-121.

[9] Haykin, S. *Neural Networks*. Upper Saddle River; NJ: *Prentice Hall*, 1998.

[10] Vapnik, VN. The Nature of Statistical Learning Theory. Berlin: *Springer-Verlag*, 1995.

[11] Suzuki, K; Horiba, I; Sugie, N; Ikeda, S. Improvement of image quality of x-ray fluoroscopy using spatiotemporal neural filter which learns noise reduction; edge enhancement and motion compensation. In: *Proc. Int. Conf. Signal Processing Applications and Technology (ICSPAT)*; 1996 October; Boston; MA; 1996, 1382-1386.

[12] Suzuki, K; Horiba, I; Sugie, N. Neural edge enhancer for supervised edge enhancement from noisy images. IEEE Transactions on *Pattern Analysis and Machine Intelligence,* 2003, 25(12), 1582-1596.

[13] Suzuki, K; Horiba, I; Sugie, N; Nanki, M. Extraction of left ventricular contours from left ventriculograms by means of a neural edge detector. IEEE *Transactions on Medical Imaging,* 2004, 23(3), 330-339.

[14] Suzuki, K; Abe, H; Li, F; Doi, K. Suppression of the contrast of ribs in chest radiographs by means of massive training artificial neural network. In: *Proc. SPIE Medical Imaging* (SPIE MI); 2004 May; *San Diego*, CA; 2004, 1109-1119.

[15] Oda, S; Awai, K; Suzuki, K; Yanaga, Y; Funama, Y; MacMahon, H; et al. Performance of radiologists in detection of small pulmonary nodules on chest radiographs: effect of rib suppression with a massive-training artificial neural network. *AJR Am J Roentgenol,* 2009, 193(5), W397-402.

[16] Arimura, H; Katsuragawa, S; Suzuki, K; Li, F; Shiraishi, J; Sone, S; et al. Computerized scheme for automated detection of lung nodules in low-dose computed tomography images for lung cancer screening. *Academic Radiology,* 2004, 11(6), 617-629.

[17] Suzuki, K; Shiraishi, J; Abe, H; MacMahon, H; Doi, K. False-positive reduction in computer-aided diagnostic scheme for detecting nodules in chest radiographs by means of massive training artificial neural network. *Academic Radiology,* 2005, 12(2), 191-201.

[18] Suzuki, K; Li, F; Sone, S; Doi, K. Computer-aided diagnostic scheme for distinction between benign and malignant nodules in thoracic low-dose CT by use of massive training artificial neural network. IEEE *Transactions on Medical Imaging,* 2005, 24(9), 1138-1150.

[19] Suzuki, K; Doi, K. How can a massive training artificial neural network (MTANN) be trained with a small number of cases in the distinction between nodules and vessels in thoracic CT? *Academic Radiology,* 2005, 12(10), 1333-1341.

[20] Li, Q; Li, F; Suzuki, K; Shiraishi, J; Abe, H; Engelmann, R; et al. Computer-aided diagnosis in thoracic CT. *Semin Ultrasound CT MR,* 2005, 26(5), 357-63.

[21] Li, F; Arimura, H; Suzuki, K; Shiraishi, J; Li, Q; Abe, H; et al. Computer-aided detection of peripheral lung cancers missed at CT: ROC analyses without and with localization. *Radiology*, 2005, 237(2), 684-90.

[22] Suzuki, K; Yoshida, H; Nappi, J; Dachman, AH. Massive-training artificial neural network (MTANN) for reduction of false positives in computer-aided detection of polyps: Suppression of rectal tubes. *Medical Physics,* 2006, 33(10), 3814-3824.

[23] Suzuki, K; Yoshida, H; Nappi, J; Armato, SG; 3rd; Dachman, AH. Mixture of expert 3D massive-training ANNs for reduction of multiple types of false positives in CAD for detection of polyps in CT colonography. *Med Phys*, 2008, 35(2), 694-703.

[24] Suzuki, K; Zhang, J; Xu, J. Massive-training artificial neural network coupled with Laplacian-eigenfunction-based dimensionality reduction for computer-aided detection of polyps in CT colonography. IEEE *Transactions on Medical Imaging,* in press.

[25] Suzuki, K; Rockey, DC; Dachman, AH. CT colonography: Advanced computer-aided detection scheme utilizing MTANNs for detection of "missed" polyps in a multicenter clinical trial. *Med Phys*, 2010, 30, 2-21.

[26] Zhang, W; Doi, K; Giger, ML; Nishikawa, RM; Schmidt, RA. An improved shift-invariant artificial neural network for computerized detection of clustered microcalcifications in digital mammograms. *Med Phys,* 1996, 23(4), 595-601.

[27] Zhang, W; Doi, K; Giger, ML; Wu, Y; Nishikawa, RM; Schmidt, RA. Computerized detection of clustered microcalcifications in digital mammograms using a shift-invariant artificial neural network. *Med Phys*, 1994, 21(4), 517-24.

[28] Lo, SC; Lou, SL; Lin, JS; Freedman, MT; Chien, MV; Mun, SK. Artificial Convolution Neural Network Techniques and Applications to Lung Nodule Detection. IEEE *Transactions on Medical Imaging,* 1995, 14(4), 711-718.

[29] Lin, JS; Hasegawa, A; Freedman, M; Mun, SK. Differentiation between nodules and end-on vessels using a convolution neural network architecture. *Journal of Digital Imaging*, 1995, 8, 132-141.

[30] Lo, SC; Li, H; Wang, Y; Kinnard, L; Freedman, MT. A multiple circular path convolution neural network system for detection of mammographic masses. IEEE *Trans Med Imaging*, 2002, 21(2), 150-8.

[31] Rumelhart, DE; Hinton, GE; Williams, RJ. Learning representations by back-propagating errors. *Nature*, 1986, 323, 533-536.

[32] Rumelhart, DE; Hinton, GE; Williams, RJ. Learning internal representations by error propagation. *Parallel Distributed Processing*, 1986, 1, 318-362.

[33] Funahashi, K. On the approximate realization of continuous mappings by neural networks. *Neural Networks*, 1989, 2, 183-192.

[34] Barron, AR. Universal approximation bounds for superpositions of a sigmoidal function. IEEE Transactions on Information *Theory*, 1993, 39(3), 930-945.

[35] Suzuki, K; Horiba, I; Ikegaya, K; Nanki, M. Recognition of coronary arterial stenosis using neural network on DSA system. *Systems and Computers in Japan*, 1995, 26(8), 66-74.

[36] Murray, CJ; Lopez, AD. Mortality by cause for eight regions of the world: Global Burden of Disease Study. *Lancet*, 1997, 349(9061), 1269-1276.

[37] Goodman, GE. Lung cancer, 1, prevention of lung cancer. Thorax 2002, 57(11), 994-999.

[38] Austin, JH; Romney, BM; Goldsmith, LS. Missed bronchogenic carcinoma: radiographic findings in 27 patients with a potentially resectable lesion evident in retrospect. *Radiology*, 1992, 182(1), 115-122.

[39] Shah, PK; Austin, JH; White, CS; Patel, P; Haramati, LB; Pearson, GD; et al. Missed non-small cell lung cancer: radiographic findings of potentially resectable lesions evident only in retrospect. *Radiology*, 2003, 226(1), 235-241.

[40] Glocker, R; Frohnmayer, W. Uber die rontgenspektroskopische Bestimmung des Gewichtsanteiles eines Elementes in Gememgen und Verbindungen. *Annalen der Physik*, 1925, 76, 369-395.

[41] Jacobson, B; Mackay, RS. Radiological contrast enhancing methods. Advances in *Biological and Medical Physics*, 1958, 6, 201-261.

[42] Ishigaki, T; Sakuma, S; Horikawa, Y; Ikeda, M; Yamaguchi, H. One-shot dual-energy subtraction imaging. *Radiology*, 1986, 161(1), 271-273.

[43] Stephane, GM. A theory for multiresolution signal decomposition: the wavelet representation. *IEEE Transactions on Pattern Analysis and Machine Intelligence*, 1989, 11(7), 674-693.

[44] Akansu, AN; Haddad, RA. *Multiresolution signal decomposition.* Boston: Academic Press, 1992.

[45] Doi, K. Computer-aided diagnosis in medical imaging: Historical review; current status and future potential. *Comput Med Imaging Graph*, 2007, 31(4-5), 198-211.

[46] Giger, ML. Update on the potential role of CAD in radiologic interpretations: are we making progress? *Acad Radiol*, 2005, 12(6), 669-70.

[47] Giger, ML; Suzuki, K. Computer-Aided Diagnosis (CAD). In: Feng DD; editor. Biomedical Information *Technology: Academic* Press, 2007, 359-374.

[48] Suzuki, K. A supervised 'lesion-enhancement' filter by use of a massive-training artificial neural network (MTANN) in computer-aided diagnosis (CAD). *Phys Med Biol*, 2009, 54(18), S31-45.

[49] Suzuki, K; Horiba, I; Sugie, N. Linear-time connected-component labeling based on sequential local operations. *Computer Vision and Image Understanding*, 2003, 89(1), 1-23.

[50] Wu, K; Otoo, E; Suzuki, K. Optimizing two-pass connected-component labeling algorithms. Pattern *Analysis and Applications,* 2009, 12, 117-135.

[51] He, L; Chao, Y; Suzuki, K; Wu, K. Fast connected-component labeling. *Pattern Recognition*, 2009, 42, 1977-1987.

[52] He, L; Chao, Y; Suzuki, K. A run-based two-scan labeling algorithm. IEEE *Trans Image Process*, 2008, 17(5), 749-56.

[53] Armato, SG; 3rd; Giger, ML; MacMahon, H. Automated detection of lung nodules in CT scans: preliminary results. *Medical Physics*, 2001, 28(8), 1552-1561.

[54] Armato, SG; 3rd; Li, F; Giger, ML; MacMahon, H; Sone, S; Doi, K. Lung cancer: performance of automated lung nodule detection applied to cancers missed in a CT screening program. *Radiology*, 2002, 225(3), 685-692.

[55] Egan, JP; Greenberg, GZ; Schulman, AI. Operating characteristics; signal detectability; and the method of free response. *Journal of the Acoustical Society of America*, 1961, 33, 993-1007.

[56] Suzuki, K; Shiraishi, J; Abe, H; MacMahon, H; Doi, K. False-positive reduction in computer-aided diagnostic scheme for detecting nodules in chest radiographs by means of massive training artificial neural network. *Acad Radiol*, 2005, 12(2), 191-201.

[57] Suzuki, K; Yoshida, H; Nappi, J; Dachman, AH. Massive-training artificial neural network (MTANN) for reduction of false positives in computer-aided detection of polyps: Suppression of rectal tubes. *Med Phys*, 2006, 33(10), 3814-24.

In: Focus on Artificial Neural Networks
Editor: John A. Flores, pp. 151-170
ISBN: 978-1-61324-285-8

Chapter 7

AN INVERSE NEURAL NETWORK MODEL OF DISC BRAKE PERFORMANCE AT ELEVATED TEMPERATURES

Dragan Aleksendrić
University of Belgrade Faculty of Mechanical Engineering, Serbia

ABSTRACT

The demands imposed on a braking system, under wide range of operating conditions, are high and manifold. Improvement and control of automotive braking systems' performance, under different operating conditions, is complicated by the fact that braking process has stochastic nature. The stochastic nature of braking process is determined by braking phenomena induced in the contact of friction pair (brake disc and disc pad) during braking. Consequently, the overall braking system's performance has been also affected especially at high brake interface temperatures. Temperature sensitivity of motor vehicles brakes has always been an important aspect of their smooth and reliable functioning. It is particularly related to front brakes that absorb a major amount (up to 80%) of the vehicle total kinetic energy. The friction heat generated during braking application easily raises temperature at the friction interface beyond the glass transition temperature of the binder resin and often rises above decomposition temperature. The gas evolution at the braking interfaces because of pyrolysis and thermal degradation of the material results in the friction force decreasing. At such high temperatures, friction force suffers from a loss of effectiveness. This loss of effectiveness (brake fading) cannot be easily predicted due to subsequent thermo-mechanical deformation of disc and disc pad (friction material) which modifies the contact profile and pressure distribution, altering the frictional heat. The instability of the brake's performance after a certain number of brake applications is common and depends on braking regimes represented by application pressure, initial speed, and brake interface temperature. Therefore, the most important issue is related to investigation of possibilities for control of brake performance, especially at elevated temperatures, in order to be stabilized and kept on some level. The control of motor vehicle brakes performance needs a model of how braking regimes, before all application pressure, affecting their performance for the specific friction pair characteristics. Analytical models of brakes performance are difficult, even impossible to be obtained due to complex and highly nonlinear phenomena involved during braking. That is why, in this chapter artificial neural network abilities have been used for modelling of the disc brake performance (braking torque) against synergy of influences of application pressure, initial speed, and brake interface temperature. Based on that, an inverse model of the disc brake performance has been developed able to predict the value of brake's application pressure, which, for current values of brake interface temperature and initial speed, provides wanted braking torque. Consequently, the brake's application pressure could be adjusted to keep the disc brake performance (braking torque) on some wanted level and prevent its decreasing during braking at elevated temperatures.

INTRODUCTION

Automotive braking systems were always given the highest importance concerning active safety of motor vehicles [1]. The demands imposed to the automotive braking system, under wide range of operating conditions, are high and manifold [1,2,3,4]. The braking system performance is mostly determined by brakes performance [5]. It is expected that a braking torque should be relatively high but also stable. The basic requirements imposed to the automotive brakes are related to the values and stability of braking torque versus different brake's operation conditions defined by changing of application pressure and/or sliding speed and/or brake interface temperature [5]. The ever-increasing demands imposed to very complex systems such as a braking system of passenger car, require highly sophisticated controllers to ensure that high performance can be achieved and maintained under adverse conditions. To address the control demands of such complex system it is possible to enhance today's control methods using intelligent techniques [5,6,7,8]. From active safety point of view, intelligent control of braking systems should be based on following abilities: (i) learning from previous experience of braking system's operation and, accordingly (ii) modelling and predicting of braking system's performance.

Actual sophisticated systems for controlling of vehicle's braking system performance operate in that way to a posterior correct its output performances in order to solve problems that may occur during braking process. These systems, like ABS, ESP, or EBD, are electronically controlled "add-on" systems that improve braking system operation primarily by increasing the speed of signal transmission from driver to ECU and other electronically controlled devices. Although without intelligent abilities, electronically controlled braking system operation offers possibilities for further innovative solutions directed to improving of braking system's performance. Improving of the braking system performance should be based on intelligent using of the same data that are measured or computed by mentioned electronically controlled add-on systems.

Taking into consideration that the automotive brake's performance results from the complex processes occurred in the contact of the friction pair, overall performance of a braking system strongly depends on the brakes operation and in turn, performance of their friction pairs. In the automotive industry, the friction system of brake pad against a cast iron disc has an enormous technical significance [9]. The nature of tribological contact between the disc and disc pad in automotive brakes is manifold and intricate. As it is already well known, brakes' performance, i.e. braking systems' performance primarily depends on the interaction of the friction pair components at their sliding interfaces [10]. Therefore, tribological processes occurring in the contact of friction pair unite questions from different physical fields, such as the mechanics, thermodynamics and chemistry [11,12]. Moreover, compared to the total contact area, the size of the real contact area between the pads and the disc is very small and highly dependent on changes of application pressure, temperature, deformation, and wear [9,13,14]. As it is the case in most sliding situations, contact is made over just a few isolated junctions who receive substantial amounts of thermal energy [14,15,16]. The temperature distribution at the friction interface generated during braking process is a complex phenomenon, which directly affects the brake performance [17,18]. Brake fade is the term used to indicate a loss in braking effectiveness at elevated temperatures due to reduction of friction coefficient (μ) which leads to reduction of braking torque realized by the brake. The return to acceptable levels of friction i.e. braking torque value, after subsequent brake applications, can be referred as recovery, which is essential for reliable braking system performance [19]. The level and stability of the motor vehicle brakes performance at elevated temperature is further influenced by vehicle's weight, axle load distribution, high of centre of gravity, maximum speed, braking system characteristics and especially the brake's friction pair properties.

The complexity of the friction pair's contact is mainly affected by physical and chemical characteristics of friction materials components, their manufacturing technology in synergy with brake operating regimes [20,21,22]. The most complicated phenomenon is related to the friction film generation between the friction pair because of temperature increasing at brake interface. The friction film or third body layer is produced incessantly by maintaining a certain thickness and it is composed of carbonaceous reaction products, unreacted constituents, oxide from metallic ingredients etc. [23,24,25]. Friction characteristics of the transfer layer between a brake disc and pads developed during subsequent braking influence the braking effectiveness at elevated temperature. Due to highly nonlinear phenomena involved by different friction characteristics of the transfer layer generated in the contact of friction pair, it is very difficult obtain analytical models of brake performance and particularly how the brake performance was affected by its operation conditions. The sensitivity of the automotive brakes performance on the transfer layer generation can be especially seen at elevated brake interface temperatures.

Minimizing of brakes performance sensitivity versus operation condition, particularly during braking at elevated temperatures could be done on at least two ways. Firstly, inherent properties of the friction pair should be improved i.e. friction material and brake disc capabilities need to provide stable own as well as stable synergy of their own performance during braking. It is especially related to stability of their performance at elevated brake interface temperatures. On the other hand, besides improving the friction pair characteristics on the part level, brakes performance could be better controlled on the system level. It needs that the braking torque changes, as an output of brake operation, should be better controlled versus different brake operation conditions [26]. Therefore, an appropriate model of brake performance should be developed versus brake operation conditions. It requires that the braking system operation should be monitored in order to brake input/output relationship be modelled. Monitoring of the braking system and accordingly brakes operation has been already provided by introduction of the electronically controlled braking systems.

Since braking process has a stochastic nature, artificial intelligence could be involved in the control of brakes performance. Accordingly, the basic step in this process is related to investigation of possibilities for modelling of braking torque changes against influence of application pressure in the whole range of initial speed and brake interface temperature changes. If this model exists, braking torque could be better controlled because the influence of brake application pressure versus braking torque variation could be predicted. Based on this model, brake application pressure could be adjusted on the level that provide wanted braking torque for current values of speed and brake interface temperature. It means that an inverse model of brake operation should be developed. Instead of modelling of influence of application pressure, initial speed, and brake interface temperature against braking torque changes, an inverse model of the disc brake performance has been developed in this chapter. The inverse model of the disc brake performance is able to predict value of brake application pressure against current values of braking toque, speed, and brake interface temperature. Taking into consideration that braking torque stability is particularly affected at elevated brake interface temperature; this approach could be especially useful under this braking regimes.

It is clear that developing of the inverse model of disc brake operation is not an easy task, especially if classical mathematical methods are used to. In contrast to analytical approaches, artificial neural networks can be used for modelling of these complex and non–linear influencing factors versus brake output performance. As a tool for systematic parameter studies based on its parallel processing property [27], artificial neural networks are much more favourable from the point of their inherent abilities to learn from experimental data in comparison with classical analytical models. Because of highly non–linear phenomena involved in the problem of tribological interactions during braking, intelligently based technique has been introduced in this chapter in order to develop the inverse model of disc brake performance installed on the front axle of passenger car.

ARTIFICIAL NEURAL NETWORK MODELLING

Due to a complex synergy of phenomena occurred during braking, artificial intelligence abilities should be embedded into the process of control of automotive brake's performance. Generally, intelligence can be defined as the ability to learn and understand, to solve problems and to make decisions [28]. That is why, brain-like information processing is needed to create artificial environment for development of intelligent decision support system. One of the most powerful tools of artificial intelligence has emerged under the name of artificial neural networks, which mimic the function of the human brain [29]. In the last decade, artificial neural networks have emerged as attractive tools for modelling of non-linear process, especially in situations where the development of phenomenological or conventional regression models becomes impractical [30]. Artificial neural networks could be adequately characterized as a computer modelling approach with particular properties such as the ability to adapt to a changing environment or learn from examples through iterations without requiring a prior knowledge of relationships of process parameters [30]. Moreover, artificial neural networks have great capabilities to generalize, to cluster or organize data, to deal with uncertainties, noisy data, and non-linear relationships [31]. Its abilities for "transformation" of experimental data and experience into rules and knowledge could be used on different ways. That is why, artificial neural networks are good candidates for data mining due to their capabilities to represent complex non-linear behaviour based on learning from experimental data. In order to knowledge be extracted from experimental data, artificial neural networks have to be trained so that a particular input leads to a specific target output [31,32,33]. Training of artificial neural networks needs sufficient number of experimental data in order that input/output relationship may be learned and well generalized.

Artificial neural network is trained to perform a particular function by adjusting the values of the connections (weights) between elements (artificial neurons). Artificial neural network resembles the brain in two respects: (i) knowledge is acquired through a learning process, and (ii) connection strengths between neurons, known as synaptic weights, are used to store the knowledge. The artificial neurons are grouped into layers. Input layer receives data from outside of the network while output layer contains data representing the network predictions. Layers between these two kinds of layers are called hidden layers. When using multilayer neural networks for solving a problem, number of neurons in the hidden layers is one of the most important issues. It is known that insufficient number of neurons in the hidden layers leads to the inability of neural networks to solve the problem. On the other hand, too many neurons lead to overfitting and decreasing of network generalization capabilities due to increasing of freedom of network more than it is required.

Artificial neural networks are composed of simple elements operating in parallel, called artificial neurons [34,35]. As in nature, the connections between artificial neurons (connection weights) largely determine the network function [35]. To describe an artificial neural network's architecture adequately, it is necessary to specify how many layers it has, each layer's transfer function, the number of neurons in each of them and to characterize how layers are interrelated [36,37]. Since artificial neural network can be categorized as a parallel processor i.e. computer model, previously learned artificial neural network is often called artificial neural model or simply neural model. The quality of artificial neural models depends on a proper setting of neural network architecture and other influential parameters (training algorithm, transfer functions, the number and distribution of input/output data etc.). The process of a neural model developing, especially in the case when very complex functional relationships need to be modelled, includes resolving of very important issues. The following steps are necessary to be performed [35,38,39,40]: (i) identification of input and output data, (ii) select data generator, (iii) data generation, (iv) data pre-processing, (v) selection of neural network architectures and training algorithms, (vi) training of neural networks, (vii) accuracy evaluation, and (viii) neural model testing. The synergistic influence of above-mentioned parameters needs to be analysed, in order to make a proper selection of the neural network parameters corresponding the problem to be solved. Identification of neural model's input and output, which functional relationship needs to be modelled, represents the first important step in the neural model developing. It primarily depends on model objectives and choice of the data generator [36,37,38,39,40]. Taking into consideration that inverse functional relationship needs to be established between braking regimes of the disc brake and its performance, inputs parameters are defined by disc brake initial speed, brake interface temperature, and realized braking torque. On the other side, application pressure has been taken as an output parameter (see figure 1).

The general concept of an artificial neural network model of disc brake performance is shown in Figure 1. Artificial neural network has to be trained with corresponding data in order to learn the functional relationships between input/output data pairs. In order to be trained, the data for artificial neural network training should be generated. The type of data generator depends on application and the availability. In this case, as a data generator, single-end full-scale inertia dynamometer has been used. Testing of the disc brake has been performed under strictly controlled conditions related to change of application pressure, initial speed, brake interface temperature, and inertia of revolving masses. The role of data generator is important from the point of view repeatability of testing conditions in order to establish required relationships between input and output parameters space. In order to provide data for artificial neural networks training and testing, the disc brake has to be tested according to the adopted testing methodology. The testing methodology should be chosen to provide data covering the wanted ranges of input/output parameters change.

As it can be seen from Figure 2, the DC motor (1) drives, via carrier (2) and elastic coupling (3), a set of six flywheels (4) providing in such way different inertia from 10 to 200 kgm^2, and independently mounted on the driving shaft (5). The flange (6) firmly jointed to the shaft (5), bears rotating part of the tested brake (disc) while immobile flange (7), being firmly connected to the foundation (8) is used for mounting stationary parts of the tested brake (calliper). The temperature near the contact of friction pair has been measured by thermocouple sensor, mounted in the the outer brake pad, 0,5 mm from the contact surface. The full–scale inertia dynamometer is equipped by data acquisition system of all measured parameters at a sampling rate of 50Hz. The disc brake, used in this case, was designed for mounting on the front axle of passenger car with static load of 730 kg, effective disc radius of 101 mm, floating calliper (piston diameter 48 mm), friction surface area of 32,4 cm^2, and thickness of 16,8 mm.

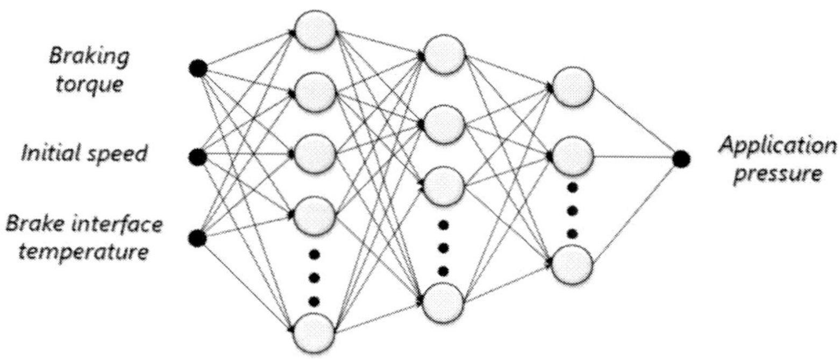

Figure 1. An inverse neural model of disc brake performance.

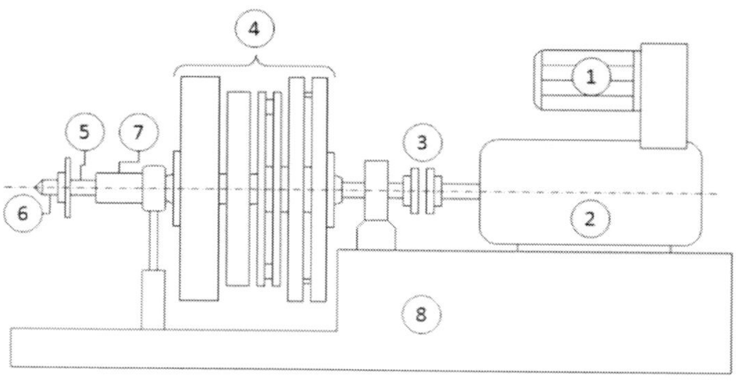

Figure 2. Single–end full–scale inertia dynamometer.

Table 1. Testing methodology

Test	Application pressure (bar)	Initial speed (km/h)	Temperature (°C)	№. brake application
Fading	Correspond to the 3 m/s^2 deceleration on first braking	90	Open	180
Recovery	20, 40, 60, 80, 100	20, 40, 60, 80, 100	Reached after 3 min cooling under 300 o/min after fading test	300

The testing methodology used for data acquisition in this case is shown in table 1. According to table 1, the disc brake has been tested under conditions specified in the tests fading and recovery. Training of artificial neural networks needs that the ranges of all input/output parameters be specified. The testing methodology of the disc brake fade and recovery performance defines not only the ranges of data that are going to be collected but also their distribution across selected ranges. Data distribution across of a training data set is important from a training strategy point of view. It is related to learning accuracy of artificial neural networks in some parts of the specified training data ranges. Regarding table 1, it can be seen that the disc brake has been tested under conditions that produce brake interface temperature rising. In the fade test, the disc brake has been subjected to 15 subsequent applications from initial speed of 90 km/h until stop and pause between each brake application of 45 s. This test is characterized by substantial brake interface temperature increasing which causing that braking torque has been decreased. The total number of braking applications performed under those conditions was 180. Recovery performance tests have been established for identification of the disc brake capability to preserve performance under high thermal load, reached after 3 min of cooling of the disc brake under 300°/min after a fading test. In the recovery test, the disc brake has been actuated by different application pressure (20 – 100 bar) at elevated brake interface temperatures when initial speed was changed in the range of 20 – 100 km/h. The number of the disc brake applications within recovery test was 300.

INVERSE MODEL OF DISC BRAKE PERFORMANCE

As it is shown in Figure 1, developing of the inverse neural model of disc brake performance at elevated temperatures has been performed by predicting influence of three input parameters related to initial speed, braking torque, and brake interface temperature versus one output parameter related to brake application pressure. Artificial neural networks are supposed to learn and generalize this complex relationship in the wide range of brake interface temperature, initial speed, and braking torque change. Since artificial neural networks are composed of simple elements operating in parallel, a computational neural model has been developed able to transform an input into output parameter space. As it was mentioned, the quality of artificial neural network models mostly depends on a proper setting of neural network architecture i.e. learning algorithm, transfer functions, range and distribution of data used for training, validation, and testing, etc. The artificial neural network architecture, which represents potential of the model prediction abilities in synergy with others network parameters, is unknown in advance. That is why the number of different architectures should be analyzed. Therefore, it is necessary properly determine the neural network architecture together with appropriate learning algorithm that shows the best prediction results. Different variations of network architectures can be obtained through varying of the number of hidden layers and the number of neurons in each of them. According to [2,3,4,36,37,41,42], it is important that the balance between the size of training, validation, and test data set, be optimally resolved.

The artificial neural network performance related to its learning and generalization capabilities strongly depends on the total amount of data presented to network. This data can be generally divided into three sets: training data, validation, and test data set. The number of training data pairs has significant influence on artificial neural network's generalization capabilities. The number of training data should be at least several times larger than the network's capacity. Therefore, it is necessary to determine the artificial neural network's architecture with appropriate learning algorithm which synergy providing the best prediction results for available data. In this chapter, collected input/output data pairs were divided into two sets (training and test data set – without validation data set). To ensure improving of the artificial neural network's performance, sufficient input/output data pairs were stored into the training data set. According to adopted testing methodology, 480 input/output data pairs were available for training and testing of different artificial neural network architectures. The total amount of input/output data pairs has been divided into two sets: 440 data pairs for the artificial neural networks training and 40 data pairs for testing of their prediction capabilities.

Data used for the artificial neural network training and testing have not been subjected to pre-processing. It means that the inverse neural model of disc brake performance at elevated temperature being developed able to deal with the real input data in the specified ranges of their change. In order to find the best neural network characteristics for establishing the functional relationships between input and output data space, a trial and error method has been used. The following 15 different artificial neural network architectures have been investigated in this chapter. The artificial neural network architectures have been denoted using the following mark $i\ [n]_l\ o$. In this mark i – indicates the number of inputs, o – indicates the number of outputs, n – indicates number of neurons in each of the hidden layers, and l – indicates the number of hidden layers.

The artificial neural network architectures have been divided according to the number of hidden layers as follows:

- one–layered network architectures: $3\ [2]_1\ 1$, $3\ [3]_1\ 1$, $3\ [5]_1\ 1$, $3\ [8]_1\ 1$, $3\ [10]_1\ 1$;
- two–layered network architectures: $3\ [2–2]_2\ 1$, $3\ [3–2]_2\ 1$, $3\ [5–2]_2\ 1$, $3\ [5–3]_2\ 1$, $3\ [8–2]_2\ 1$, $3\ [10–4]_2\ 1$;
- three–layered network architectures: $3\ [2–2–2]_3\ 1$, $3\ [3–3–2]_3\ 1$, $3\ [5–4–3]_3\ 1$, $3\ [8–5–3]_3\ 1$.

Each one of these 15 different neural network architectures have been trained with 6 different types of training algorithm: Levenberg–Marquardt (LM), Bayesian Regulation (BR), Resilient Backpropagation (RP), Scaled Conjugate Gradient (SCG), Gradient Descent (GDX), and Quasi–Newton (BFG), respectively.

A sigmoid transfer function has been used between the input and hidden layers as well as within the hidden layers (see expression (1)).

$$f(x) = \frac{1}{1+e^{-x}} \tag{1}$$

To avoid limiting the output to a small range, a linear activation function $f(x) = x$ has been employed between the hidden and the output layer.

After training of the selected artificial neural network architectures, the developed inverse neural models of disc brake performance have been tested against their ability for prediction of application pressure of the disc brake versus speed – temperature - braking torque change. Since 15 different artificial neural networks have been trained by 6 training algorithms, 90 inverse neural models have been developed. These neural models have been tested by the same data stored in the test data set in order to evaluate their capabilities for predicting the brake application pressures versus change of braking torque, speed, and brake interface temperature. The quality of prediction has been evaluated taking into consideration the difference between predicted (artificial neural network outputs) and real (experimentally obtained) values of the application pressure (expressed in percentage). The six error intervals have been established (0–5%; 5–10%, 10–15%; 15–20%; 20–25%; 25–30%) for that purpose. Based on the calculated errors between predicted and real values of the application pressure, the number of predicted results, which belong to each of these error intervals, has been calculated and expressed as a fraction of the test data set. The prediction results of the best neural models obtained after training and testing of the artificial neural networks are shown in Figure 3. It is evident from Figure 3 that different artificial neural network architectures (one-layered, two-layered, and three-layered) reached the best generalization capabilities versus used training algorithms. Furthermore, these network's architectures have been consisted by different number of neurons in the hidden layers (from 5 to 16). Complex synergy of the number of neurons in hidden layers and used training algorithms can be seen from Figure 3. The general trend of influence of the artificial neural network architectures on their prediction capabilities was that increasing of the number of neurons in the hidden layers may increase the network prediction capabilities but often, as it can be noticed in Figure 3, the prediction error can be also increased. It confirms that larger artificial neural network provide better learning abilities but it does not simultaneously provide better generalization properties of the artificial neural network. That is why different artificial neural network architectures should be trained and tested.

In order to demonstrate how artificial neural network architectures influence their prediction abilities in the case when the same training algorithm, training and test data sets were used, influence of Bayesian regulation algorithm on the artificial neural network prediction abilities is shown in Figure 4. According to Figure 4, there is a big difference in the network prediction abilities versus their architectures. If the first error interval is taken into consideration, the worst prediction result was reached by the neural model denoted as BR 3 $[10–4]_2$ 1 with only 24% of predicted results belong this interval. On the other hand, if the neural model based on three-layered neural network has been used, denoted as BR 3 $[8–5–3]_3$ 1, the percentage of predicted results in the first error interval was 48. Obviously, from Figure 4, increasing the number of neurons in synergy with increasing the number of hidden layers provided increasing the network's prediction abilities in the case when Bayesian regulation algorithm was used.

It is important to emphasize that a training algorithm has significant influence on the network prediction abilities. For instance, influence of the training algorithm on the networks prediction abilities is illustrated in Figure 5. Using the same training and testing data set for learning of the same artificial neural network architectures but in this case when the other training algorithm (Levenberg-Marquardt) was used, see Figure 5, the networks prediction abilities have been completely changed. It is evident from Figure 5 that in this case increasing of the number of neurons in the hidden layers in synergy with increasing the number of hidden layers caused generally decreasing of the networks prediction abilities in the first error interval. Comparing influence of these two training algorithms, it is evident different effect of the used training algorithms on the networks prediction results.

The best inverse neural models obtained by using mentioned training algorithms have been employed for prediction of the disc brake application pressure. The neural model prediction capabilities to predict the value of brake application pressure in fading and recovery tests were shown in Figure 6. According to Figure 6, although trained and tested by the same data, the neural models based on different neural network architectures and training algorithms have predicted the brake application pressure with different accuracy. The best prediction results were shown by the neural model denoted as BR 3 $[3–3–2]_3$ 1 obtained by training of the three–layered neural network architecture (3 $[3–3–2]_3$ 1) with Bayesian Regulation algorithm.

In order to better illustrate significance of the inverse neural model of disc brake performance and its ability to predict the brake application pressure, prediction of the brake application pressure has been shown in Figure 7. Figure 7 shows change of the real braking torque versus wide change of brake interface temperature in the fading test when initial speed was 90 km/h and brake application pressure took a constant value of 34 bar during all 15 brake applications. According to Figure 7, the real braking torque has shown instability versus brake interface temperature changing in the specified range. The key issue, which determines significance of the inverse neural model, is related to the neural model ability to predict the brake application pressure that provides the wanted value of braking torque. As it is shown in Figure 7, the wanted braking torque is set on 530 Nm.

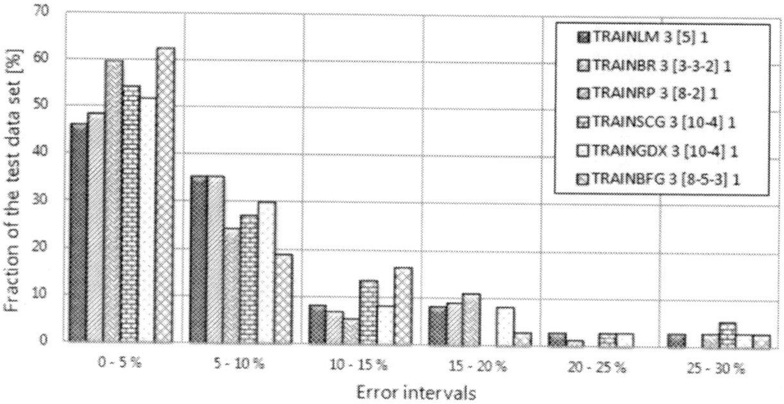

Figure 3. Prediction capabilities of artificial neural networks vs. training algorithm

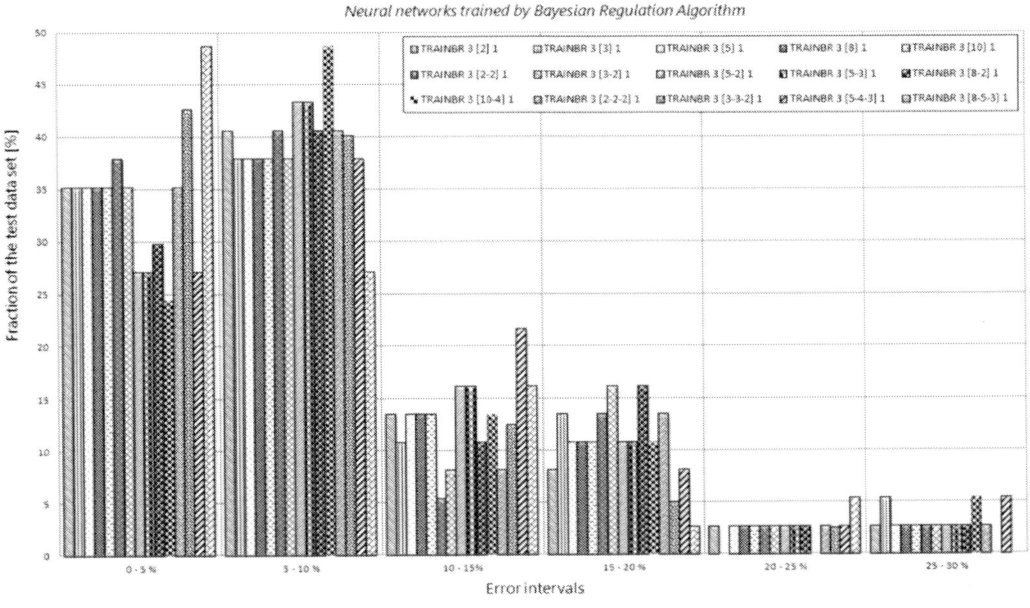

Figure 4. Influence of Bayesian Regulation algorithm on the networks prediction capabilities

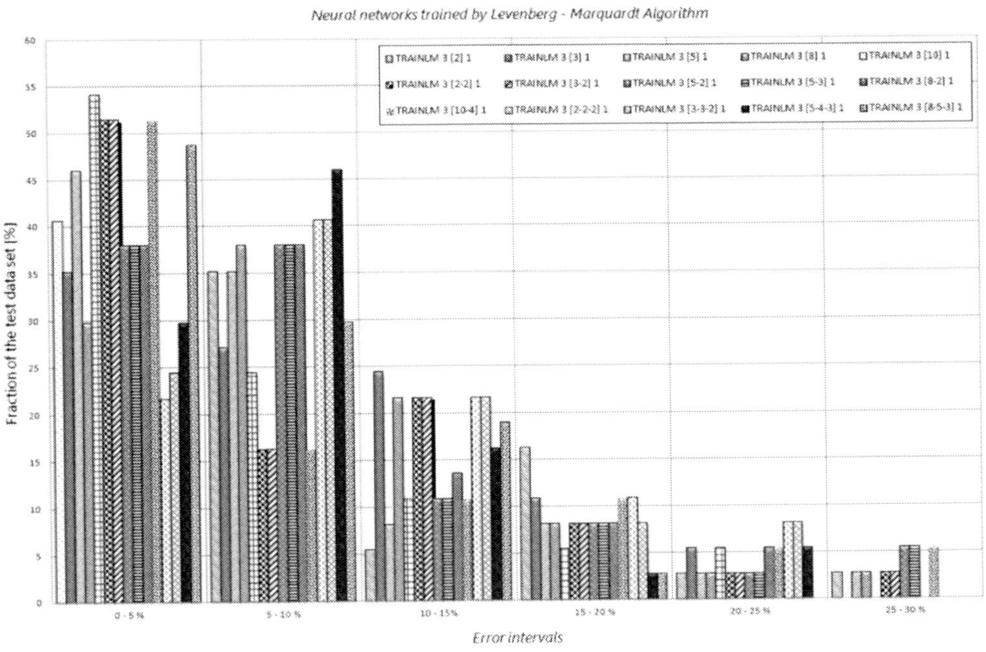

Figure 5. Influence of Levenberg - Marquardt algorithm on the networks prediction capabilities

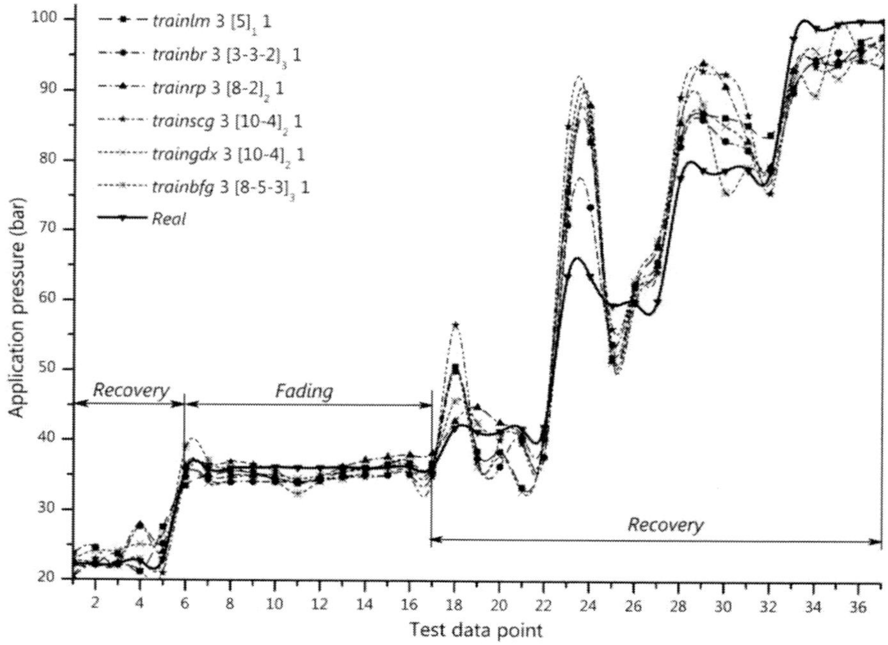

Figure 6. Comparison between the real and predicted pressure level by neural models.

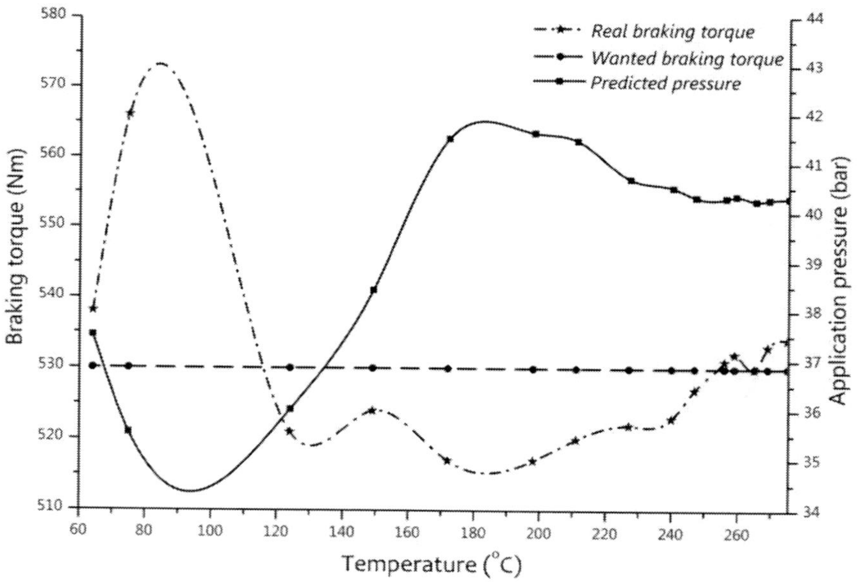

Figure 7. Prediction of brake application pressure versus wanted braking torque value.

In order to reach this value of braking torque and keep its value constant during all 15 brake applications, the brake application pressure has been modulated versus the real braking torque changes caused by brake interface temperature increasing. The inverse neural model of disc brake performance has been tested against those braking conditions. Instead of to be constant during all 15 brake applications, the brake application pressure has adjusted, as it is shown in Figure 7, in order to provide wanted value of braking torque. Obviously, in the case when the real braking torque is decreased, the inverse neural model predicted that the brake application pressure needs to be increased on the level which providing wanted braking torque value. This ability of the inverse neural model to control of brake application pressure versus braking regimes changes could be used for intelligent control of the disc brake performance especially at elevated brake interface temperatures.

The inverse neural model capabilities to predict the brake application pressure at elevated temperature has been further tested under braking conditions described in the recovery test. The recovery test has been performed after corresponding fading test at brake interface temperature reached after 3 min cooling of the disc brake under 300 °/min. The recovery test has been set up with main goal to investigate influence of high brake interface temperature load on its performance when initial speed and application pressure have been also changed. In this case, application pressure has been uniformly distributed between 20 – 100 bar (20, 40, 60, 80, and 100 bar) as well as initial speed between 20 – 100 km/h. The brake interface temperature was randomly distributed, after the fading test, depending on the disc brake braking regimes. Under these conditions, the inverse neural model has been tested to predict the disc brake application pressure, in synergy with other influencing factors, which causing braking torque change measured on the single-end full-scale brake dynamometer.

Figure 8 illustrates a comparison between the real and predicted values of application pressure versus synergistic influence of initial speed, braking torque, and the brake interface temperature changes between 208–230°C. As it can be seen, the inverse neural model (artificial neural network architecture 3 [3–3–2]$_3$ 1 trained by Bayesian Regulation algorithm) shows good capabilities for predicting of the brake application pressure, in the range of initial speed change between 20 – 100 km/h. From Figure 8 it can be seen that change of real brake application pressure was more linear than predicted one at lower initial speeds between 20 – 60 km/h. However, the inverse neural model of disc brake performance very well learned and generalized the relationships between the input and output parameters under these braking conditions.

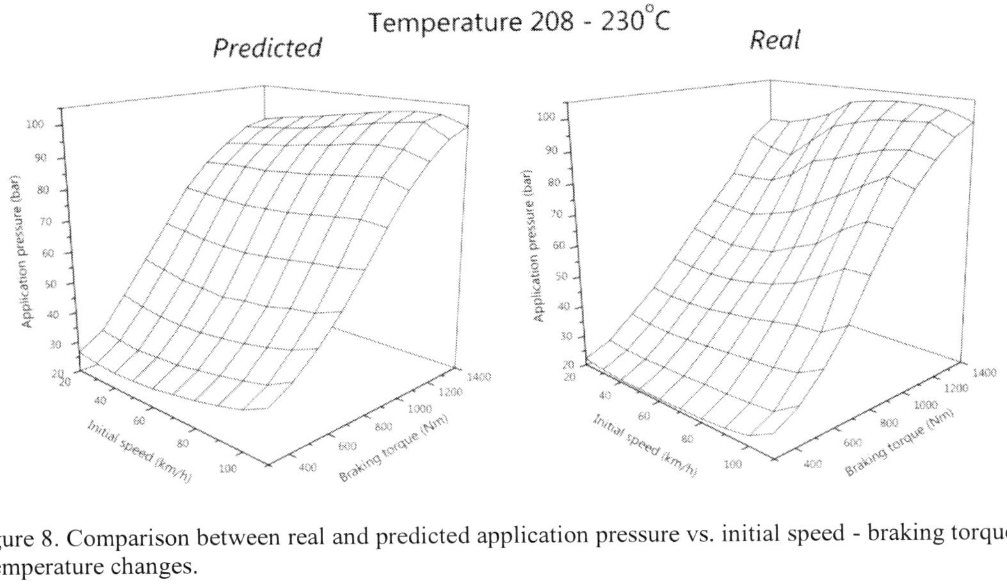

Figure 8. Comparison between real and predicted application pressure vs. initial speed - braking torque - temperature changes.

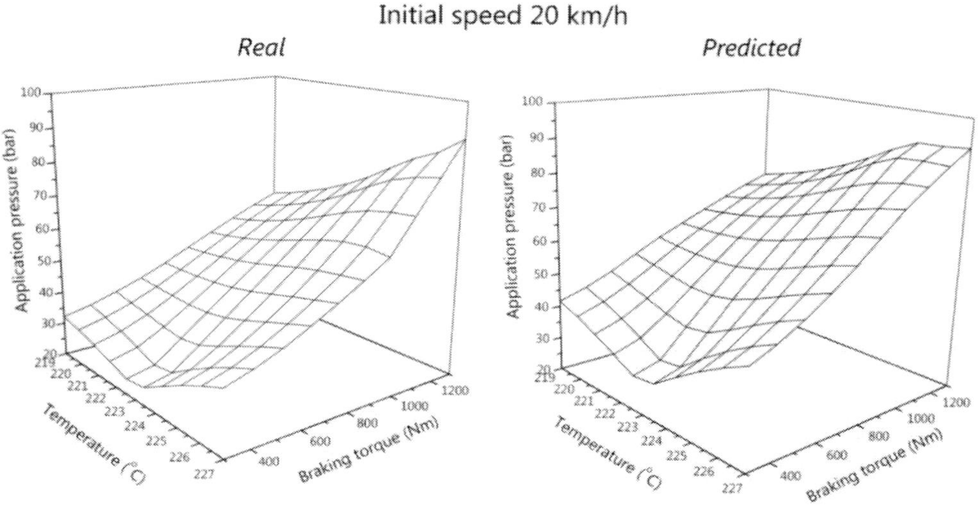

Figure 9. Comparison between real and predicted application pressure vs. braking torque and temperature changes (speed 20 km/h).

Since Figure 8 illustrates the neural model ability to predict the disc brake application pressure in the whole range of change of initial speed, braking torque, and brake interface temperature, from the neural model accuracy point of view it was interesting to analyse its prediction abilities for each value of initial speed in the tested range. Figure 9 shows a comparison between the real and predicted values of application pressure versus brake interface temperature and braking torque change when initial speed was 20 km/h. From Figure 9 is evident not only influence of the brake application pressure but also brake interface temperature on the maximum braking torque values. For instance, the maximum braking torque at brake interface temperature of 219°C can be reached for application pressure of 65 bar but at brake interface temperature of 227°C it was almost 90 bar. The neural model well predicted the disc brake performance under these braking conditions although the predicted values have been slightly higher than real ones. Furthermore, the neural model well predicted braking conditions (application pressure) for providing the minimum braking torque value around 223°C.

If Figure 10 is taken consideration, it can be seen that 3D profile of the disc brake performance related to influence of the brake application pressure has been changed in the case when initial speed was increased to 40 km/h. The higher fluctuation of the brake application pressure was requested in order to compensate a complex synergy of influences of brake interface temperature and initial speed versus braking torque, as it is shown in Figure 10. It is interesting that the general trend of influence of the brake application pressure against the disc brake performance was different from previous one. It can be noticed from Figure 10 that that brake application pressure should be decreased versus brake interface temperature increasing. The neural model well learned and accordingly recognized this 3D profile of the disc brake performance under these braking conditions. It can be seen, from Figure 10, that the predicted values, done by the inverse neural model, still slightly higher than real ones. Furthermore, the neural model well generalized at which braking conditions the specific value of braking torque needs maximum or minimum value of brake application pressure.

Moreover, the neural model well predicted the trend of brake application pressure change versus synergy of influences of initial speed-brake interface temperature.

Further increasing of initial speed from 40 to 60 km/h again changed 3D profile of the disc brake application pressure change required for reaching the braking torque values in the range between 200 – 1400 Nm (see Figure 11). It can be noticed from Figure 11 that increasing of the brake interface temperature required significant increasing of the brake application pressure in order to maximum braking torque be realized. According to Figure 11, the neural model did not predict substantial increasing of the brake application pressure in the case when maximum braking torque of 1400 Nm had to be realized (see Figure 11) versus real one. The neural model shown higher fluctuation of the brake application pressure for the braking torque value around 400 Nm than it was the case with the real braking torque in the range of brake interface temperature change. However, it can be seen in Figure 11 that the neural model again well predicted the general trend of disc brake application pressure changes needed for providing the braking torque specified in Figure 11.

The neural model has been also tested for initial speeds of 80 and 100 km/h (see figures 12 and 13). According to Figure 12 it can be seen that, for initial speed of 80 km/h, the range of brake application pressure increasing between the minimum and maxim braking torque values, at lower brake interface temperature around 211°C, was shortened versus previous situation when initial speed was 60 km/h.

Figure 12 shows substantial decreasing of the brake application pressure with brake interface temperature increasing at braking torque values around 400 Nm. That fluctuation of the brake application pressure was significantly smaller at maximum braking torque values around 1400 Nm. The neural model well predicted such 3D profile of the disc brake application change to reach the specified braking torque values (see figure 12).

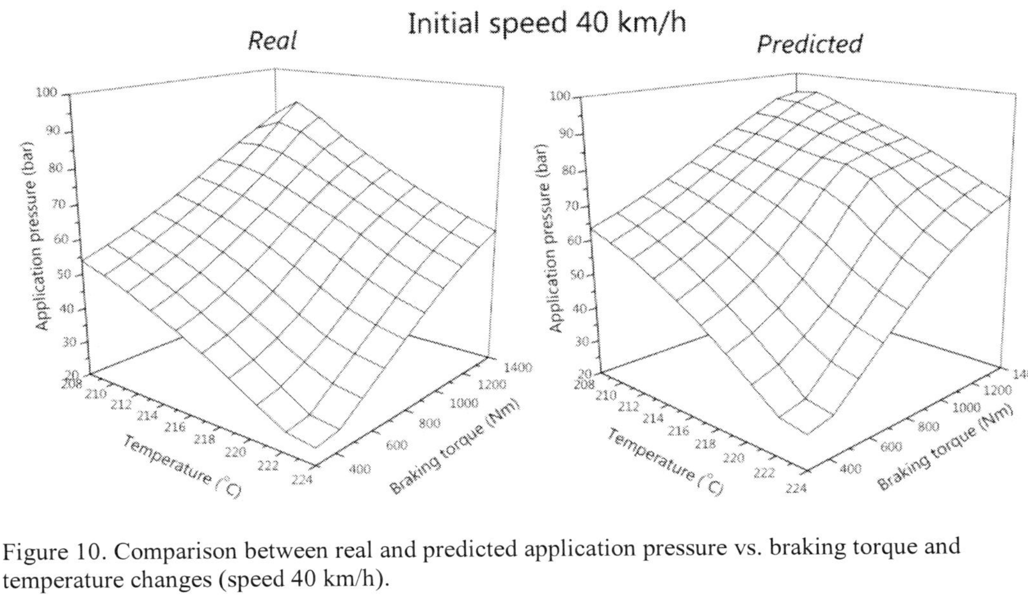

Figure 10. Comparison between real and predicted application pressure vs. braking torque and temperature changes (speed 40 km/h).

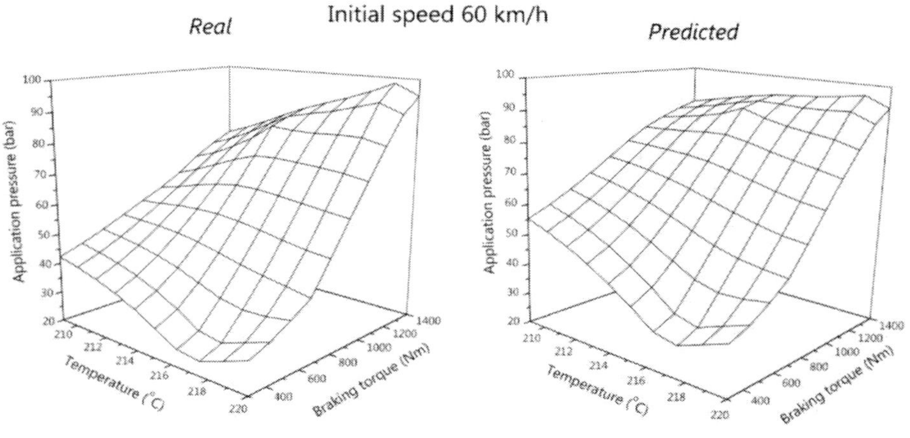

Figure 11. Comparison between real and predicted application pressure vs. braking torque and temperature changes (speed 60 km/h).

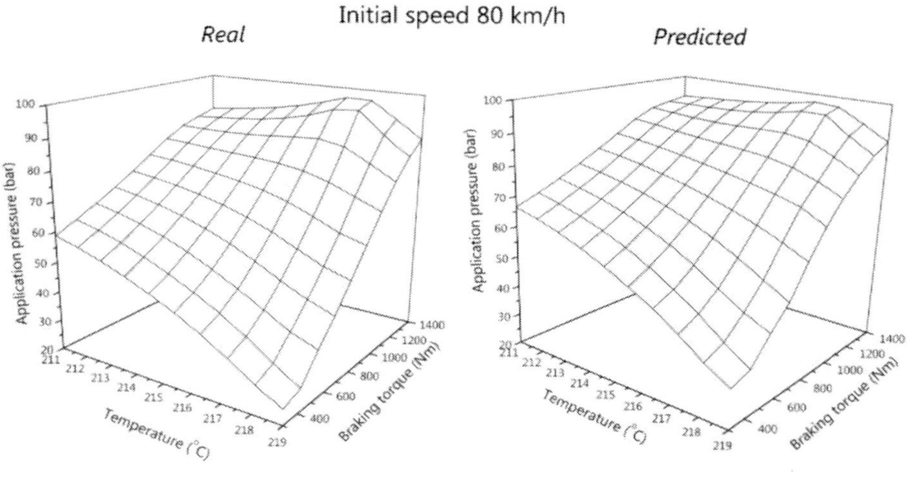

Figure 12. Comparison between real and predicted application pressure vs. braking torque and temperature changes (speed 80 km/h).

The trend of influence of brake application pressure on its performance change has been again changed when the disc brake initial speed increased to 100 km/h (see Figure 13). The disc brake, at this initial speed, has become more sensitive with brake interface temperature increasing. The span of the brake application pressure, at brake interface temperature around 210°C, between the minimum and maximum braking torque value, has been increased versus higher brake interface temperature around 230°C (see Figure 13). Although change of the disc brake interface temperature has been slightly changed, it required significant brake application pressure modulation. The neural model generalized such changes of the disc brake performance under these braking regimes. The general trend of influence of the disc brake application pressure on its performance has been learned and well predicted.

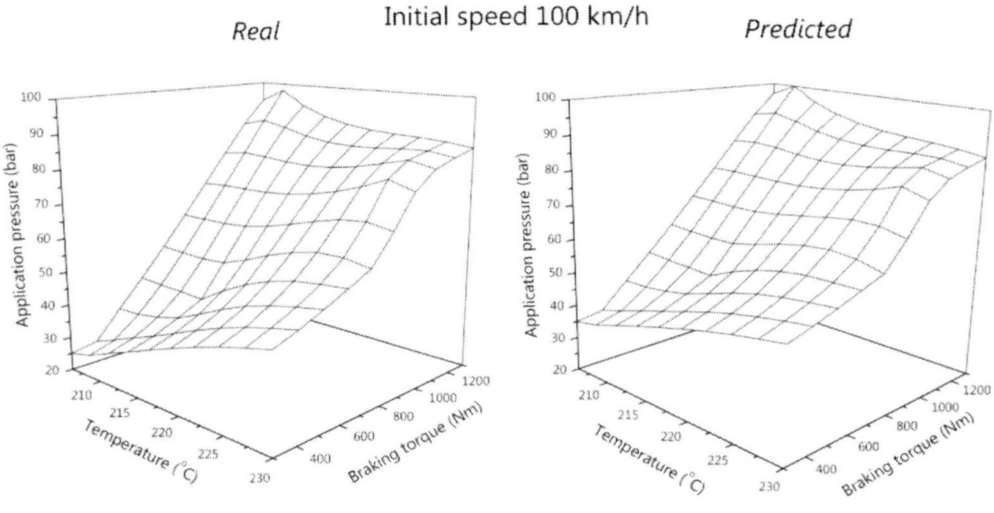

Figure 13. Comparison between real and predicted application pressure vs. braking torque and temperature changes (speed 100 km/h).

Conclusion

It was shown that the disc brake performance has a stochastic nature and that its performance is differently influenced by braking regimes. The braking regimes which causing significant change of the disc brake performance conditions can be represented by application pressure, initial speed, and brake interface temperature for the used friction pair. It is especially evident at elevated brake interface temperature when the disc brake performance suffers form a loss of effectiveness. The disc brake performance cannot be easily predicted particularly at elevated temperature due to complex interrelated influences. Furthermore, the classical mathematical model of the disc brake operation is impossible to be established. That is why, in this chapter, the technique of artificial intelligence has been used in order to model complex influence of braking regimes on change of the disc brake performance. In this chapter has been shown that the brake application pressure can be used for control of the disc brake performance. Since the brake application pressure can be used for better control of the disc brake performance, the inverse neural model of the disc brake performance has been developed. Inversion in this chapter was related to modeling and prediction of the brake application pressure versus initial speed, brake interface temperature, and braking torque changes. It is shown that braking torque, speed, and brake interface temperature can be correlated with the brake application pressure using artificial neural networks. Moreover, it is shown that the inverse neural model of the disc brake operation can be developed with inherent abilities to predict the value or required change of the brake application pressure for reaching of the specific braking torque values at different braking conditions. The developed inverse neural model has abilities to predict 3D profile of the brake application pressure change for different values of the braking torque in the range of initial speed change between 20 – 100 km/h and brake interface temperature between 60 – 275°C. The methodology of the inverse neural model development indicates that this approach can be used for introducing of intelligent control of the disc brake performance based on modulation of brake application pressure according to wanted value of braking torque.

REFERENCES

[1] Aleksendrić, D. *Intelligent Control of Commercial Vehicles Braking System Function,* FISITA Yokohama, Japan, 2006.

[2] Aleksendrić, D. *Neural networks in automotive brakes' friction material development,* PhD thesis, Faculty of Mechanical Engineering University of Belgrade, (In Serbian), 2007.

[3] Aleksendrić, D; Duboka, C. Prediction of automotive friction material characteristics using artificial neural networks-cold performance, *Wear*, 2006, 261(3-4), 269-282.

[4] Aleksendrić, D; Duboka, C. Fade performance prediction of automotive friction materials by means of artificial neural networks, *Wear*, 2007, 262(7-8), 778-790.

[5] Aleksendrić, D; Duboka, C. A Neural Model of Automotive Cold Brake Performance, *FME Transactions*, 2007, 35, 9-14.

[6] Hagan, MT; Demuth, HB; De Jesus, O. An introduction to the use of neural networks in control systems, *International Journal of Robust and Nonlinear Control*, 2002, 12 (11), 959-985.

[7] Demuth, H; Beale, M. Neural network toolbox for use with MATLAB, Users guide ver. 4.0., *The 9. Mathworks*. Inc. 1998.

[8] Antsaklis, PJ; Passino, KM. *An Introduction to intelligent and autonomous control,* Kluwer Academic Publishers, Norwell, MA, USA, 1993.

[9] Eriksson, M; Bergman, F; Jacobson, S. On the nature of tribological contact in automotive brakes, *Wear*, 2002, 26–36.

[10] Xiao, G; Zhu, Z. Friction materials development by using DOE/RSM and artificial neural network, *Tribology International,* 2010, 43 (1-2), 218-227.

[11] Müller, M; Ostermeyer, GP. A Cellular Automaton model to describe the three–dimensional friction and wear mechanism of brake systems, *Wear*, 2007, 1175–1188.

[12] Zhang, SY; Qu, SG; Li, YY; Chen, WP. Two–body abrasive behaviour of brake pad dry sliding against interpenetrating network ceramics/Al–alloy composites, *Wear*, 2010, 268 (7-8), 939-945.

[13] Eriksson, M; Jacobson, S. Tribological surfaces of organic brake pads, *Tribology International*, 2000, 33(12), 817-827.

[14] Eriksson, M; Bergman, F. Surface characterization of brake pads after running under silent and squealing condition, *Wear*, 1999, 232(2), 163-167.

[15] Ray, S; Chowdhury, SKR. Prediction of contact temperature rise between rough sliding bodies: An artificial neural network approach, *Wear*, 2009, 1029–1038.

[16] Zhang, S; Wang, F. Comparison of friction and wear performances of brake materials containing different amounts of $ZrSiO_4$ dry sliding against SiC_p reinforced Al matrix composites, *Materials Science and Engineering*, A 2007, 443 (1–2), 242–247.

[17] Qi, HS; Day, A. J. Investigation of disc/pad interface temperatures in friction braking, *Wear*, 2007, 505–513.

[18] Gurunath, PV; Bijwe, J. Friction and wear studies on brake–pad materials based on newly developed resin, *Wear*, 2007, 1212–1219.

[19] Gopal, P; Dharani, LR; Blum, FD. Load, speed and temperature sensitivities of a carbon–fiber–reinforced phenolic friction material, Wear 1995 181–183, Part 2, 10th International *Conference on Wear of Materials*, 913-921.

[20] Aleksendrić, D; Duboka, Č; Ćirović, V. Intelligent Control of disc brake operation, 26th Annual Brake Colloquium 2008, SAE *Paper*, 2008-01-2570, 2008, Texas, USA.

[21] Aleksendrić, D; Barton, DC. Modelling of brake friction materials performance at elevated temperatures, *Braking*, 2009, June 9-12, York, United Kingdom.

[22] Aleksendrić, D. *Prediction of brake friction materials speed sensitivity*, 27th Annual Brake Colloquium 2009, SAE Paper 2009-01-3008, Florida, USA.

[23] Satapathy, BK; Bijwe, J. Performance friction materials based on variation in nature of organic fibers- Part I. Fade and recovery behavior, *Wear*, 2004 257, 573-584.

[24] Myshkin, NK. Friction transfer film formation in boundary lubrication, *Wear*, 2000, 245, 116-124.

[25] Jintang, G. Tribochemical effects in formation of polymer transfer film, *Wear*, 2000, 245, 100-106.

[26] Ćirović, V; Aleksendrić, D. Intelligent control of passenger car braking system, FISITA 2008 World Automotive *Congress*, F2008-SC-046, 14-19 Sep. Munich, Germany.

[27] Xiao, G; Zhu, Z. Friction materials development by using DOE/RSM and artificial neural network, *Tribology International*, 2010, 43(1-2), 218-227.

[28] Voracek, J. Introduction to knowledge-base intelligent systems, *Pearson Education*, 2002.

[29] Lee, JW; Oh, JH. Time delay control of non-linear systems with neural network modelling, *Mechatronics*, 1997, 7(7), 613-640.

[30] Lahiri, SK; Ghanta, KC. Artificial Neural Network Model with Parameter Tuning Assisted by Differential Evolution Technique - Study of Pressure Drop of Slurry Flow in Pipeline, *Chemical Industry & Chemical Engineering,* 2009, Quarterly 15(2), 103−117.

[31] Krose, D; Smagt, P. *An Introduction to Neural Networks*, The University of Amsterdam, Eighth edition, November, 1996.

[32] Issa, RA; Fletcher, D. Neural Networks in Engineering Applications, Proceedings of the 29th Annual Conference Colorado State University - Fort Collin's, *Colorado*, April 15-17, 1993, 177-186.

[33] Demuth, H; Beale, M. Neural network toolbox for use with MATLAB, Users guide ver. 6.0.1, *The Mathworks*., Inc. 2006.

[34] Larose, DT. *Discovering Knowledge in Data-An Introduction to Data Mining*, John Willey & Sons, 2005.

[35] Devabhaktuni, VK; Yagoub, MCE; Fang, Y; Xu, J; Zhang, QJ. *Neural Networks for Microwave Modeling: Model Development Issues and Nonlinear Modeling Techniques*, John Wiley & Sons, Inc. 2001.

[36] Aleksendrić, D; Barton, DC. Neural network prediction of disk brake performance, *Tribology International*, 2009, 42 (7), 1074-1080.

[37] Aleksendrić, D. Neural network prediction of brake friction materials wear, *Wear*, 2010, 268(1-2), 117-125.

[38] Aleksendrić, D; Duboka, C. Artificial technologies in sustainable braking system development, *Int. J. Vehicle Design,* 2008, 46(2), 237-249.

[39] Aleksendrić, D; Duboka, C; Mariotti, GV. Neural modelling of friction material cold performance, Proc. IMechE Part D: *J. Automobile Engineering,* 2008 222(7), 1021-1029.

[40] Aleksendrić, D; Barton, DC; Vasic, B. Prediction of brake friction material recovery performance using artificial neural networks, *Tribology International*, 2010, 43(11), 2092-2099.

[41] Ćirović, V; Aleksendrić, D. Development of neural network model of disc brake operation, *FME Transactions*, 2010, 38, 29-38.

[42] Aleksendrić, D; Ćirović, V. *Effect of brake friction material manufacturing conditions on its wear*, 28th Annual Brake Colloquium 2010, SAE Paper 2010-01-1679, Arizona, USA.

In: Focus on Artificial Neural Networks
Editor: John A. Flores, pp. 171-189
ISBN: 978-1-61324-285-8

Chapter 8

ARTIFICIAL NEURAL NETWORKS; DEFINITION, PROPERTIES AND MISUSES

Erkam Guresen [*] [1] ***and Gulgun Kayakutlu*** [2]
[1]Lecturer, Okan University, Department of Business Administration, Turkey,
[2]Asst. Prof. Dr., Istanbul Technical University, Department of Industrial Engineering, Turkey

SUMMARY

There are no such clear and good definitions of ANNs in the literature. Many of the definitions refer to the figures instead of well explaining the ANNs. That is why many weighted graphs (as in shortest path problem networks) fit the definition of ANN. This study aims to give a clear definition that will differentiate ANN and graphs (or networks) by referring to biological neural networks. Although there is no input choice limitation or prior assumption in ANN, sometimes researchers compare ANN achievements with the results of other methods using different input data and make comments on these results. This study also gives examples from literature to misuses, unfair comparisons and evaluates the underlying reasons which will guide researchers.

1. WHAT IS AN ARTIFICIAL NEURAL NETWORK (ANN)?

There are no such clear and good definitions of ANN in the literature. Many of the definitions refer to the figures instead of well explained networks. That is why many weighted graphs (as in shortest path network) fits the definition of ANN. Even some preferable definitions consider ANN as distributed processing elements.

A good definition of ANN is given by Haykin (1999) as follows;

[*] Corresponding author: Email: erkamguresen@gmail.com

> "A neural network is a massively parallel processor made up of simple processing units, which has a natural propensity for storing experiential knowledge and making it available for use. It resembles the brain in two respects:
>
> 1. Knowledge is acquired from the environment through a learning process run in the network.
> 2. Interneuron connection strengths, known as synaptic weights, are used to store the acquired knowledge."

Eberhart and Shi (2007), Gurney (2003), Rojas (1996), Muller and Reinhardt (1990) also mention about Processing Elements (PE) and learning algorithm in defining ANN. Haykin (1999), Eberthart and Shi (2007), Gurney (2003), Rojas (1996) referred learning as modifying synaptic weights to capture information. Haykin (1999), Gurney (2003) also added that ANNs can modify their own topology.

Eberhart and Shi (2007) defined output of a PE as a function of function, in which a summation is performed to combine inputs than an activation function is used to calculate the output. In a similar way Haykin (1999) identified three basic elements of a PE as; *synaptic weights*, *summing function* to combine inputs with respect to their weights, and an *activation function* to produce an output. These PEs are inspired by the existing neurons in animal nerve system. Real neurons get stimulus and change them via synaptic weights, combine them, and lastly produce a single response (output) different from combination.

Rojas (1996) noted that we still do not fully understand the computing mechanism of a biological neuron so we prefer PEs (or computing units) instead of artificial neuron. Similar to Haykin (1999), Rojas (1996) added that four structures (dendrites, synapses, cell body and axon) of biological neuron are the minimal structure we would adopt from biological models.

Principe et al. (1999) used a pragmatic definition for ANNs as follows;

> "ANNs are distributed, adaptive, generally nonlinear learning machines built from many different processing elements (PEs). Each PE receives connections from other PEs and/or itself. The interconnectivity defines the topology. The signals flowing on the connections are scaled by adjustable parameters called weights, w_{ij}. The PEs sum all these contributions and produce an output that is a nonlinear (static) function of the sum. The PEs' outputs become either system outputs or are sent to the same or other PEs."

To make things more comprehensible, definition of graph is taken as the starting point. Geometrically a *graph* is a set of points (vertices or nodes) in space which are interconnected by a set of lines (edges or links) (Gibbons, 1999). A graph with no self-loops and no parallel edges is called *simple graph* (Gibbons 1999). And the *weighted graph* can be defined as follows; a *weighted graph* is a graph in which a number is assigned to for any edge *e* (Gibbons, 1999). A *directed graph* or shortly *digraph* is a graph whose edges have a direction and edges are represented by arrows showing the direction (Gibbons, 1999). And lastly we should define *connectivity* in digraphs as follows: v_i is connected to v_j if there is a path from v_i to v_j (Gibbons, 1999).

Haykin (1999) mentioned about ANNs as directed graphs, in which three graphical representations are used for defining a neural network:

1. With a block diagram to describe the network functionally,
2. With signal-flow graph to describe signal flow in the network,

3 With the architectural graph to describe the network layout.

Haykin's (1999) description of ANN as directed graphs is not complete as long as it excludes the learning process, input and output set (number of input or output records and number of attributes of inputs and outputs) and parallel structure.

An interesting but the most mathematical definition we found in literature was from Muller's studies (Muller et al. 1996). Muller et al. (1996) define ANN from the point of graphs as follows:

"A neural network model is defined as a directed graph with the following properties:

1. A state variable n_i is associated with each node *i*,
2. A real valued weight w_{ik} is associated with each link (*ik*) between two nodes *i* and *k*,
3. A real valued bias v_i is associated with each node *i*,
4. A transfer function $f_i[n_k, w_{ik}, v_i, (i \neq k)]$ is defined for each node i, which determines the state of the node as a function of its bias, of the weights of its incoming links, and the states of the nodes connected to it by these links. "

Muller et al. (1996) defines input nodes as the nodes with no incoming link and output nodes as the nodes with no outgoing links. This definition contains some problems like not fitting to recurrent neural networks, in which output of each neuron can be its input. Thus Muller et al.'s (1996) definition cannot point out input and output nodes clearly.

Another problem with Muller et al.'s definition is that it does not contain input and output nodes (or layers) in the definition of ANN. This creates the confusion with other graphs. Clearly an ANN should have some input neurons and some output neurons with specific features that are not applicable for all other graphs. Muller et al.'s definition does not refer to parallel distribution of nodes and learning process. This will also cause confusion with other graphs.

In literature many studies can be found which compare biological neural networks with ANN like Haykin (1999), DARPA Report (1992), Rojas (1996) Braspenning et al. (1995) and Muller et al. (1996). The inspirations of similarities are summarized in Table 1. Note that receptors are specialized neurons for gathering specific information from environment, neural net generally refers to brain and the effectors are the specialized neurons for evoking the specific tissues.

Table 1. Similarity between biological neural networks and artificial neural networks

Biological Neural Networks	Artificial Neural Networks
Stimulus	Input
Receptors	Input Layer
Neural Net	Processing Layer(s)
Neuron	Processing Element
Effectors	Output Layer
Response	Output

Table 2. Similarities of neurons and processing elements (PEs)

Neurons	Processing Elements (PEs)
Synapses	Weights
Dendrites	Summing Function
Cell Body	Activation Function
Axon	Output
Threshold value	Bias

Activities of biological neurons and processing elements of ANN can be compared as in Table 2. Briefly, synapses act like a weight of the incoming stimulus and inspired the weights of ANN; dendrites that accumulates the incoming weighted stimulus, inspired the summing function of ANN; cell body, that causes conversion of summed stimulus in to a new stimulus, inspires activation function; axon, which distributes the new stimulus to the corresponding neurons, inspires the output and output links; and lastly, threshold value with a role of activating or inactivating increase an decrease of the stimulus, inspires the bias. All four structures mentioned by Rojas (1996) (dendrites, synapses, cell body and axon) are necessarily contained in PEs.

In the light of above analysis on definitions and inspirations we can enrich the definitions by mentioning a network which is made up of massively parallel processors with connections. A clear definition of processors will differentiate an artificial neural network with unique features. In general, nodes in a graph could be considered as PEs with identity function, which returns the same input as output. A complete definition of ANN from the point of graphs is suggested to include the features give in the following definitions.

Definition 1. A directed simple graph is called an *Artificial Neural Network (ANN)* if it has

- at least one *start node* (or *Start Element; SE*),
- at least one *end node* (or *End Element; EE*),
- at least one *Processing Element* (*PE*),
- all the nodes representing must be *Processing Elements* (*PEs*), except *start nodes* and *end nodes*,
- a state variable n_i associated with each node i,
- a real valued weight w_{ki} associated with each link (ki) from node k to node i,
- a real valued bias b_i associated with each node i.
- at least two of the multiple PEs connected in parallel,
- *a learning algorithm* that helps to model the desired outputs for given inputs.
- a flow on each link (k_i) from node k to node i, that carries carrying exactly the same flow which equals to n_k caused by the output of node k,
- each start node is connected to at least one end node, and each end node is connected to at least one start node.

The definition of Artificial Neural Networks will be complete when we define *Start Element* (*SE*), *End Element* (*EE*), *Processing Element* (*PE*) and *Learning Algorithm*;

Definition 2. *Start Element* (*SE*) *i* is a node in a directed graph, which gets an input I_{ij} from the input matrix $I=\{I_{ij};\ i=1,2,\ldots,n\ ,\ j=1,2,\ldots,m\}$ of *n* attributes of *m* independent records, and starts a flow in the graph.

Definition 3. *End Element (EE) i* is a node in a directed graph, which produces an output O_{ij} from the output matrix $O=\{O_{ij};\ i=1,2,\ldots,n\ ,\ j=1,2,\ldots,m\}$ of *n* desired outputs of *m* independent input records and ends a flow in the graph.

Definition 4. Let *G* be a directed simple graph with the following properties;

1. A state variable n_i is associated with each node *i*,
2. A real valued weight w_{ki} is associated with each link (*ki*) from node *k* to node *i*,
3. A real valued bias b_i is associated with each node *i*.

Let $f_i[n_k, w_{ki}, b_i, (i \neq k)]$ be the following function in graph *G* for node *i*;

$$f_i(u_i, b_i) = n_i = \varphi(u_i + b_i)$$

where $\varphi(.)$ is the *activation function* and u_i is as follows;

$$u_i = \sum_{j=1}^{m} w_{ji}\, n_j \qquad \text{Eq.1}$$

where *j* is the node which has a link to node *i* and hence node *i* is called a *Processing Element* (*PE*).

Corollery 1. In a directed graph for each node can be considered as PE with (if not specially assigned) $w_{ki} = 1$, $b_i = 0$ and $\varphi(.) = I(.)$, where *I(.)* is the identity function. With these properties flow does not change at nodes.

We can shortly explain Definition 4 (PEs of ANN) as nodes with functions, constructed by the state of the node, weights of its incoming links, and the bias of weights.

Definition 5. *Learning Algorithm* in an *ANN* is an algorithm which modifies weights of the ANN to obtain desired outputs for given inputs.

Hint 1. Desired outputs can be exactly known values, number of desired classes or some pattern expectations certain input sets. Therefore, "desired output" term contains output of supervised, unsupervised or reinforced learning.

Hint 2. Note that every element *k* (SE, PE and EE) can generate only and only one output value at a time. But every element *k* can send the same output value to other element *i* with no restrictions if there is a link (edge) from *k* to *i*.

The suggested definition an ANN can be differentiated from any other kind of graph and is strong enough to avoid previous issues. First of all, it is a network which has specific starting and ending nodes. This new start and end node (element) definitions do not contradict with recurrent neural networks as in Muller et al. (1996). By differentiating the nodes as PEs, SEs or EEs, components of an ANN are clarified. By describing the variables and parameters associated with each node and link contained in a graph, confusions are void.

Besides, massively parallel structure makes it more biological based than computer based. Structures containing some SEs and EEs with one or more PEs which connected serially cannot be mentioned as an ANN because it will lose the power of parallel computing and starts to act more like existing computers than a brain. Good explanations of parallel and serial computing can be found in Haykin (1999). Shortly it can be said that parallel computing is powerful in complex calculations and mappings; serial computing is powerful for arithmetic calculations (Haykin, 1999). Also serial structure cannot contain fault tolerance property (will be discussed in detail in further sections). Thus a damage or corruption in serial structure will cause catastrophic failures but in biological neural networks death of a neuron does not cause catastrophic failures of the neural network.

A graph which cannot be taught (or corrected through its weights) cannot be mentioned as an ANN, because it cannot contain **adaptivity** property (will be discussed in detail in further sections). If a graph cannot be taught, environmental changes cause us to build a new graph to represent the environment instead of modifying the existing graph through learning (or updating synaptic weights).

The flow in an ANN is also specific since every outgoing link carries the same flow n_i produced as output of the node i. For this reason proposed definition contains details of flow to avoid confusion. For example in an ANN, when a flow comes to node k (can be SE, EE or PE k), node k generates an output n_k. The output n_k is send to the all nodes i if an edge exists from node k to node i. On other words each edge from node k to node i, duplicates the output value n_k and carries as a flow. But in many other graphs each edge from node k to node i, carries some portion of the output value n_k as a flow, such a way that their sum is equal to n_k.

And the last feature in the proposed definition is about connectivity of the ANN, in which they have to map inputs to outputs.

2. Common Properties of ANN

Artificial neural network is a parallel distributed processer, which stores information through learning. This parallel structure gives many advantages to ANN. Haykin (1999) listed benefits of ANN as follows:

(a) Nonlinearity
(b) Input-Output Mapping
(c) Adaptivity
(d) Evidential Response
(e) Contextual Information
(f) Fault Tolerance
(g) Very Large Scale Integrated (VLSI) Implementability

(h) Uniformity of Analysis and Design
(i) Neurobiological Analogy

In addition to listed benefits, parallel distributed processing and generalization ability must be mentioned as primary properties.

Parallel distributed processing in ANN comes from biological analogy and it is the main property of ANN. Many properties such as nonlinearity, input-output mapping, fault tolerance and VLSI implementability, naturally comes from parallel distributed processing.

Generalization ability is the property of extracting general information from data. ANN can deal with this if it is trained well enough. Assume that the given data is like the dots in figure XX (a). General behavior of the data is a curve like given in (a). An ANN can extract this general behavior from data but if it over-trained, it will memorize the points, like given in (b). The difference is if (a) occurs, model will work correctly with unseen data but if (b) occurs model will not correctly with unseen data. In (b) model will work correctly with only the trained data.

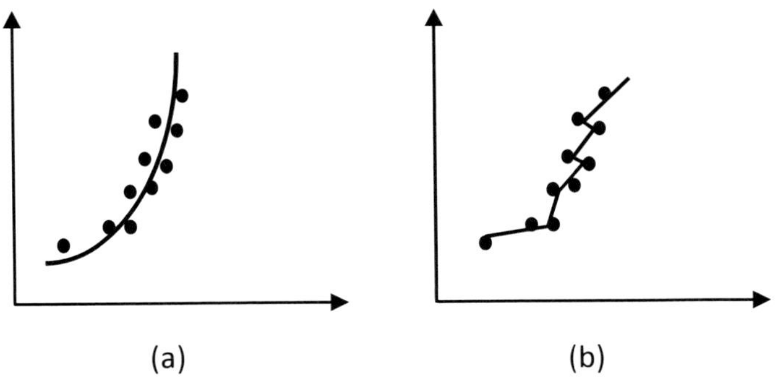

Figure 1. Data set and curve (a), data set and over-train zigzags (b).

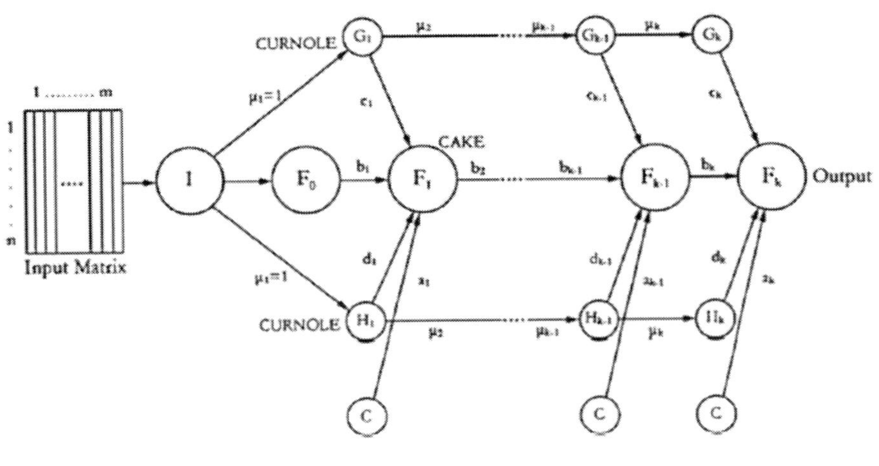

Figure 2. The DAN2 Network Architecture (Ghiassi and Saidane 2005).

Input-output mapping property refers to ability of using any input to obtain any output without any prior assumption or any restriction. ANN can find ***nonlinear*** relations in the data even if it is formed from linear functions (Haykin, 1999). ***Adaptivity*** is the ability to adapt changes in environment. Adaptivity refers to retraining of ANN with only the new data set when it is available. Thus once an ANN model is formed, no need to built a new ANN model when new data is available (it can be referred as environmental change).

Fault tolerance ability refers to the results, that will contain some errors but not at a catastrophic level, when a hardware implementation of an ANN is damaged (Haykin, 1999). Fault tolerance is the result of parallel distributed structure since it distributes learned information to the PEs through weights (Gurney, 2003). Thus when a PE is damaged instead of losing the whole information, only the information that stored in the corresponding weights is lost.

VLSI implementability refers that ANNs can deal with large complex data sets (Haykin, 1999) and can be used for data mining. Since ANNs has its own notation and architectures, that can be understand by other researchers, which offers ***Uniformity of Analysis and Design.*** Haykin (1999) referred ***Neurobiological Analogy*** as the living proof of ANNs. Detailed explanations about general properties of ANN and biological neural networks can be found in Haykin (1999) and Muller et al. (1995).

3. Misuses of ANNs

3.1. New Architectures That Are Not Actually ANN

New ANN architectures are developed by academicians with the aim of improving results of ANNs or in a search of more understandable architectures instead of accepting "black box" assumption. But many times practice with new ANN architectures is limited by the applications of the developers, so they are not fully evaluated before excepting as a working ANN architecture.

Dynamic Architecture for Neural Networks (DAN2) is one of the newly developed architectures, which was first introduced by Ghiassi and Saidane (2005). Figure 1 shows the structure of DAN2. In this structure there is an "I" node, which represents the Input node, for the input layer. For the hidden (and if it is the last layer than is called output) layers there are 4 nodes; one "C", one CAKE ("F"), and two CURNOLE nodes ("G" and "H"). CURNOLE stands for "Current Residual Nonlinear Element" and CAKE stands for "Current Accumulated Knowledge Element". C node is a constant value obtained from linear regression.

In DAN2 structure, learning is obtained by an algorithm based on linear regression and teaching the remaining residuals by adding a new hidden layer(s) and again using linear regression. This layer adding procedure is continued until desired level of learning is obtained. If the desired level is reached at first layer, it can be said that the mapping from inputs to outputs is linear and do not contain any non-linearity. For non-linear relations additional layer(s) is added to architecture.

Every node in each layer has a duty; such as CAKE node accumulates the knowledge gained by architecture and CURNOLE nodes transfers the remaining non-linearity and adds it to the next layers CAKE node.

The learning algorithm steps given by Ghiassi and Saidane (2005) are as follows:

For input matrix $X = \{X_i; i = 1,2,\ldots,n\}$ as *n* independent records of *m* attributes let $X_i = \{x_{ij}; j = 1,2,\ldots,m\}$, and the reference vector $R = \{r_j = 1,2,\ldots,m\}$

1. The initial linear layer:

$$F_0(X) = a_0 + \sum_j b_{0j} x_{ij} \quad \text{Eq.2}$$

2. Subsequent hidden layers' CAKE node at iteration *k*:

$$F_k(X_i) = a_k + b_k F_{k-1}(X_i) + c_k G_k(X_i) + d_k H_k(X_i) \quad \text{Eq.3}$$

3. The CURNOLE node's input and transfer function at iteration *k* (*k*=1,2,…,*K*; where *K* is the maximum sequential iterations or number of hidden layers) is defined as:

(a) Specify a random set of m constant representing the ''reference'' vector *R* (default r_j=1 for all *j*=1, 2,…, *k*, *m*).

(b) For each input record X_i , compute the scalar product:

$$R \times X = \sum_j r_j x_{ij} \quad \text{Eq.4}$$

(c) Compute the length (norm) of the vector R and a record vector

$$X_i: \|R\| = \sqrt{\sum_j r_j^2}; \|X_i\| = \sqrt{\sum_j x_{ij}^2} \quad \text{Eq.5}$$

(d) Normalize $R \times X$ to compute

$$(R \times X_i)_N = (R \times X_i) = \frac{(R \times X_i)}{\|R\| \times \|X_i\|} \quad \text{Eq.6}$$

Recall that:

$$(R \times X_i)_N = (\|R\| \times \|X_i\|) \times \cos(angle(R, X_i)) \quad \text{Eq.7}$$

thus,

$$\cos(angle(R.X_i)) = \frac{(R \times X_i)}{\|R\| \times \|X\|_i} = (R \times X_i)_N \quad \text{Eq.8}$$

(e) For i=1,2,…, n; compute

$$angle(R \times X_i) = \arccos(R \times X_i)_N = \alpha_i \quad \text{Eq.9}$$

(f) Compute the transferred nonlinear component of the signal as:
$G_k(X_i) = \cos(\mu_k \times \alpha_i), H_k(X_i) = \sin(\mu_k \times \alpha_i)$ μ_k is a constant multiplier for iteration k.

(g) Replacing $G_k(X_i)$ and $H_k(X_i)$ in Equation 3 will result

$$F_k(X_i) = a_k + b_k F_{k-1}(X_i) + c_k \cos(\mu_k \times \alpha_i) + d_k \sin(\mu_k \times \alpha_i) \quad \text{Eq.10}$$

Data normalization in DAN2 can be represented by the trigonometric function$\cos(\mu_k \times \alpha_i + \theta)$ suggested by Ghiassi and Saidane (2005). At each layer vector R is rotated and shifted to minimize the resulting total error.

If the model training stops too early, the network is said to be under-trained or under-fit. An under-trained model often has high SSE values for either or both the training and validation data sets. Under-training often occurs when there are insufficient data for model fitting. DAN2 uses $\varepsilon_1 = \frac{(SSE_k - SSE_{k-1})}{SSE_k} \leq \varepsilon_1^*$ to assess existence or absence of under-training in the models (Ghiassi and Saidane, 2005).

Over-training or over-fitting is a more common problem in neural net modeling. A neural network modeler considered over-fitted (over-trained) when the network fits the in sample data well but produces poor out-of-sample results. To avoid over-fitting, Ghiassi and Saidane, (2005) divide the available in-sample data into the training and validation data sets. At each iteration k, (k>1), they compute MSE values for both the training (MSET) and validation (MSEV) sets and they use $\varepsilon_2 = |MSE_T - MSE_V|/MSE_T \leq \varepsilon_2^*$ to guard against over-fitting. The modeler should consider fully trained when the user specified accuracy criteria and the over fitting constraint are both satisfied. The accuracy levels ε_1^* and ε_2^* are problem dependent and should be determined experimentally (Ghiassi and Saidane, 2005).

According to Ghiassi and Saidane (2005), Ghiassi et al. (2005), Ghiassi et al. (2006), Ghiassi and Nangoy (2009), Ghiassi and Burnley (2010), Gomes et al. (2006), Guresen and Kayakutlu (2008a and 2008b); DAN2 gives better results than the compared models. But when the architecture is evaluated with respect to the existing ANN definitions or reported ANN properties it is easy to see that this new architecture is not and ANN and does not behave like an ANN.

First off all, the PEs (CURNOLE and CAKE) in DAN2 are not same PEs as in ANN. Actually they are a kind of divided PE whose summing function is the CAKE and the activation function is the CURNOLE.

Another problem is C nodes. C nodes have no incoming links, so such nodes must be input nodes that start a flow in ANN but they do not. They do not use any incoming flow to produce an output. For this reason C nodes are only the threshold values.

With C node as a threshold value, CAKE node as a summing function and CURNOLE nodes as the activation functions, each hidden layer is only one PE of an ANN. Thus DAN2 contradicts with the massively parallel structure because in terms of PE DAN2 is a serial structure. This causes another problem: ***fault tolerance***. As explained previously, dead of a neuron in biological neural network or failure of a PE in ANN will not cause a catastrophic failure. But what happens if a CAKE node fails to work? Clearly DAN2 cannot produce meaningful results because obtained knowledge up to that CAKE node will not go any further. The failure of a CURNOLE node will not be that catastrophic but still cause serious errors in results since they some part of the nonlinear relation in data will not be transmitted to the next layers.

When learning algorithm of DAN2 is evaluated, it is easy to see that learning in DAN2 does not modify existing architecture. DAN2 always adds new layers and calculates weights for new links, but never modifies existing weights. Besides contradicting with our new definition of ANN, DAN2 do not have ***adaptivity*** property of an ANN. As considered above an ANN can be adapted to changes in the environment like biological neural networks. Since DAN2 do not modify existing weights in the architecture, a small change in environment will turn DAN2 model into salvage. And a new DAN2 model is needed for the new environment. To explain this we will give an example as follows:

Consider a price forecasting problem of a stock. Let's have two models, a standard ANN model and a DAN2 model. Both of them are trained to forecast the future prices of the stock using daily closing values. Assume that training is done in all fair conditions and both of them gives results with desired accuracy. After a month we will have about 20 new values and probably stock price will be moved. The standard ANN model can be adjusted by retraining only with new data and afterward can be used again. But since DAN2 cannot modify existing weights, it cannot be used after a month with retraining, so a new DAN2 model must be constructed.

Flow in DAN2 is different from other ANN models since CURNOLE nodes sends to different outputs, a matrix of input vectors to the next CURNOLE node and an output value to the corresponding CAKE node. This structure contradicts with both ANNs and biological neural networks.

As mentioned above DAN2 starts with a special node that calculates the linearity with multiple linear regression. And for the hidden layers, output value of CAKE node is calculated by similar way; multiple linear regression. Thus in each linear regression equation restricts the input selection due to prior assumptions of linear regression. But normally there is no prior assumption on data for ANNs, so anything can be used as input. Not considering multiple linear regression's limitations and using anything as input with the excuse of "ANNs are black boxes" will not produce meaningful results. DAN2 uses multiple linear regression as a sub model and multiple linear regression has to produce meaningful results. So inputs of DAN2 architecture must obey the assumptions of multiple linear regression which clearly contradicts with ***input-output mapping*** property of ANNs.

Another problem with the DAN2 is that each model can only give us one output value. This is also caused by multiple linear regression because multiple linear regression equations have only one dependent variable (y). Each layer of DAN2 has only one CAKE node so from each DAN2 model we can only get one output. This is also contradicts with ***input-output mapping*** property of ANNs since ANNs can map n inputs onto m outputs. This is also noticed by Ghiassi and Burnley (2010) so they suggest a procedure for classification problems. Each time they used a new DAN2 model to decide an input whether fits to a specific class or not. By using a hierarchical procedure given in figure 2 and figure 3, they guarantee that procedure deals with fewer classes in each consequent step.

It can be observed that DAN2 structure clearly contradicts with the graph structures given in definitions and the underlying reasons such as, the necessity of PEs, the parallel distribution of them and the necessity of a learning algorithm which modifies weights.

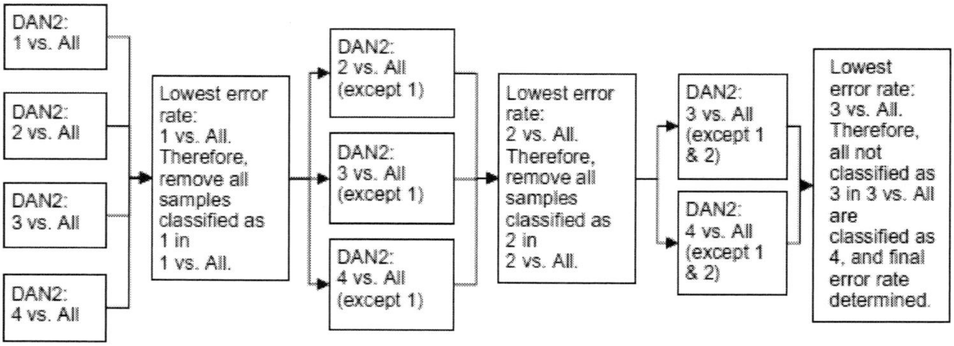

Figure 3. Hierarchical DAN2 procedure example for a four class data set. (Ghiassi and Burnley 2010).

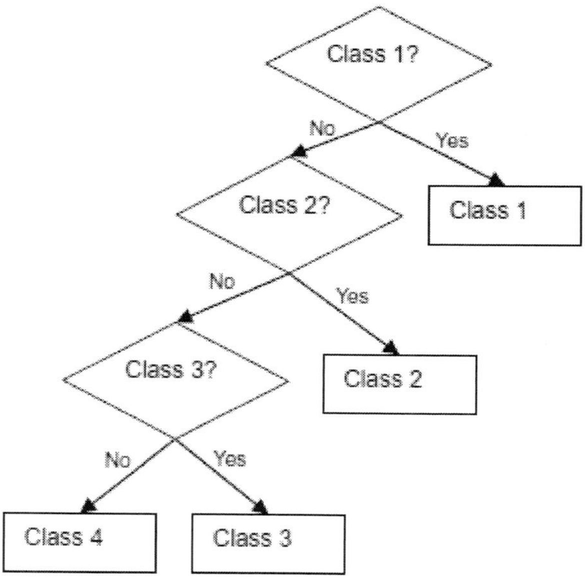

Figure 4. Tree form of hierarchical DAN2 procedure's results for a four class data set. (Ghiassi and Burnley 2010).

Why researchers do not notice problems about DAN2? There may be several answers for this question but we can summarize in the following headlines:

- There was no such clear mathematical definition in literature. Existing definitions were not aim to differentiate ANN from other methods, so many of them use pragmatic definitions. Researchers should focus to make clear definitions which can clearly differentiate defined object from other objects.
- Overall behavior of the new architecture was not fully evaluated as if it behaves like an ANN or not. Only structure or steps of algorithms are considered. Researches should evaluate overall properties of new architectures.
- Weaknesses of the new algorithm did not fully evaluated. Thus researches assumed that they can apply this structure to every kid of problem. Actually applying to new kinds of problems uncovers structural problems like given example above. Researchers should be in a search of new architecture's limitations. Finding limitations will give researchers two opportunities. First, they can focus on overcoming the limitation within the existing architecture, this will result a better architecture. Or researchers can find where the existing architecture works and where not, thus new research are areas will be revealed. In both cases they can expand the existing literature into a new level.
- Existence of computers enabled use of more calculations in algorithms. But researchers should calculate new algorithms by hand on the way of algorithm perfection. By this way researchers can easily catch algebraic problems in the steps of algorithm.

3.2 .Unfair Comparisons with Other Methods

3.2.1. Evaluation with respect to computation time

For comparing new methods or algorithms, mainly two fundamental criteria is used in literature. First one is based on comparing methods with respect to errors or with respect to the global optimum solution. The second fundamental criterion is working time on a computer. In the second one, coding perfectness will clearly effects the computation time. How can we be sure if the algorithm is perfectly coded for both alternatives? It is known that an unnecessary "for" cycle can turn the algorithm form O(n) time algorithm to O(n^2) time algorithm. Hence, evaluation with respect to computation time is a critical subject.

Complexity of an algorithm generally refers to computational step which is needed to turn the inputs into results (Gibbons, 1999). This complexity is represented by a function O(.) which is the order of the function of calculating time. Required calculating time for an algorithm is a function of quantity of the input data. In this function the biggest order term is much more important that lower order terms of the polynomial can be ignored, thus O(x) always uses the term with the biggest order in the polynomial as x. For example for a function $f(n)=3n^2+5n+6$, algorithmic complexity is denoted with O(n^2). Table 3 shows the comparison of time complexities for given problem sizes.

Table 3. Computation step required with respect to time complexity for given input size (Gibbons, 1999)

		Problem size (*n*)			
		2	8	128	1024
Time-complexity O(.)	n	2	8	128	1024
	n.logn	2	24	896	10240
	n^2	4	64	16384	1048576
	n^3	8	512	2097152	2^{30}
	2^n	4	256	2^{128}	2^{1024}
	n!	2	40320	$\sim 5x2^{714}$	$\sim 7x2^{8766}$

As it can be seen in the Table 3, we never want an algorithm to have O(n!) complexity since even for *n*=8, it takes 40320 steps to produce results. Also $O(2^n)$ is not preferable since for *n*=128, it needs 2^{128} steps to calculate results while 2^{30} steps required for an $O(n^3)$ algorithm for *n*=1024.

As mentioned above common a comparison method is based on computing time for given problems. What about time-complexity of the algorithms? Researchers should also evaluate their algorithm's time complexity. Many researchers do the coding of the algorithm by themselves. This contains a risk; computation time will increase due to lack of coding experience. Consider the following pseudo code:

```
for i=1 to i=input length, do
CALCULATION STEPS;
i=i+1;
end for;
```

The code above is a *for cycle*, which does calculation steps, until counter unit i reaches the length of inputs. This kind of code works in O(n) time (assume not depending on calculation steps) since it only goes over inputs only once. But take look at the following pseudo code:

```
for i=1 to i=input length, do

    for j=1 to j=input length, do

            CALCULATION STEPS;
            j=j+1;

    end for;

    i=i+1;
end for;
```

This code still does the calculation steps with a difference; it takes two inputs systematically and then does the calculation steps. This kind of code takes an input and for that input it takes whole input set once again which means it takes *n*x*n* inputs. Thus this code will work in $O(n^2)$ time. At this point researchers must be careful not to make such critical coding mistakes.

Researchers should pay attention to make comparison as fair as possible to get objective results. If the researcher is not much experienced with coding with the corresponding programming language, researcher should leave the coding of the algorithm to experts.

3.2.2. Input data

Some statistical models have prior assumption like *linearity*, *independency* etc. If the input data does not fit to the prior assumptions, some of the inputs will be removed from input data or the results of the model must be questioned.

Although there is no input choice limitation or prior assumption in ANN, sometimes researchers compare ANN achievements with the results of other methods using different input data and make comments on these results. An example to this kind of misuse can be Leu et al. (2009)'s research for exchange rates forecasting.

Leu et al. (2009) used three main models; Random Walk (RW), an ANN architecture Radial Basis Function Neural Network (RBFNN) and Distance-Based Fuzzy Time Series (DBFTS) to forecast NTD/USD (New Taiwan Dollar / USA Dollar).

For RW model, Leu et al. (2009) used the following general formula of random walk:

$$S_t=S_{t-1}+\varepsilon_t \qquad \text{Eq.11}$$

where S_t is the exchange rate at time *t* and ε_t is the white noise. Briefly exchange rate at time *t* is equal to exchange rate at time *t*-1 plus a white noise; further details can be found at Leu et al. (2009).

For RBFNN model, Leu et al. (2009) used statistical software package R. Leu et al.'s (2009) RBFNN model had three layers an input layer with 3 nodes, a hidden layer with 4 nodes and an output layer with one node. Although they did not mention how they choose the number of nodes in hidden layer, starting weights and learning rate not to stick on a local optimum, they mention that they use values of NTD/USD exchange rates at time *t*-3, *t*-2 and *t*-1 to predict exchange rate at time *t*.

For the DBFTS Leu et al. (2009) used eight steps to construct the model. At first step they chose candidate variables; JPY/USD, KRW/USD, CNY/USD exchanges and TAIEX (Taiwan Stock Exchange Capitalization Weighted Stock Index). They test the correlation coefficient between candidate variables and NTD/USD. Since the entire candidate variables have significant coefficients, in step two Leu et al. (2009) used principle component analysis to construct a new *factor*. Their first *factor* was the historical data of the NTD/USD exchange rate. In third step they divide universe of first and second factor into equal length intervals, whose union is the corresponding universe. Leu et al. (2009) define fuzzy sets on the universe with respect to the equal interval at step four. Then they adjust number of equal length intervals of the second factor's universe. At step six Leu et al. (2009) fuzzify the historical data set of first and second factor. Then at step seven they construct fuzzy logic relation (FLR) database with the following formulation:

$$FLR_i\text{: } A_kB_xA_lB_yA_mB_z \rightarrow A_n \qquad \text{Eq.12}$$

where FLR_i is the ith fuzzy logic relation; $A_{k,}$, A_l, A_m and A_n are the fuzzy values of the ith, $(i+1)^{th}$, $(i+2)^{th}$ and $(i+3)i^{th}$ days exchange rates respectively. B_x, B_y and B_z are the fuzzy values of the ith, $(i+1)^{th}$, $(i+2)^{th}$ days corresponding second factor value respectively. Thus they relate a day's exchange rate to the previous 3-days exchange rate and previous 3-days second factor value. By this way Leu et al. (2009) used six inputs; at time t-3, t-2, t-1; fuzzy values exchange rate and at time t-3, t-2, t-1; fuzzy values of the second factor to forecast exchange rate at time t. Leu et al. (2009) compare the result of the models with respect to mean square error (MSE) and directional symmetry (DS) and conclude that proposed DBFTS model outperformed the RW and RBFNN models.

At this point there is a comparison problem since DBFTS model used six inputs while RBFNN uses only three. And RBFNN has no restriction to use of inputs. To make objective comparison one of two things can be done. First one is about using FLR formulation as follows:

$$FLR_i: A_k A_l A_m \rightarrow A_n \qquad \text{Eq.13}$$

where FLR_i is the ith fuzzy logic relation; $A_{k,}$, A_l, A_m and A_n are the fuzzy values of the ith, $(i+1)^{th}$, $(i+2)^{th}$ and $(i+3)i^{th}$ days exchange rates respectively. Thus both DBFTS and RBFNN models will be using equivalent inputs; only previous 3-days exchange rates.

Second option is leaving the FLR formulation as it is and increasing RBFNN input nodes to six. To make inputs equivalent, previous 3 days second factor values can be used as additional inputs. Only by this way we can get an objective comparison. Of course different methods will require different types of processing on input data, like fuzzifying as in the example, but this will never be an excuse for using different types of inputs while there is no restriction on selection of input data. Researchers should pay more attention to make comparisons in fair conditions, or else results would not be reliable.

3.3. Under Evaluated Outputs

Some researchers use same input types to train the ANN and use statistical methods to determine the effect of inputs on output. However, missing application of factor analysis causes every input to have a small stand alone effect on the output. This result misleads for difference in importance of inputs. An example of this is Roh's (2007) study about forecasting KOSPI (Korea Composite Stock Price Index) 200. Roh (2007) used one ordinary ANN model and three hybrid-ANN models. In hybrid models Roh (2007) extracted new input variables from GARCH (Generalized Autoregressive Conditional Heteroscedasticity), EGARCH (Exponential GARCH) and EWMA (Exponentially Weighted Moving Average). Roh (2007) calculated statistical meanings of extracted input variables and expressed by relative contribution allocation. Roh's (2007) findings are given in Table 4.

Findings in Table 4 will lead researchers misunderstanding the results, such as the most important inputs are the new extracted ones since 14th and 15th inputs for NNEWMA has % 32, 16th and 17th inputs for NNGARCH has % 31 and 18th, 19th, 20th inputs for NNEGARCH has % 35 effect on results. But on the other side, remaining % 68, % 69 and % 65 are explained by other inputs. Although none of the remaining inputs have an effect bigger than

% 10 but with a closer look it is easy to see that some of the inputs are highly relative such as KOSPI200 yield and KOSPI200 yield square; 3-Month government bond yield and 3-Month government bond price; 1-Year government bond yield and 1-Year government bond price; KOSPI200 at *t*-1 and KOSPI 200 at *t* etc. At this point researchers should do a statistical factor analysis to clearly see which type of inputs has how much impact on results for forecasting stock indexes.

Another issue with the evaluating outputs is that researchers generalize their findings too quickly. They claim that one method clearly outperforms another one without repeating the tests many times and on many other types of data. And after many repetitions, statistical tests such as ANAVO and t-tests should be done to compare the means and variances of the errors (or results) if the methods produce statistically different results from each other.

Table 4. Input variables and relative contribution factors (Roh, 2007)

Input variables	NN	NNEWMA	NNGARCH	NNEGARCH
KOSPI200 yield square	0.0518	0.0532	0.0567	0.0393
Promised volume	0.0545	0.0452	0.0421	0.0422
KOSPI200 at *t*-1	0.0568	0.0489	0.0504	0.0516
KOSPI200 yield	0.0583	0.0626	0.0602	0.0534
3-Month government bond price	0.0596	0.0555	0.0567	0.0614
1-Year government bond yield	0.0605	0.0519	0.0561	0.0489
Open interest volume	0.0633	0.0555	0.0564	0.0528
Premium average	0.0654	0.0623	0.0693	0.0606
Contract volume	0.0667	0.0591	0.0593	0.0566
1-Year government bond price	0.0674	0.0593	0.0623	0.0576
3-Month government bond yield	0.0685	0.0556	0.0549	0.0558
KOSPI 200 at *t*	0.0731	0.0722	0.0648	0.0680
σ_{t-1}^2	0.2542	-	-	-
$\sigma_{t-1}^{2\prime} = 0.97\sigma_{t-1}^2$	-	0.2244	-	-
$\varepsilon_{t-1}^{2\prime} = 0.03\varepsilon_{t-1}^2$	-	0.0946	-	-
$\sigma_{t-1}^{2\prime} = 0.836611\,\sigma_{t-1}^2$	-	-	0.2144	-
$\varepsilon_{t-1}^{2\prime} = 0.064785\,\varepsilon_{t-1}^2$	-	-	0.0963	-
$\ln \sigma_{t-1}^{2\prime} = 0.900850 \ln \sigma_{t-1}^2$	-	-	-	0.2176
LE(leverage - effect)	-	-	-	0.0778
L(leverage)	-	-	-	0.0565
Total	1.000	1.000	1.000	1.000

4. Conclusion and Recommendations

This study is driven by a research targeting for dynamic forecasting of time series. One of the widely accepted methods in this field is ANN, which is used as a black box by the majority of the researchers. In depth analysis of literature on ANN forecasting has lead us through the advantages of this method with respect to the statistical or heuristic methods.

Nevertheless, we were confused by conflicting utilization and inconsistent findings based on different properties of ANN. In order to continue research a robust definition of all the features had to be accumulated and the observations were to be clearly demonstrated.

This chapter is prepared as a guide to researchers who would like to use ANN mathematics; benefit advantages of this method in forecasting and compare this method with other methods. That is why; each property is handled in detail by comparing the definitions by distinguished scientists. Integrating a variety of definitions features of ANN are demonstrated to included graph theory, mathematics and statistics.

Once the concept is clearly defined, examples of casual utilization or misuses were explained with references. We highly recognize the research in chosen examples and would like to add more value by the suggestions based on the ANN properties.

Scientists work on comparison of different methods in order to demonstrate respective improvements and lead young researchers for the best fit tools. In order to achieve these goals comparisons must be applied on the same data analyzed with the same statistical tools. In the field of ANN forecasting such mistreatment is also observed again based on conflicts in definition. This study points out examples of such exploitation.

Research for a dynamic forecasting model of ANN to be used for time series continues. Progressions will allow us to avoid myths and high expectations from ANN which will give freedom for improvements in hybrid methods.

It is our hope that respected ANN researchers benefit the definition and selected examples.

REFERENCES

Braspenning, P. J., Thuijsman, F. & Weijters, A. J. M. M. (1995). *Artificial Neural Networks*, Germany, Springer-Verlag Berlin Heidelberg.

DARPA *Neural Network Study* (1992). USA, AFCEA International Press.

Eberhart, R. & Shi, Y. (2007). Computational Intelligence, USA, Morgan Kaufmann.

Haykin, S. (1999). *Neural Networks: A Comprehensive Foundation*, New Jersey, USA, Prentice Hall.

Gibbons, A. (1999). *Algorithmic Graph Theory*, USA, Cambridge University Press.

Ghiassi, M. & Burnley, C. (2010). Measuring effectiveness of a dynamic artificial neural network algorithm for classification problems, *Expert Systems with Applications*, *37*, 3118-3128.

Ghiassi, M. & Saidane, H. (2005). A dynamic architecture for artificial neural networks. *Neurocomputing*, *63*, 397-413.

Ghiassi, M., Saidane, H. & Zimbra, D. K. (2005). A dynamic artificial neural network model for forecasting time series events, *International Journal of Forecasting*, *21*, 341-362.

Ghiassi, M., Zimbra, D. K. & Saidane, H. (2006). Medium term system load forecasting with a dynamic artificial neural network model, *Electric Power Systems Research*, *76*, 302-316.

Ghiassi, M. & Nangoy, S. (2009). A dynamic artificial neural network model for forecasting nonlinear processes, *Computers & Industrial Engineering, Volume 57*, Issue 1, Pages 287-297.

Gomes, G. S. S., Maia, A. L. S., Ludermir, T. B., Carvalho, F. A. T. & Araujo, A. F. R. (2006). Hybrid model with dynamic architecture for forecasting time series, *International Joint Conference on Neural Networks*, Vancouver, Canada.

Guresen, E. & Kayakutlu, G. (2008a). Forecasting stock exchange movements using artificial neural network models and hybrid models. In Proceedings of 5th *IFIP International conference on Intelligent Information Processing* (Vol. 288, pp. 129-137). Intelligent Information Processing IV; Zhongzhi Shi, E. Mercier-Laurent, D. Leake. Boston: Springer.

Guresen, E. & Kayakutlu, G. (2008b). Evaluating Artificial Neural Network Approaches Used for Time Series Forecasting, *Informs Annual Meeting*, Washington, USA.

Gurney, K. (2003). *An Introduction to Neural Networks*, London, UK, CRC Press.

Leu, Y., Lee, C. P. & Jou, Y. Z. (2009). A distance-based fuzzy time series model for exchange rates forecasting, *Expert Systems with Applications*, *36*, 8107-8114.

Muller, B., Reinhardt, J. & Strickland, M. T. (1995). *Neural Networks an Introduction,* Germany, Springer-Verlag Berlin Heidelberg.

Principe, J. C., Euliano, N. R. & Lefebvre, W. C. (1999). *Neural and Adaptive Systems: Fundamentals Through Simulations,* New York, USA, John Wiley & Sons.

Roh, T. H. (2007). Forecasting the volatility of stock price index, *Expert Systems with Applications*, *33*, 916–922.

Rojas, R. (1996). *Neural Networks: A Systematic Introduction, Germany, Springer-Verlag Berlin Heidelberg.*

In: Focus on Artificial Neural Networks
Editor: John A. Flores, pp. 191-207

ISBN: 978-1-61324-285-8

Chapter 9

EVIDENCES OF NEW BIOPHYSICAL PROPERTIES OF MICROTUBULES

***Rita Pizzi*[1], *Giuliano Strini*[2], *Silvia Fiorentini*[1], *Valeria Pappalardo*[3] *and Massimo Pregnolato*[3]**

[1] Department of Information Technology,- Via Bramante 65, Università degli Studi di Milano, 26013 Crema - Italy

[2]Department of Physics – Via Celoria 16, Università degli Studi di Milano, 20133 Milano - Italy

[3]QuantumBiolab- Pharmaceutical Chemistry Department – Viale Taramelli 12, Università degli Studi di Pavia, 27100 Pavia - Italy.

ABSTRACT

Microtubules (MTs) are cylindrical polymers of the protein tubulin, are key constituents of all eukaryotic cells cytoskeleton and are involved in key cellular functions. Among them MTs are claimed to be involved as sub-cellular information or quantum information communication systems. MTs are the closest biological equivalent to the well known carbon nanotubes (CNTs) material. We evaluated some biophysical properties of MTs through two specific physical measures of resonance and birefringence, on the assumption that when tubulin and MTs show different biophysical behaviours, this should be due to the special structural properties of MTs. The MTs, as well as CNTs, may behave as oscillators, this could make them superreactive receivers able to amplify radio wave signals. Our experimental approach verified the existence of mechanical resonance in MTs at a frequency of 1510 MHz. The analysis of the results of birefringence experiment highlights that the MTs react to electromagnetic fields in a different way than tubulin.

INTRODUCTION

Microtubules (MTs) are cylindrical protein polymers and are key constituents of all eukaryotic cells cytoskeleton. They are involved in the regulation of essential cellular functions such as the transport of materials within the cell, the movement of cytoplasmic organelles or vesicles and the cell division [1]. These filaments are constructed from αβ-tubulin heterodimers that through a process of polymerization and depolymerization will arrange to form a slightly distorted hexagonal lattice. This dynamic nature makes MTs sensitive to several pharmacological agents, i.e. some classes of anticancer agents that are able to destroy or stabilize their structure.

Several MTs biophysical characteristics have been studied in the last decade and are increasing mainly due to the close analogy that exists between MTs and carbon nanotubes (CNTs).

CNTs display a wide range of physical effects among them electronic properties are particularly attractive. In the case of MTs suitable experiments are more difficult to be performed and required expertises in both biological and physical disciplines.

The purpose of this research project is the study and evaluation of some biophysical properties of MTs through two specific physical measures of birefringence and resonance, on the assumption that when tubulin and MTs show different biophysical behaviours, this should be due to the special structural properties of MTs.

Tubulins and Microtubules

MTs are stiff cytoskeletal filaments characterized by a tubelike structure, they are also relatively fragile and more liable to break than microfilaments or intermediate-filaments. The building block of a MT is a 110-kDa heterodimeric protein said tubulin, that is the association product of two different subunits, designated α and β tubulin [2,3] and encoded by separate genes. The word tubulin always refers to the αβ heterodimer, that is usually considered as one unit, although the association is only due to non-covalent interactions. Each monomer of α and β tubulin is a compact ellipsoid of approximate dimensions 46 x 40 x 65 A° (width, height, and depth, respectively); while dimensions of αβ-heterodimer are 46 x 80 x 65 A°. Both α- and β- tubulin is composed of approximately 450 amino acids and, in spite of their sequence identity (approximately 40%), slight folding difference can be seen. The two tubulins exhibit homology with a 40,000-MW bacterial GTPase, called FtsZ, a ubiquitous protein in eubacteria and archeobacteria. Like tubulin, this bacterial protein has the ability to polymerize and participates in cell division. Perhaps the protein carrying out these ancestral functions in bacteria was modified in the course of evolution to fulfill the diverse roles of MTs in eukaryotes [4].

While many questions remain about tubulin, in 1998 Nogales et al. obtained the structure of the αβ-heterodimer at 3,7 Å resolution by electron crystallography of zinc-induced crystalline sheets of tubulin stabilized with taxol [5]. In 2001 this structures has been refined [6]. The core of each monomer contains two β-sheets of 6 and 4 strands, that are surrounded by α-helices, and a pair of globular domains set on either side of a central (core) helix H7. The monomer is a very compact structure and can be divided into three functional and sequential domains. The larger globular domain comprises the N-terminal half of the polypeptide that include the binding site for the guanosine nucleotide. The second globular domain has a binding site for Taxol on the opposite side from its contact with the nucleotide base and a predominantly helical carboxy-terminal region which probably constitutes the binding surface for motor proteins.

Calculations of the potential energy displayed that tubulin is quite highly negatively charged at physiological pH and that much of the charge is concentrated on the C-terminus of each tubulin monomer. The C-terminal end forms two long helices (H11 and H12) connected by a U-turn while the final 13 residues of α-tubulin and 9 residues of β-tubulin are too disordered in the 2D crystals to show up as electron density but are assumed to project out into the solution [7]. A detailed map of the electric charge distribution on the surface of the tubulin dimer showed that the C-termini, which extend outward, carry a significant electric charge [8]. In physiological conditions (neutral pH), the negative charge of the carboxy-terminal region causes it to remain extended due to the electrostatic repulsion within the tail. Under more acidic conditions, the negative charge of this region is reduced by association of hydrogen ions. The effect is to allow these tails to acquire a more compact form by folding.

Each tubulin heterodimers binds two molecules of guanine nucleoside phosphates (GTP) and exhibits GTPase activity that is closely linked to assembly and disassembly of MTs. One GTP-binding site is located in α-tubulin at the interfaces between α- and β- tubulin monomers; in this site GTP is trapped irreversibly and it is not hydrolyzable. The second site is located at the surface of the β-tubulin subunit; in this site GTP is bound reversibly and it is freely hydrolyzable to GDP. The GTP bound to β-tubulin modulates the addition of other tubulin subunits at the ends of the MT.

Recently important information about tubulin conformational changes during the MTs polymerization have been obtained through X-ray crystallography [9].

The general structure of MTs has been established experimentally [10,11]. MTs have been considered as helical polymers and they are built by the self-association of the αβ-heterodimer. In those polymers tubulin subunits are arranged in a hexagonal lattice which is slightly twisted, resulting in different neighboring interactions among each subunit. The polymerization occurs in a two-dimensional process that involves two types of contacts between tubulin subunits. The first process involve head-to-tail binding of heterodimers and it results in polar protofilaments that run along the length of the MT. The second process involve lateral interactions between parallel protofilaments and it complete the MT wall to form a hollow tube [12]. The longitudinal contacts along protofilaments appear to be much stronger than those between adjacent protofilaments [13].

The head-to-tail arrangement of the α- and β-tubulin dimers in a protofilament confers an overall polarity on a MT. All protofilaments in a MT have the same orientation. One end of a MT is ringed by α-tubulin and it is designed minus end because here the GTP is not exchangeable. The opposite end is ringed by β-tubulin, it is designed plus end because here the nucleotide is exchangeable. The longitudinal interactions between tubulin subunits in the protofilament seem to involve exclusively heterologous (α-β) subunits. In contrast, the lateral interactions involve predominantly homologous subunits (α-α, β-β) but heterologous interactions (α-β) occur also. When all or most lateral interactions are α-β the lattice is known as the A-lattice; instead, when all lateral contacts are α-α or β-β the lattice is known as the B-lattice.

Assembly mechanism of α- and β- tubulin gives rise *in vitro* to a variety of cylindrical structures that differ by their protofilament and monomer helix-start numbers [14-19]. In contrast, most MTs assembled *in vivo* seem to be composed of 13 protofilaments, although many exceptions have been noted in different species and cell types; for example in neurons of the nematode *Caenorhabditis elegans* some specialized MTs have 15-protofilaments [20,21]. The lengths of MTs vary but commonly reach 5-10 μm dimensions; and their diameter depends on the protofilament number. For example in the case of 13 protofilaments the tube has an outer diameter of 23 nm and an inner diameter of roughly 15 nm.

Microtubules Quantum Theories

In the last decade many theories and papers have been published concerning the biophysical properties of MTs including the hypothesis of MTs implication in coherent quantum states in the brain evolving in some form of energy and information transfer.

The most discussed theory on quantum effects involving MTs has been proposed by Hameroff and Penrose that published the OrchOR Model in 1996 [22,23].

They supposed that quantum-superposed states develop in tubulins, remain coherent and recruit more superposed tubulins until a mass-time-energy threshold, related to quantum gravity, is reached (up to 500 msec). This model has been discussed and refined for more than 10 years, mainly focusing attention to the decoherence criterion after the Tegmark critical paper of 2000 [24, 25] and proposing several methods of shielding MTs against the environment of the brain [26-28]. In the Hameroff model MTs perform a kind of quantum computation through the tubulins working like a cellular automata. The MTs interior works as an electromagnetic wave guide, filled with water in an organized collective states, transmitting information through the brain [29].

In the same years Nanopoulos et al adopted the string theory to develop a so called QED-Cavity model predicting dissipationless energy transfer along MTs as well as quantum teleportation of states at near room temperature [30-33].

The Tuszynski approach is based on the biophysical aspects of MTs. Tubulins have electric dipole moments due to asymmetric charges distribution and MTs can be modeled as a lattice of orientated dipoles that can be in random phase, ferroelectric (parallel-aligned) and an intermediate weakly ferroelectric phase like a spin-glass phase [34-36]. The model has been sustained by Faber et al [37] who considered a MT as a classical subneuronal information processor.

In 1994 Jibu and Yasue suggested that the Fröhlich dynamics of ordered water molecules and the quantizated electromagnetic field confined inside the hollow MTs core can give rise to the collective quantum optical modes responsible for the phenomenon of superradiance by which any incoherent molecular electromagnetic energy can be transformed in a coherent photon inside the MTs. These photons propagate along the internal hollow core as if the optical medium were transparent and this quantum theoretical phenomenon is called "self-induced transparency".

A decade before, applying quantum field theory (QFT), Del Giudice et al [38,39] reported that electromagnetic energy penetrating into cytoplasm would self-focus inside filaments whose diameter depend on symmetry breaking (Bose condensation) of ordered water dipoles. The diameter calculated was exactly the inner diameter of MTs (15 nm).

In any case, all phenomena occurring within the brain, both at macroscopic or microscopic level, can be related to some form of phase transition and a number of authors [40,41] pointed out the inconsistence of a quantum mechanical framework based only on traditional computational schemata. It is to be recalled, in this regard, that these schemata have been introduced to deal with particles, atoms, or molecules, and are unsuitable when applied to biological phenomena. In particular Pessa suggested that adopting a wider framework of QFT and, in particular, the dissipative version of it, relying on the doubling mechanism, we could achieve a generalization of QFT able to account for change phenomena in the biological world [42-44].

Carbon Nanotubes and Microtubules

The time required to process and transfer information faster has reached the point at which quantum effects can no longer be neglected. The electronics industry will evolve from the technology based on silicon towards innovative materials with new physical properties. These new materials include the carbon nanotubes which currently represent one of the most promising alternatives to overcome the current limits of silicon.

Currently, with a large commitment of academic and industrial scientists, the research is developing nanotubes with extremely advanced and useful properties, as they can act both as semiconductors and as superconductors. Thanks to the structure of these nanoscale materials, their properties are not restricted to classical physics, but presents a wide range of quantum mechanical effects. These may lead to an even more efficient tool for information transfer.

Quantum transport properties of CNTs has been reviewed by Roche et al [45] both from a theoretical and experimental view. Recently has been described the low-temperature spin relaxation time measurement in a fully tunable CNT double quantum dots. This is an interesting study for new microwave-based quantum information processing experiments with CNTs [46].

According to Pampaloni et al. [47] CNTs are the closest equivalent to MTs among the known nanomaterials. Although their elastic moduli are different, MTs and CNTs have similar mechanical behaviours. They are both exceptionally resilient and form large boundless with improved stiffness. Nanobiotechnology can move towards a next generation of materials with a wide range of functional properties. As suggest by Michette et al, MTs associated with carbon chemistry will allow to build complex macromolecular assemblies for sharing the exciting electronic properties of semi- and super-conductors [48].

Resonance Experiment on Microtubules

Antennas are devices capable to transform an electromagnetic field into an electrical signal, or to radiate, in the form of electromagnetic field, the electrical signal they are fed by. When powered by an electrical signal to their ends, antennas absorb energy and return it in the surrounding space as electromagnetic waves (transmitting antenna), or absorb energy from an electromagnetic wave and generate a voltage to their ends (receiving antenna). On theoretical bases any conductive object acts as an antenna, regardless of the electromagnetic wave frequency they are hit or the signal that is fed by. In particular, any tubular conductor cable, resonating mechanically, acts as a cavity antenna. The magnitude of the effect becomes significant when the frequency corresponds to the resonance frequency and in this case the output voltage can be used for receiving and transmitting radio waves. The resonance is a physical condition that occurs when a damped oscillating system is subjected to a periodic solicitation with a frequency equal to the system oscillation. A resonance phenomenon causes a significant increase in the extent of the oscillations that corresponds to a remarkable accumulation of energy within the oscillator.

Recent observations and experiments on CNTs have led to the development of an array of CNTs able to act as antennas [49]. These, instead to transmit and receive radio waves (measured in meters), due to their scale capture wavelengths at the nanoscale (measured in nanometers). In the study of the physical properties of MTs compared with those of CNTs, it is desired to search and analyze a possible reaction to microwaves, observing any ability of MTs to absorb or emit like antennas. The MTs, as well as CNTs, may behave as oscillators, this could make them superreactive receivers able to amplify the signals.

Our experimental approach was intended to verify the existence of mechanical resonance in MTs, in analogy with the CNTs, at the frequency that amplifies the wave.

Birefringence Experiment on Microtubules

Birefringence is an optical property of materials that arises from the interaction of light with oriented molecular and structural components [50]. Birefringence is the decomposition of a beam of light into two rays that occurs when the light crosses specific anisotropic media depending on the polarization of the light. The interaction between light and magnetic field in a medium results in the rotation of the plane of polarization proportional to the intensity of the magnetic field component in the direction of the beam of light (Faraday effect). By means of a polarized light and a suitable detection apparatus, it is possible to observe the associated birefringence and, therefore, the index of orientation of MTs subjected either to transverse electric fields and to transverse and longitudinal magnetic fields [51].

We performed *in vitro* experiment on different samples of MTs and tubulins, in stabilizing buffer solution, and measured the polarization under controlled conditions in order to determine different effects in the interaction of almost static electromagnetic fields. For our comparative experiments the variation of the refraction index is important because it is a function of the wavelength of the electromagnetic radiation and the nature of the crossed material.

Behavioural differences observed between samples of tubulin and MTs, would lead us to understand weather the cavity structure in the MT reacts in a peculiar way in response to specific stimuli or not.

Materials and Methods

Materials

Stabilized microtubules (MTs, #MT001-A), tubulin (#TL238), taxol (# TXD01), GTP (#BST06) and General Tubulin Buffer (# BST01) are supplied by Cytoskeleton Inc. Denver, CO. USA.

Preparation of buffer MT: MTs resuspension buffer is obtained by adding 100 μl of 2mM taxol stock in dry DMSO to 10 ml of room temperature PM buffer (15 mM PIPES pH 7.0, 1 mM MgC12). It is important to make sure that PM buffer is at room temperature as taxol will precipitate out of solution if added to cold buffer. Resuspended taxol should be stored at -20 °C.

Preparation of buffer T: GTP stock solution (100mM) is added to General Tubulin Buffer (80 mM PIPES pH 6.9, 2 mM MgC12, 0.5 mM EGTA) at a final concentration of 1mM GTP. The buffer T will be stable for 2-4 hours on ice.

Microtubules Reconstitution. 1 ml of buffer MT is added to 1 mg of lyophilized MTs and mixed gently. Resuspended MTs are left at room temperature for 10–15 minutes with occasional gentle mixing. The MTs are now ready to use. They are at a mean length of 2 μm and the tubulin concentration is 1mg/ml. MTs will be stable for 2-3 days at room temperature, although it should be noted that the mean length distribution will increase over time. MTs can be snap frozen in liquid nitrogen and stored at -70 °C.

Tubulin Reconstitution. 1 mg of lyophilized tubulin is resuspended in 1 ml of buffer T at 0-4 °C (final tubulin concentration is 1 mg/ml). The reconstituted tubulin solution is not stable and needs to be used soon after its preparation.

Microwave Generator. The bench for the MTs resonance experiment consisted of two ¼ wave dipole custom antennas centered on a frequency of 1.5 GHz. The antennas have been placed on the same horizontal plane and spaced 1.6 in. The test-tube containing the solution was placed between the antennas. The system was placed in a Mu-metal container in order to shield the measurement system from any external signal. The first antenna was connected with a shielded cable to a Polarad mod. 1105 Microwave Signal Generator (Figure 1), generating frequencies between 0.8 GHz and 2.5 GHz. The second antenna shielded cable was connected with an Avantest mod. TR4131 Spectrum Analyzer. The experiment displays changes in the resonance reference peak of the tested material. If the peak is lower the analyzed sample is absorbing, if higher it is emitting electromagnetic energy.

Polarimeter specifications. For the measurement a polarimeter was prepared. In a classic polarimeter a monochromatic source radiates a beam of light (initially not polarized) that is sent on a pair of polarized filters (normally Nicol prisms) oriented so as to polarize light. In the following, the beam of polarized light crosses a cuvette containing the test solution which, if optically active, rotates both polarization planes of light. Finally, the beam passes through a polarized filter, the analyzer, whose main section is rotatable. A more descriptive schema is depicted in the following (Figure 2).

The light source consists of a Hughes 3222HP Helium-Neon Laser, 633 nm, power 5 mW. The magnetic field is 18 Gauss RMS for the 632 Hz test cuvette and 9.8 Gauss RMS for the 610.1 Hz cuvette, while the applied electric field (632 Hz) is 1 Volt/cm RMS.

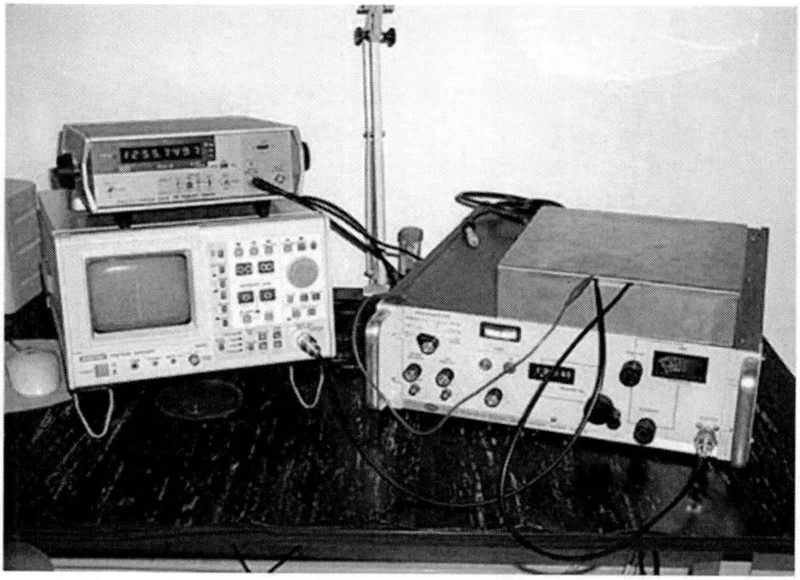

Figure 1. Microwave Signal Generator.

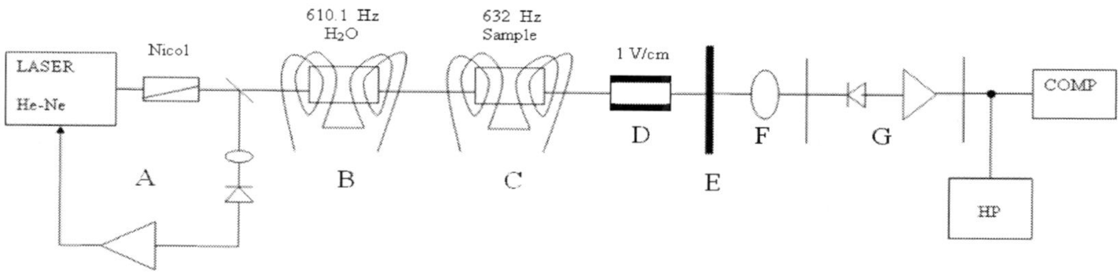

Figure 2. Scheme of the polarimeter
A: Elio-Neon Laser (Hughes 3222H-P, 633 nm; 5 mW max; Polarizing Nicol; beam splitter
B : cuvette and 610.1 Hz coil for the reference cell
C: cuvetta and 632 Hz coil for the sample
D: electric field cell
E: analyzer filter
F: lens that focuses the beam on the photodiode
G: photodiode and amplifier
HP : spectrum analyzer (HP 3582A) for on-line check
COMP : data acquisition system

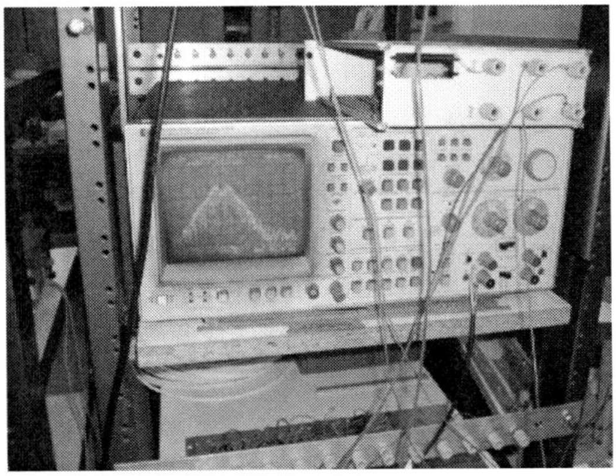

Figure 3. Spectrum analyzer HP 3582A

The cuvettes used for the magnetic field measured 15 mm, while that for the electric field was 23 mm long. The transverse electric field was achieved with simple aluminium electrodes, 3 mm far and 5 mm high. The magnetic field (longitudinal or transverse) was obtained by a pair of Helmholtz coils powered by sinusoidal generators. Electric field and transverse magnetic field were oriented according to the horizontal and the first polarizer was oriented at 45 degrees with respect to the direction of the transverse fields. The laser beam after the cuvette was examined by a polarization analyzer oriented at 45 degrees with respect to the first polarizer and finally sent to the photodiode: with this orientation the maximum signal is achievable by modulation due to the Faraday effect (longitudinal magnetic field). The photodiode was a HP 5082-4220 and the spectrum analyzer was an HP 3582A; the signal was sampled at 8000 samples/sec (Figure 3).

Signals analysis software. The analysis with Hamming windowing was performed using home-made analysis software written in Fortran at the Department of Physics (University of Milan). Other tests have been performed using the Sigview® SignalLab software and have exploited Hann and Hamming windowing, with or without Hann smoothing.

Methods

Resonance experiment. We compared the responses of samples of MTs, tubulin and buffer solutions without proteins when subjected to high frequency electromagnetic stimulations.

1. *Tubulin analysis.* The tubulin sample was prepared as previously described (see: *Materials; Tubulin Reconstitution*). 1 ml of tubulin solution was placed in a plastic test tube positioned between the transmitting and receiving antennas. In order to detect possible resonances on specific frequencies, we carried out a frequency scan between 800 MHz and 2500 MHz using a radiofrequencies generator and checking the presence of an absorption resonance, visible by means of a difference in the peak amplitude, with an Avantest TR-3130 spectrum analyzer.

2. *Microtubules analysis.* The MTs sample was prepared as previously described (see: *Materials; Microtubules Reconstitution*). 1 ml of MTs solution has been analyzed as described in the previous section (Tubulin Analysis).
3. *Microtubule buffer without MTs analysis* (see: *Materials; Preparation of Buffer MT*). 1 ml of Buffer MT been analyzed as described in the previous section (Tubulin Analysis).

Birefringence experiment. The tests were performed on solutions of tubulin and MTs, each in its own stabilizing buffer. Then we repeated the tests with tubulin in MTs buffer and with the buffer alone as control.

TUBT. *Tubulin in T buffer analysis*
The tubulin sample was prepared resuspending 0.5 mg of lyophilized tubulin in 1 ml of T buffer at 0-4 °C (final tubulin concentration is 0.5 mg/ml).

TUBMT. *Tubulin in MT buffer analysis*
The tubulin sample was prepared resuspending 0.5 mg of lyophilized tubulin in 1 ml of MT buffer at 0-4 °C (final tubulin concentration is 0.5 mg/ml).

MT. *MT Buffer analysis* (see: Materials; Preparation of MT Buffer). We analyzed 1 ml of MT buffer.

MTMT. *Microtubules in MT buffer analysis*
The MT sample was prepared as previously described (see: *Materials; Microtubules Reconstitution*) by using 0.5 mg of lyophilized MTs (final MT concentration is 0.5 mg/ml). We analyzed 1 ml of MT solution.

Each sample solution was submitted to 4 tests:

(a) Transverse electric field (1 volt/cm)
(b) Transverse magnetic field
(c) Longitudinal magnetic field
(d) No field

For each test the value displayed on the polarimeter measures directly the current in the photodiode, expressing the intensity of the laser beam after passing through the cuvette. In presence of a strong scattering, the intensity decreases. The cuvette for the magnetic field was 15 mm long, whereas that for the electric field was 23 mm long. To minimize spurious effects, the windows of the cuvettes are made of coverslip glass about 18 microns thick. The spectrum analyzer window was set to see a width of 50 Hz, within which range the frequencies of the two samples are included, the distilled water 610 Hz reference and the analyzed 632 Hz solution.

We used two cells simultaneously, a first cell was always present with a low intensity longitudinal magnetic field at 610.1Hz frequency and filled with distilled water. This allowed a reference signal in all the various measures on the second cell, excited at a 632 Hz frequency. The choice of almost static fields permitted the highest sensitivity. The frequency (632 Hz) is sufficiently low to exclude dynamic effects. An important point is that for longitudinal magnetic fields a strong Faraday effect is present due to the water contained in the analyzed solution and producing a consistent background noise.

Results and Discussion

Resonance of Microtubules

In the tubulin analysis of tubulin no significant changes have been detected in the amplitude of the signal received by the spectrum analyzer; while in the MTs analysis we observed at 1510 MHz a 0.3 dB lowering of the reference peak (absorption), and between 2060 MHz and 2100 MHz a small lowering of the reference peak (absorption). The Microtubule buffer without MT analysis gave no evidence of absorption.

The outcome of the last analysis is important; the fact that the MT buffer did not cause changes in the reference peak means that the fluctuation found in the test tube with microtubules and MT buffer depends only on the protein assembling in the tubelike structure typical of MTs.

Birefringence Results

Already at an early stage we noticed a strong response to the longitudinal magnetic field of all samples submitted to a frequency of 632 Hz, due at least in large part to the Faraday effect, while without field no reaction peaks were visible.

FFT Analysis of the Acquired Signals

In Table 1 we show the values obtained with different set-ups, normalized by the value of the control sample at 610 Hz [value (632 Hz) / value (610 Hz)] allowing a direct comparison between the analyses. All values have been multiplied by 10^5 factor. The 632 Hz signal is shown normalized for the presence of changes in measurements due to scattering, by comparing this value to the value of the 610 Hz signal of the control sample containing distilled water. The parameter choices were different for each of the four tests shown. Since the signal was sampled at 8000 Hz, the bandwidth per channel is 4000/131072 = 0.003052 Hz/channel and the transformed FFT was performed on 18 bits, or 262,144 points.

The Hann windowing is useful for analyzing transients longer than the length of the window and for general purpose applications. The Hamming windowing is very similar to the previous one; in the time domain it does not come so close to zero near the peak, as the Hann windowing does.

For the Hann window function analysis (HN) we did not use smoothing; we used instead a 15 pts smoothing (HNS) trying to remove noise without altering the possible relevant data.. The Hamming window function analysis (HM) had no smooth, while a 5 pts smoothing have been applied in HMS. We did not deepen the analyses on tubulin in tubulin buffer, since the different buffer would affect the possible comparison with the other samples. By comparing the results we observe that there are major differences in values over the third decimal place.

Table 1.

	HM	*HMS*	*HN*	*HNS*
Electric Field (EF)				
MTMT	0.0267	0.0249	0.0283	0.0238
TUBMT	0.0177	0.0175	0.0197	0.0169
MT	0.0099	0.0089	0.0123	0.0083
TUBT	0.0025			0.0018
Transverse Magnetic Field (TMF)				
MTMT	0.0810	0.0781	0.0837	0.0766
TUBMT	0.0996	0.0966	0.1018	0.0946
MT	0.0925	0.0893	0.0953	0.0872
TUBT	0.0895			0.0849
Longitudinal Magnetic Field (LMF)				
MTMT	1.828	1.7717	1.8480	1.7320
TUBMT	2.327	2.2544	2.3567	2.2025
MT	2.336	2.2628	2.3654	2.2115
TUBT	2.311			2.1883
No Field (NF)				
MTMT	0.00860	0.01069	N P	0.00389
TUBMT	0.00285	0.00135	N P	0.00088
MT	0.00585	0.00353	N P	0.00245
TUBT	0.00353			0.00112

NP: No Peak in 632 Hz
HN: Hann window function
HNS: Hann window function (smooth 15 pts)
HM: Hamming window function
HMS: Hamming window function (smooth 5 pts)
MTMT: Microtubules in microtubule buffer
TUBMT: Tubuline in microtubule buffer
MT: microtubule buffer alone
TUBT: Tubuline in tubuline buffer

Considering the relationship between the responses of the solutions in each context, we note that for all the analyses the MTs solution gave higher responses. There is a significant difference between the readings of the solution without protein, that gives values about ten times lower than that of the solution with MTs, which suggests a degree of response due to the proteins itself.

The MTs solution always shows higher values than the tubulins solution when crossed by electric field. The tubulins solution always shows larger values than the control solution when an electric field is applied. Tests with buffer alone show values equal to the tests with proteins, this suggests that there was no significant response for MTs and tubulins subjected to transverse magnetic field.

The comparison among the same tests with different windowing and smoothing highlighted the difference in the response of the MTs samples, while for the other solutions the values are virtually identical. The MTs solution has always lower value of both the tubulins solution and the solution alone when crossed by a longitudinal magnetic field. We can also observe that the solution with MTs has always a higher value if compared with the solution with tubulins and the solution alone in absence of electromagnetic field. The value of the tubulins solution results to be lower than the value of the solution alone in the cases of longitudinal magnetic field and no field.

It should be noted that the various parameterizations lead to small differences in absolute value, but substantially retain the ratio values. The uniformity of the different analysis suggests that these differences are not random or due to noise and, given this correlation, we do not need to evaluate a best choice among possible parameterizations.

Statistical Analysis

Below the statistical analysis is reported to verify possible significances.

With 8000 samples / sec run for 32 seconds, we provided more than 262,000 entries for each set-up. The analysis was performed using the paired t-test. Given the substantial equivalence between parameterizations, the analysis was performed on the significance of data processed with Hamming windowing and Hamming smoothing (5 pts). Comparisons were made on the most interesting portion of data, that includes the frequencies from 600 Hz to 650 Hz.

We compared with Paired T test the data where we had observed different behaviours (Table 2).

Table 2.

	95% CI	T-Value (P-Value)
MTMT(EF) ; TUBMT(EF)	(-1,1188; -0,9555)	-24,91 (0,000)
MTMT(EF)* ; TUBMT(EF)*	(0,000733; 0,000873)	22,53 (0,000)
MTMT(EF) ; MT(EF)	(-2,2282; -2,0130)	-38,66 (0,000)
MTMT(EF)* ; MT(EF)*	(0,000680; 0,000827)	20,12 (0,000)
TUBMT(EF) ; MT(EF)	(-1,2012; -0,9658)	-18,06 (0,000)
TUBMT(EF)* ; MT(EF)*	(-0,000105; 0,000006)	-1,76 (0,078)
MTMT(LMF) ; TUBMT(LMF)	(-0,5861; -0,3924)	-9,91 (0,000)
MTMT(LMF)* ; TUBMT(LMF)*	(0,000570; 0,000724)	16,56 (0,000)
MTMT(LMF) ; MT(LMF)	(-2,0424; -1,7779)	-28,33 (0,000)
MTMT(LMF)* ; MT(LMF)*	(0,000427; 0,000593)	12,07 (0,000)
MTMT(NF) ; TUBMT(NF)	(0,5588; 0,7656)	12,56 (0,000)
MTMT(NF)* ; TUBMT(NF)*	(0,001982; 0,002171)	43,08 (0,000)
MTMT(NF) ; MT(NF)	(-0,7297; -0,4794)	-9,47 (0,000)
MTMT(NF)* ; MT(NF)*	(0,001831; 0,002027)	38,74 (0,000)
TUBMT(NF) ; MT(NF)	(-1,3829; -1,1508)	-21,41 (0,000)
TUBMT(NF)* ; MT(NF)*	(-0,000204; - ,000091)	-5,14 (0,000)

* Normalized at 610 Hz

CI: confidence interval for mean difference

T-Value: T-Test of mean difference = 0 (vs not = 0)

EF: Electric Field ; LMF: Longitudinal Magnetic Field ; NF: No Field

Among all the tests just the Paired T for TUBMT (Electric Field) normalized at 610 Hz and MT (Electric Field) normalized at 610 Hz, which compares tubulin in microtubules buffer and buffer without cellular matter, both subjected to electric field, shows a value above the 5% threshold.

All the other comparisons show a good statistical significance, for which the P-Value is always <0.0005, suggesting that the already highlighted differences in the behaviour, allow us to draw some conclusions on the achieved results.

Conclusions and Future Developments

In this work we described the results and analysis of data collected from experiments on MTs and tubulin subjected to electromagnetic stimulations in order to observe possible different behaviours.

In the electromagnetic resonance experiment we identified a difference in the peak amplitude of the solution with MTs at a frequency of 1510 MHz, whereas the solution with tubulin and the control solution did not show any reaction. The lack of response in tubulin and control can be considered a hint that the peculiar structure of microtubules could be the cause of the observed signal.

Considering the nanoscopic size of MTs, the resonance analysis would be more effective if carried out on much higher frequencies (up to 100 GHz), with suitable instrumentation. The presence of a small but sharp resonance effect at a low frequency could be the hint of a much evident effect at higher frequencies.

The analysis of the results of birefringence experiment highlights that the MTs react to electromagnetic fields in a different way than tubulin. In particular, electric field and longitudinal magnetic field show opposite effects in the two types of proteins. Anyway in spite of the effect under electric field is the same as with no field, an unexpected and interesting effect is shown in the case of longitudinal magnetic field. The achieved results, supported by statistical significance, suggest that the tubular structure of MTs might be responsible for the different behaviour in respect to free tubulins. These preliminary positive results encourage us to continue our experimental research. In particular we will carry out a replication of the already performed tests on MTs and tubulins interacting with different ligands. It will be necessary to assess the statistical significance of possible differences in value. The experimental results will be coupled with a three-dimensional simulation of the protein folding binding different ligands, to study the emerging conformational differences. These studies would support hypotheses on the origin of the different biophysical behaviour in relationship with conformational changes. This work aims to deepen the knowledge on the behaviour of MTs and tubulin and to deduce a number of reasonable assumptions on the function of MTs as information or quantum information communication structures.

Acknowledgments

We are indebted to Dr. Fabio Costanzo and to Eng. Danilo Rossetti (Department of Information Technology, Università degli Studi di Milano) for their valuable contribution.

REFERENCES

[1] Hyams, J. S. & Lloyd, C. W. (1994). editors. *Microtubules.* Wiley-Liss, New York.

[2] Postingl, H., Krauhs, E., Little, M. & Kempf, T. (1981). Complete amino acid sequence of α-tubulin from porcine brain. *Proc. Natl. Acad. Sci.*, USA, *78*, 2757–61.

[3] Krauhs, E., Little, M., Kempf, T., Hofer-Warbinek, R., Ade, W. & Postingl, H. (1981). Complete amino acid sequence of β-tubulin from porcine brain. *Proc. Natl. Acad. Sci. USA.*, *78*, 4156–60.

[4] Lowe, J. & Amos, L. A. (1998). Crystal structure of the bacterial cell-division protein FtsZ. *Nature.*, *391*, 203–6.

[5] Nogales, E. (1998). Structure of the αβ-tubulin dimer by electron crystallography. *Letters to nature*, *391*, 192–203.

[6] Lowe, J., Li, H., Downing, K. H. & Nogales, E. (1998). Refined Structure of αβ-Tubulin at 3.5 A° Resolution. *J. Mol. Biol.*, *313*, 1045–57.

[7] Amos, L. A. & Schlieper, D. Microtubules and MAPs. *Advances in chemistry*, *Vol. 71*, 257.

[8] Tuszynski, J. A., Brown, J. A., Crawford, E. & Carpenter, E. J. (2005). Molecular Dynamics Simulations of Tubulin Structure and Calculations of Electrostatic Properties of Microtubules. *Mathematical and computer modeling.*

[9] Ravelli, R., Gigant, B., Curmi, P. A., Jourdain, I., Lachkar, S., Sobel, A. & Knossow, M. (2004). Insight into tubulin regulation from a complex with colchicine and a stathmin-like domain. *Letters to Nature*, *428*, 198–202.

[10] Amos, L. A., Amos, W. B. (1991). *Molecules of the Cytoskeleton*, First Edition, MacMillan Press, London.

[11] Chrétien, D. & Wade, R. H. (1991). New data on the microtubule surface lattice. *Biol Cell.*, *71(1-2)* , 161-74.

[12] Nogales, E., Whittaker M, Milligan R. A. & Downing, K. H. (1999). *High-Resolution Model of the Microtubule. Cell.* , 96, 79-88.

[13] Mandelkow, E. M., Mandelkow, E. & Milligan, R. A. (1991). Microtubules dynamics and microtubules caps: a time-resolved cryoelectron microscopy study. *J. Cell Biol.*, *114*, 977–91.

[14] Binder, L. I. & Rosenbaum, J. L. (1978). The in vitro assembly of flagellar outer doublet tubulin. *J. Cell Biol.*, *79*, 500-15.

[15] Burton, P. R. & Himes, R. H. (1978). Electron microscope studies of pH effects on assembly of tubulin free of associated proteins. *J. Cell Biol.*, *77(1)*, 120–33.

[16] Chrétien, D., Metoz, F., Verde, F., Karsenti, E. & Wade, R. H. (1992). Lattice defects in microtubules: Protofilament numbers vary within individual microtubules, *J. Cell Biol.*, *117(5)*, 1031-40.

[17] Linck, R. W. & Langevin, G. L. (1981). Reassembly of flagellar B (αβ) tubulin into singlet microtubules: consequences for cytoplasmic microtubule structure and assembly. *J. Cell Biol.*, *89*, 323–37.

[18] Pierson, G. B., Burton, P. R. & Himes, R. H. (1978). Alterations in number of protofilaments in microtubules assembled in vitro. *J. Cell Biol.*, *76*, 223–8.

[19] Chrétien, D. (2000). Microtubules Switch Occasionally into Unfavorable Configuration During Elongation. *J. Mol. Biol.* , *298*, 663–76.

[20] Savage, C., Hamelin, M., Culotti, J. G., Coulson, A., Albertson, D. G. & Chalfie, M. (1989). MEC-7 is a β-tubulin gene required for the production of 15-protofilament microtubules in Caenorhabditis elegans. *Genes Dev.*, *3*, 870–81.

[21] Fukushige, T., Siddiqui, Z. K., Chou, M., Culotti, J. G., Gogonea, C. B., Siddiqui, S. S. & Hamelin, M. (1999). MEC-12, an α-tubulin required for touch sensitivity in C. elegans. *Journal of Cell Science*, *112*, 395–403.

[22] Hameroff, S. R., Penrose, R. Orchestrated reduction of quantum coherence in brain microtubules: A model for consciousness. *Mathematics and Computers in Simulation*, *40*, 453–80.

[23] Hameroff, S. R., Penrose, R. (1996). Conscious events as orchestrated space-time selection. *Journal of Consciousness Studies*, *3*, 36–53.

[24] Tegmark, M. (2000). The importance of quantum decoherence in brain processes. *Phys. Rev. E.*, *61(4)*, 4194–206.

[25] Tegmark, M. (2000). Why the brain is probably not a quantum computer. *Information Science*, *128(3-4)*, 155-79.

[26] Woolf, N. J., Hameroff, S. R. (2001). A quantum approach to visual consciousness. *Trends in Cognitive Science*, *5(11)*, 472–78.

[27] Hagan, S., Hameroff, S. R. & Tuszynski, J. A. (2002). Quantum computation in brain microtubules: Decoherence and biological feasibility. *Physical Review E.*, *65*, 061901–11.

[28] Hameroff, S. R. (2007). The brain is both neurocomputer and quantum computer. *Cognitive Science*, *31*, 1035–45.

[29] Hameroff, S. R. (2007). Orchestrated Reduction of Quantum Coherence in Brain Microtubules. *NeuroQuantology*, *5(1)* , 1–8.

[30] Nanopoulos, D. V. (1995). *Theory of Brain Function, Quantum Mechanics And Superstrings*, http://arxiv.org/abs/hep-ph/9505374.

[31] Nanopoulos, D. V. & Mavromatos, N. (1996). *A Non-Critical String (Liouville) Approach to Brain Microtubules: State Vector Reduction, Memory Coding and Capacity.*, http://arxiv.org/abs/quant-ph/9512021.

[32] Mavromatos, N. (2000). *Cell Microtubules as Cavities: Quantum Coherence and Energy Transfer?* , http://arxiv.org/pdf/quant-ph/0009089.

[33] Mavromatos, N., Mershin, A. & Nanopoulos, D. V. (2002). *QED-Cavity model of microtubules implies dissipationless energy transfer and biological quantum teleportation,* http://arxiv.org/pdf/quant-ph/0204021.

[34] Tuszynski, J., Hameroff, S. R., Satarić, M. V., Trpišová, B. & Nip, M. L. A. (1995). Ferroelectric behavior in microtubule dipole lattices: implications for information processing, signaling and assembly/disassembly. *J. Theor. Biol.*, *174*, 371–80.

[35] Tuszynski, J. A., Trpišová, B., Sept, D. & Satarić, M. V. (1997). The enigma of microtubules and their self-organization behavior in the cytoskeleton. *Biosystems*, *42*, 153-75.

[36] Tuszynski, J. A., Brown, J. A. & Hawrylak, P. (1998). Dielectric Polarization, Electrical Conduction, Information Processing and Quantum Computation in Microtubules. Are They Plausible? Phil. Trans. *The Royal Society. Lond. A.*, 356(1743). 1897–926, 1998.

[37] Faber, J., Portugal, R. & Rosa, L. P. (2006). Information processing in brain microtubules. *Biosystems*, *83(1)*, 1–9.

[38] Del Giudice, E., Doglia, S., Milani, M. & Vitiello, G. (1983). Spontaneous symmetry breakdown and boson condensation in biology. *Phys. Rev. Letts*, *95A*, 508–10.
[39] Del Giudice, E., Doglia, M. & Milani, M. (1982). Self-focusing of Fröhlich waves and cytoskeleton dynamics. *Phys. Letts*, *90A*, 104–06.
[40] Pessa, E. (2007). Phase Transition in Biological Matter. Physics of Emergence and Organization. Licata I, Sakaji A, editors. *World Scientific*; 165-228.
[41] Alfinito, E., Viglione, R. & Vitiello, G. (2001). The decoherence criterium, *Modern Physics Letters B 15,127-136.*
[42] Vitiello, G. (1995). Dissipation and memory capacity in the quantum brain model. Int. *J. Mod Phys. B.*, *9(8)*, 973–89.
[43] Pessa, E. & Vitiello, G. (2004). Quantum noise induced entanglement and chaos in the dissipative quantum model of brain. *Int. J. Mod Phys. B*, *18(6)*, 841–58.
[44] Freeman, W. J. & Vitiello, G. (2006). Nonlinear brain dynamics as macroscopic manifestation of underlying many-body field dynamics, *Phys. of Life Reviews*, *3(2)*, 93–118.
[45] Roche, S., Akkermans, E., Chauvet, O., Hekking, F., Issi, J. P., Martel, R. & Montambaux, G. (2006). Poncharal Ph.. Transport Properties. Understanding Carbon Nanotubes. Loiseau A, Launois P, Petit P, Roche S, Salvetat J-P, editors. *Springer*, *677*, 335–437.
[46] Sapmaz, S., Meyer, C., Beliczynski, P., Jarillo-Herrero, P. & Knowenhoven, L. P. (2006). Excited State Spectroscopy in Carbon Nanotube Double Quantum Dots. *Nano Lett*, *6(7)*, 1350–55.
[47] Pampaloni, F. & Florin, E. L. (2008). Microtubule architecture: inspiration for novel carbon nanotube-based biomimetic materials. *Trends in Biotechnology*, *26(6)*, 302–10,.
[48] Michette, A. G., Mavromatos, N., Powell, K., Holwill, M. & Pfauntsch, S. J. (2004). Nanotubes and microtubules as quantum information carriers. *Proc. SPIE*, *522*, 5581,.
[49] Wang, Y., Kempa, K., Kimball, B., Carlson, J. B., Benham, G., Li, W. Z., Kempa, T., Rybczynski, J., Herczynski, A. & Ren, Z. F. (2004). Receiving and transmitting light-like radio waves: Antenna effect in arrays of aligned carbon nanotubes. *Applied physics letters*, *85(13)*, 2607–9.
[50] Huang, X. R. & Knighton, R. W. (2005). Microtubules Contribute to the Birefringence of the Retinal Nerve Fiber Layer. *Invest Ophthalmol Vis Sci.*, *46(12)*, 4588–93, 2005.
[51] Oldenbourg, R., Salmon, E. D. & Tran, P. T. (1998). Birefringence of Single and Bundled Microtubules. *Biophysical Journal.*, *74*, 645–54.

In: Focus on Artificial Neural Networks
Editor: John A. Flores, pp. 209-229
ISBN: 978-1-61324-285-8

Chapter 10

FORECASTING STREAM TEMPERATURE USING ADAPTIVE NEURON-FUZZY LOGIC AND ARTIFICIAL NEURAL NETWORK MODELS

Goloka Behari Sahoo *

University of California at Davis, Department of Civil and Environmental Engineering, and Tahoe Environmental Research Center, Davis, California, USA

ABSTRACT

All biological processes in water are temperature dependent. The plunging depth of stream water and its associated pollutant load into a lake/reservoir depend on stream water temperature. Lack of detailed datasets and knowledge on physical processes of the stream system limits the use of a phenomenological model to estimate stream temperature. Rather, empirical models have been used as viable alternatives. In this study, models using artificial neural networks (ANN) were examined to forecast the stream water temperature from available solar radiation and air temperature data. Observed time series data were non-linear and non-Gaussian, thus the method of time delay was applied to form the new dataset that closely represented the inherent system dynamics. Mutual information function indicates that optimum time lag was approximately 3 days. Micro-genetic algorithms were used to optimize the ANN geometry and internal parameters. Results of optimized ANN models showed that the prediction performance of four layer back propagation neural network was highest to those of other models when data are presented to the model with one-day to three-day time lag. Air temperature was found to be the most important variable in stream temperature forecasting; however, the prediction performance efficiency was somewhat higher if short wave radiation was included.

Keywords: back propagation neural network, radial basic function neural network, neuron-fuzzy logic, genetic algorithms, mutual information, multiple regression analysis, stream temperature

* Corresponding author: E-mail:gbsahoo@ucdavis.edu, Telephone: (530) 752-1755

1. INTRODUCTION

Rivers carry large amount of nutrients, suspended sediments, and contaminants into lakes, reservoirs, and ocean. Particularly, this problem is more acute when rivers empty into deep lakes. A high-density river inflow (i.e. low stream water temperature) that plunges into the hypolimnion of a lake results in different ecological consequences than a low-density river inflow (i.e. high stream water temperature) that spreads out on the lake surface [Dallimore et al., 2004]. Thus predicting the fate of a river inflow is an important concern for water quality management. Importantly, water temperature is considered as the metabolic indicator for the water body since most biological processes are temperature dependent [Mackey and Berrie, 1991]. For these reasons, several authors have investigated ways of predicting water temperatures of streams [Mackey and Berrie, 1991; Krasjewski, et al., 1982; Ozaki et al, 2003; Kinouchi et al. 2007].

A variety of approaches such as statistical models [Mackey and Berrie, 1991; Ozaki et al, 2003; Kinouchi et al. 2007; Mohseni and Stefan, 1999; Benyahya et al., 2007] and deterministic models [Lowney, 2000; Bogan et al, 2006; Caissie, et al., 2007] have been used to predict stream water temperature. The common approach for deterministic modeling is to use the equilibrium temperature concept [Mohseni and Stefan, 1999; Edinger et al. 1968; O'Driscoll et al., 2006]. Equilibrium temperature is the temperature that water reaches in response to a particular set of flow and atmospheric heat fluxes. In most cases where detailed information on the stream flow and meteorology are not available, investigators have found that air temperature is a good index of stream temperature in statistical models.

Estimation of stream water temperature at the stream mouth is important because it (1) determines the stream load plunging depth in the lake, (2) influences the lake water temperature after mixing, and (3) is the metabolic indicator of the water body. In absence of adequate field data, assumptions need to be made for the construction of conceptual or deterministic models, which can often lead to poor estimates [Maier and Dandy,1996, 1997; Sahoo and Ray, 2008a; Jain and Srinivasulu, 2004]. Therefore, there is a need to develop modeling techniques that can predict stream temperature using only the cause-effect data (i.e. causing variables: air temperature and solar radiation; and affected variable: stream temperature).

Artificial neural networks (ANNs) have been used as a viable alternative approach to physical models [Maier and Dandy, 1997; Sahoo and Ray, 2008a; Jain and Srinivasulu, 2004; Birikundavyi, 2002]. The objectives of this paper are to (1) predict daily stream water temperature using readily available air temperature and short-wave radiation by developing an adaptive neuron-fuzzy inferences system (ANFIS), micro-genetic algorithms optimized ANN models, and multiple regression analysis (MRA), and (2) demonstrate the use of optimized method for future prediction purposes.

For achieving the above objectives, three types of ANN models: back propagation neuronal network (BPNN), radial basis function network (RBFN), and adaptive neuron-fuzzy inference system (ANFIS) are employed to forecast stream temperature. ANN's optimum predictive performances highly depend on selection of network's geometry and values of modeling parameters [Sahoo and Ray, 2008b]. Thus, the BPNN and RBFN network geometry and internal parameters are optimized using a micro-genetic algorithms (μGA) technique. ANN models were developed using MATLAB version 7.1 [The MathWorks, Inc, 2005].

2. METHODOLOGY

The μGA model generates a set of solutions (model geometry and parameters) which is passed to ANN (Figure 1). ANN forms the architecture and trains the network using model parameters and input data subsets. ANN estimates a correlation coefficient (*R*) value based on a testing subset which is unused during training and passes onto μGA. Since, the objective is to reduce the difference between measured and ANN-estimated value, maximization of the *R*-value is used as the objective function. The interchange of information between μGA and ANN, named here as μGA-ANN, occurs to evolve the optimum solution set.

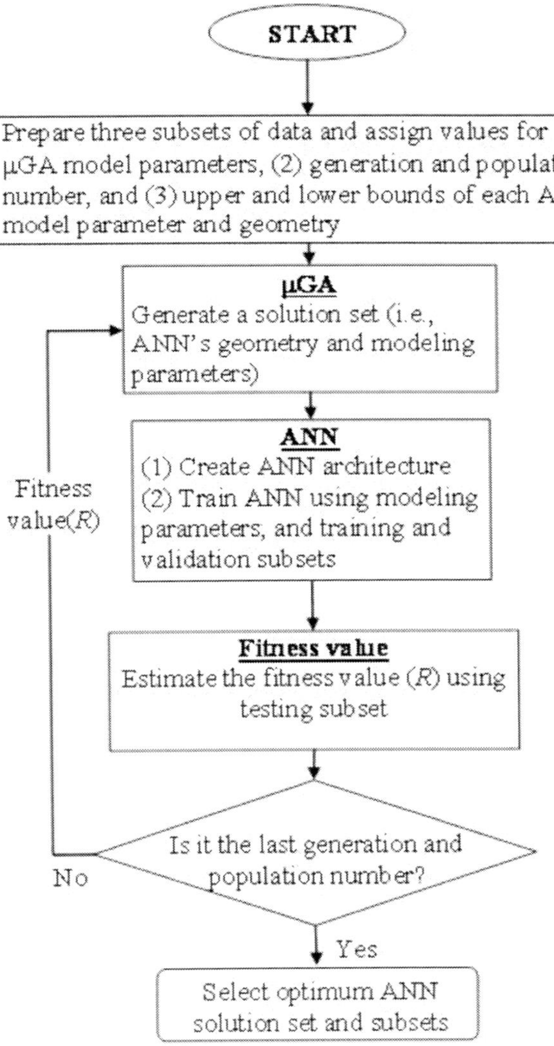

Figure 1. Flowchart for optimization of the ANN's geometry and modeling parameters.

2.1. Adaptive Neuron-Fuzzy Inferences System (ANFIS)

Adaptive neuron-fuzzy inference system combines the fuzzy qualitative approach with the neural network adaptive capabilities to achieve a desired outcome. Fuzzy set theory and fuzzy logic were established by Zadeh [1965] in order to deal with vagueness and ambiguity associated with human thinking, reasoning, cognition and perception. In this study, we used the fuzzy inference system (FIS) coupled with ANN to forecast the stream temperature. The modeling process based on ANFIS can broadly be classified in three steps.

Step 1: System Identification

The first step in system modeling is the identification of input and output variables of the ANFIS model. Fuzzy IF-THEN rules based on the Takagi-Sugeno-Kang (TSK) model [Takagi and Sugeno, 1985; Sugeno and Kang, 1988] are then formed. Fuzzy inference is a method that interprets the values in the input vector and assigns values to the output by means of some set of fuzzy "IF-THEN" rules: IF x is A THEN y is B, where A and B are fuzzy sets, e.g., "low", "high". Each fuzzy set is characterized by appropriate membership functions that map each element to a membership value between 0 and 1. The IF part (antecedent) and THEN part (consequent) of a rule can have multiple parts linked by Boolean operators (AND, OR) which have counterpart fuzzy operations (MIN, MAX).

Step 2: Determining the network structure

Once the input and output variables are identified, the neuron-fuzzy system is constructed using a six-layered network as shown in Figure 2a. The input (layer 1), output (layer 6), and node functions of each layer are explained in the subsequent paragraphs [Bateni et al, 2008].

- Layer 2 (fuzzification layer): The fuzzification layer describes the membership function of each input fuzzy set $\nu(x)$. We examined the predictive performance of ANFIS model using Gaussian, Triangular, Trapezoidal, S-shaped membership functions. ANFIS model using Gaussian membership function produced highest predictive performance than that of others. The Gaussian shaped membership function is specified by a set of two fitting parameters $\{a, b\}$ as

$$\nu(x) = \exp\left[-\left(\frac{x-a}{b}\right)^2\right] \quad \textbf{(1)}$$

  where a is the parameter determining the center, and b is the parameter determining the width of the membership function. The desired shape of Gaussian function can be obtained by proper selection of the parameters. The parameters in this layer are referred to as premise parameters.
- Layer 3 (inference layer): Each node in this layer is a fixed node and represents the IF part of a fuzzy rule. This layer aggregates the membership grades using any fuzzy intersection operators which can perform fuzzy AND operation.
- Layer 4 (normalization layer): The ith node of this layer is also a fixed node and calculates the ratio of the ith rule's firing strength in the inference layer to the sum of all the rules' firing strengths

$$\overline{w}_i = \frac{w_i}{\sum_{i=1}^{R} w_i} \tag{2}$$

where i = 1, 2,..., R and R is total number of rules. The outputs of this layer are called normalized firing strengths.

- Layer 5 (defuzzification layer): This layer aggregates the qualified consequents to produce a crisp output. The single node in this layer is a fixed node. It computes the weighted average of output signals of the output layer as

$$O = \sum_{i=1}^{R} O_i = \sum_{i=1}^{R} \overline{w}_i f_i = \frac{\sum_{i=1}^{R} w_i f_i}{\sum_{i=1}^{R} w_i} \tag{3}$$

- Layer 6 (output layer): This layer represents the THEN part (i.e. the consequent) of the fuzzy rule. The output of the model is computed as

$$O_i = \overline{w}_i f_i$$

where $\overline{w}_i$ is a normalized firing strength from layer 3 and f_i is a linear function of input variables.

Step 3: Learning algorithm and parameter tuning

The ANFIS model fine-tunes the parameters of membership functions using either the back propagation learning algorithm [Rumelhart et al., 1986] or hybrid learning rule. The back-propagation algorithm is an error-based supervised learning algorithm. The hybrid learning rule is a combination of a least squares estimator and the gradient descent method used in the back propagation method. Examining both the learning rules, we found that ANFIS using back propagation learning algorithms has advantage over other for (1) predictive performance is higher and (2) computational processing time is lower for the same epoch numbers.

2.2. Artificial Neural Network (ANN)

ANN uses a multilayered approach that approximates complex mathematical functions to process data. An ANN is arranged into discrete layers each layer consisting of at least one neuron (i.e. node). Each node of a layer is connected to nodes of preceding and/or succeeding layers but not to nodes of the same layer with a connection weight (Figure 2b). Thus, as the number of layers and nodes in each layer increases, the process becomes more complex demanding more computational effort. In general hydrologic and environmental problems are complex and require a complex ANN structure for prediction purposes. The number of layers and nodes in each layer is problem specific and needs to be optimized [Maier and Dandy, 2000; Sahoo and Ray, 2006a, 2006b].

Figure 2b depicts the flow of information from input layer to output layer (feed forward). The value of connection weight between two nodes determines the strength of the node in estimating the target. The connection weights from input-layer neuron to hidden-layer neuron and from hidden-layer neuron to output-layer neuron are w_{ij} and w_{jk}, respectively. The input- and output-layer neurons are fixed according to the input and output parameter(s) of the specific problem. The weight vectors w_{ij} and w_{jk} are randomly generated in the range between -1 and 1. Total input signal at the jth hidden-neuron is

$$v_j(n) = \sum_{i=1}^{N_I} y_i \times w_{ij} + b_j \tag{4}$$

where y_i is the value of the ith input parameter to hidden-layer neurons, b_j is the bias for the jth hidden-layer neuron, and N_I is the total input neurons. The total input signal received by the jth hidden-layer neuron, v_j, is converted to an output signal using an activation function φ_j. The output signal of the jth hidden-layer neuron is $\varphi_j[v_j(n)]$ for the nth pattern of the training dataset. The output neuron receives signals from all hidden-layer neurons and converts them to a single signal as output using an activation function $\varphi_k[.]$. Thus, the input and output signal of the kth output neuron are

$$v_k = \sum_{j=1}^{N_h} \varphi_j \left[y_j(n) \right] \times w_{jk} + b_k \tag{5}$$

And

$$y_k(n) = \varphi_k[v_k(n)] \tag{6}$$

respectively, where b_k is the bias, and N_h is the total hidden layer neurons.

The estimated output ($y_k(n)$) for nth pattern is compared with the measured output ($o_k(n)$) and an error ($e_k(n) = o_k(n) - y_k(n)$) is estimated. If the average mean square error ($1/N \sum_{n=1}^{N} 0.5\{e_k(n)\}^2$) of all patterns is greater than specified error goal [Haykin, 1999, Sahoo and Ray, 2007], then the weights are changed by an amount proportional to the difference between the desired output and the actual output. N is the total number of patterns in the training dataset. Depending on the algorithms to adjust weights, different neural networks have been developed. The process of adjusting the weights to minimize the differences between the estimated value and the actual output is called training the network. The two most commonly used neural networks: back propagation neural network (BPNN) and radial basis function network (RBFN) [Haykin, 1999; ASCE Task Committee, 2000] are used in this study. The details of BPNN and RBFN can be found in Hagan et al. [1996], Haykins [1999] and Sahoo and Ray [2007]. Brief details on functions and subroutines used or developed or modified in the current study are described below.

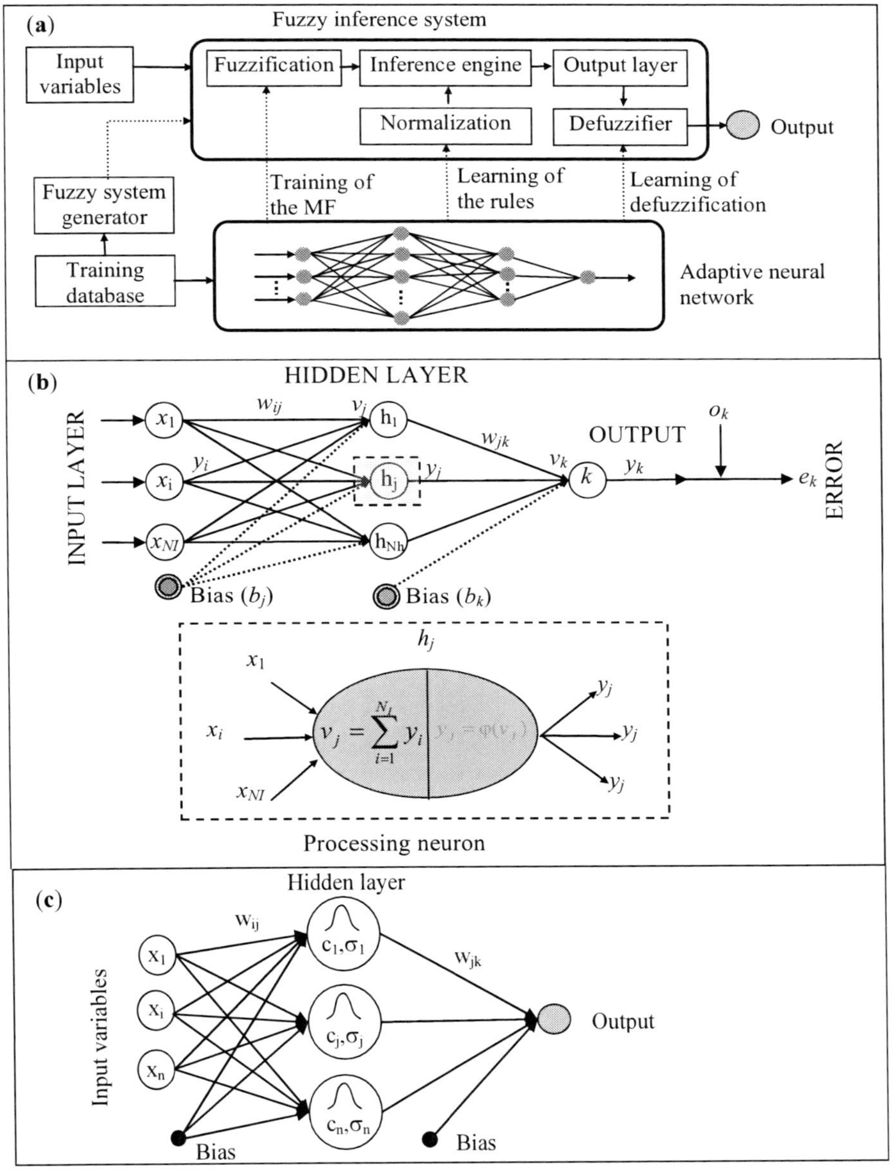

Figure 2. Topology of a (a) conceptual ANFIS (Adaptive network-based fuzzy inference system), (b) schematic of a three-layer feed-forward neural network architecture, and (c) radial basis function neural network model. For the feed-forward neural network, the connection weights from input-layer neuron to hidden-layer neuron and from hidden-layer neuron to output-layer neuron are w_{ij} and w_{jk}, respectively. The input- and output-layer neurons are fixed according to the input and output parameter(s) of the specific problem. b_j and b_k are the biases for the input and hidden layer, respectively. y_i, y_j, and y_k are the output signals of i, j, and kth nodes of input, hidden and output layer, respectively. An error e_k is measured between measured values o_k and y_k.

2.2.1. Back propagation neural network (BPNN)

The back propagation algorithm is essentially a gradient descent technique that minimizes the network error function [Haykin, 1999; ASCE Task Committee, 2000]. It involves two steps: a forward pass and a backward pass. In the forward pass, the output is calculated based on inputs and the connection weights (w_{ij} and w_{jk}) as described above. An error (e_k) is estimated at the output neuron. In the backward pass, the error at the hidden nodes is calculated by back-propagating the error at the output units through the connection weights and new connection weights, $w(\gamma+1)$, on the hidden nodes are estimated using the equation

$$w(\gamma+1) = w(\gamma) + \Delta w(\gamma+1) \tag{7}$$

where, γ is the epoch (i.e. iteration) number, Δw is the increment in the weight vector computed so as to move the weight in the direction of the negative gradient of the cost function $(-\partial F(w)/\partial w)$.

The Levenberg–Marquardt (LM) algorithm was selected for the back propagation training (i.e. estimation of Δw) because (1) it converges faster than the conventional gradient descent algorithms [Principe et al., 1999; El-Bakyr, 2003; Kisi, 2004; Cigizoglu and Kisi, 2005; Alp and Cigizoglu, 2007], (2) it does not need a learning rate and momentum factor like the gradient descent algorithms, and (3) in many cases it converges when other back propagation algorithms fail to converge [Hagan and Menhaj, 1994]. Samani et al. [2007] reported that using the Levenberg-Marquardt (LM) algorithms instead of the gradient descent algorithms for BPNN training helped reduce the convergence criterion from 10^{-3} to 10^{-9}. The optimum number of hidden nodes and hidden layers depend on the complexity of the modeling problem and the threshold value for training error goal [Fausett, 1994].

For each data pair to be learned a forward pass and a backward pass is performed. One complete forward and backward process is referred to as an iteration (also called an epoch). The process is repeated either until the error between the predicted and the measured values falls below the pre-specified error goal (a value of 10^{-20} in this study) or until the number of epochs reaches a pre-determined maximum value.

The number of nodes in each layer and epoch number were optimized using micro-genetic algorithms (μGA). Both three- or four-layered BPNN structure were examined. The transfer functions, most commonly used for hidden layer(s), are the logarithmic sigmoid (Figure 3a) and the hyperbolic tangent sigmoid (Figure 3b) [ASCE Task Committee, 2000; Maier and Dandy, 2000]. For the output layer, a linear transfer function (Figure 3c) is used so that the output range is between $-\infty$ and ∞. This avoids remapping of the outputs.

2.2.2. Radial basis function network (RBFN)

Both RBFN and BPNN are feed-forward networks, where the primary difference is in the hidden layer and training algorithms. RBFN is a three-layered network (see Figure 2c) and RBF hidden layer units have a receptive field which has a center (see Figure 3d): that is, a particular input value at which they have a maximal output.

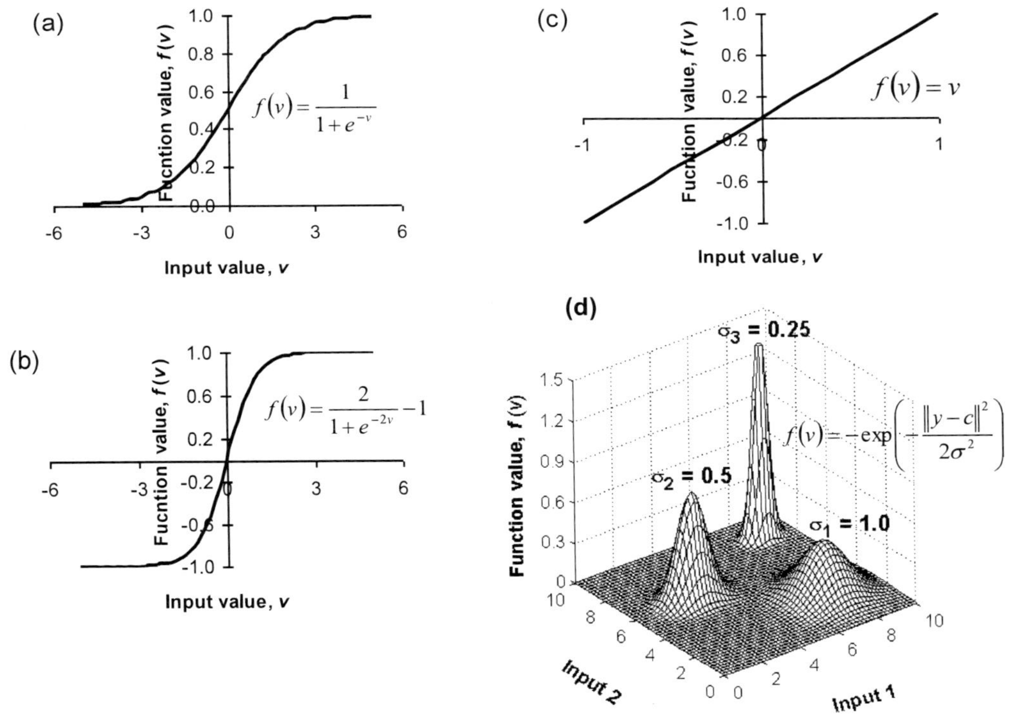

Figure 3. Transfer functions (a) logarithmic sigmoid (0 to 1) (b) hyperbolic tangent sigmoid (-1 to 1) (c) linear ($-\infty$ to $+\infty$), and (d) Gaussian type radial basis function. Shown are the responses of the RBF to three different values of spread (i.e. standard deviation) $\sigma_1 = 1.0$, $\sigma_2 = 0.50$ and $\sigma_3 = 0.25$ for the inputs value ranging between 0 to 10 at centers $c_1 = (7.5, 3)$, $c_2 = (3, 6)$, and $c_3 = (8.5, 8.5)$. Shown are the mathematical expressions of activation functions in each subfigure. $\left\| \cdot \right\|$ represents the Euclidean distance between the input value and the RBF center, c.

Of the several radial basis functions (RBFs), the most commonly used is the Gaussian RBF [ASCE Task Committee, 2000; Govindaraju and Rao, 2000; Haddadnia et al., 2003; Chang and Chen, 2003], that is described by

$$R_j = -\exp\left(-\frac{\left\| y_i - c_j \right\|^2}{2\sigma_j^2} \right) \tag{8}$$

where c_j is the center of the jth RBF neuron, R_j [i.e. $\varphi_j(\cdot)$ for the general equation 5] is the radially symmetric basis activation function, y_i is the input vector, $\left\| \cdot \right\|$ denotes a *norm* that is usually the Euclidean distance, and σ_j is the spread or the radial distance from the center of the jth RBF neuron. The standard deviation or spread (σ_j) of a radial basis neuron determines the width of an area in the input space to which each neuron responds. The function value R_j is highest at the center, c_j, and drops off rapidly to zero as the argument, y_i, moves away from the center. Thus, the neurons in the RBFN have localized receptive fields [Hagan et al., 1996]. Figure 3d shows the width of input space for three σ - values; for better vision three RBF centers are shown. The spread of a radial basis neuron determines the width of an area in the input space to which each neuron responds. Thus, the spread should be large enough so that neurons respond strongly to the overlapping regions of the input space, although too large a spread can cause numerical instability [ASCE Task Committee, 2000; Govindaraju and Rao, 2000]. It is also important that the spread of each RBF neuron should not be so large that all the neurons respond essentially in the same manner. If the spread is too large, the slope of the RBF surface becomes smoother, leading to a large area around the input vector. As a result, several neurons may respond to an input vector. On the other hand, if the spread is too small, the RBF surface becomes steep so that neurons with the weight closest to the input will have a larger output than other neurons [ASCE Task Committee, 2000; Govindaraju and Rao, 2000]. When input patterns fall within close proximity to each other, the RBF may have overlapping receptive fields. Therefore, it is important to determine the optimum value of the spread for an RBFN.

The weighted sum of the inputs at the output layer is transformed to the network output using a linear activation function. The output y_k of the RBFN is computed using the following equation [Haykin, 1999; ASCE Task Committee, 2000; Chang and Chen, 2003; Haddadnia et al., 2003]:

$$y_k = \varphi_k\left(\sum_{j=1}^{N_0} w_{jk} R_j + b_k\right) \tag{9}$$

where b_k is the bias, w_{jk} is the connection between the hidden neuron and output neuron, and N_0 is the total number of RBFN centers. First term is the weighted sum of all inputs. Since each RBFN center must respond to at least one input pattern, N_0 is always less than or equal to the total number of input patterns (N). Therefore, the total number of RBF centers (N_0) is always less than or equal to the total number of input patterns (N). Thus, setting a large value for N_0 (i.e., close to N) does not mean that the network will produce a good result because the mean square error (MSE) of the predicted and measured values during training may be low because the network might be overtrained.

RBF networks are trained by deciding on how many hidden units there should be depending on their centers and the sharpness (standard deviation) of their Gaussians training up the output layer. It starts with a minimal network (i.e. one hidden node) and grows during training by adding new hidden nodes one by one. The training cycle is divided into two phases. First, the output nodes are trained to minimize the total output error. Second, a new node is inserted in the hidden layer and connected to every node the input layer and the output layer. After a new node is added, the network is trained and an error is estimated. The addition of new node in the hidden layer is continued until either the sum of squared error falls beneath a pre-determined value (a value of 10^{-20} in this study) or N_0 equals to N. In this study the number of centers (N_0) and standard deviations (σ_j) are optimized using a μGA model. The output layer weights are then trained using the Least Mean Square (LMS) method [see Haykin, 1999; Principe et al. 1999, Sahoo and Ray, 2007].

2.3. Micro Genetic Algorithms (μGA)

We used the modified μGA [Carroll, 1999] for optimization of network geometry and internal parameters. Krishnakumar [1989] and Caroll [1996, 1999] suggested using a population size of 5. Abu-Lebdeh and Benekohal (1999) reported that μGA performs best for a population size around or above the square root of the string length. Note that each bit represents either 0 or 1 for the binary μGA. We used the population size equal to 10. Abu-Lebdeh and Benekohal [1999] reported using binary tournament selection, 0.5 uniform probability crossover rate (P_{cross}) and no mutation. Carroll [1999] suggested using a binary tournament selection with shuffling and a uniform crossover rate of 0.5, whereas Krishnakumar [1989] recommended using a crossover rate of 1.0 and a mutation rate equal to 0. Ines and Honda [2005] reported using a binary tournament selection with shuffling, a uniform P_{cross} equal to 0.5, a probability of creep mutation (P_{creep}) equal to 0.1, and 150 generations. However, Wardlaw and Sharif [1999] pointed out that the value of a uniform P_{cross} is problem specific. Thus, this study uses a binary tournament selection with shuffling. In this study, μGA was found to be robust for P_{cross} and P_{creep} equal to 0.5 and 0.1, respectively.

2.4. Multiple Regression Analysis (MRA)

The goal of multiple regression analysis is to evaluate the relationship between several independent or predictor variables and a dependent or criterion variable. This is done by fitting a straight line to a number of data points. Specifically, a line is produced so that the squared deviations of the observed points from that line are minimized. Thus this procedure is generally referred to as least squares estimation. Mathematically, if stream water temperature ($T_{w,\,t}$) is dependent on the current day air temperature ($T_{a,\,t}$) and solar radiation ($Q_{SW,\,t}$), and the previous day air temperature ($T_{a,\,t\text{-}1}$) and solar radiation ($Q_{SW,\,t\text{-}1}$) then a multiple regression model is

$$T_{w,t} = a_0 + a_1 Q_{SW,t} + a_2 T_{a,t} + a_3 Q_{SW,t-1} + a_4 T_{a,t-1} \tag{10}$$

where a_0, a_1, a_2, a_3, and a_4 are unknown coefficients that are estimated by the least squares method. MRA has been the traditional approach utilized in water resources hydrology for several decades since the last Century. Some recent applications appears in Chiang et al. [2002] and Leclerc and Ouarda [2007].

3. DATA USED

The data used in these analyses comes from the stream that flow into Lake Tahoe, CA-NV, U.S.A (Figure 4). In total, 64 streams discharge directly into Lake Tahoe, only one outflow, the Truckee River. Flows from ten streams (Upper Truckee River, Ward Creek, Trout Creek, Third Creek, Logan House Creek, Incline Creek, Glenbrook Creek, General

Creek, Edgewood Creek, and Blackwood Creek) are regularly monitored as part of the Lake Tahoe Interagency Monitoring Program (LTIMP); these 10 tributaries account for up to 40% of the total annual stream discharge into the lake. For many years temperature and water quality data were only available on an event basis with sampling frequency on the order of 25-30 times per year [Rowe et al. 2002]. However, as part of this long-term monitoring program, the U.S. Geological Survey (Carson City, NV) was able to measure time series water temperature at every 10 minutes in five streams: the Upper Truckee River, Trout Creek, Incline Creek, Glenbrook Creek, and Blackwood Creek for a shorter, 4-5 year period. Since lake clarity model needs water temperature on daily basis, the 10 minutes water temperature data were averaged to daily time scale. There is potential data gap in the time series water temperature data. In this study, continuous daily time series data from 1/1/1999 to 9/29/2002 of Upper Truckee were used for the models training, validating and testing, and comparison of their forecasting ability.

All models need at least two subsets of time series data: training (popularly known as calibration) and testing. However, ANN models need three subsets of data: training, validating and testing. The training and validating subsets are used for the network training and the testing subset is used to measure the predictive ability of the ANN model. The validating subset is used to prevent the ANN model being overtrained. The training of ANN is important for it acquires all the information from the presented dataset. Thus, the training subset should embrace the range of the whole dataset. Further, it should contain the lowest and maximum values of the dataset since ANNs are not very good predictors for the new dataset whose samples are far wide from those used during training [ASCE Task Committee, 2000]. These subsets should include information of the modeling domain (i.e. all seasons). For ANN, three subsets: training, validating, and testing were prepared using the data from 2000 to 2001, 1999, and 2002, respectively. Water temperature data for the year 2002 was used as testing subset to measure the predictive performance efficiency of all models.

4. Estimation of ANN Performance Efficiency

The performance efficiency of the network was estimated by comparing the measured and ANN-estimated values. The ANN performances used in this study are the correlation coefficient (*R*), root mean square error (RMSE), and mean square error (MSE). The mathematical expressions of *R*, RMSE, and MSE can be found in any statistics book and Haykin [1999]. In brief, the ANN predictions are optimum if *R*, RMSE, and MSE are found to be close to 1, 0, and 0, respectively. In the present study, MSE is used only for the estimation of network training performance, whereas *R* and RMSE are used to measure the prediction performance of ANN on the testing dataset, which is independent of ANN network training and validation.

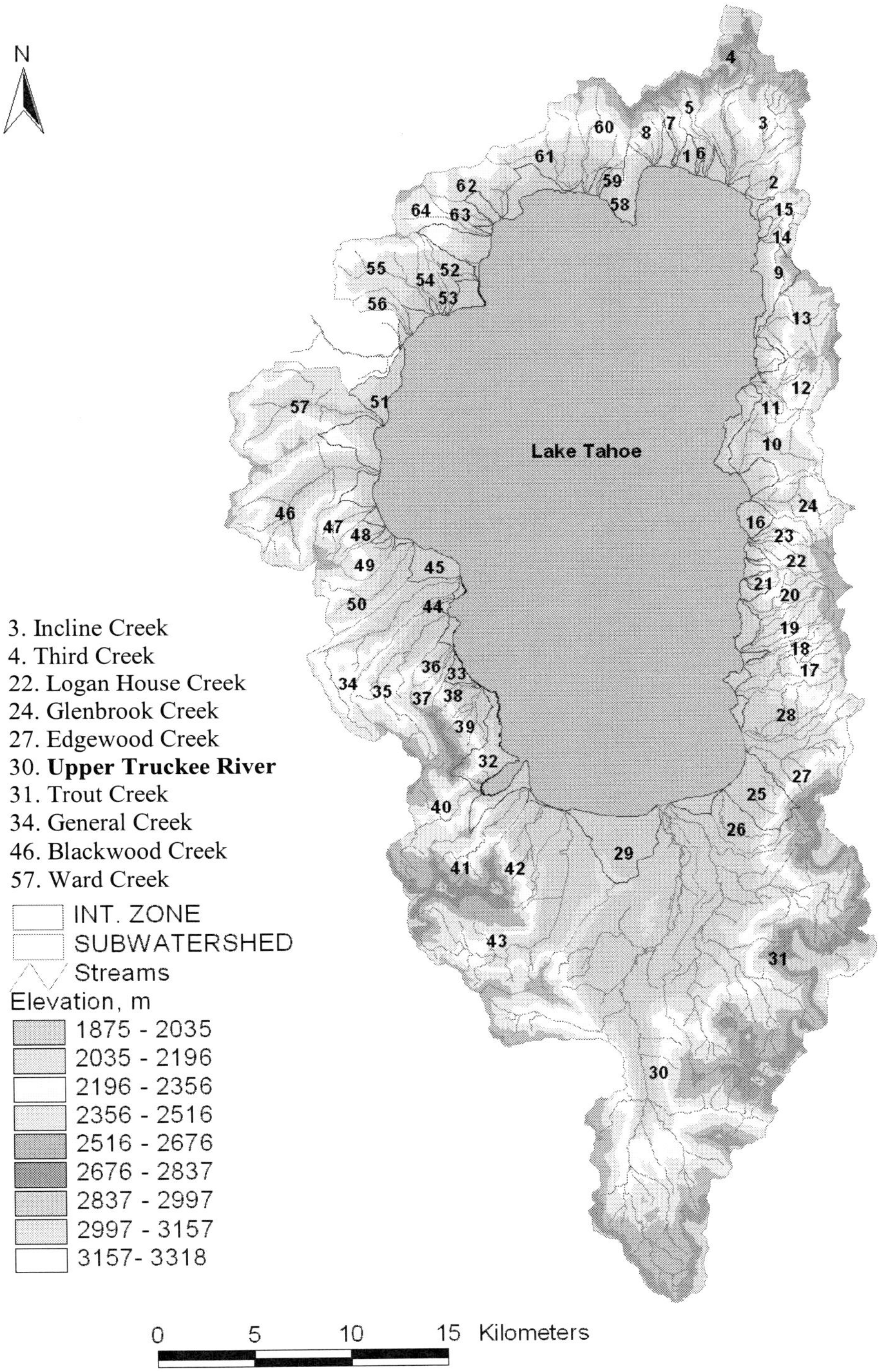

Figure 4. Lake Tahoe and locations of the all 64 streams. Name and corresponding map ID of 10 LTIMP streams are shown at the left-hand side. The stream temperature analysis was carried out on streams in bold texts. The stream without a map ID represents Truckee River, the only outflow to Lake Tahoe.

5. Results and Discussions

5.1. Data Preprocessing

Data pre-processing techniques are a powerful means of pre-structuring the problem setting of function approximation through an adaptive training procedure. In particular, integral transforms may change the nature of the training problem significantly without loss of generality if carefully selected. This provides an excellent opportunity to incorporate additional knowledge about the process in the training dataset that closely represents the inherent physical processes of the system. For example, stream temperature does not change instantaneously in response to the change of current day air temperature. Although, current day heat inputs are significant contributor for the change in water temperature, the previous days' heat inputs are also carried over to current day temperature to some extent. However, the number of days whose heat inputs influence on the current day stream temperature is not known *a priori*. Thus, data analysis was carried out using mutual information (MI).

Figure 5 presents the variation of MI against the lag time. The MI function exhibits the initial rapid decay up to a lag time of around 2 – 4 days before attaining the near saturation. It is clear that the selection of the minimum MI is difficult, as except the initial decay, the function appears decreasing very slowly as the lag time increases. As observed from Figure 5, it is difficult to select the saturation point. Thus, the initial rapid decay was selected as the lag time. However its superiority over other values is also verified by running the models for all cases.

5.2. Sensitivity Analysis of lag-time on ANN Predictive Performance

The four-layered BPNN (referred as 4BPNN), three-layered BPNN (referred as 3BPNN), and RBFN are optimized using μGA and referred here as μGA-4BPNN and μGA-RBFN, respectively. Results of three models: ANFIS, μGA-4BPNN, and μGA-RBFN are compared in Figure 6. Result of a multiple regression analysis (MRA) was compared in Figure 6 for comparison. Prediction performance efficiency of MRA was found to be the lowest to ANN models. The overall prediction performance efficiency in terms of R and RMSE of μGA-4BPNN is higher than the other three ANN models (i.e. μGA-3BPNN, μGA-RBFN, and ANFIS) and MRA for time lag less or equal to 3 days. The overall prediction performance efficiencies of μGA-4BPNN for time lag equal to 4 days were lower than those of time lag equal to 3 days. All three models predict stream temperature with R greater than 0.95. Figure 6 showed that R-value and RMSE do not significantly improve after time lag equals to 3 days.

5.3. Sensitivity Analysis of Input Parameters

Sensitivity analysis on the two input variables (air temperature and short wave radiation) influencing stream water temperature was carried out by deleting one input variable from the input data set to measure the importance of that variable over other. Since μGA-4BPNN produced the highest predictive performance efficiency, it was used to carry out the sensitivity analysis. The results presented in Figure 7 show that the predictive performance efficiency using only shortwave radiation was very low. The RMSE values of μGA-4BPNN using only shortwave radiation as input are nearly double than those of using air temperature as input. The predictive performance efficiency of the 1-day time lag increased substantially than those of the case of no time lag. In all cases, the *R*-values were found to be increasing as time lag (days) increased. However, *R*-values using only air temperature were found to be above 0.96 in all cases for time lag equal to 1 day or higher. The overall predictive performance efficiency of all streams was found to be optimum for time lag equal to 3 days. The general concept is that water temperature does not change instantaneously with a change in air temperature. Effects of previous days' heat inputs into water were carried to current day. Since stream flow travel time (headwaters to the inlet to the lake) of Upper Truckee River is nearly 2 to 3 days, applying a time lag of nearly two days was justified.

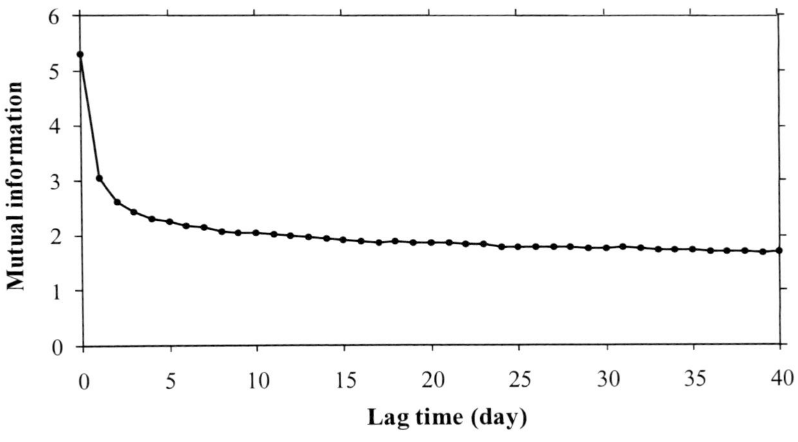

Figure 5. Mutual information of daily water temperature with air temperature.

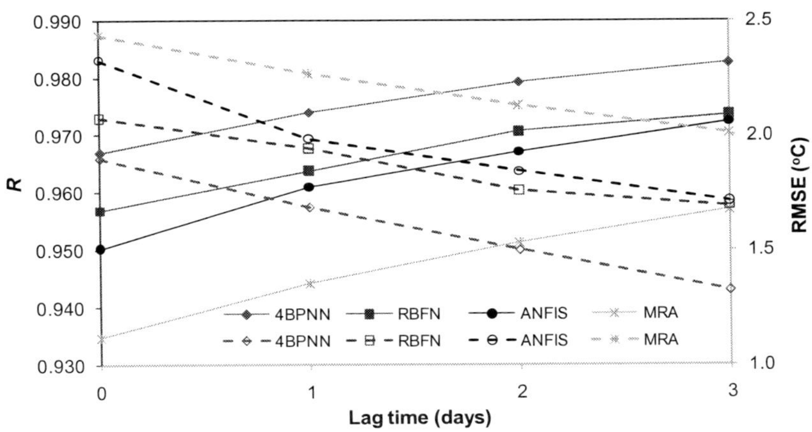

Figure 6. Comparison of prediction performance efficiency of μGA-4BPNN, μGA-RBFN, ANFIS, and multiple regression analysis (MRA) using test dataset. The solid line with solid symbol and dashed line with hollow symbol represent the *R*-values and RMSE (°C) values, respectively.

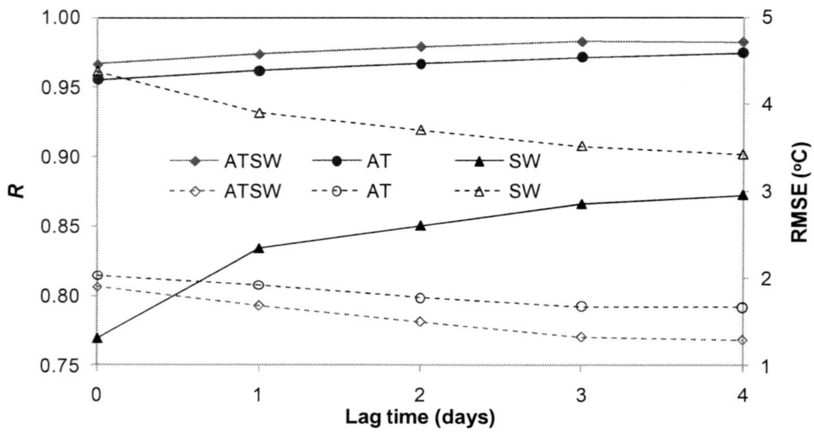

Figure 7. Prediction performance efficiency of μGA-4BPNN for different time lags (days) and for input variables: only air temperature (AT), only shortwave radiation (SW), and both air temperature and shortwave radiations (ATSW). The solid line with solid symbol and dashed line with hollow symbol represent the R-values and RMSE (°C) values, respectively.

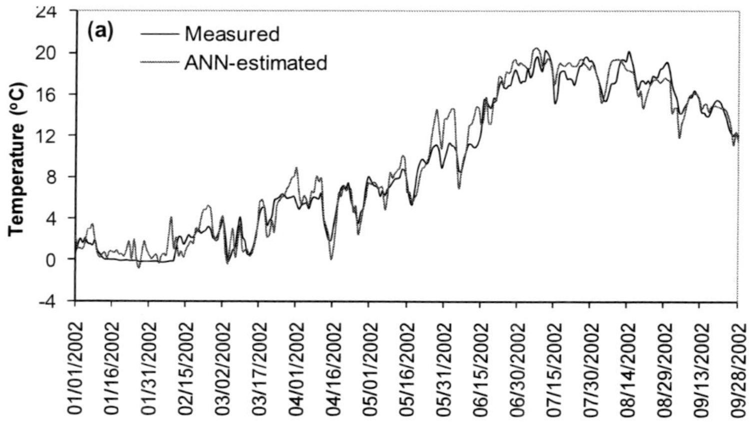

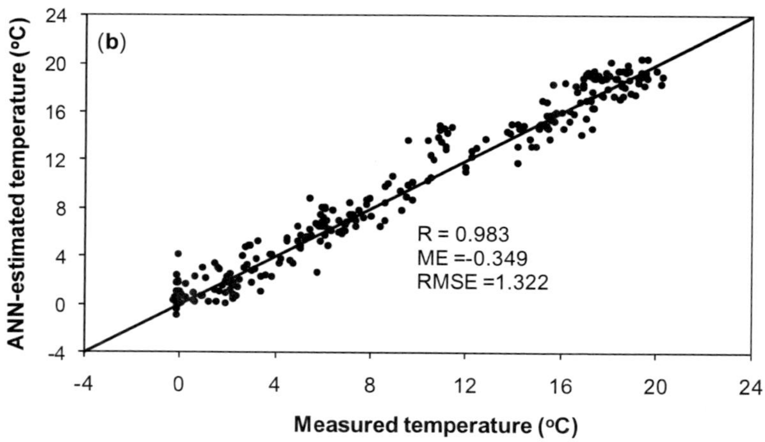

Figure 8. (a) Close of view of μGA-4BPNN estimated and measured testing data set and (b) the performance efficiency and ANN-estimated values on the 1:1 line.

The finding that air temperature alone had significant effect on stream water temperature is similar to the findings of other investigators [e.g., Mackey and Berrie, 1991; Mohseni and Stefan, 1999]. The reason for air temperature being the primary factor in stream temperature estimation is the tendency of stream water temperature to attain the equilibrium temperature. Predictive performance efficiency is higher for input dataset containing both air temperature and shortwave radiation. The results of μGA-4BPNN for time lag equal to 3 days are compared with measured stream water temperature in Figure 8. As is seen in Figure 8, the predicted values are in good agreement with the measured values.

6. Conclusions

Mutual information shows that water temperature data are highly dependent and correlated with air temperature for no time lag and the dependency exponentially decreases during initial 2-4 days as time lag increases. Results of the μGA-4BPNN show that the prediction performance is optimum for time lag equal to 3 days; however, a higher time lag (greater than 3 days) appears to be good choice to obtain higher prediction efficiency using μGA-RBFN. Prediction performance efficiency of multiple regression analysis was found to be lower than those of ANN models.

Overall, the prediction performance efficiency of μGA-4BPNN was found to be the highest among all algorithms. Comparing the predicted values of μGA-4BPNN with corresponding measured values it was found that predicted values from the μGA-4BPNN analysis were in better agreement with measured values. A sensitivity analysis showed that air temperature was found to be the most important parameter in stream temperature estimation; however, the inclusion of short wave radiation did enhance performance efficiency somewhat. Shortwave radiation alone could not predict stream temperature with acceptable accuracy.

References

Abu-Lebdeh, G. & Benekohal, R. F. (1999). Convergence variability and population sizing in Micro-Genetic Algorithms. *Computer-Aided Civil and Infrastructure Engineering, 14(5)*, 321–334.

Alp, M. & Cigizoglu, H. K. (2007). Suspended sediment load simulation by two artificial neural network methods using hydrometeorological data. *Environmental Modelling and Software, 22(1)*, 2-13.

ASCE Task Committee, 2000. Artificial neural network in hydrology. *Journal of Hydrologic Engineering, 5(2)*, 124–144.

Bateni, S. M., Mortazavi, S. M. & Jeng, D. S. (2008). Runoff forecasting using an adaptive neuron-fuzzy approach. IN: *New Topics in Water Resources Research and Management,* H.M. Andreassen, Nova Publishers, Inc, New York, 15-27.

Benyahya, L., St-Hilaire, A., Ouarda, T. B. M. J., Bobée, B. & Ahmadi-Nedushan, B. (2007). Modeling of water temperatures based on stochastic approaches: case study of the

Deschutes River. *Journal of Environmental Engineering and Science, 6*, 437–448. doi:10.1139/S06-067.

Birikundavyi, S., Labib, R., Trung, H. T. & Rousselle, J. (2002). Performance of Neural Networks in Daily Streamflow Forecasting. *Journal of Hydrologic Engineering, 7(5)*, 392–398.

Bogan, T., Othmer, J., Mohseni, O. & Stefan, H. (2006). Estimating extreme stream temperatures by the standard deviate method. *Journal of Hydrology*, *317*, 173–189.

Boughton, C., Rowe, T., Allander, K. & Robledo, A. (1997). Stream and groundwater monitoring program, *Lake Tahoe Basin, Nevada and California.* U.S. Geological Survey Fact Sheet, FS–100–97, 7.

Caissie, D., Satish, M. G. & El-Jabi, N. (2007). Predicting water temperatures using a deterministic model: Application on Miramichi River catchments (New Brunswick, Canada). *Journal of Hydrology*, *336*, 303–315.

Carroll, D. L. (1996). Genetic algorithms and optimizing chemical oxygen–iodine lasers, in Developments in Theoretical and Applied Mechanics 18, 411–424, edited by H. Wilson, R. Batra, C. Bert, A. Davis, R. Schapery, D. Stewart, and F. Swinson, *School of Engineering*, University of Alabama.

Carroll, D. L. (2006). *Fortran Genetic Algorithm (GA)* Driver version 1.7.0. Available from http://www.cuaerospace.com/carroll/ga.html (1999) (accessed during January).

Chang, F. & Chen, Y. (2003). Estuary water-stage forecasting by using radial basis function neural network. *Journal of Hydrology*, *270*, 158–166.

Chen, H. & Kim, A. S. (2006). Prediction of permeate flux decline in crossflow membrane filtration of colloidal suspension: a radial basis function neural network approach. *Desalination*, *192*, 415–428

Chiang, S., Tsay, T. & Nix, S. J. (2002). Hydrologic Regionalization of Watersheds. II: Applications. *Journal of Water Resources Planning and Management 128(1)*, 12–20, doi: 10.1061/~ASCE!0733-9496~2002!128:1~12.

Cigizoglu, H. K. & Kisi, O. (2005). Flow prediction by three back propagation techniques using k-fold partitioning of neural network training data. *Nordic Hydrology*, *36(1)*, 49-64.

Dallimore, C. J., Imberger, J. & Hodges, B. R. (2004). Modeling a plunging underflow. *Journal of Hydraulic Engineering*, *130(11)*, 1068–1076.

Edinger, J. E., Duttweiler, D. W. & Geyer, J. C. (1968). The response of water temperatures to meteorological conditions. *Water Resources Research*, *4(1)*, 1137–1143.

El-Bakyr, M. Y. (2003). Feed forward neural networks modeling for K-P interactions. Chaos, *Solitons & Fractals*, *18(5)*, 995–1000.

Fausett, L. (1994). *Fundamentals of neural networks*. Prentice-Hall, Englewood Cliffs, NJ.

Fleenor, W. E. (2001). Effects and Control of Plunging Inflows on Reservoir *Hydrodynamics and Downstream Releases*. PhD dissertation, UC Davis, CA, USA

Goldberg, D. E. (1989). *Genetic algorithms in search, optimization, and machine learning*, Addison-Wesley-Longman, Reading, Mass, USA.

Govindaraju, R. S. & Rao, A. R. (2000*). Artificial Neural Networks in Hydrology*. Kluwer Academic Publishers, Dordecht.

Haddadnia, J., Faez, K. & Ahmadi, M. (2003). A fuzzy hybrid learning algorithm for radial basis function neural network with application in human face recognition. *Pattern Recognition, 36*, 1187–1202.

Hagan, M. T. & Menhaj, M. B. (1994). Training feed forward techniques with the Marquardt algorithm. IEEE *Transactions on Neural Networks*, *5(6)*, 989–993.

Hagan, M. T., Demuth, H. P. & Beale, M. (1996). *Neural Network Design*. PWS Publishing, Boston.

Haykin, S. (1999). *Neural Networks: A Comprehensive Foundation*. Macmillan, New York.

Hejazi, M. I., Cai, X. & Ruddell, B. L. (2008). The role of hydrologic information in reservoir operation – learning from historical releases. *Advances in Water Resources*, *31*, 1636 – 1650.

Ines, A. V. M. & Honda, K. (2005). On quantifying agricultural and water management practices from low spatial resolution RS data using genetic algorithms: A numerical study fro mixed-pixel environment. *Advances in Water Resources*, *28*, 856–870.

Islam, M. N. & Sivakumar, B. (2002). Characterization and prediction of runoff dynamics: a nonlinear dynamical view. *Advances in Water Resources*, *25*, 179–190.

Jain, A. & Srinivasulu, S. (2004). Development of effective and efficient rainfall-runoff models using integration of deterministic, real-coded genetic algorithms and artificial neural network techniques. *Water Resources Research 40*, W04302, doi:10.1029 /2003WR002355.

Jingyi, Z. & Hall, M. J. (2004). Regional flood frequency analysis for the Gan-Ming River Basin in China. *Journal of Hydrology*, *296*, 98–117.

Kantz, H. & Schreiber, T. (2004). *Nonlinear Time Series Analysis,* Second Edition, Cambridge University Press.

Kim, H. S., Eykholt, R. & Salas, J. D. (1998). Delay Time Window and Plateau Onset of the Correlation Dimension for Small Data Sets. *Physica Review E*, *58(5)*, 5676-5682.

Kim, H. S., Eykholt, R. & Salas, J. D. (1999). Nonlinear Dynamics, Delay Times, and Embedding Windows. Physica D Nonlinear Phenomena, *Physica*, D127, 48-60.

Kinouchi, T., Yagi, H. & Miyamoto, M. (2007). Increase in stream temperature related to anthropogenic heat input from urban wastewater. *Journal of Hydrology, 335*, 78-88.

Kisi, Ö. (2004). Multi-layer perceptrons with Levenberg–Marquardt training algorithm for suspended sediment concentration prediction and estimation, *Hydrological Sciences Journal, 49(6)*, 1025–1040.

Krasjewski, W. F., Kraszewski, A. K. & Grenney, W. J. (1982). A graphical technique for river water temperature predictions. *Ecological Modelling*, *17*, 209–224.

Krishnakumar, K. (1989). Micro-genetic algorithms for stationary and non-stationary function optimization, SPIE: *Intelligent Control and Adaptive Systems*, *1196*, 289–296, Philadelphia, PA.

Leclerca, M. & Ouarda, T. B. M. J. (2007). Non-stationary regional flood frequency analysis at ungauged sites. *Journal of Hydrology 343*, 254–265, doi:10.1016/j.jh ydrol.2007.06.021.

Lowney, C. L. (2000). Stream temperature variation in regulated rivers: Evidence for a spatial pattern in daily minimum and maximum magnitudes. *Water Resources Research*, *36(10)*, 2947–2955.

Mackey, A. P. & Berrie, A. D. (1991). The prediction of water temperatures in chalk streams from air temperatures. *Hydrobiologia*, *210*, 183–189.

Maier, H. R. & Dandy, G. C. (1996). The use of artificial neural networks for the prediction of water quality parameters. *Water Resources Research*, *32*, 1013–1022.

Maier, H. R. & Dandy, G. C. (1997). Author's reply to comments by Fortin, V., Ouarda, T.B.M.J. and Bobee, B. on "The use of artificial neural networks for the prediction of water quality parameters" by Maier, H.R., Dandy, G.C.. *Water Resources Research, 33(10)*, 2425–2427.

Maier, H. R. & Dandy, G. C. (2000). Neural networks for the prediction and forecasting of water resources variables: a review of modelling issues and applications. *Environmental Modelling and Software, 15(1)*, 101–124.

Manrique, D., Ríos, J. & Rodríguez-Patón, A. (2006). Evolutionary System for Automatically Constructing and Adapting Radial Basis Function Networks. *Neurocomputing, 69*, 2268–2283.

Marquardt, D. (1963). An algorithm for least squares estimation of non-linear parameters, *J. Soc. Ind. Appl. Math., 11(2)*, 431–441.

Massei, N., Dupont, J. P., Mahler, B. J., Laignel, B., Fournier, M., Valdes, D. & Ogier, S. (2006). Investigating transport properties and turbidity dynamics of a karst aquifer using correlation, spectral, and wavelet analysis. *Journal of Hydrology, 329*, 244-257, doi:10:10.1016/j.jhydrol.200602.021.

Mohseni, O. & Stefan, H. G. (1999). Stream temperature/air temperature relationship: a physical interpretation. *Journal of Hydrology, 218*, 128–141.

O'Driscoll, M. A. & DeWalle, D. R. (2006). Stream-air temperature relations to classify stream-ground water interactions in a karst setting, central Pennsylvania, USA. *Journal of Hydrology, 329*, 140–153.

Ozaki, N., Fukushima, T., Harasawa, H., Toshiharu, K., Kawashima, K. & Ono, M. (2003). Statistical analyses on the effects of air temperature fluctuations on river water qualities. *Hydrological Processes, 17*, 2837–2853.

Poole, G. C. & Berman, C. H. (2001). An ecological perspective on in-stream temperature: Natural heat dynamics and mechanisms of human-caused thermal degradation. *Environmental Management, 27(6)*, 787–802.

Principe, J. C., Euliano, N. R. & Lefebvre, W. C. (1999). Neural and adaptive systems: fundamentals through simulations. John Wiley & Sons, New York.

Ramírez, M. C. V., Velho, H. F. C. & Ferreira, N. J. (2005). Artificial Neural network technique for rainfall forecasting applied to the São Paulo region. *Journal of Hydrology, 301*, 146–162.

Ray, C. & Klindworth, K. K. (2000). Neural Networks for agrichemical vulnerability assessment of rural private wells, *Journal of Hydrologic Engineering, 5(2)*, 162–171.

Roberts, D. M. & Reuter, J. E. (2007). *Lake Tahoe total Maximum Daily Load, Technical Report CA-NV*, California Regional Water Quality Control Board, Lahontan Region, CA, USA.

Rowe, T. G., Saleh, D. K., Watkins, S. A. & Kratzer, C. R. (2002). Streamflow and Water Quality Data for Selected Watersheds in the Lake Tahoe Basin, California and Nevada, through September 1998. U.S. *Geological Survey Water Resources Investigations Report*, 02–4030, Carson City, NV. 117.

Rumelhart, D. E., Hinton, G. E. & Williams, R. J. (1986). Learning representations by back-propagation error. *Nature, 323*, 533-536.

Sahoo, G. B. & Ray, C. (2006a). Flow forecasting for a Hawaiian stream using rating curves and neural networks. *Journal of hydrology, 317*, 63–80.

Sahoo, G. B. & Ray, C. (2006b). Predicting Flux Decline in Cross-Flow Membranes using Artificial Neural Networks and Genetic Algorithms. *Journal of Membrane Science*, *283*, 147-157.

Sahoo, G. B. & Ray, C. (2007). Reply to comments made by W. Sha on "Flow forecasting for a Hawaii stream using rating curves and neural networks", *Journal of Hydrology*, *340*, 122–127, doi:10.1016/j.jhydrol.2007.04.004.

Sahoo, G. B. & Ray, C. (2008a). Flow Forecasting Using Artificial Neural Network and a Distributed Hydrological Model, MIKE SHE. *New Topics in Water Resources Research and Management* (Henrik M. Andreassen (eds), NOVA Publishers, New York, 315–333.

Sahoo, G. B. & Ray, C. (2008b). Micro-genetic algorithms and artificial neural networks to assess minimum data requirements for prediction of pesticide concentrations in shallow groundwater on a regional scale. *Water Resources Research*, *44*, W05414, doi:10.1029/2007WR005875

Samani, N., Gohari-Moghadam, M. & Safavi, A. A. (2007). A simple neural network model for the determination of aquifer parameters. *Journal of Hydrology*, *340*, 1–11.

Shi, D., Yeung, D. S. & Gao, J. (2005). Sensitivity Analysis Applied to the Construction of Radial Basis Function Networks. *Neural Networks*, *18(7)*, 951–957.

Sugeno, M. & Kang, G. T. (1988). Structure identification of fuzzy models. *Fuzzy Sets System*, *28*, 15-33.

Swift, T. J., Perez-Losada, J., Schladow, S. G., Reuter, J. E., Jassby, A. D. & Goldman, C. R. (2006). Water Quality Modeling in Lake Tahoe: linking suspended matter characteristics to Secchi Depth. *Aquatic Sciences*, *68*, 1–15.

Tahoe Environmental Research Center, (2007). *State of the Lake Report*, Tahoe Environmental Research Center, UC Davis, CA, USA.

Takagi, M. T. & Sugeno, M. (1985). Fuzzy identification of systems and its applications to modeling and control, *IEEE Transactions on Systems, Man, and Cybernetics*, *15*, 116-132.

Takens, F. (1981). Detecting strange attractors in turbulence. In: DA; Rand, LS. Young, (Eds.). *Dynamic Systems and Turbulence. Lecture Notes in Mathematics 898*. Springer, Berlin, 366-381.

The MathWorks, Inc., (2005). *MATLAB version 7.1*, 3 Apple Hill Drive, Natick, Massachusetts, USA.

The MathWorks, (2005). Inc.: *MATLAB version 7.1, 3* Apple Hill Drive, Natick, Massachusetts, USA.

Wang, W., Pieter, H. A. J. M., Gelder, V., Vrijling, J. K. & Ma, J. (2006). Forecasting daily streamflow using hybrid ANN models. *Journal of Hydrology*, *324*, 383–399, doi:10.1016/j.jhydrol.2005.09.032.

Wardlaw, R. & Sharif, M. (1999). Evaluation of Genetic Algorithms for Optimal Reservoir System Operation. *Journal of Water Resource Planning and Management*, *125(1)*, 25–33.

Zadeh, L. A. (1965). Fuzzy sets. *Information Control, 8*, 338-353.

In: Focus on Artificial Neural Networks
Editor: John A. Flores, pp. 231-256
ISBN: 978-1-61324-285-8

Chapter 11

NEURAL NETWORK APPLICATIONS IN MODERN INDUCTION MACHINE CONTROL SYSTEMS

Dinko Vukadinović and Mateo Bašić
University in Split, Faculty of Electrical Engineering, Mechanical Engineering and Naval Architecture, Croatia

ABSTRACT

This chapter shows an overview of neural network applications in modern induction machine control systems. Induction motors have been used as the workhorse in industry for a long time due to their being easy to build, highly robust, and having generally satisfactory efficiency. In addition, induction generators play an important role in renewable energy systems such as energy systems with variable-speed wind turbines. The induction machine is a nonlinear multivariable dynamic system with parameters that vary with temperature, frequency, saturation and operating point. Considering that neural networks are capable of handling time varying nonlinearities due to their own nonlinear nature, they are suitable for application in induction machine systems. In this chapter, the use of artificial neural networks for identification and control of induction machine systems will be presented. An overview of neural network applications in induction machine control systems will be the focus:

1. Drive feedback signal estimation,
2. Inverter control,
3. Identification of machine parameters,
4. Neural network based approaches for the efficiency improvement in induction machine systems,
5. Neural network implementations by digital signal processors and ASIC chips.

1. INTRODUCTION

Induction motors play an important part in industry. Their popularity is due to their being easy to build, highly robust, and having generally satisfactory efficiency [1]. In addition, squirrel cage induction motors are, by far, the most commonly used type of induction motor and, hence, are the subject of our analysis. Induction motors are often the preferred choice in variable-speed drives. Furthermore, these drives mandatorily include power electronic inverters. Voltage source inverters (VSIs) based upon insulated gate bipolar transistors (IGBTs) have a dominant position in the AC drives market, in the power range up to 200 kW. In addition, VSIs are unavoidable in control systems of stand-alone induction generators. In contrast, thyristor-based current source inverters (CSIs) continue to have a role for single induction motor drives up to 3 MW [2].

When an induction machine operates as a motor, the rotor speed, ω_r, is lower than the synchronous speed, ω_e. The difference of these speeds, called a slip and given by

$$\omega_s = \omega_e - \omega_r \tag{1}$$

is then positive.

An equivalent circuit that is normally adopted for analyzing the motor speed characteristics at steady state is represented in Figure 1.

The speed-torque relationship of the induction motor can be found using the steady-state equivalent circuit shown in Figure 1. The developed torque that controls the behavior of the motor under different operating conditions is given by

$$T_e = \frac{3}{2} p \frac{k_s^2}{\sigma L_r T_r^{'}} \left(\frac{U_s}{\omega_e} \right)^2 \frac{\omega_s}{\left(\frac{1}{T_r^{'}} + \frac{\omega_s}{\omega_e T_s^{'}} \right)^2 + \left(\omega_s - \frac{\sigma}{\omega_e T_s^{'} T_r^{'}} \right)^2} \tag{2}$$

where:

$$k_s = \frac{L_m}{L_s}, T_s^{'} = \frac{\sigma L_s}{R_s}, T_r^{'} = \frac{\sigma L_r}{R_r},$$

p is the number of pole pairs, σ is the total leakage factor, L_s, L_r, L_m are, respectively, the stator, rotor and mutual inductances, the stator resistance is R_s and the rotor resistance is R_r.

The electromagnetic torque in an induction motor is dependent on the supply voltage (U_s) and the motor parameters. The rotational speed of the induction motor ω_r, as well as its electromagnetic torque T_e, may be controlled by several techniques, such as scalar, vector or direct torque, to meet the load requirements.

For constant torque operation, the voltage/frequency ratio (U_s/f_e) is maintained almost constant from starting to the maximum rated synchronous speed. In order to ensure constant pull-out torque at low speed, the inverter voltage is boosted to compensate for the stator winding's ohmic losses and to establish the required starting torque at low frequency [1]. This type of control is usually referred to *as scalar control*.

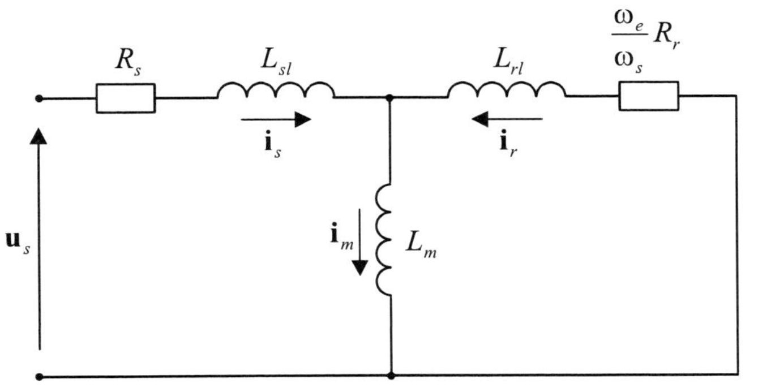

Figure 1. Per-phase equivalent circuit of induction motor with iron losses neglected.

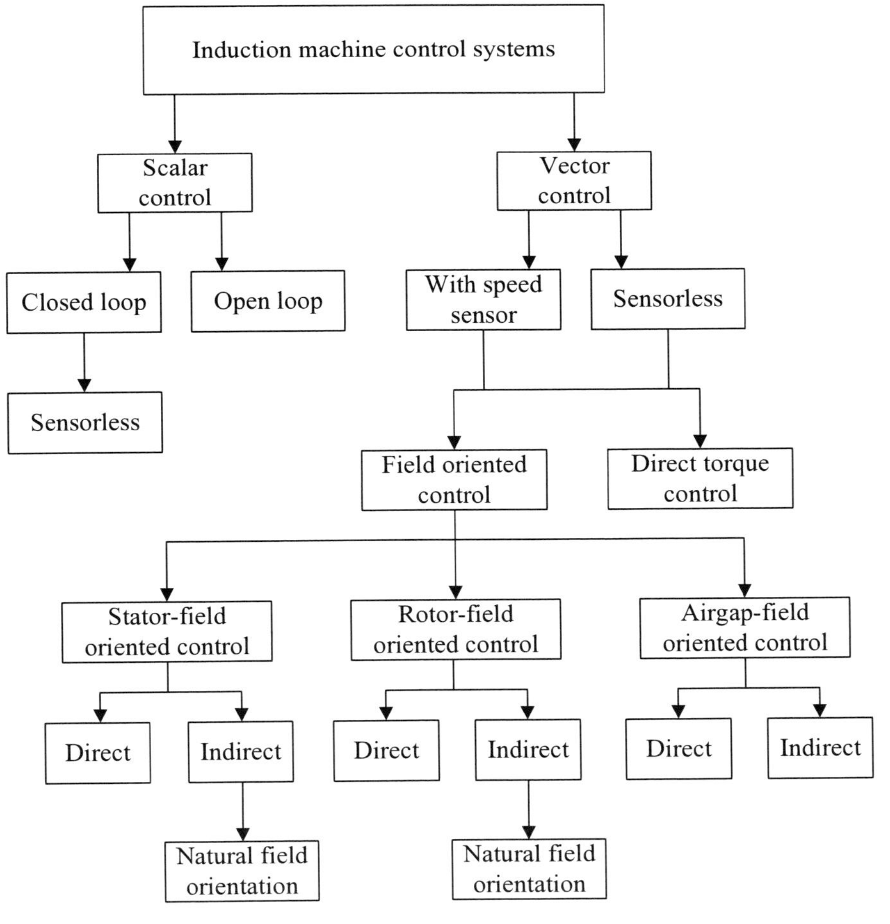

Figure 2. Induction machine control systems.

Field-oriented or vector control techniques use the space vector model of the induction motor to accurately control the speed and torque both in steady-state operation and in dynamic operation. The dynamic performance achieved by field-oriented techniques equals the dynamic performance offered by DC motor drives. In fact, field-oriented techniques ensure decoupled flux and torque control of an induction machine by controlling the instantaneous phase current values, with respect to the chosen flux linkage space-vector. The class of vector control strategies encompasses field-oriented control methods and direct torque control methods. Field-oriented control methods use the rotor field-oriented reference frame, the airgap field-oriented reference frame or the stator field-oriented reference frame (see Figure 2). In each case, the reference frame real axis (*d* axis) is oriented along the direction indicated by the corresponding magnetic flux. The rotor field-oriented vector control is the most commonly applied technique because it has the simplest structure and generates very fast transient response.

In classical induction machine control systems, knowledge of the controlled system in the form of a set of algebraic and differential equations is required. This set of equations, written in a synchronously rotating (d,q) reference frame is as follows [3]:

$$\mathbf{u}_s = \mathbf{i}_s R_s + \frac{d\boldsymbol{\psi}_s}{dt} + j\omega_e \boldsymbol{\psi}_s \tag{3}$$

$$0 = \mathbf{i}_r R_r + \frac{d\boldsymbol{\psi}_r}{dt} + j(\omega_e - \omega_r)\boldsymbol{\psi}_r \tag{4}$$

$$J\frac{d\omega_r}{dt} = \frac{3}{2}\frac{p^2 L_m}{L_r}\left|\boldsymbol{\psi}_r \times \mathbf{i}_s\right| - pt_l - R_\omega \omega_r \tag{5}$$

$$\boldsymbol{\psi}_s = L_s \mathbf{i}_s + L_m \mathbf{i}_r \tag{6}$$

$$\boldsymbol{\psi}_r = L_m \mathbf{i}_s + L_r \mathbf{i}_r \tag{7}$$

where $\mathbf{u}_s$, $\mathbf{i}_s$, $\mathbf{i}_r$, $\boldsymbol{\psi}_s$, $\boldsymbol{\psi}_r$ denote space-vectors of the stator voltage, stator current, rotor current, stator flux, and rotor flux, respectively, the friction coefficient is R_ω and J is the drive's moment of inertia.

Taking into account (7), (4) becomes

$$0 = \frac{1}{T_r}\boldsymbol{\psi}_r - \frac{L_m}{T_r}\mathbf{i}_s + s\boldsymbol{\psi}_r + j(\omega_e - \omega_r)\boldsymbol{\psi}_r, \tag{8}$$

where T_r is the rotor time constant and s is the Laplace operator (=*d/dt*).

In Eqs. (3)-(8) the space-vectors are denoted with a bold face. Equation (3), expressed in terms of the stator current space-vector and rotor flux space-vector (from (6) and (7)), can be written as follows:

$$\mathbf{u}_s = (R_s + (s + j\omega_e)\sigma L_s)\mathbf{i}_s + (s + j\omega_e)\frac{L_m}{L_r}\psi_r \tag{9}$$

Equation (8) is well known as the current model, and Eq. (9) represents the voltage model of the induction machine. Equations (5), (8) and (9) present the state-space model of the induction motor with rotor speed, stator currents and rotor fluxes as the state variables of the system. However, the mathematical models of any induction machine control system, described by Eqs. (3)-(7) or (8)-(9), are complex, rely on many assumptions, contain parameters which are difficult to identify or change significantly during operation, and sometimes such mathematical models cannot be applicable. These problems can be overcome by using neural network-based control techniques, especially when the analytical models are not known. Moreover, ANN-based techniques can be less sensitive to parameter variation (more robust) than classical control systems.

Applications of artificial neural networks (ANNs) in induction motor drives are still emerging industry, but their growth potential is tremendous. In the literature, most publications on the application of ANNs in induction motor drives focuse on speed or position controller applications, where an existing conventional controller is replaced by an ANN-based controller [4-6]. Although this is an important application, there are further possibilities for a much wider range of ANN-based applications in variable speed AC drives [7]. Just a few of these are briefly described in this chapter, such as: ANN-based inverter control, applications of ANNs in waveform processing and delayless filtering, identification of machine parameters based on neural concepts and ANN-based approaches for the efficiency improvement in induction machine systems. In addition, this chapter the sensorless rotor field-oriented control system where the rotor speed measurement is replaced by the ANN-based speed estimator is presented.

2. Drive Feedback Signal Estimation

a) Delayless filtering

Consider a three-phase induction motor supplied by a three-phase SVPWM inverter, which switching frequency is usually higher than 1 kHz. This inverter outputs distorted three phase voltages. When these voltages are utilized for stator/rotor flux linkage estimation (upon Eqs. (6)-(9)), then they need to be filtered. The fundamental frequency and the amplitude of the phase voltage can be variable. A conventional low-pass filter (LPF) causes frequency-sensitive phase delay and attenuation at the output. In this case, a neural network can be used for delayless filtering [1] of a distorted wave as illustrated in Figure 3. The voltages are initially filtered by the three identical LPFs with the time constant T in order to convert them into continuous waves before processing through the ANN. The fundamental voltages at zero phase-shift angle, obtained by the neural network, permit correct calculation of feedback signals for a vector drive.

u_a, u_b, and u_c denote the phase voltage waves of the inverter (only u_a with the fundamental voltage u_{af} is shown in Figure 4).

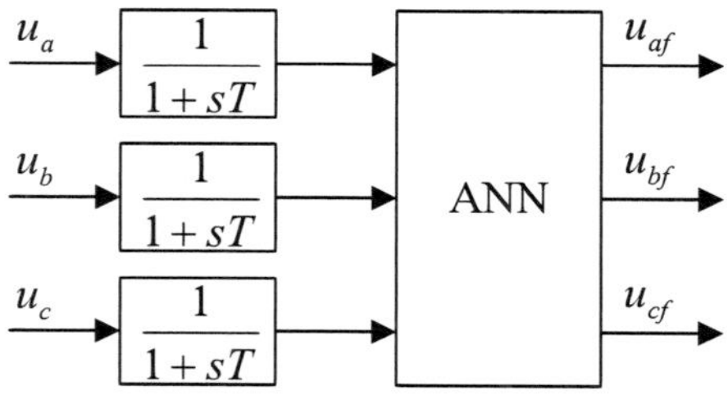

Figure 3. ANN-based delayless filtering of induction machine phase voltages

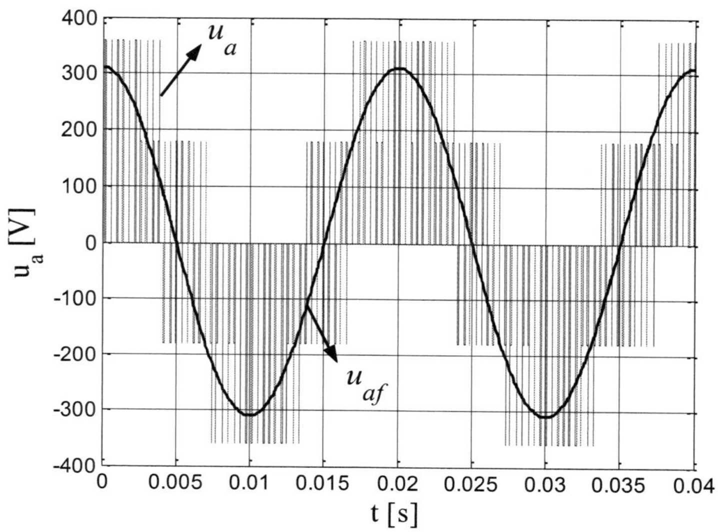

Figure 4. The phase voltage of the induction motor and corresponding filtered voltage

The ANN needs to be trained at variable-frequency and variable-magnitude voltages as encountered in variable-frequency drives. This implicates that an ANN of a large structure might have to be used for this purpose (e.g. four layer), which, in turn, implicates utilization of a high computing power digital signal processor (DSP). A small deviation in frequency and distortion from the trained wave produces a tolerable error.

b) Signal estimation

The capability of ANNs to provide signal estimation for an induction motor drive will be demonstrated. Figure 5 shows a neural network used for estimation of motor variables such as electromagnetic torque, rotor speed, stator flux linkage space-vector and rotor flux linkage space-vector. Simultaneous estimation theoretically can be done for all of the mentioned variables or for some of these [8]. The DSP receives the three-phase machine voltages and currents and outputs quantities: the rotor/stator flux linkage space-vector (ψ_r/ ψ_s), the rotor speed (ω_r) and the electromagnetic torque (T_e). For a three-phase machine with isolated neutral point only two phase currents are sufficient.

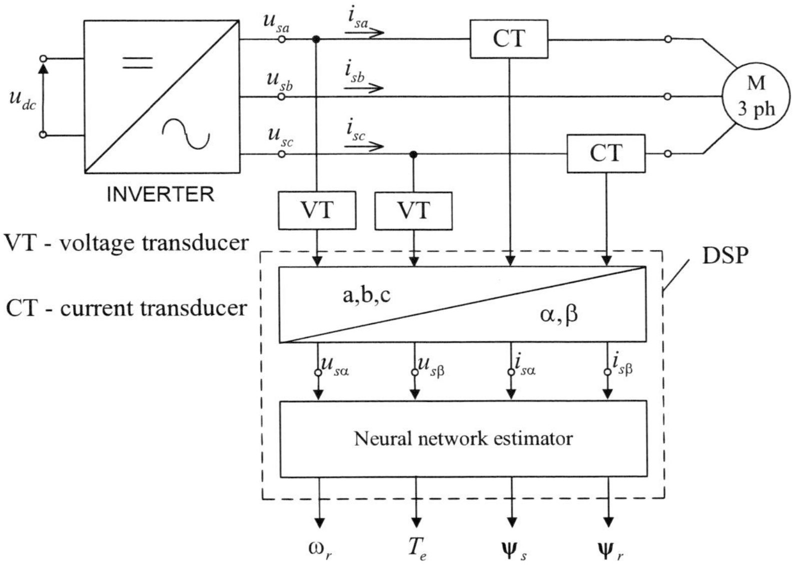

Figure 5. ANN-based state variable estimation.

For the ANN-based transient analysis of an induction machine a multilayer ANN should to be considered. In the case when only the rotor speed needs to be estimated, a four-layer ANN can be used. The inputs to the ANN can be the stator voltages and stator currents expressed in the stationary reference frame at two instants ($u_{s\alpha}(k)$, $u_{s\alpha}(k\text{-}1)$, $u_{s\beta}(k)$, $u_{s\beta}(k\text{-}1)$, $i_{s\alpha}(k)$, $i_{s\alpha}(k\text{-}1)$, $i_{s\beta}(k)$, $i_{s\beta}(k\text{-}1)$). Similarly, when only electromagnetic torque estimation is needed, the ANN structure with four layers satisfies. In this case, the electromagnetic torque is first calculated conventionally, for example by Eqs. (5) and (8) at each time instant of a simulated operating regime. The obtained numerical results can be used for the backpropagation training of the ANN whose structure must be specified *a priori*.

A much more complex ANN needs to be applied to obtain four quantities on its outputs, namely: the rotor speed, the electromagnetic torque and components of the stator/rotor flux linkage space-vector. Satisfying results will be given when a four-layer ANN is applied. However, in this case, the number of neurons in the hidden layers would need to be increased, compared to the ANN that outputs only the rotor speed. From this viewpoint, practical applications of this ANN are limited.

Figures 6 and 7 show the simultaneous estimation and corresponding actual values of the rotor speed and electromagnetic torque of the induction motor rated at 0.75 kW. A transient response of the scalar controlled (volt/hertz control) induction motor drive is presented.

At time 0.6 s, the previously unloaded motor is loaded with 5 Nm. During the initial transient, up to 0.6 s, the stator frequency is equal to 10 Hz. In addition, at time 1 s, the stator frequency rises linearly up to 50 Hz. The volt/hertz ratio retains its nominal value. A multilayer feedforward ANN with the structure 8-20-2 (8 input nodes, 20 hidden nodes in the hidden layer and 2 output nodes) has been used, since it has given correct results. This type of the ANN structure is chosen by trial and error. The sum of the squared errors during training was $2\text{x}10^{-5}$. The obtained results show the successful implementation of the ANN for the rotor speed and electromagnetic torque estimation.

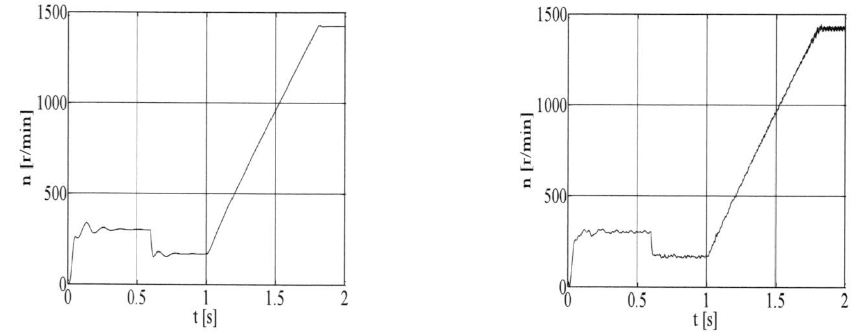

(a) (b)

Figure 6. Actual (a) and estimated rotor speed (b) in scalar controlled drive.

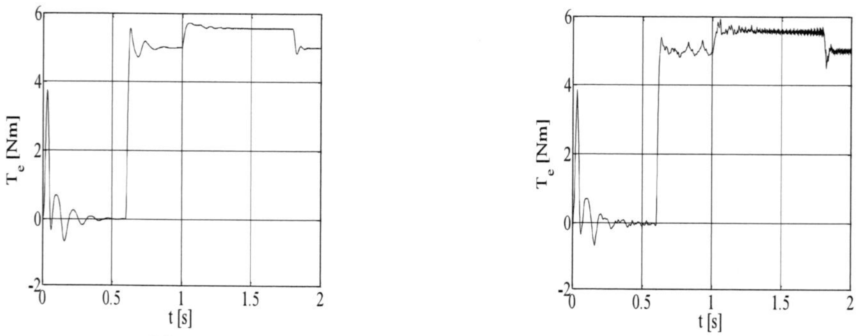

(a) (b)

Figure 7. Actual (a) and estimated electromagnetic torque (b) in scalar controlled drive.

With developments in DSPs and power electronics, induction motors can now be used in high-performance variable-speed drives. Some of these drives are based on the indirect rotor field-oriented method. The indirect rotor field-oriented drive is shown in Figure 8.

The goal is to achieve speed-sensorless operation of the drive, e.g. the ANN replaces the digital encoder [9-11]. Simulation-based data or measured data can be used for training. Once the ANN is trained and tested, it replaces the digital encoder. To obtain good estimation accuracy, the inputs to the network are the present and past values of the stator voltage and current components in the stationary reference frame. The final structure of the neural network used is a multilayer net with three layers. The training algorithm of the neural network speed observer is as follows [9]:

1. Initially randomize the weights,
2. Obtain the stator currents and voltages,
3. Calculate the error between real and observed speeds,
4. Adjust the weights of the neural network,
5. Calculate the output of the neural network,
6. Go to 2nd step until the stipulated error is reached.

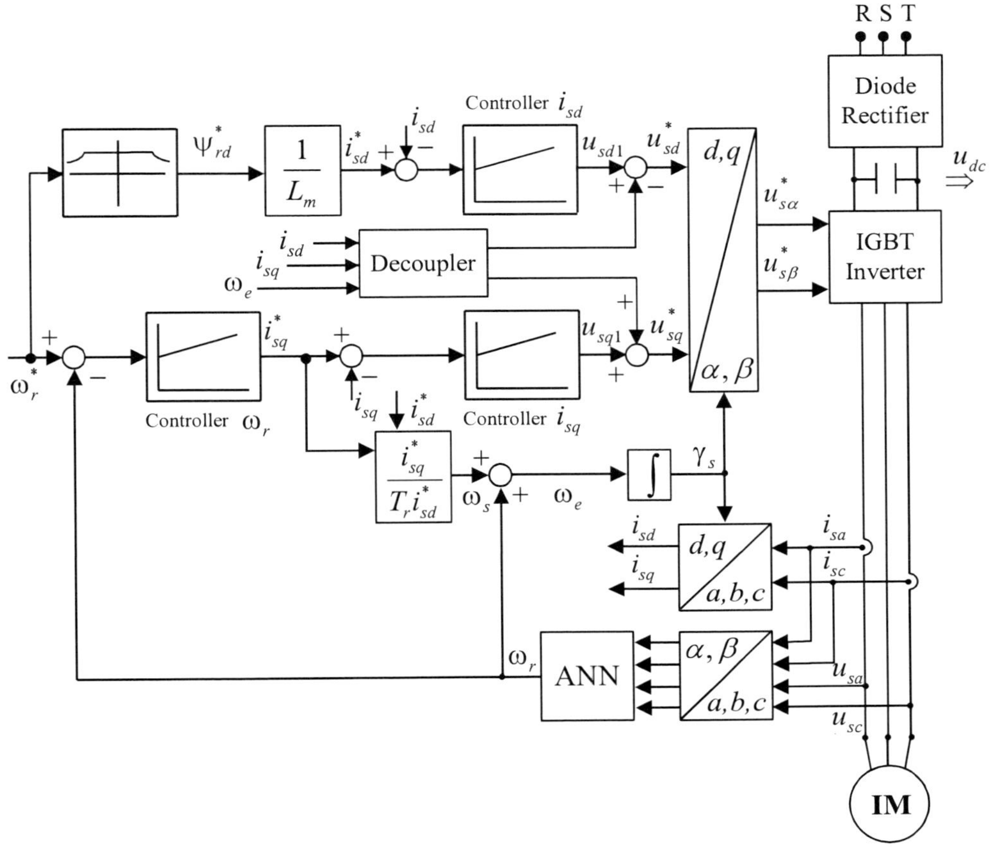

Figure 8. Indirect rotor field-oriented control system.

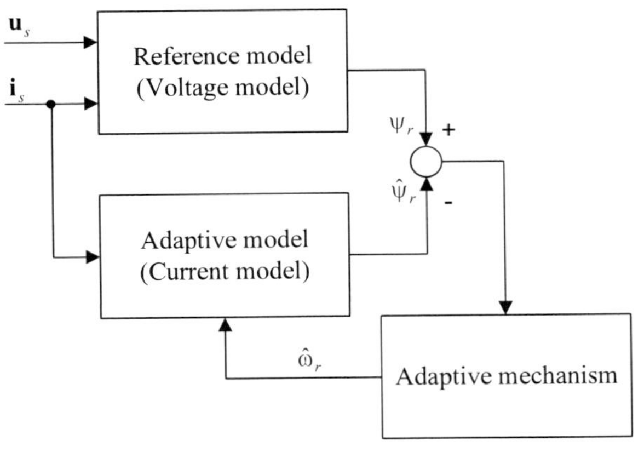

Figure 9. Classical rotor flux MRAS speed observer.

Another approach to obtain speed-sensorless operation of the drive can be based on the model reference adaptive system (MRAS) [12, 13]. The basic idea is that one quantity (vector or scalar) can be calculated in two different ways (by using the voltage model and the current model of the induction machine). The difference between the two quantities is an error signal, whose existence is assigned to the error in the estimated rotor speed. The error signal is applied to drive an adaptive mechanism (proportional-integral or integral) which provides correction of the rotor speed. The classical rotor flux MRAS speed observer structure, shown in Figure 9, consists of the reference model, the adaptive model and an adaptation scheme which generates the estimated speed.

The classical MRAS observer suffers from several disadvantages as follows:

1. The rotor flux estimation based on the voltage model needs open loop integration for the flux calculation.
2. Parameter sensitivity. Since the speed estimation is based on the machine models, it is highly sensitive to machine parameter variations. Especially, the stator resistance variation with machine temperature is a problem over a low speed region.
3. Inverter nonlinearity. The stator voltages are pulse-width modulated and include high harmonic components.

To overcome these problems, the voltage model can be replaced by an ANN. Figure 10 shows the modified MRAS observer.

A three layer feedforward ANN can be used to estimate the rotor flux linkage components in the stationary reference frame. Using the ANN scheme for rotor flux linkage estimation eliminates the need for pure integration, with less sensitivity to stator resistance variations [12].

3. Inverter Control

Current controlled VSIs with PWM are widely used in induction machine drives. The fast development of the switching capabilities of power electronics devices requires faster and simpler modulation techniques. There are many types of PWM techniques, but the space-vector modulation in power converters is the most often used technique in power converters. Usually, implementation of this modulation is powered by DSPs. Figure 11 shows a conventional hysteresis current controlled VSI for induction motor supply. In this case, when the stator current becomes greater or less than the corresponding reference by the hysteresis $\pm\delta$ in the hysteresis comparator, then the inverter leg is switched to the negative or positive side of the DC link. The conventional current controller outputs the switching signals sA, sB and sC (binary signals 0 or 1).

If an ANN-based current controller is used (Figure 12), the goal is to generate a non-linear function which maps the analogue current error signals ε_A, ε_B and ε_C into binary signals. For this purpose, it is also possible to use a multilayer feedforward ANN. The number of hidden layers can be selected by trial and error and the ANN can be trained by the backpropagation technique. The goal of the training is to adjust the weights so that the current error is minimized. However, this ANN-based current controller has a disadvantage. Considering a three-phase symmetrical motor, the current errors are never of the same sign and, hence, the ANN would never output the switching states sA=1, sB=1 and sC=1 (or sA=0, sB=0 and sC=0) at the same time. This means that the three switching devices would never be all turned-off or turned-on. In addition, higher harmonic content of the stator currents would be appearing. This disadvantage can be avoided by using an improved ANN-based current controller. In this technique, the ANN outputs are sampled with a phase shift [8].

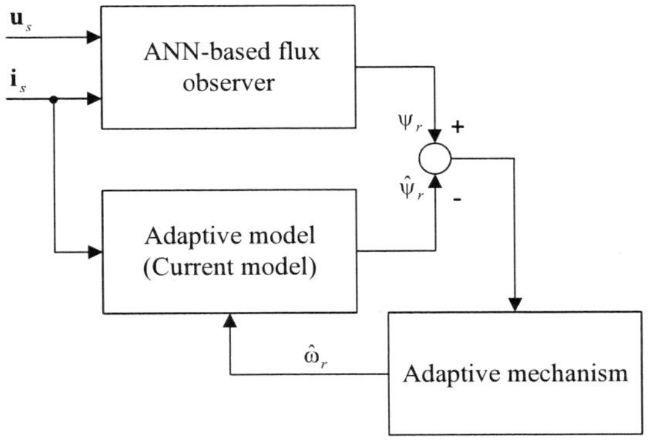

Figure 10. Modified MRAS observer.

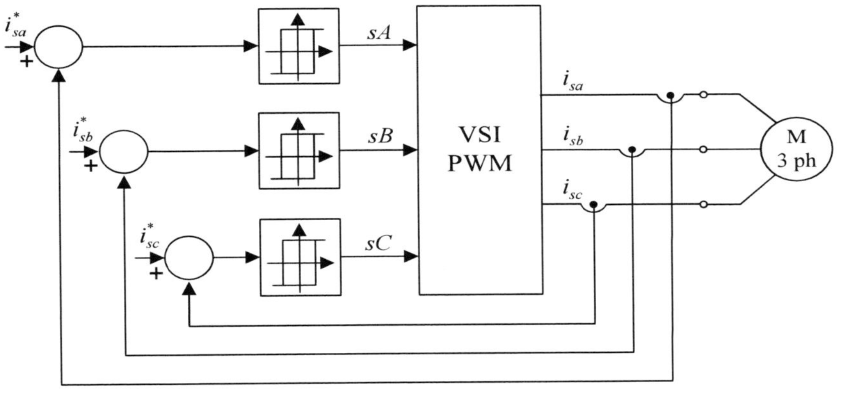

Figure 11. Current-controlled voltage source PWM inverter with three hysteresis controllers.

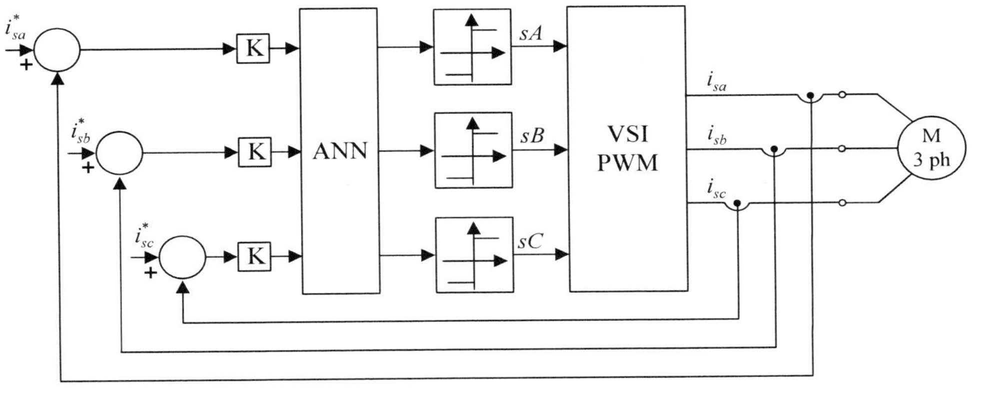

Figure 12. Current-controlled voltage source PWM inverter with ANN-based controller.

In addition, the required computations for the pulse times and overall executing time limit the sampling time. To reduce the software complexity and overall executing time, an ANN-based space-vector modulator, briefly described in this section, can be used.

A three-phase 2-level inverter with the dc link outputs the phase voltages (line-to-neutral voltages) determined by the eight possible inverter switching states. Each inverter switching state generates a voltage space-vector ($\mathbf{u}_1$ to $\mathbf{u}_6$ active vectors, $\mathbf{u}_7$ and $\mathbf{u}_8$ zero voltage vectors) in the space-vector plane (Figure 14). The magnitude of each active vector ($\mathbf{u}_1$ to $\mathbf{u}_6$) is 2/3 u_{dc} (DC bus voltage). The six non-zero phase voltages switching space-vectors ($\mathbf{u}_k$, k=1,2,…,6) and the two zero vectors ($\mathbf{u}_7$, $\mathbf{u}_8$) can be expressed as:

$$\mathbf{u}_k = \begin{cases} \frac{2}{3} u_{dc} e^{j(k-1)\frac{\pi}{3}}, k = 1,2,...,6 \\ 0, \qquad\qquad k = 7,8 \end{cases} \tag{10}$$

where u_{dc} is the DC link voltage.

The eight switching states with the corresponding switching voltage space-vectors are shown in Figure 13. Six switching space-vectors are evenly distributed at 60 intervals with the length of 2/3 u_{dc} and form a hexagon.

If the reference stator voltage space-vector is located in any of the six sectors of the hexagon, then pulse times of the two adjacent and zero space-vectors can be determined as follows:

$$t_{k+1} = mT_s \sin\vartheta \tag{11}$$

$$t_k = mT_s \sin(60^0 - \vartheta) \tag{12}$$

$$t_0 = T_s - t_k - t_{k+1}, \tag{13}$$

where ϑ is the angle between the reference voltage space-vector and the closest space-vector, and m is the modulation index defined by

$$m = \sqrt{3}\left|\mathbf{u}_{sref}\right| / u_{dc} \tag{14}$$

The conventional SVPWM requires utilization of Eqs. (11)-(13). Because of its real time implementation, i.e. storing the sine function in a look-up table, the SVPWM has the three main disadvantages, as follows:

1. Utilization of a look-up table implies the need for additional DSP memory
2. Interpolation achieved by a look-up table leads to increased harmonics in the PWM waveforms
3. Utilization of a look-up table demands additional computing time, limiting the maximum inverter switching frequency

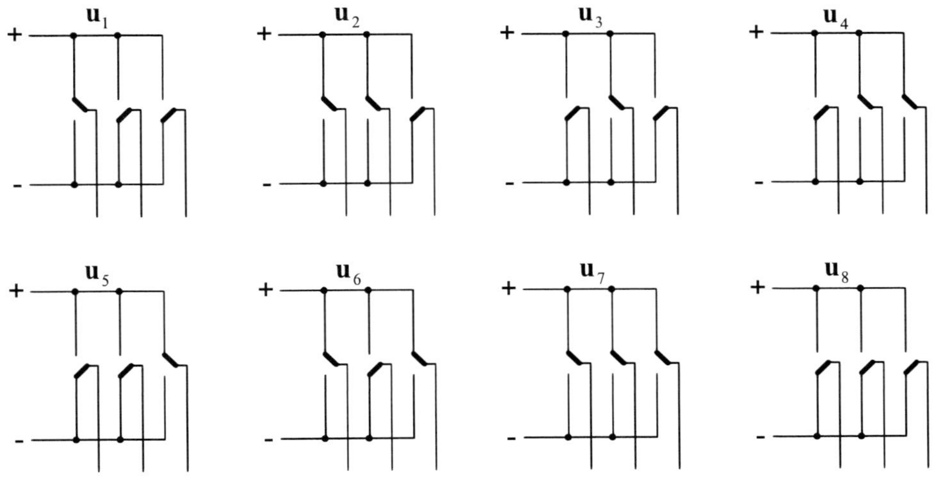

Figure 13. Eight switching states and corresponding voltage space-vectors.

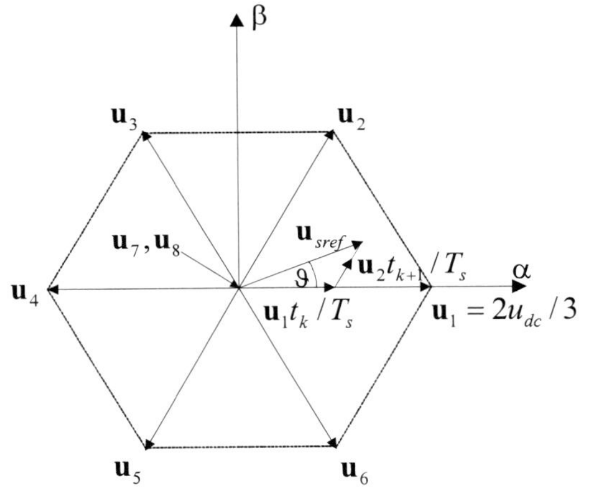

Figure 14. Determination of switching sequences in the three-phase VSI.

As an alternative to the conventional space-vector modulation technique, an ANN-based technique can be used, avoiding in that way the direct on-line computation of trigonometric functions [8, 14].

Space-vector modulation technique based on an ANN

As mentioned above, the SVPWM requires the utilization of the switching vectors adjacent to the reference voltage vector and the angle ϑ between this vector and the closest clockwise switching vector. For this purpose, first the sector where the reference voltage space-vector is positioned must be determined. The pulse times upon Eqs. (11)-(13) can then be determined by using the sine function. Except of the sine function, the pulse times can be determined by computing the cosine of the angle ϑ. This can be realized by computing the real parts of the products of the voltage reference space-vector and the six non-zero switching vectors, and finally selecting the two largest values. These values are proportional to $\cos\vartheta$

and $\cos(60^0 - \vartheta)$. The ANN shown in Figure 15, which resembles a conventional competitive ANN, can be used to implement the SVPWM [8].

This ANN consists of the two layers (input layer containing the source nodes and an output or competitive layer). The inputs are the components of the stator voltage reference space-vector expressed in the stationary reference frame, $u_{s\alpha}$ and $u_{s\beta}$. These are followed by a layer of six neurons which output n_1, n_2, …, n_6. These outputs correspond to the real part of the product of the reference voltage space-vector $\mathbf{u}_s^*$ and one of the six normalized switching vectors computed by

$$\mathbf{u}_k = e^{j(k-1)\frac{\pi}{3}}, k = 1,2,...,6 \tag{15}$$

Taking into account Eq. (15), the outputs n_i (i=1,2,…,6) can be written as:

$$n_k = \mathrm{Re}(\mathbf{u}_k \cdot \mathbf{u}_s^*) = \cos\left[(k-1)\frac{\pi}{3}\right]u_{s\alpha}^* + \cos\left[(2k+1)\frac{\pi}{6}\right]u_{s\beta}^*, k = 1,2,...,6 \tag{16}$$

The pulse times t_i and t_{i+1} can be computed from the largest values of the ANN. Taking into account following equations:

$$n_i = \left|\mathbf{u}_s^*\right|\cos\vartheta \tag{17}$$

$$n_{i+1} = \left|\mathbf{u}_s^*\right|\cos(60^0 - \vartheta) \tag{18}$$

and by considering the trigonometric relationships

$$\cos\vartheta = \frac{2}{\sqrt{3}}(\sin(60^0 - \vartheta) + \frac{\sin\vartheta}{2}) \tag{19}$$

$$\cos(60^0 - \vartheta) = \frac{2}{\sqrt{3}}(\frac{\sin(60^0 - \vartheta)}{2} + \sin\vartheta) \tag{20}$$

and also Eqs. (11) and (12), the pulse times can be written as

$$t_i = \frac{T_s}{u_{dc}}(2n_i - n_{i+1}) \tag{21}$$

$$t_{i+1} = \frac{T_s}{u_{dc}}(2n_{i+1} - n_i) \tag{22}$$

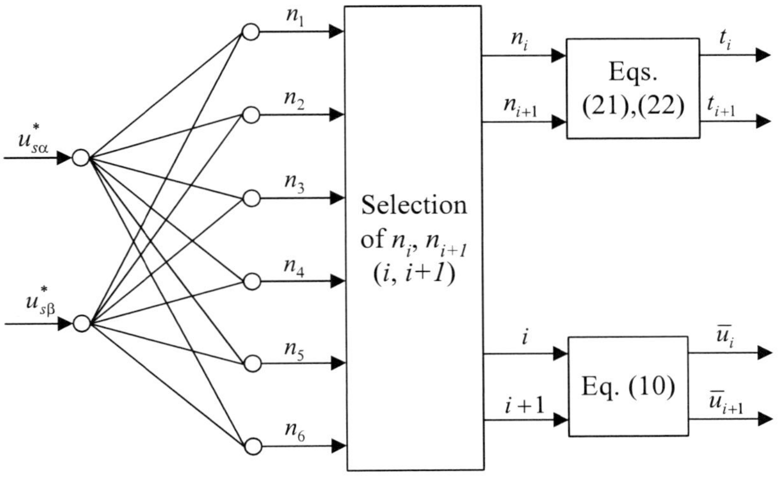

Figure 15. Space-vector modulation implemented by a competitive ANN.

In addition, the time t_0 can be computed by Eq. (13).

In summary, the SVPWM based on the competitive ANN consists of the following steps:

1. The ANN values n_1, n_2, …, n_6 are computed by using Eq. (16)
2. The two largest values n_i, n_{i+1} of the six values and the corresponding indexes are selected by the ANN
3. The pulse times t_i and t_{i+1} are obtained by using Eqs. (21) and (22)
4. The switching vectors $\mathbf{u}_i$ and $\mathbf{u}_{i+1}$ are determined according to the values of i and $i+1$.

Similar procedure is described in [8], but it contains several errors in Eqs. (19)-(22) which have been corrected here.

4. Identification of Machine Parameters

Induction motor control systems are known to be extremely non-linear control systems, because of induction motor parameter variability under different conditions. Heating of motor windings depends on stator and rotor currents leading to variability of stator and rotor resistances. Variable mutual inductance is a consequence of different flux levels of the motor. This is very important from the viewpoint of field oriented control systems. Most types of field oriented control systems are sensitive to errors resulting from non-constant parameters and furthermore, do not give an accurate representation of the machine under consideration.

In the past, many methods have been developed for estimations of induction machines parameters. Some of these methods are based on neural networks that replace the adaptive model of an induction machine. The basic structure of an adaptive scheme for stator or rotor resistance identification is shown in Figure 16. This scheme is based on the model reference adaptive system (MRAS) [15].

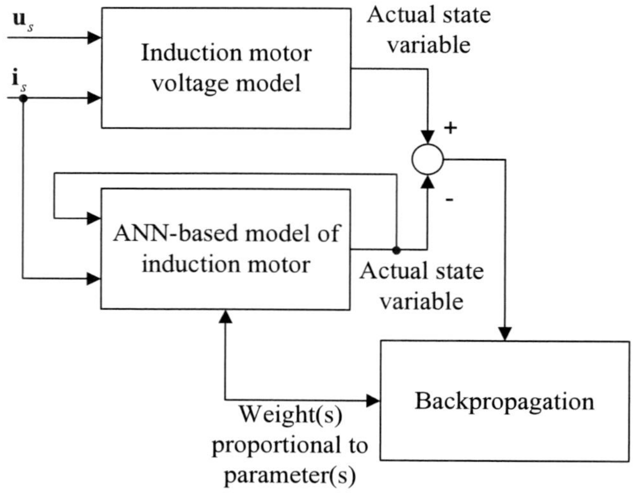

Figure 16. ANN-based parameter identification..

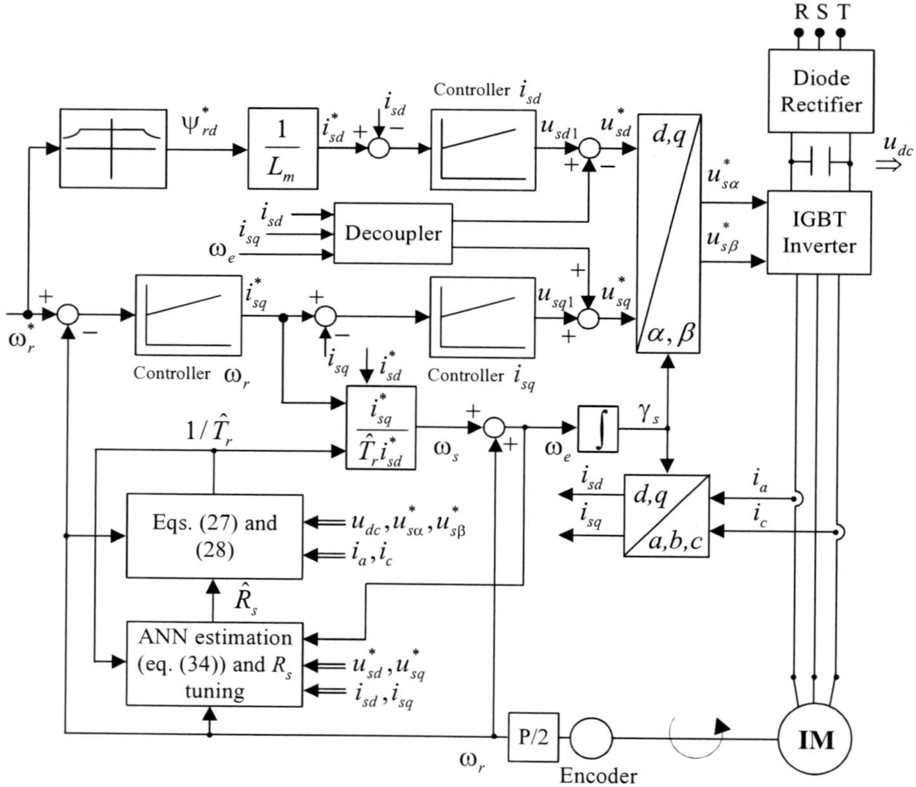

Figure 17. ANN-based indirect rotor-field control system.

The following example shows the stator resistance identification in the IRFO control system shown in Figure 17 [16].

The MRAS theory, as described, is utilized in order to estimate the rotor speed of induction motor. The rotor flux space-vector is estimated in the d,q reference frame by the voltage model (reference model) and by the ANN-based model (adaptive model) of the induction motor. Conventionally, the current model is used as the adaptive model because it is the rotor speed-dependent one. The difference between flux space-vectors estimated using the two ways is then used in an adaptive mechanism that outputs the estimated value of the rotor speed and adjusts the adaptive model until good performances are obtained. The inputs to the reference model are the direct- and quadrature-axis stator voltages and currents of the induction motor and the angular stator frequency ω_e. The outputs of the reference model are the components of the rotor flux space-vector in the d, q reference frame, which can be obtained from Eq. (9) as follows:

$$\frac{d\psi_{rd}}{dt} = \frac{L_r}{L_m}\left\{u_{sd} - \hat{R}_s i_{sd} - \sigma L_s \frac{di_{sd}}{dt} + \omega_e \sigma L_s i_{sq}\right\} + \omega_e \psi_{rq} \tag{23}$$

$$\frac{d\psi_{rq}}{dt} = \frac{L_r}{L_m}\left\{u_{sq} - \hat{R}_s i_{sq} - \sigma L_s \frac{di_{sq}}{dt} - \omega_e \sigma L_s i_{sd}\right\} - \omega_e \psi_{rd} \tag{24}$$

Equations (23) and (24) determine the reference model as shown in Figure 18. These equations do not contain the rotor speed. However, Eq. (8) contains the rotor flux space-vector and the rotor speed as well. This is the equation for the adaptive model. Rewriting (8) to give the rotor flux components in the d,q reference frame yields

$$\frac{d\hat{\psi}_{rd}}{dt} = \frac{1}{\hat{T}_r}\left(L_m i_{sd} - \hat{\psi}_{rd}\right) + \left(\omega_e - \omega_r\right)\hat{\psi}_{rq} \tag{25}$$

$$\frac{d\hat{\psi}_{rq}}{dt} = \frac{1}{\hat{T}_r}\left(L_m i_{sd} - \hat{\psi}_{rq}\right) - \left(\omega_e - \omega_r\right)\hat{\psi}_{rd} \,. \tag{26}$$

Equations (25) and (26) contain the rotor speed, which is generally changing, and the intent is to estimate this speed by using an ANN. Consequently, Eqs. (25) and (26) can be implemented by a single-layer ANN, which contains variable weights, and the variable weights are proportional to the rotor speed.

When there is no mismatch between the actual and identified parameters of the induction motor, then the errors ε_d and ε_q (Figure 18) are zero in the steady state. In this case, the rotor speed estimated by the ANN must be the same as the actual rotor speed. During transient states, there is a difference between the actual rotor speed and the speed estimated by the ANN, even if there is no mismatch between the actual and identified parameters of the induction motor. In these cases, the errors ε_d and ε_q are not zero, and they are used to adjust the weights of the ANN.

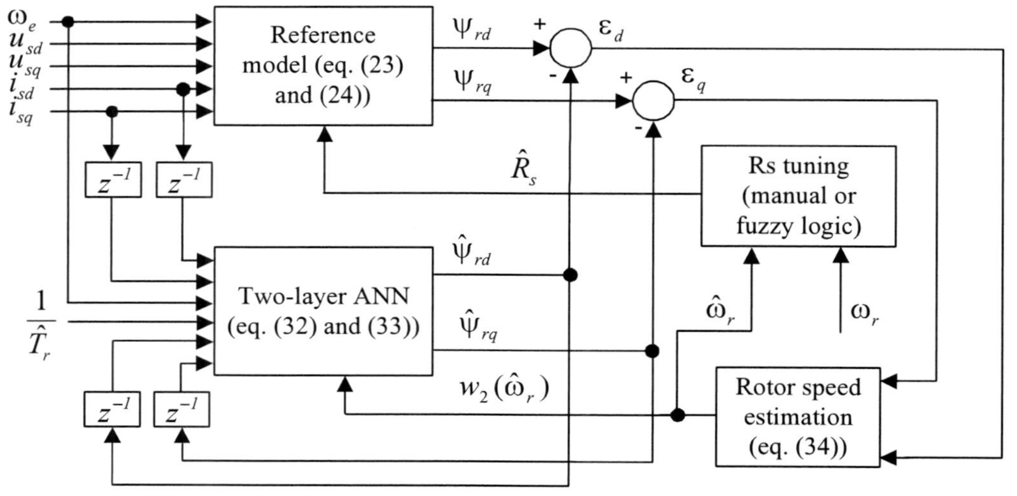

Figure 18. Stator resistance tuning based on MRAS theory and ANN.

Otherwise, when there is any mismatch between the actual and identified parameters of the induction motor, then the errors ε_d and ε_q are not zero in the steady state. Consequently, the actual rotor speed is different from the estimated rotor speed. Taking into account the constant magnetizing level of the induction motor (constant mutual inductance), the difference between the actual and the estimated rotor speed can be caused by the following two reasons:

(a) incorrect rotor resistance identification (incorrect inverse rotor time constant), and
(b) incorrect stator resistance identification.

The stator resistance is an important parameter for inverse rotor time constant identification, especially in the low speed region. When the stator resistance is incorrectly identified, then the inverse rotor time constant is incorrectly identified as well. As a result, there is a mismatch between the actual rotor speed and the estimated rotor speed in the steady state. When the stator resistance is correctly identified, then there is no mismatch between the actual rotor speed and the estimated rotor speed, and the inverse rotor time constant is correctly identified. As a result, the stator resistance tuning can be done either by a manual tuning procedure observing the difference between the actual rotor speed and the estimated rotor speed or by an automated fuzzy logic principle as will be described.

There are many methods for estimation of the (inverse) rotor time constant. One group of online rotor time constant adaptation methods is based on the principles of MRAS. This is the approach with relatively simple implementation requirements. Replacing the actual rotor flux space-vector $\boldsymbol{\psi}_r$ with the estimated rotor flux space-vector $\hat{\boldsymbol{\psi}}_r$, in Eq. (8), and rewriting Eqs. (8) and (9) in the α,β reference frame (ω_e=0), yields

$$0 = \frac{1}{\hat{T}_r}\hat{\boldsymbol{\psi}}_r - \frac{L_m}{\hat{T}_r}\mathbf{i}_s + s\hat{\boldsymbol{\psi}}_r - j\omega_r\hat{\boldsymbol{\psi}}_r \tag{27}$$

$$\mathbf{u}_s = (\hat{R}_s + s\sigma L_s)\mathbf{i}_s + s\frac{L_m}{L_r}\psi_r + K_1(\psi_r - \hat{\psi}_r), \quad (28)$$

where K_1 is an observer gain.

A hat above a symbol in (27) and (28) denotes identified parameters. Equation (27) gives an estimation of the rotor flux space-vector based upon easily measured stator currents and rotor speed. This estimation mainly depends on the accuracy of the inverse rotor time constant identification. Equation (27) presents an *adaptive model* of the rotor flux estimation. On the other hand, (28) gives an estimation of the rotor flux space-vector based upon measured stator currents and the reconstructed voltage space-vector from the measured DC link voltage and the inverter driving signals. Equation (28) is independent of the inverse rotor time constant and, accordingly, can be used as the reference model of the rotor flux space-vector estimation. This estimation mainly depends on the accuracy of the stator resistance identification. The error signal of the rotor flux magnitude of the two estimators is applied to drive an adaptive mechanism (PI) which provides correction of the inverse rotor time constant.

The overall identification procedure of the inverse rotor time constant identification is shown in Figure 19.

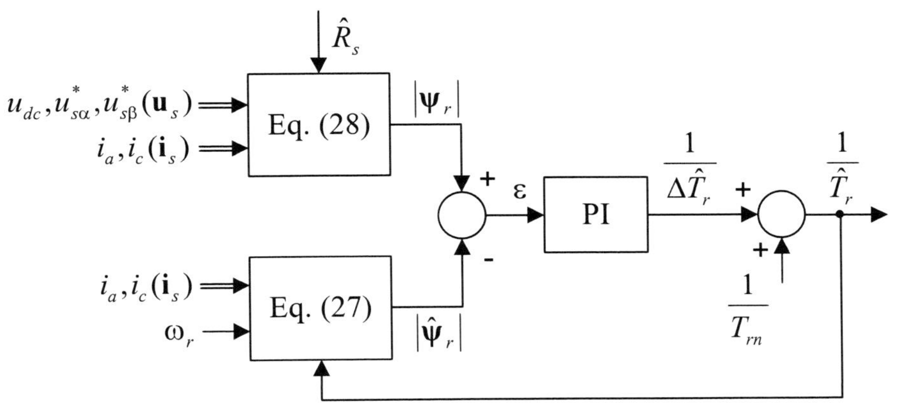

Figure 19. Inverse rotor time constant identification.

To obtain the weight adjustment in the ANN, the sampled data forms of Eq. (25) and (26) are derived. The actual rotor speed is then replaced by the estimated rotor speed. By using the backward difference method, the rotor flux components at the kth sampling instant can be described in the recursive form as follows:

$$\hat{\psi}_{rd}(k) = \hat{\psi}_{rd}(k-1)\left(1 - \frac{T}{\hat{T}_r}\right) + (\omega_e - \hat{\omega}_r)T\hat{\psi}_{rq}(k-1) + \frac{T}{\hat{T}_r}L_m i_{sd}(k-1) \quad (29)$$

$$\hat{\psi}_{rq}(k) = \hat{\psi}_{rq}(k-1)\left(1 - \frac{T}{\hat{T}_r}\right) - (\omega_e - \hat{\omega}_r)T\hat{\psi}_{rd}(k-1) + \frac{T}{\hat{T}_r}L_m i_{sq}(k-1) \quad (30)$$

where T is the sampling rate.

In Eqs. (29) and (30) the following weights are introduced:

$$w_1 = 1 - \frac{T}{\hat{T}_r}, w_2 = -(\omega_e - \hat{\omega}_r)T, w_3 = \frac{T}{\hat{T}_r}L_m. \tag{31}$$

It can be seen that w_2 is a variable weight and, in addition, is proportional to the speed. From the viewpoint of the ANN training procedure, the weights w_1 and w_3 do not depend on the ANN training, but rather on the inverse rotor time constant identification.

Equations (29) and (30) can be expressed in the following forms:

$$\hat{\psi}_{rd}(k) = w_1\hat{\psi}_{rd}(k-1) - w_2\hat{\psi}_{rq}(k-1) + w_3 i_{sd}(k-1) \tag{32}$$

$$\hat{\psi}_{rq}(k) = w_1\hat{\psi}_{rq}(k-1) + w_2\hat{\psi}_{rd}(k-1) + w_3 i_{sq}(k-1). \tag{33}$$

Equations (32) and (33) are shown in Figure 20 in the form of the single-layer ANN. There are four input nodes (denoted by the small circuit) and two output nodes. The connections between the nodes are represented by weights.

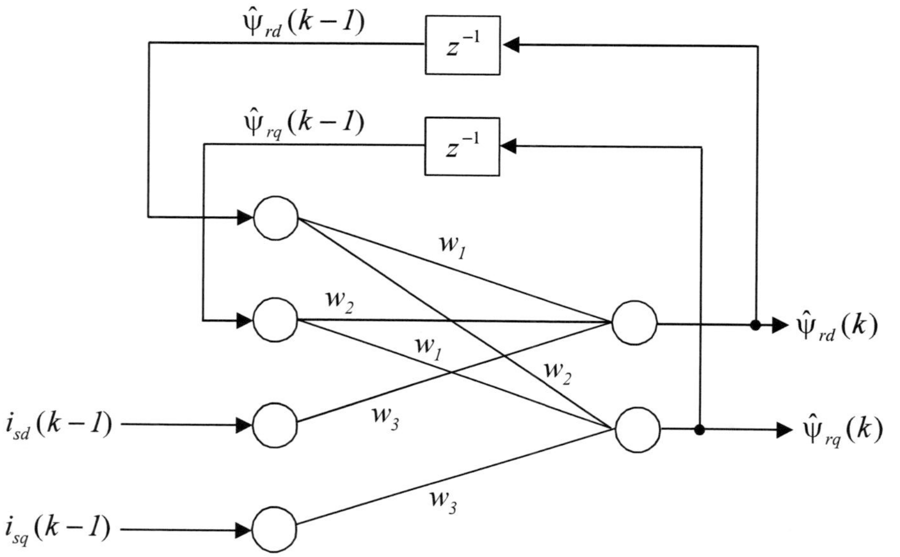

Figure 20. Rotor flux estimation by ANN.

During the training procedure of the ANN, the adaptive weights are adjusted so that the total network error is minimized. Consequently, the estimated rotor speed can be obtained as follows:

$$\hat{\omega}_r(k) = \omega_e(k) + \frac{1}{T} w_2(k-1) + \\ + \frac{\eta}{T}\left\{-\left[\psi_{rd}(k) - \hat{\psi}_{rd}(k)\right]\hat{\psi}_{rq}(k-1) + \left[\psi_{rq}(k) - \hat{\psi}_{rq}(k)\right]\hat{\psi}_{rd}(k-1)\right\} + \frac{\alpha}{T}\Delta w_2(k-1) \tag{34}$$

where η is the learning rate and α is the momentum constant.

Equation (34) presents the simple algorithm of the rotor speed estimation by the ANN shown in Figure 4. This simple single-layer ANN does not require an off-line learning procedure since the learning takes place during the on-line rotor speed estimation process. The results obtained by this method can be found in [16].

5. Neural Network Based Approaches for the Efficiency Improvement in Induction Machine Systems

Induction machine drives are high efficiency drives when operating close to the rated torque and speed. However, at light loads, their efficiency is considerably reduced. In the steady-state, if the induction machine electromagnetic torque is below the rated torque, the induction machine efficiency may be increased by applying the optimum rotor flux linkage, which depends on motor torque and speed. In addition, the optimized flux can be applied in a field-oriented drive. Implementation of ANNs is also possible in this area, although deviations of induction machine parameters can deteriorate their performances [17].

The overall induction machine losses can be calculated as follows:

$$P = \frac{2}{3}\left(\frac{T_e}{p\psi_r}\right)^2 \left(\left|\frac{Z_r + Z_m}{Z_m}\right|^2 R_s + R_r + \frac{|Z_r|^2}{R_m} + K_a\omega_r^2\right) + K_\omega\omega_r^2 \tag{35}$$

where K_a and K_ω present the coefficients of stray and mechanical losses, respectively, and R_m is the magnetic resistance representing the iron losses,

$$Z_r = \frac{R_r}{s} + j\omega_e L_{rl},\ Z_m = \frac{j\omega_e L_m R_m}{j\omega_e L_m + R_m}.$$

The optimum rotor flux which ensures efficiency optimization is given by

$$\psi_r^{opt} = \arg\left[\left.\frac{\partial P(\psi_r, \omega_r, T_e, R_r, R_s)}{\partial \psi_r}\right|_{\omega_r, T_e, R_r, R_s}\right] = 0 \tag{36}$$

The optimum rotor flux, taking into account deviations of the stator and rotor resistance, can be derived by using Eq. (36). This equation determines the optimum rotor flux for each operating point defined by the rotor speed, the electromagnetic torque and the thermal variation of the stator and rotor resistances. The described determination of the optimum rotor flux would be very cumbersome and too slow for real-time control implementation. In order to avoid utilization of Eq. (36) for real-time implementation, it is possible to carry out extensive computer simulations, and, subsequently, implement the numerical results obtained by Eq. (36) in a feedforward ANN that is suitable for real-time implementation. This network can be a 3-3-1 feedforward neural network as proposed in [17], with three inputs (torque, speed and rotor resistance) and two layers (hidden and output layer). The hidden layer neurons have a nonlinear sigmoid transfer function and the only output layer neuron has a linear transfer function.

The following example shows the utilization of an ANN for determining the maximum power output of an induction generator driven by a wind turbine [18]. The wind turbine developed torque is a function of its rotational speed at different wind speeds. For a particular wind speed, the turbine speed is varied so that it can deliver the maximum power. For this purpose, a closed-loop wind-turbine emulator, using the radial basis function (RBF) ANN, can be designed to produce a maximum-power command for the induction generator system at varying wind speeds [18]. The RBF ANN is implemented within the rotor field-oriented induction generator system (Figure 21). The rotor speed of the induction generator, the DC link voltage and current of the power converter are detected simultaneously to yield maximum power output of the converter through DC link power control.

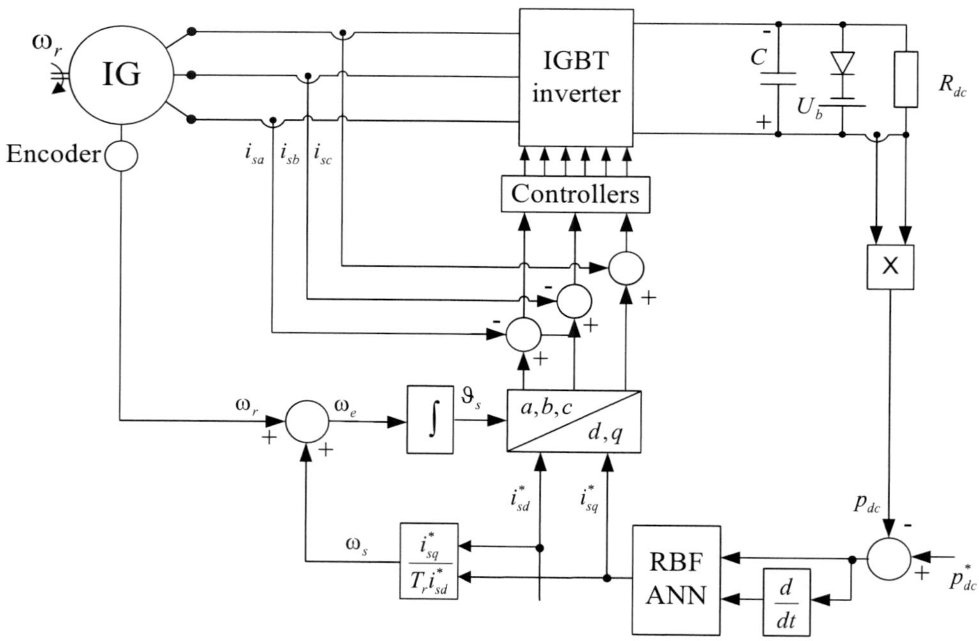

Figure 21. Rotor field-oriented induction generator system.

The RBF ANN performs function mappings and converges faster than a multilayer perceptron neural network because of a simpler network structure. Additionally, the RBF ANN has features similar to a fuzzy-logic system [19].

6. Neural Network Implementations by Digital Signal Processors and ASIC Chips

Modern control systems with induction machines are commonly implemented by using a microcomputer or DSP software, dedicated digital hardware, or a combination of both. Before the DSP genesis, control systems were controlled by hardwired analog devices and digital circuits. The majority of systems today are controlled digitally because digital control has a number of advantages. Since the creation of the first digital computer in the early 1940s (the Atanasoff–Berry Computer), the price of digital computers has dropped considerably. Except of this, digitally controlled systems (or microcomputer-based systems) are easy adaptable and digital components are much less prone to environmental conditions than capacitors, inductors, etc. DSP technology is advancing very fast with higher processing speeds and more functional integration. Generally, digital controllers may be classified into the following two categories: VLSI (very large scale integration) controllers and discrete IC (integrated circuit). In a VLSI, a very large number of devices are integrated in a chip to provide great simplification of hardware. A VLSI chip may use digital, analog, or mixed signals. VLSI chips can be designed either for general purposes (such as DSPs) or for a particular application, i.e., application specific IC (ASIC).

Examples of ASIC chips are numerous. In this chapter, just a few are specified as follows [20]:

1. The AD2S100/AD2S110 (made by Analog Devices) vector controller that performs the Clark and Park transformations, usually required for implementing field-oriented control of induction machines.
2. NLX230 fuzzy controller (made by American Neurologix) is a fully configurable fuzzy logic engine containing a 1-of-8 input selector, 16 fuzzyfiers, a minimum comparator, a maximum comparator and a rule memory. Up to 64 rules can be stored in the on-chip, 24-bit-wide rule memory. The NLX230 can perform 30 million rules per second.
   The same company has introduced NLX420 Neural Processor Slice. This is a digital chip with 16 processing elements (PE's). The "slice" architecture allows for building of multi-chip configurations. Using time multiplexing, one NLX420 configured for 16-bit inputs, can emulate a maximum of 1048576 neurons, each having 64k synaptic inputs. Weights are stored in external RAM. Transfer functions are implemented with user loaded piecewise continuous approximation. Input data can be in form of 1, 4, 8, and 16-bit integer values. The ADS420 Neural Processor Slice Development System includes a PC AT board, with up to 4 chips, and software.
3. Intel's 80170NX ETANN (Electrically Trainable Analogue Neural Network) was the first large analog ASIC chip, and it was introduced to the market in 1991. Analog neural network with 64 inputs, 16 internal biases, and 64 neurons with sigmoidal transfer functions (external gain control available). Two layer feedforward networks can be implemented with 64 inputs, 64 hidden neurons, and 64 output neurons using the two 80x64 weight matrices. It was recently withdrawn from the market because it had an analog signal drift problem.

4. Philips L-Neuro chips. Philips has previously offered the L-Neuro 1.0 chip and has now announced development of the follow-on L-Neuro 2.3 chip. L-Neuro 1.0 contains 16 processing elements with 16-bit registers. There is a 1 kByte weight memory buffer on-chip to provide 1024 8-bit or 512 16-bit weights. The transfer functions are implemented off-chip so that multiple chips can be cascaded together. L-Neuro 2.3 is a second generation that builds on the L-Neuro 1.0 experience. The chip was developed at Laboratoires d'Electronic Philips (LEP), France.
5. The ZISC 036 (made by IBM) is a digital chip with 64 8-bit inputs and 36 radial basis function neurons. Multiple chips can be easily cascaded together to create networks of arbitrary size.
6. The Signetics HEF4752V motor control circuit utilizing LOCMOS (Local Oxidation Complementary Metal-Oxide Semiconductor) technology for the control of three-phase pulse width modulated inverters in induction machine control systems. A pure digital waveform generation is used for synthesizing three 120° out of phase signals, the average voltage of which varies sinusoidally with time in the frequency range from 0 to 200 Hz.

Digital chips are also available from Siemens, Hitachi, Mitsubishi etc. A disadvantage of ASICs in induction machine control systems is the lack of flexibility to modify or to adapt the design to different types of motor drives, once the chip is built. The high development and fabrication cost for an ASIC can thus only be justified in large volume production.

In small-volume production and in prototyping stages, field programmable gate arrays (FPGAs) offer a realistic alternative to full gate arrays design to implement specific motion control functions of high complexity requiring up to a million gates.

The few dedicated circuit examples, together with the general modern trend towards 'systems-on-a-chip' integration in electronics, illustrate the need for further complex ASIC/FPGA designs for drives and power systems.

Since the 2000s, except of the ASIC/FPGA controllers, DSP-based digital controllers play a very important role. DSPs are practically mandatory in the research stage of any induction machine control system. Modern DSPs yield greater performance. For example, Technosoft announces the new MSK28335 DSP Motor Control Kits family based on the TMS320F28335 floating-point DSP [21]. The MSK28335 DSC boards use the 150 million instructions per second (MIPS) computational power of the DSP, combined with a manager able to drive up to 18 PWM and 16 A/D converters. The embedded CAN interface may be used to connect the board to multiple-axis structures. The kits can be connected to a PC via an RS232 interface to download, execute and debug the software applications without the need of other hardware devices. The basic MSK28335 kits are designed for users who already have the power module and motor, and want to develop their motion control software application for educational purposes. These kits allow for quick debugging and implementation of neural network algorithms.

7. CONCLUSION

Induction motors have a unique and important role in industry and electricity generation. Their main advantage is the elimination of all sliding contacts, resulting in a very simple and rugged construction. Induction machines are built in a variety of designs with ratings from a few watts to tens of megawatts. Because of their nonlinear nature, induction motors are somewhat difficultly controlled. Neural networks (and fuzzy neural networks as well) are very suitable for solving such problems. Although numerous scientific papers have been written in this area, we have found just two manufacturers of induction machine converters: ABB and Fuji, who specify neural network applications on their websites. Other reputable manufacturers (such as Hitachi, Yaskawa, Danfoss, Lenze, Siemens, etc.), on the other hand, do not specify such applications on their websites.

Nowadays, intelligent control and estimation, particularly neural network-based, have brought a new dimension to induction machine applications. An increasing number of papers in this field open up new perspectives on power electronics and induction machine control systems.

REFERENCES

[1] Modern, BK. *Power Electronics and AC Drives*; Prentice Hall PTR: Upper Saddle River, NJ, 2002.

[2] Vas, P; Drury, W. Electrical Machines and Drives: Present and Future; *Melecon*, 1996, Vol. 1, 404-408.

[3] Novotny, DW; Lipo, TA. *Vector Control and Dynamics of AC Drives; Monographs in Electrical and Electronic Engineering 41*; Oxford University Press Inc.: New York, NY, 1996.

[4] Vas, P. *Artificial-Intelligence-Based Drives*; Rashid, M. H; Ed; Academic Press Series in Engineering; Academic Press: San Diego, CA, 2001, 769-774.

[5] Kim, S; Han, W. Induction Motor Servo Drive using Robust PID-like Neuro-Fuzzy Controller; *Control Eng. Pract.*, 2006, Vol. 14, 481–487.

[6] Uddin, MN; Wen, H. Development of a Self-Tuned Neuro-Fuzzy Controller for Induction Motor Drives; *IEEE Trans. Ind. Appl.*, 2007, Vol. 43, 1108-1116.

[7] Cirstea, MN; Dinu, A; Khor, JG; *McCormick, M. Neural and Fuzzy Logic Control of Drives and Power Systems; Electronic Engineering*; Newnes: Oxford, UK, 2003.

[8] Vas, P. Electrical Artificial-Intelligence-Based Machines and Drives, Application of Fuzzy, Neural, Fuzzy-Neural and Genetic-Algorithm-Based Techniques; *Monographs in Electrical and Electronic Engineering*; Oxford University Press Inc.: New York, NY, 1999.

[9] Mouna, BH; Lassaâd, S. Neural Networks for Controlled Speed Sensorless Direct Field Oriented Induction Motor Drives; *JEE* 2008, Vol. 8, 88-99.

[10] Tsaji, C. CMAC-Based Speed Estimation Method for Sensorless Vector Control of Induction Motor Drive; *Electr. Power Compon. Syst.* 2006, Vol. 34, 1213–1230.

[11] Zaky, MS; Khater, M; Yasin, H; Shokralla, S. Review of Different Speed Estimation Schemes for Sensorless Induction Motor Drives; *JEE* 2008, Vol. 8, 100-138.

[12] Gadoue, SM; Giaouris, DJ; Finch, W. Sensorless Control of Induction Motor Drives at Very Low and Zero Speed using Neural Network Flux Observers; *IEEE Trans. Ind. Electron.*, 2009, Vol. 56, No. 8, 3029-3039.

[13] Ben-Brahim, L; Tadakuma S; Akdag, A. Speed Control of Induction Motor without Rotational Transducers; *IEEE Trans. Ind. Appl.*, 1999, vol. 35, 844–850.

[14] Pinto, JOP; Bose, BK; da Silva, LEB; Kazmierkowski, MP. A Neural-Network-Based Space-Vector PWM Controller for Voltage-Fed Inverter Induction Motor Drive; *IEEE Trans. Ind. Appl.*, 2000, Vol. 36, 1628-1636.

[15] Karanayil, B; Rahman, MF; Grantham, C. Online Stator and Rotor Resistance Estimation Scheme using Artificial Neural Networks for Vector Controlled Speed Sensorless Induction Motor Drive; *IEEE Trans. Ind. Electron.*, 2007, Vol. 54, 167-176.

[16] Vukadinovic, D; Basic, M; Kulisic, Lj. Stator Resistance Identification Based on Neural and Fuzzy Logic Principles in an Induction Motor Drive; *Neurocomputing,* 2010, Vol. 73, 602-612.

[17] Pryymak, B; Moreno-Eguilaz, JM; Peracaula, J. Neural Network Flux Optimization using a Model of Losses in Induction Motor Drives; *Math. Comput. Simul.*, 2006, Vol. 71, 290-298.

[18] Lin, FJ; Teng, LT; Shieh, PH; Li, YF. Intelligent Controlled-Wind-Turbine Emulator and Induction-Generator System using RBFN; *IEE Proc–Electr. Power Appl.,* Vol. 153, 2006, 608-618.

[19] Barazane, L; Khwaldeh, A; Krishan, MM; Ouiguini, R. Optimization by Gaussian Radial Basis Function Neural Network of the Performance of Induction Motor System Based on New Linguistic Fuzzy Model, *IREE* 2008, Vol. 3, 344-354.

[20] Lindsay, CS; Denby, B; Lindblad, T. (1998). *Neural Network Hardware.* http://neuralnets.web.cern.ch/NeuralNets/nnwInHepHard.html, accessed on April 30th, 2010.

[21] *http://www.technosoftmotion.com/products/TOOLS_MSK.htm*, accessed on April 30th, 2010.

In: Focus on Artificial Neural Networks
Editor: John A. Flores, pp. 257-275
ISBN: 978-1-61324-285-8

Chapter 12

WAVELET NEURAL NETWORKS: A RECENT STRATEGY FOR PROCESSING COMPLEX SIGNALS APPLICATIONS TO CHEMISTRY

Juan Manuel Gutiérrez[1,2], Roberto Muñoz[2], and Manel del Valle[1], *

[1]Sensors & Biosensors Group, Department of Chemistry, Universitat Autònoma de Barcelona, 08193 Bellaterra, Spain
[2]Department of Electrical Engineering, Bioelectronics Section, CINVESTAV, 07360 Mexico D.F., Mexico

ABSTRACT

In the last three decades, Artificial Neural Networks (ANNs) have gained increasing attention due to their wide and important applications in different areas of knowledge as adaptive tool for processing data. ANNs are, unlike traditional statistical techniques, capable of identifying and simulating non-linear relationships without any a priori assumptions about the data's distribution properties. Furthermore, their abilities to learn, remember and compare, make them useful processing tools for many data interpretation tasks in many fields, for example in chemical systems or in the analytical field.

Nevertheless, the development of new analytical instruments producing readouts of higher dimensionality and the need to cope with each time larger experimental data sets have demanded for new approaches in data treatment. All this has lead to the development of advanced experimental designs and data processing methodologies based on novel computing paradigms, in order to tackle problems in areas such as calibration systems, pattern recognition, resolution and recovery of pure-components from overlapped spectra or mixtures.

This chapter describes the nature and function of Wavelet Neural Networks (WNNs), with clear advantages in topics such as feature selection, signal pre-processing, data meaning and optimization tasks in the treatment of chemical data. The chapter focuses on

---

* Corresponding author: Manel del Valle: Phone +34 935813235; fax: +34 935812379. E-mail address: manel.delvalle@uab.es

the last applications of WNNs in analytical chemistry as one of its most creative contributions from theoretical developments in mathematical science and artificial intelligence. Specifically, recent contributions from our laboratory showing their performance in voltammetric electronic tongue applications will be outlined and commented.

INTRODUCTION

In practice, chemical properties of compounds and mixtures are determined using some measuring process in order to obtain raw experimental data. Traditionally, classical laboratory experiments involved the intervention of a specialist with trained eye to determine the cause-effect relationship between the dependent/independent variables that give rise to the phenomenon measured. However, in the case of multivariable systems (more common in reality) is absolutely essential to use appropriate tools to find multidimensional inner relationships between the variables within the measured or theoretical data. To explore and to validate these data sets are often the first actions needed for their interpretation. Currently, this chemometrics stage is a mandatory step in the analysis of chemical data to explore their information content. Normally, this will involve to carry out tasks such as reducing the size of the data and extracting relevant features aimed to discard irrelevant and/or redundant information, in order to build adequate interpretation models and to reduce computational costs (Gutiérrez et al. 2008).

Nowadays, the extraction of information from chemical data is an active area of research in multidisciplinary groups around the world, particularly those related to electroanalytical methods where feature selection is a difficult problem because of the non-idealities present, the existence of interference effects, the colinearities among measured variables and the presence of noise. Electroanalysis has long benefited from the well-established techniques such as potentiometry, coulometry, polarography, voltammetry and electrochemical impedance spectroscopy. Novel working principles, such as electronic noses and electronic tongues, have enlarged the scope of application of the above techniques (Ni and Kokot, 2008).

Electronic tongues are analytical chemical systems which employ electrochemical sensors in a novel way, having as main foundation the sense of taste in animals. A widely accepted definition of electronic tongue entails an analytical instrument comprising an array of non-specific, poorly selective, chemical sensors with cross-sensitivity to different compounds in a solution, and an appropriate chemometric tool for data processing (Holmberg et al., 2004). The low selectivity of sensors is what precisely produces complex cross-response signals containing information about different compounds plus other features. Due to the lack of selectivity, an important part in a multisensor approach is the signal processing stage.

During the last years, the electroanalytical methods have taken advantage of different data processing methods to improve tasks such as the interpretation, correlation and modeling of measured chemical responses based on properties of a set of samples, e.g., concentration, intensity, potential, mass, spectra, etc. (Wold, 1995). Thus, the development of computational paradigms and advanced algorithms to face proper handling of the information, have encouraged the establishment of novel computational models with predictive capabilities and improved performance.

By tradition, classical experimentation involves the use of modeling methods to find data meaning; these models are mainly classified in linear and non-linear methodologies. Regardless of the nature of the methods, their use is often employed in applications such as multivariate calibration and resolution. On the one hand, multivariate calibration entails relating the concentration or any other measured chemical value with a response obtained from a multicomponent mixture. On the other hand, resolution focuses on the recovery of pure-components from an overlapped spectra or electrochemical signal (Wold and Sjöström, 1998).

The most common methods used for modeling include Multiple Linear Regression (MLR), Partial Least Squares (PLS) and Principal Components Regression (PCR). Its implementation is often associated to solve problems where there are a large number of variables and relatively few data. In this way, the obtained prediction models are not suitable to discriminate new information, performing good results particularly when the relationship between variables is linear (Gutiérrez et al., 2009). An alternative to solve this inconvenience is the use of Artificial Neural Networks (ANN). These mathematical models which mimic the human brain, have shown great ability to interpret and model linear and nonlinear data relationships, making them useful for modeling and calibrating complex analytical signals (Despagne and Massart, 1998).

A common feature in published papers using ANN, especially if the measured records are too complex, is the need of a pre-processing stage. Methods such as Principal Components Analysis (PCA), Discrete Fourier Transform (DFT) or Wavelet Transform (WT) have proved to be appropriate to extract important features of the information. WT is specially useful because of its ability to analyze non-stationary signals with higher performance (Shao et al., 2003). Analytical techniques such as chromatography, infrared spectroscopy, mass spectrometry, nuclear magnetic resonance spectroscopy, ultraviolet-visible spectroscopy and electroanalysis, have also been benefited from the properties of wavelet processing for data compression, noise removal, base-line correction, zero crossing, regression and multi-resolution of overlapping signals (Leung et al., 1998; Nie et al., 2001).

The use of WT and ANN has been adopted as a strategy for non-stationary signal processing, especially when they have very large data sets. The main purpose of the WT is obtaining a new representation of the signals by a set of wavelet coefficients. These coefficients are used to construct an ANN that accomplishes the quantitative calibration of chemical species. Recently, some authors used this processing scheme in order to identify individual components present in voltammetric signals of mixtures (Cocchi et al., 2003; Palacios-Santander et al., 2003; Moreno-Barón et al., 2005). This methodology may be considered as an initial approach of the Wavelet Neural Network (WNN). Under this scheme, to select configuration details such as the mother wavelet, the level of compaction and the selection of ANN architecture, represents an exhaustive work which aims to get an appropriate combination of WT and ANN with the best performance.

True WNN models are obtained by merging the two data-processing techniques into a single one; these networks have the main feature that the transfer function in neurons of the hidden layer is a mother wavelet (Zhang and Benveniste, 1992). WNNs allowed in a single step, to perform feature extraction of chemical signals and to build the multivariate calibration model, making them powerful chemometric tools (Gutés et al., 2006).

In this chapter, we will attempt to describe the nature of WNNs to build models able to interpret and decode the complexity of chemical signals. Our research group has pioneered the use of WNN as a chemometric tool in the processing used for voltammetric electronic tongues. The knowledge gained will be summarized along the different sections to give the reader a comprehensive and specialized information on the use of this novel networks. For this purpose, a review of the most recent contributions of the different analytical chemistry fields, especially those related to sensor systems such as electronic tongues is also included.

FUNDAMENTALS OF WNN

Theory

As mentioned before, the idea of combining wavelets with neural networks resulted in a successful synthesis of theories that generated a new class of networks called WNNs or simply wavelet networks (Zhang and Benveniste, 1992). The innovation of these networks is the use of wavelet functions as hidden neuron activation functions in the ANN. Applying theoretical features of WT, different network construction methods can be proposed. The foundation of a basic WNN model is evident by observing the similarity between Strömberg's definition and the hidden layer of a Multi-layer Perceptron (MLP) (Akay, 1992; Meyer, 1994). Thus, wavelet decomposition can be seen as a neuronal model assuming that the scale and translation (s,t) of wavelets can be replaced by a new index $i = 1,..., k$, as is defined in the equation:

$$y = \sum_{i=1}^{k} w_i \psi_i (x) \qquad (1)$$

where w_i represent the wavelet coefficients of the decomposition of y, and ψ_i the daughter wavelets.

This approximation represents an important relation between the single output of a one-hidden layer MLP and one-dimensional signal decomposition using simple scaling and translations of a mother wavelet. A graphic interpretation of this concept can be seen on Figure 1.

For developing a WNN, the use of $1+1/2$ network's architectures (one input layer, one hidden layer and a single output) is enough to approximate any arbitrary and continuous function, using an appropriate family of functions in the hidden layer (Heil and Walnut, 1989; Hornik et al., 1989; Scarselli and Chung-Tsoi, 1998). The final accuracy in the approach depends of the used family function characteristics as well as the network's training error.

The essence of WNN is to find an appropriate family of wavelets in a representative space in order that the complex relationships contained in the original signals might be properly expressed (Daubechies, 1992). In this sense, there are two possibilities to construct a complete and useful family of wavelets for its inclusion in a neuronal model. The first, involves an orthogonal decomposition; the obtained daughter wavelets are linearly independent and orthogonal to each other so that there are not any redundant functions (Kaiser, 1994). The second is the use of wavelet frames which can be constructed by simple operations of translation and dilation without fulfilling stringent orthogonal conditions (Heil and Walnut, 1989; Akay, 1992; Gutés et al., 2006). Although many wavelet applications work well using orthogonal wavelet basis, many others work better with redundant wavelet families. The redundant representation offered by wavelet frames has demonstrated to be good both in signal denoising and in compaction (Daubechies et al., 1986; Daubechies, 1992).

In this manner, a signal $f(x)$ can be approximated by generalizing a linear combination of daughter wavelets $\psi_{s,t}(x)$ derived from its mother wavelet $\psi(x)$, this family of functions is defined as:

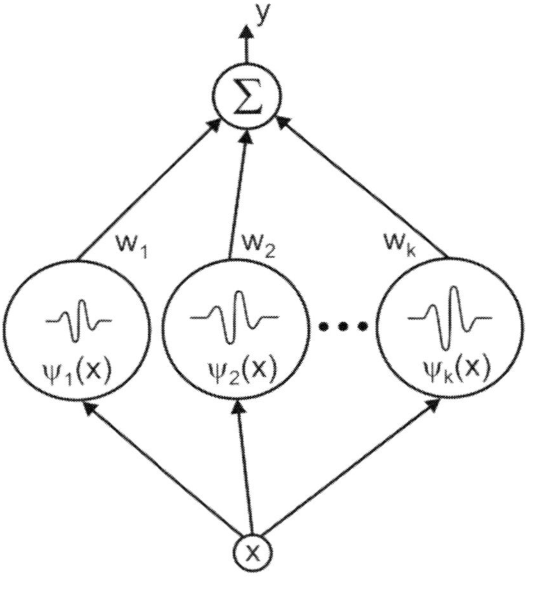

Figure 1. Wavelet expansion shown as neural network.

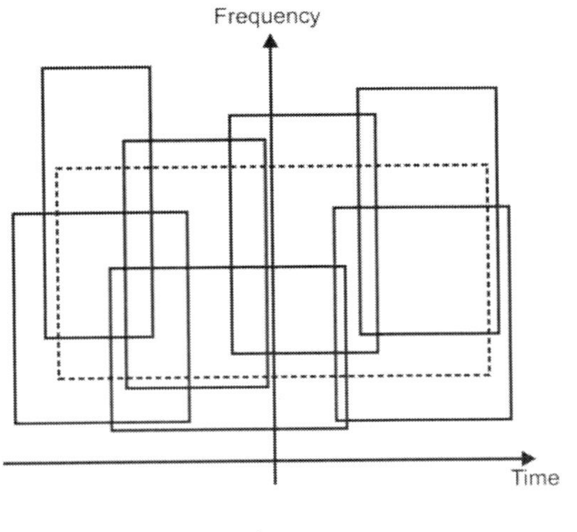

Figure 2. Continuous Wavelet Frames representation.

$$M_c = \left\{ \frac{1}{\sqrt{|s_i|}} \psi\left(\frac{x - t_i}{s_i}\right), \quad t_i, s_i \in \Re, \quad s_i > 0 \right\} \tag{2}$$

where the translation t_i and the scaling s_i are real numbers in $\Re$.

The family of functions M_c is known as a continuous wavelet frame of $L^2(\Re)$ if there exist two constants A and B that fulfills with (Kugarajah and Zhang, 1995):

$$A\|f(x)\|^2 \le \sum_{s,t} |\langle f(x), \psi_i(x)\rangle|^2 \le B\|f(x)\|^2 \text{ with } A > 0, B < +\infty \tag{3}$$

Figure 2 shows schematically the generation of functions M_c (represented by solid lines) covering the space where it is located the function to approximate (box with dotted lines).

Nevertheless for multi-variable model's applications it is necessary to use multidimensional wavelets. Families of multidimensional wavelets can be obtained from the product of P monodimensional wavelets, $\psi(a_{ij})$, of the form:

$$\Psi_i(x) = \prod_{j=1}^{P} \psi(a_{ij}) \text{ where } a_{ij} = \frac{x - t_{ij}}{s_{ij}} \tag{4}$$

Where now t_i and s_i are the translation and scaling vectors respectively.

WNN Algorithm

The most widespread WNN architecture is shown in Figure 3. This model corresponds to a one-hidden layer feedforward MLP with single output. The output y^n (where n is an index, not a power) depends on the connection weights c_i between the output of each neuron and the output of the network, the connection weights w_j between the input data and each output, an offset value $b_0(r)$ useful when adjusting functions that has a mean value other than zero, the n-th input vector x^n and the wavelet function Ψ_i of each neuron. The approximated signal of the model y^n can be represented by the next equation:

$$y^n = \sum_{i=1}^{K} c_i \Psi_i(x^n) + b_0 + \sum_{j=1}^{P} w_j x_j^n \text{ with } \{i, j, K, P\} \in Z \tag{5}$$

where subindexes i and j stand for the i-th neuron in the hidden layer and the j-th element in the input vector, x^n, respectively, K is the number of wavelet neurons and P is the length of input vector, x^n. With this model, a P-dimensional space can be mapped to a monodimensional space $(R^P \rightarrow R)$, allowing the network to predict a value for the output y^n when the n-th signal x^n is input to the trained network.

The basic neuron will be a multidimensional wavelet $\Psi_i(x^n)$ which is built using the definition (4), where scaling s_{ij} and translation t_{ij} coefficients are the adjustable parameters of the i-th wavelet neuron. With this mathematical model for the wavelet neuron, the network's output becomes a linear combination of several multidimensional wavelets (Gutés et al., 2006).

Training Stage

Different learning methods have been implemented in WNN training: some authors use stochastic gradient algorithm (Zhang and Benveniste, 1992), conjugate gradient method (Szu et al. 1996), backpropagation algorithm (Oussar and Dreyfus, 1998) and most recently the use of sampling theory (Zhang, 2007). There are also cases where backpropagation is combined with Orthogonal Least Square-Backward Elimination (Zhang and Benveniste, 1992; Zhang, 1997), mainly used in the selection of network structures. Nevertheless, backpropagation remains the most popular algorithm in the training process. There are several reasons for the backpropagation method is still widely used. Some of them are listed below.

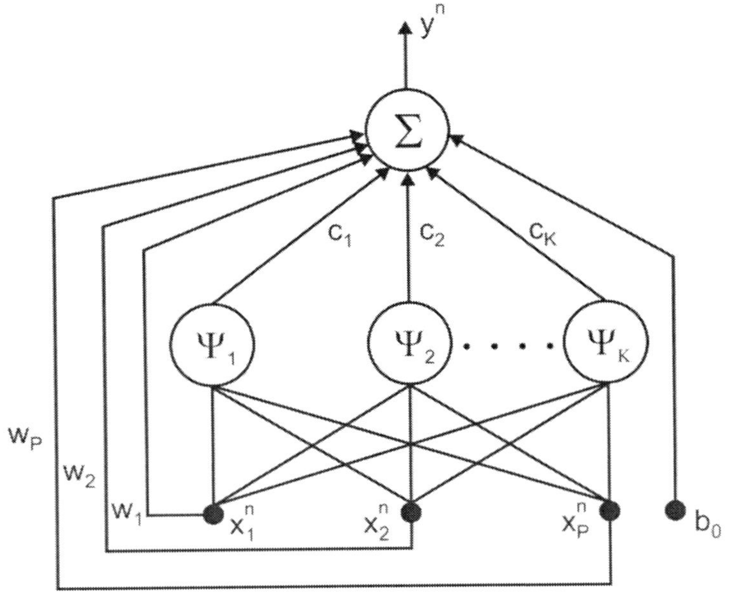

Figure 3. WNN topology architecture.

- It is easy to understand and implement.
- It has been shown that it can approximate'well-behaved' functions and thus is often qualified as a generic learning mechanism. In particular, it can learn nonlinearities in separable pattern sets.
- It involves only few parameters and these are not very critical for the final training result.
- Although the algorithm is often found to be slow and has as critical point how to find the appropriate the number of neurons in the hidden layer, its implementation with WNN it is quite useful because the number of neurons required in these models is usually reduced.

On the other hand, many algorithms are suggested to initialize the weights (Oussar et al., 2000), adjust the structures of wavelet networks (Yevgeniy et al., 2005) and accelerate the convergence when the error backpropagation is applied to the training algorithms (Gutés et al., 2006).

Other algorithms besides backpropagation have been also proposed to optimize the configuration of the wavelet network, e.g., Kalman filter (Qingmei and Yongchao, 2003), genetic algorithms (ChangGyoon et al., 2004) and immune algorithms (Huang and Cui, 2005). In parallel to this, to have a sufficiently large training set (at least 2 samples per connection weight) is a must when developing sound and meaningful response models (Despagne and Massart, 1998).

An unsolved problem in the design of WNN is related with the *curse of dimensionality* which brings some difficulties in applying this kind of networks to high-dimensional problems. In this sense, the determination of the number of hidden neurons and the correct initialization is an important issue. As in conventional ANN models, an appropriate architecture is still important to promote generalization and avoid overfitting.

WNNs and Their Applications to Chemistry

Although the development of the WNNs and their benefits in signal processing have been reported for more than 20 years, there are not many studies showing their application in chemistry. Some important contributions tend to use them in prediction tasks or quantitative structure-properties relationship (QSPR). For example the prediction of inclusion complexation constants of α- cyclodextrins (Guo et al., 1998) for a considered molecule, departing from a substituent molar index, its hydrophobic constant and its Hammet constant. The compounds considered were benzene derivatives, mono and 1,4-disubstituted, demonstrating a good agreement for more than 40 different compounds. Later, the same authors repeated a similar work, now employing β-cyclodextrin ligands (Liu and Guo, 1999), with close results.

Another similar property prediction study was proposed in which a WNN bettered a conventional back-propagation ANN in the prediction of gas chromatographic retention times using the programmed-temperature technique, a variant usually adopted in the petrochemical industry (Zhang et al., 2001). The chemical compounds considered were naphtas (94 different compounds), and the departure information was their density plus the isothermal retention index. A curious QSPR study employing WNNs was performed for the prediction of gas chromatography retention indexes of methyl-substituted alkanes produced by insects (Atabati and Zarei, 2008). A number of structure and molecular geometry descriptors were the inputs for the WNN model, achieving an average relative error of prediction of 2.2%.

A number of interesting works with WNNs originated from research groups in Iran. The first recorded work is a QSPR study in order to predict the solubility of polycyclic aromatic hydrocarbons in supercritical CO_2 (Khayamian and Esteki, 2004). The WNN model departed from six descriptors of the chemical considered: temperature, pressure, volume of the molecule, highest occupied molecular orbital, dipole moment and number of double bonds. A similar work developed a QSPR relationship to predict the critical micelle concentrations of 94 Gemini surfactants (Kardanpour et al., 2005), in this case from different molecular descriptors related to geometry of the molecule and their electronegativity. Performance of the WNN was shown to be superior to Multivariate Linear Regression (MLR). Two additional works did the solubility prediction of anthraquinone dyes (Tabaraki et al., 2006) or azo dyes (Tabaraki et al., 2007) at different temperatures and pressures in supercritical CO_2. A number of descriptors related to the molecular structure and properties were initially tested and reduced down to six in both cases. A similar work compared the abilities of ANNs and WNNs in the preditction of three gasoline properties (density, benzene and ethanol content), departing from near infrared spectroscopy data (Balabin et al., 2008), where the algorithm employing wavelets demonstrated to be more efficient. In 2007, another QSPR study employed ANNs and WNNs to predict solvent polarity (Zarei et al., 2007). A set of 69 compounds, including unsaturated hydrocarbons, solvents containing halogen, cyano, nitro, amide, sulfide, mercapto, sulfone, phosphate, ester, ether, etc. was used, and again WNN outperformed ANN. The application of WNN in chemical process control has also been considered, in this case to perform real-time dynamic fault diagnosis (Zhao et al. 1998).

Moreover, WNNs have been employed to build response models for multicomponent determinations using spectrophotometric techniques. A first contribution is the multivariate spectrophotometric determination of Cu^{2+}, Fe^{2+} and Al^{3+} using pyrocatechol violet as chromogenic reagent (Khayamian et al., 2005). This work developed a WNN response model from the the principal components of the UV-VIS spectra (obtained from PCA), a somewhat redundant operation, given WNNs are already performing a feature extraction process. A similar work was that of Zarei and Atabati, who determined Fe^{2+}, Ni^{2+} and Co^{2+} using pyridylazo naphtol in micellar media (Zarei and Atabati, 2006), again from the UV-VIS spectra preprocessed with PCA. A related work was the simultaneous kinetic determination of thiocyanate and sulfide from the time transient absorbance record at 349 nm, corresponding to the decoloration reaction of the reagent iodine by these compounds (Ensafi et al., 2007). As each considered substance showed a differentiated transient, with different time constant, the resolution of the mixture was attainable. In this case, the response model used a WNN with the absorbance transient pre-processed employing PCA as input.

In a specific instrument study, the WNN signal processing variant demonstrated a 10-fold improvement over traditional signal processing methods in the detection limit of various nitrogen and phosphorus compounds from the output of a thermionic detector attached to a gas chromatograph (Fu et al., 2005). With results obtained from this analytical technique, the WNN treatment had the potential to extract signals for at least 10 times lower signal-to-noise ratios than standard filtering and averaging techniques.

The first study related to electroanalysis was the oscillographic chronopotentiometric determination of mixtures of Pb^{2+}, In^{3+} and Zn^{2+} where it was used a discrete version of wavelet network (Zhong et al., 2001). Given the experimental signal is a highly overlapped trace, WNNs were employed as a chemometric stage to resolve the three-species mixture, with recoveries close to 100% for the three considered heavy metals. Khayamian in Isfahan (Iran) showed the extension of the calibration range for Cu^{2+} using adsorptive stripping voltammetry and xylenol orange as ligand (Khayamian et al., 2006), when a WNN response model was employed instead of the classical linear regression fit. The range was extended from 1-50 ng mL^{-1} to 1-1500 ng mL^{-1}, obtaining better linearity with the WNN model than a equivalent ANN one. Nevertheless, there are few applications dealing with multianalyte quantification in overlapped-signals of voltammetric applications: the simultaneous quantification of oxidizable compounds (Gutés et al., 2006) the resolution of mixtures of three phenolic compounds (Gutiérrez et al., 2008) and the determination of three heavy metals (Pb^{2+}, Cd^{2+} and Cu^{2+}) in presence of Tl^{+} and In^{3+} as interfering species (Gutiérrez et al., 2009).

Latest report of use of WNN is related with a QSPR work for the prediction of the half-wave reduction potential of a set of 73 aldehydes and ketones (Garkani-Nejad et al., 2010). Authors compared performance of MLR, PLS, ANN and WNN, preferring the non-linear last two methods over the two linear ones.

All the works mentioned have in common the use of wavelet networks with a single output using different wavelet activation functions. Among them, only the papers presented by Gutiérrez et al. developed multi-output models capable of decoding the multivariate information present in voltammetric signals.

WNN in a Chemical Sensing Application

In the recent years, there have been published contributions on the use of electronic tongues for environmental monitoring purposes (Legin et al., 1996; Di Natale et al., 1997; Krantz-Rülcker et al., 2001; Winquist et al., 2005; Gutiérrez et al., 2007; Mimendia et al., 2010). Our research group proposed recently the use of voltammetric electronic tongues and WNN data processing as potential monitor of pollution episodes by heavy metals (Gutiérrez et al., 2009).

The study presented here, corresponds to the direct multivariate determination of Pb^{2+}, Cd^{2+}, Cu^{2+}, Tl^{+} and In^{3+} from the complete cyclic voltammogram obtained with a laboratory made sensor electrode (consisting of a composite of graphite, epoxy and platinum). Cyclic Voltammetry (CV) has commonly been used as qualitative technique in chemistry. However, recent publications are introducing it coupled to chemometric techniques in order to quantify or semi-quantify the analytes of interest in a multivariate calibration approach (Saurina et al., 2000; Moreno-Barón et al., 2005; Kramer et al., 2007).

WNN Multi-Output Model

The developing of a multi-output WNN follows the same criteria described above, where basic hidden neurons were constructed with multidimensional wavelets employing wavelet frames. In this way, different network's outputs are weighted sums of inner products between multidimensional wavelets bases and input signal vectors. As the space of input data is represented by a new family of multidimensional wavelets, signal feature extraction is carried out by estimating the parameters in the wavelet neurons of the network whit highest similarity (scaling and translating) plus the connection weights. Multi-output model is shown in Figure 4. The approximate signal of the model $y^n(r)$ can be expressed by the equation (Gutiérrez et al., 2008):

$$y^n(r) = \sum_{i=1}^{K} c_i(r)\Psi_i(x^n) + b_0(r) + \sum_{j=1}^{P} w_j(r)x_j^n \text{ with } \{i, j, K, P\} \in Z \tag{6}$$

where $r = 1,2,...m$, with $m \in Z$, represents the number of outputs, and subindexes i and j stand for the i-th neuron in the hidden layer and the j-th element in the input vector, x^n, respectively, K is the number of wavelet neurons and P is the length of input vector, x^n. With this model, a P-dimensional space can be mapped to a m-dimensional space $(R^P \to R^m)$, allowing the network to predict a value for the output $y^n(m)$ when the n-th signal x^n is input to the trained network.

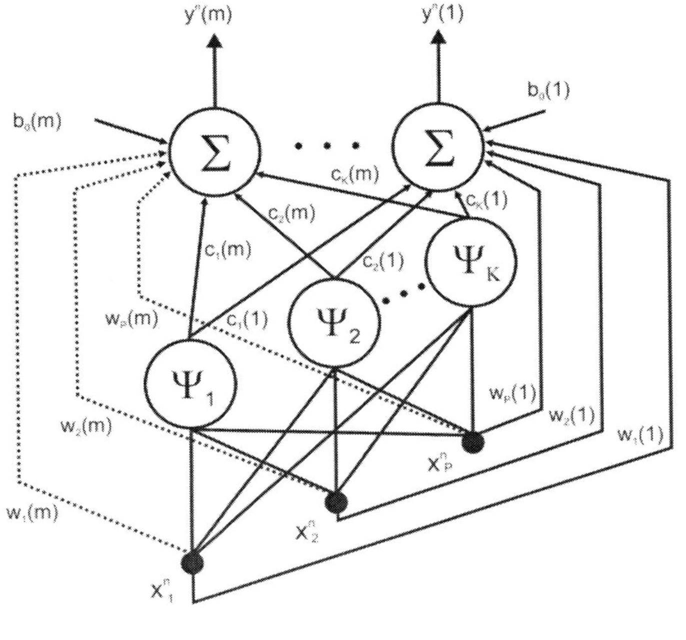

Figure 4. WNN multi-output topology.

Error backpropagation was used as learning algorithm. The difference between expected and obtained values is evaluated according to the Mean Squared Error (MSE) defined by:

$$MSE = \frac{1}{2}\sum_{n=1}^{N}\sum_{r=1}^{m}\left(y_{\exp}^{n}(r) - y^{n}(r)\right)^{2} \quad (7)$$

where $y^n(r)$ is the r-th output of the network and $y_{\exp}^{n}(r)$ is the r-th expected concentration value related to the voltammogram input vector x^n. N corresponds to the number of input vectors and m the number of outputs.

The employed WNN´s activation function corresponded to the first derivative of a Gaussian function defined by equation (8).

$$\psi(x) = xe^{-0.5x^2} \quad (8)$$

Chemical Procedure

Lead, copper, cadmium, thallium and indium stock solutions were prepared from analytical reagent grade chemicals in acetate buffer solution (0.1 M, pH 3.76). A total of 30 samples with different concentrations of metals (randomly distributed in the range of 0.01-0.1 ppm) were considered. The concentration of each sample were selected in order to generate a simplified model space without drifts and trends, which ensures that learning of WNNs is not conditioned by any previous data sample.

For voltammetric measurements, a commercial potentiostat (Autolab PGSTAT 30) with the laboratory made working electrode was used (Gutiérrez et al., 2009). The cell was completed with an Ag/AgCl reference electrode (Orion 900200) and a commercial Platinum counter electrode (model 52-67 1, Crison). All experiments were carried out without any oxygen removal from the sample, which is a common great interference in electroanalysis and here gives an interesting value to this contribution.

CV signal is obtained by applying a linear potential sweep to the working electrode; once it reaches a set potential, the sweep direction is reverted. In this way, each resulting voltammogram consisted of 534 current values recorded in a range of potentials from -1.0V to 0.3V to -1.0V in steps of 0.0048V. A sample of the obtained voltammograms is plotted in Figure 5. As can be seen, similarities between different signals and the overlapping effect difficult the direct metal determination.

Data Processing

With the 30 samples determined, the input data for WNN models is a voltammetric matrix of dimension [534,30], and the target is a [5,30] matrix (Pb^{2+}, Cd^{2+}, Cu^{2+}, Tl^{+} and In^{3+} concentrations). The processed information was normalized to an interval of [-1,1], and randomly separated in two sets; three quarters of the data were used for training and the remaining quarter for testing. Each voltammogram is represented by x^n which is mapped to a point of the five-dimensional space of concentrations identified by $y^n(m)$. WNN structures with architecture of 153x3x5 were programmed and trained to obtain the concentration of each metal. Structures with more neurons in the hidden layer were not tested because the first architectures tested allowed to obtain satisfactory results, as reported in our previous works (Gutiérrez et al., 2008; Gutiérrez et al., 2009). The expected output error was programmed to reach a value of 0.025 ppm and the learning rate was set to 0.0004.Prediction capability of the model was evaluated using k-fold cross validation method, choosing different training and testing sets, in order to validate the reliability of the model in an unconditioned way.

Obtained Results

To determine the WNN performance in quantification of metals, different comparison graphs between obtained and expected concentrations were built, both for training and for testing subsets. The linear regression of the obtained vs. expected concentration comparison graph was a measure of the goodness of the model for each metal. This fitted line, in ideal conditions, should yield the diagonal identity line (slope 1 and intercept 0). Figure 6 shows the comparative graphs between real concentrations for lead, cadmium and copper and those predicted with the WNN model for training and testing subsets. The slope (m) and intercept (b) which defines the comparison line y=mx+b that best fits the data (along with the uncertainly interval for a 95% of significance) are shown for one of the study cases in Table 1, where the ideal situation is fulfilled in all cases (at the 95% confidence level). Simultaneously, the high level of linearity accomplished permits to express the low dispersion of obtained data.

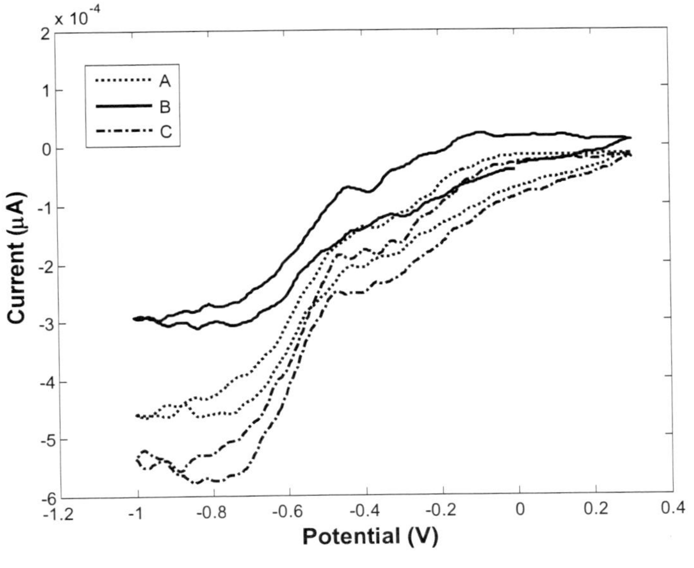

Figure 5. Some examples of CV voltammograms. Sample composition corresponds to a mixture of the five metals at equal levels of concentration: (A) 0.07 ppm; (B) 0.05 ppm; (C) 0.1 ppm.

Table 1. Linear regression parameters for interested heavy metals with the training and testing data sets using a WNN architecture whit three neurons in the hidden layer

Metal	Training					Testing				
	R	*m*	*error*	*b*	*error*	*R*	*m*	*error*	*b*	*error*
Pb^{2+}	0.986	0.994	±0.084	-3.979E-05	±0.006	0.899	1.188	±0.472	-4.795E-03	±0.034
Cd^{2+}	0.964	0.997	±0.135	-7.704E-04	±0.010	0.923	1.361	±0.462	-1.365E-02	±0.034
Cu^{2+}	0.960	0.979	±0.141	9.791E-04	±0.010	0.892	1.240	±0.514	-5.192E-03	±0.037

Table 2. Linear regression parameters for the interfering metals with the training and testing data sets using a WNN architecture whit three neurons in the hidden layer

Metal	Training					Testing				
	R	*m*	*error*	*b*	*error*	*R*	*m*	*error*	*b*	*error*
Tl^{+}	0.983	0.992	±0.093	3.434E-04	±0.007	0.745	0.654	±0.477	1.445E-02	±0.035
In^{3+}	0.949	1.011	±0.166	-1.357E-03	±0.011	0.837	1.028	±0.548	-2.419E-03	±0.037

As well as for the metals of interest, the capability of the presented model was studied for the determination of the interfering metal Tl^{+} and In^{3+}. The correlation coefficient obtained and comparison lines are shown in Table 2.

Five extra training procedures, using the same WNN architecture, were programmed with random initialization of weights to verify the final model's consistency. Employing a 10-fold cross validation method, a testing set was selected each time randomly from the total set of the available data. To compare the accuracy of the predicted information, an average Recovery Percentage (RP) was calculated for each WNN model. RP is defined by equation (9):

$$RP = \frac{\sum_{i=1}^{N} 100 \cdot \left(1 + \frac{y_i - y_{\exp_i}}{y_{\exp_i}}\right)}{N} \tag{9}$$

where y_i is the *i*-th obtained concentration value, $y_{\exp_i}$ is the *i*-th expected concentration value and N is the size of the external test set.

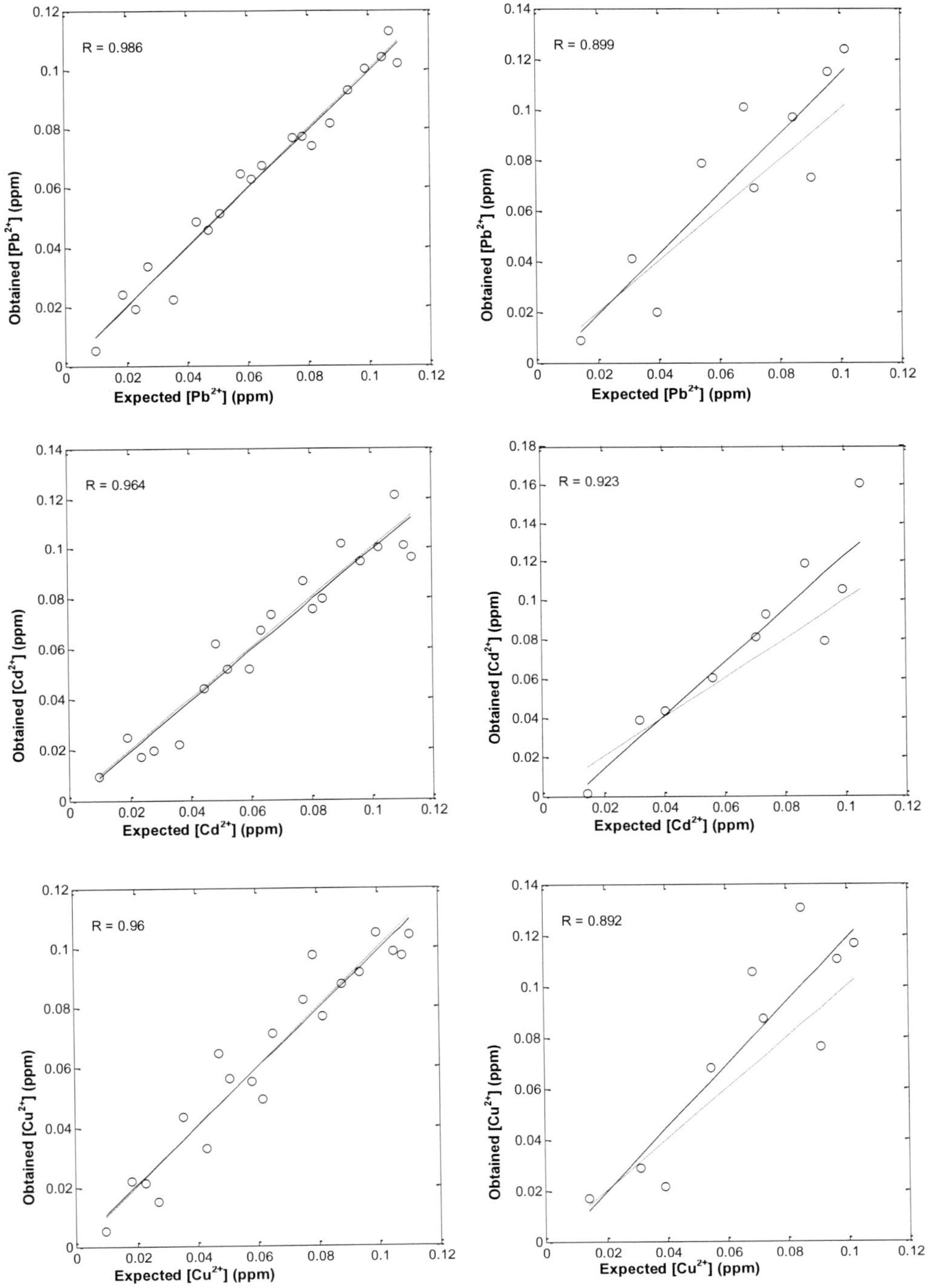

Figure 6. Comparison between obtained results and those expected using a WNN model with three neurons in the hidden layer. The dashed line corresponds to ideality (y=x) and the solid line is the regression of the comparison data (correlation coefficient R is adjoined on each graph). Plots on the left correspond to training and plots on the right to test subset.

RP value is a parameter which indicates the ability of the model to determine quantitatively a chemical analyte present in a sample; the ideal value is 100% (Gutés et al., 2006). Figure 7 summarizes the results for principal metal species. In this figure, it is possible to observe the fluctuation of the RP for each metal in the different WNN models. RP's value is a consequence of the random selection of the data, which in the worst case never exceeded the 10% of the ideal RP.

The mean recovery percentages and their standard deviations for the different tests can be seen in Figure 8. Accuracy of the predicted information is always below a 5% RSD uncertainly, and the ideal 100% situation is always attained. This ensures the models' validity regardless of the data selection used or its construction.

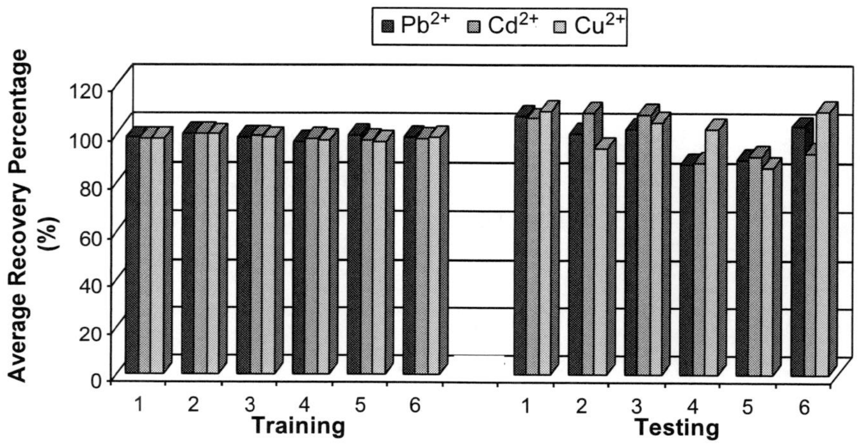

Figure 7. Average Recovery yield rated for the programmed WNN. Bars on the left correspond to each training session and bars on the right to the test set selected. The numbers below the x-axis, represent each WNN case.

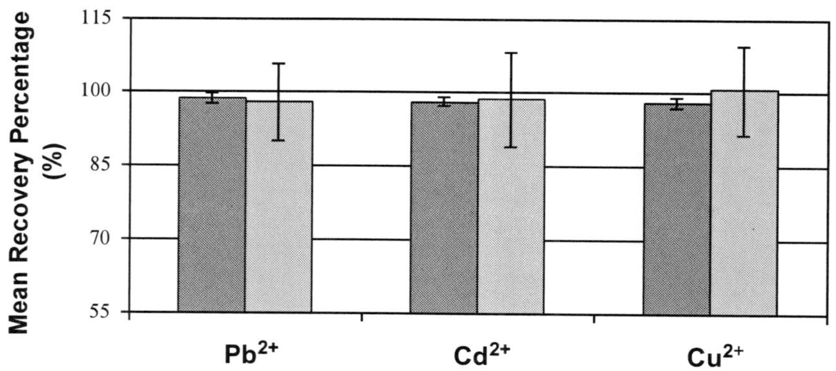

Figure 8. Mean Recovery yield and Standard Deviation obtained for the six WNN cases. In the representation, training is shown with dark grey and testing with pale grey.

Conclusion

The combination of wavelet theory and neural networks has led to the development of Wavelet Neural Networks (WNNs). Their impact on signal processing has been inherited from the rigorous mathematical foundations of both theories. WNNs have proven to be useful in tasks of modeling, classification and identification problems with some success over the most traditional data-processing methods used in analytical chemistry, e.g., multiple linear regression, partial least squares, principal components regression and artificial neural networks. The strength of WNN models lies in their capabilities to catching essential features in signals, allowing them to interpret difficult input-output relationships present in data sets coming from complex and non-stationary signals, which is especially useful for QSPR studies.

WNN architectures described in this chapter, that make use of continuous wavelet frames and back-propagation training algorithm allow to obtain feasible and effective models compared to the methods provided by conventional pre-processing stages plus artificial neural networks. Once a WNN network has been trained, the adjusted parameters represent a family of wavelet functions that matches the frequency content of signals. This family of functions represents a multi-resolution signal analysis which allows not only getting more useful information of data but also avoiding pre-processing stages such as feature extraction, data reduction and filtering.

A complete case study of signal treatment in electrochemical analysis has been shown in detail. Based on its results, it is possible to observe that the simultaneous quantitative determination of metallic species was achieved successfully employing a WNN model. Estimated concentration values, in the sub-ppm range, showed errors lower than 5% for Pb^{2+}, Cd^{2+} and Cu^{2+}; and in the ±5% interval for Tl^{2+} and In^{3+}, all values around the theoretical 100% recovery percentage. The proposed approach has demonstrated to be a proper multivariate modeling tool for voltammetric analytical signals. For its operation, the WNN adjusts the parameters for a family of wavelet functions that best fits the shapes and frequencies of the sensors signals. WNNs were able to extract meaningful information from the signal in order to estimate properly different species concentrations, even in the presence of important interfering elements as oxygen.

Acknowledgments

Financial support for this work was provided by Spanish Ministry of Science and Innovation, project TEC2007-68012-C03-02 and the Mexican National Council of Science and Technology CONACYT through Postdoctoral scholarship for J.M. Gutiérrez.

References

Akay, M. Time frequency and wavelets; In: Akay, M., (Ed.) Biomedical Signal Processing. *IEEE Press Series in Biomedical Engineering*, Wiley-IEEE Press: Piscataway NJ, 1992.

Atabati, M; Zarei, K. *J. Chinese Chem.Soc.*, 2008, 55, 732-739.

Balabin, RM; Safieva, RZ; Lomakina, EI. *Chemom. Intell. Lab. Syst.*, 2008, 93(1), 58-62.

Cocchi, M; Hidalgo-Hidalgo-de-Cisneros, JL; Naranjo-Rodríguez, I; Palacios-Santander, J. M; Seeber, R; Ulrici, A. *Talanta*, 2003, 59(4), 735-749.

ChangGyoon, L; Kangchui, K; Eungkon, K. Proceedings of the Fifteenth IASTED International Conference on Modeling and Simulation, CA, USA, 2004, 55-59.

Daubechies, I. Ten Lectures of wavelets. CBMS-NSF Regional Conference Series In *Applied Mathematics*, Vol. 61. Society for Industrial and Applied Mathematics, Philadelphia, PA, 1992.

Daubechies, I; Grossmann, A; Meyer, Y. *J. Math. Phys.*, 1986, 27(5), 1271-1283.

Despagne, F; Massart, D. *Analyst*, 1998, 123, 157R - 178R.

Di Natale, C; Macagnano, A; Davide, F; D'Amico, A; Legin, A.: Vlasov, Y; Rudnitskaya, A; Selezenev, B. *Sens. Actuators B.*, 1997, 44(1-3), 423-428.

Ensafi, A. A; Khayamian, T; Tabaraki, R. *Talanta*, 2007, 71(5), 2021-2028.

Fu, C. Y; Petrich, L. I; Daley, P. F; Burnham, A. K. *Anal. Chem.* 2005, 77, 4051-4057.

Garkani-Nejad, Z; Rashidi-Nodeh, H. *Electrochim. Acta.*, 2010, 55(8), 2597-2605.

Guo, Q. X; Liu, L; Cai, W. S; Jiang, Y; Liu, Y. C. *Chem. Phys. Lett.*, 1998, 290(4-6), 514-518.

Gutés, A; Céspedes, F; Cartas, R; Alegret, S; del Valle, M; Gutiérrez, J. M; Muñoz, R. *Chemom. Intell. Lab. Syst.*, 2006, 83(2), 169-179.

Gutiérrez, JM; Gutés, A; Céspedes, F; del Valle, M; Muñoz, R. *Talanta*, 2008, 76(2), 373-381.

Gutiérrez, JM; Moreno-Barón, L; Céspedes, F; Muñoz, R; del Valle, M. *Electroanal.*, 2009, 21(3-5), 445-451.

Gutiérrez, M; Gutiérrez, JM; Leija, L; Hernández, PR; Favari, L; Muñoz, R; del Valle, M. *Intern. J. Environ. Anal. Chem.*, 2007, 88(2), 103–117.

Heil, C; Walnut, DF. *SIAM Review*, 1989, 31(4), 628-666.

Holmberg, M; Eriksson, M; Krantz-Rülcker, C; Artursson, T; Winquist, F; Lloyd-Spetz, A; Lundström, I. *Sens. Actuators B.*, 2004, 101(1-2), 213-223.

Hornik, K; Stinchcombe, M; White, H. *Neural Netw.*, 1989, 2(5), 359-366.

Huang, M; Cui, B. A novel learning algorithm for wavelet neural networks, In: L; Wang, K; Chen, YS. Ong, (Eds), *Advances in natural computation: First international conference on natural computation ICNC 2005*, Springer-Verlag, Berlin, 2005, 1-7.

Kaiser, GA. *Friendly Guide to Wavelets.* Ed. Birkhäuser, Boston MA, 1994.

Kardanpour, Z; Hemmateenejad, B; Khayamian, T. *Anal. Chim. Acta.*, 2005, 531(2), 285-291.

Khayamian, T; Ensafi, AA; Benvidi, A. *Talanta*, 2006, 69(5), 1176-1181.

Khayamian, T; Ensafi, AA; Tabaraki, R; Esteki, M. *Anal. Lett.*, 2005, 38(9), 1477-1489.

Khayamian, T; Esteki, M. *J. Supercrit. Fluids.*, 2004, 32(1-3), 73-78.

Kramer, KE; Rose-Pehrsson, SL; Hammond, MH; Tillett, D; Streckert, HH. *Anal. Chim. Acta.*, 2007, 584(1), 78-88.

Krantz-Rülcker, C; Stenberg, M; Winquist, F; Lundström, I. *Anal. Chim. Acta.*, 2001, 426(2), 217-226.

Kugarajah, T; Zhang, Q. *IEEE Trans. Neural Netw.*, 1995, 6(6), 1552-1556.

Legin, AV; Vlasov, YG; Rudnitskaya, AM; Bychkov, EA. *Sens. Actuators B.*, 1996, 34(1-3), 456-461.

Leung, AK; Chau, F; Gao, J. *Chemom. Intell. Lab. Syst.*, 1998, 43(1-2), 165-184.

Liu, L; Guo, QX. *J. Chem. Inf. Comput., Sci.*, 1999, 39(1), 133-138.

Meyer, Y. Wavelets, *Algorithms and Applications*. Society for Industrial and Applied Mathematics 2nd Edition, SIAM, Philadelphia, PA, 1994.

Mimendia, A; Gutiérrez, JM; Leija, L; Hernández, PR; Favari, L; Muñoz, R; del Valle, M. *Environ. Modell. Softw.*, 2010, in press doi: 10.1016/j.envsoft.2009.12.003.

Moreno-Barón, L; Cartas, R; Merkoçi, A; Alegret, S; Gutiérrez, JM; Leija, L; Hernandez, PR; Muñoz, R; del Valle, M. *Analy. Lett.*, 2005, 38(13), 2189-2206.

Ni, Y; Kokot, S. *Anal. Chim. Acta.*, 2008, 626(2), 130-146.

Nie, L; Wu, S; Wang, J; Zheng, L; Lin, X; Rui, L. *Anal. Chim. Acta.*, 2001, 450(1-2), 185-192.

Oussar, Y; Dreyfus, G. *Neurocomputing*, 2000, 34(1-4), 131-143.

Oussar, Y; Rivals, I; Personnaz, L; Dreyfus, G. *Neurocomputing*, 1998, 20(1-3), 173-188.

Palacios-Santander, JM; Jiménez-Jiménez, A; Cubillana-Aguilera, LM; Naranjo-Rodríguez, I; Hidalgo-Hidalgo-de-Cisneros, JL. *Microchim. Acta.*, 2003, 142(1), 27-36.

Qingmei, S; Yongchao, G. Proceedings of the International Conference on Wavelet Analysis and Its Applications (WAA), *Chongqing, China*, 2003, 633–638.

Saurina, J; Hernández-Cassou, S; Fàbregas, E; Alegret, S., *Anal. Chim. Acta.*, 2000, 405(1-2), 153-160.

Scarselli, F; Chung-Tsoi, A. *Neural Netw.*, 1998, 11(1), 15-37.

Shao, XG; Leung, AKM; Chau, FT. *Acc. Chem. Res.*, 2003, 36(4), 276-283.

Szu, H; Telfer, B; Garcia, J. *Neural Netw.*, 1996, 9(4), 695-708.

Tabaraki, R; Khayamian, T; Ensafi, AA. *J. Mol. Graph. Model.*, 2006, 25(1), 46-54.

Tabaraki, R; Khayamian, T; Ensafi, AA. *Dyes Pigm.*, 2007, 73(2), 230-238.

Winquist, F; Bjorklund, R; Krantz-Rülcker, C; Lundström, I; Östergren, K; Skoglund, T. *Sens. Actuators B.*, 2005, 111-112, 299-304.

Wold, S. *Chemom. Intell. Lab. Syst.*, 1995, 30(1), 109-115.

Wold, S; Sjöström, M. *Chemom. Intell. Lab. Syst.*, 1998, 44(1-2), 3-14.

Yevgeniy, B; Nataliya, L; Iryna, P; Olena, V. *Expert Syst.*, 2005, 22(5), 235-240.

Zarei, K; Atabati, M. *Anal. Lett.*, 2006, 39(9), 2085-2094.

Zarei, K; Atabati, M; Ebrahimi, M. *Anal. Sci.,* 2007, 23, 937-942.

Zhang, Q. *IEEE Trans. Neural Netw.*, 1997, 8(2), 227-236.

Zhang, Q; Benveniste, A. *IEEE Trans. Neural Netw.*, 1992, 3(6), 889 - 898.

Zhang, X; Qi, J; Zhang, R; Liu, M; Hu, Z; Xue, H; Fan, B. *Comput. Chem.*, 2001, 25(2), 125-133.

Zhang, Z. *Neurocomputing*, 2007, 71(1-3), 244-269.

Zhao, J; Chen, B; Shen, J. *Comput. Chem. Eng.*, 1998, 23(1), 83-92.

Zhong, H; Zhang, J; Gao, M; Zheng, J; Li, G; Chen, L. *Chemom. Intell. Lab. Syst.*, 2001, 59(1-2), 67-74.

In: Focus on Artificial Neural Networks
Editor: John A. Flores, pp. 277-297
ISBN: 978-1-61324-285-8

Chapter 13

ROBUSTNESS VERIFICATION OF ARTIFICIAL NEURAL NETWORK PREDICTORS IN A PURPOSE-BUILT DATA COMPRESSION SCHEME

Rajasvaran Logeswaran*

Faculty of Engineering, Multimedia University, Cyberjaya, Malaysia

ABSTRACT

Artificial Neural Networks (ANN) are reputed to be error tolerant due to their massively parallel architecture, where the performance of faulty components maybe compensated by other parts of the network. However, most researchers take this for granted and do not verify the fault tolerance capabilities of their purpose-built ANN systems. This article reports on the robustness performance of various ANN architectures to the influences of noise and network failure in a block-adaptive predictor scheme developed to compress numeric telemetry data from remote sensors. Various single and multilayered feedforward and recurrent ANN architectures are tested as the predictor. For real-time adaptability, yet preventing network rigidity due to over-training, the ANN are retrained at the block level by segmenting the incoming data, providing good adaptability to even significantly varying input patterns. The results prove that while some ANN architectures in the proposed scheme do indeed provide better robustness as compared to classical schemes, this is not necessarily true for other architectures. The findings and discussions provided would be useful in determining the suitability of ANN architectures in future implementations that require sustainable robustness to influences such noise and network failures.

Keywords: Error tolerance; network resilience; fault tolerance; noise; damage; performance considerations.

* Corresponding author: Email: loges@mmu.edu.my

1. Introduction

One of the significant strengths of the biological neural system of the brain is its ability to make new connections. Although, this is mainly a feature for learning new information, the ability provides the basis for a very robust error-tolerant system that is capable of rerouting connections in the face of injury or damage. Artificial neural networks (ANN), based on a very rudimentary model of the biological system, has proven itself as a valuable artificial intelligence tool capable of self-learning and solving even non-linear problems, as in [1] and [2]. The ANN, because of its massively parallel architecture and adaptive learning, should also be capable of rerouting information by adjusting its weight coefficients, thus making it error tolerant.

This capability is often taken for granted by researchers incorporating ANN in their system. However, just because a module has certain properties, it does not necessarily follow that the implementation of a system incorporating the module also possesses the same property. For the purpose of illustration, take water as an example. Although a shapeless liquid, its physical properties change significantly in an implementation that either heats or freezes it. In the case of ANN, although it is theoretically, by virtue of its architecture, error tolerant, very little work has been done to prove this in completed systems. Some recent literature on this can be found in [3]-[5].

This paper revisits this issue in order to show that simple ANN systems are indeed tolerant, to a certain extent, to several types of typical errors. An ANN-based lossless data compression scheme developed by the author several years ago is used in this work as a sample implementation to test the hypothesis. The aim of this chapter is to prove that various popular ANN architectures are error tolerant, albeit to different degrees of robustness, to common types of errors faced in the typical use of ANN. The testing also includes expected transmission errors as a large number of ANN system interact with transmission in one way or the other. Although the test results would be given for the implementation of a specific scheme, the robustness and other advantages of the ANN highlighted in this chapter are generally applicable to a wide variety of ANN systems.

2. System Setup

As mentioned above, a sample ANN system is used to test the hypothesis. The system was originally implemented as a two-stage data compression scheme employing an ANN predictor and a lossless encoder [6]-[7]. However, for the purposes of this work, the testing will be done exclusively on the ANN predictor stage in order to provide results exclusively for the ANN implementation and not be influenced further by the characteristics of the encoders.

The ANN predictor is setup such that it uses a number (p) of past values (X_{n-i}) to predict the current value (X_n). Lossless data compression is achieved by transmitting only the residue (r_n), i.e. the result of subtracting the predicted value ($\hat{X}_n$) from the corresponding current value (X_n). Figure 1 illustrates the system overview at both the transmitter (compression) and receiver (decompression) sides. At the receiver, decompression is realized by adding the received residue (r_n) to the predicted value ($\hat{X}_n$), resulting in the reconstructed original value (X_n). For this scheme to be lossless, the prediction process has to be identical at both the transmitter and receiver, and there should be no error incurred in the transmission. Inaccuracies or errors encountered during prediction do not affect the outcome (i.e. the losslessness of the compression scheme) as long as the identical error or inaccuracy affects both predictors. The unit-time delay for feeding a past value to the predictor is represented by Z^{-1}.

Due to the amplitude constraints (finite precision) on computer architectures, floating point implementations for high precision arithmetic are likely to incur losses. This representation may be used, only if it can be proven that the predictors are well behaved and that each process and error at the transmitter can be identically reproduced and compensated for at the receiver. The nearest integer approximation technique (N_INT) [8] is used in this work to reduce the effect and influence of truncation errors, overflow errors and noise. The actual implementation discussed uses limited precision arithmetic of up to 32-bits, without suffering loss of data integrity of the satellite telemetry data used in the experiments.

3. ANN Architectures

An ANN is essentially made up of neurons or nodes. These are situated in layers where information flows from one layer to the other. Nodes with processing capabilities are known as processing element (PE). This essentially applies to all nodes except those in the input layer. Generally, a weight coefficient is assigned to each connection between a node in one layer to a node in the next layer. The weight coefficients determine the importance of the individual connections on the decision making process at the corresponding PE in the hidden and output layers. Each PE also has a bias, which is a coefficient assigned to the node itself that determines the importance of the node as a whole.

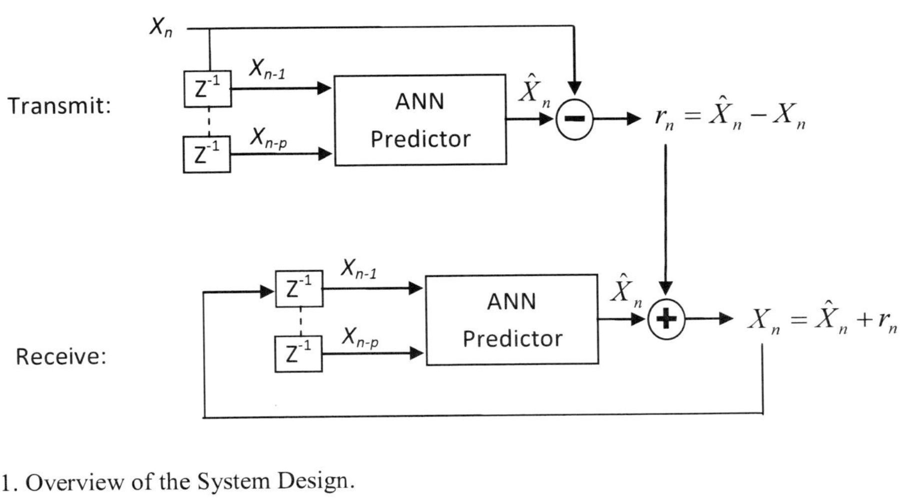

Figure 1. Overview of the System Design.

In order to provide a fair evaluation of the ANN's robustness property, various popular ANN are evaluated in this work. The tested architectures range from single-layer feedforward networks, to multilayer feedforward networks, recurrent networks, and function approximators. The configurations includes various network sizes, although the network sizes are kept small to reduce processing and resource overheads. A selection of activation functions are used with the optimum configurations determined experimentally for the test data. The ANN architectures tested n this work are shown in Figure 2 and described below.

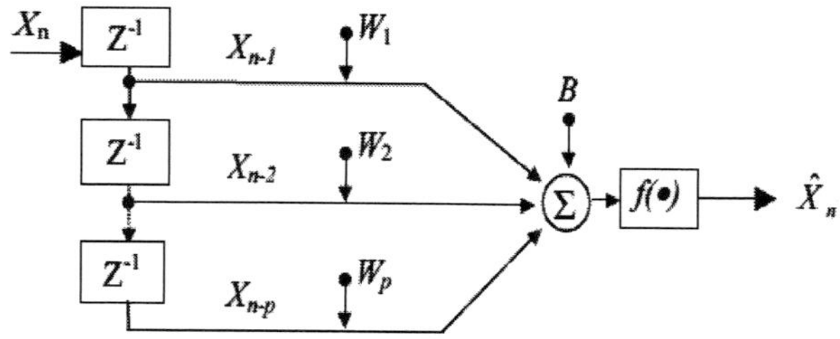

(a) RP / SLFNL

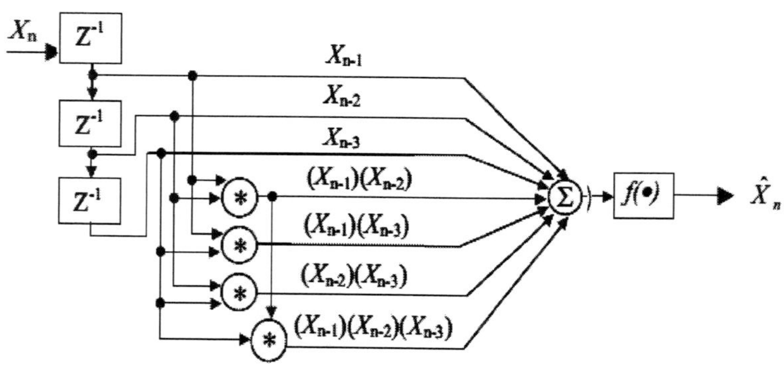

(b) SLFNH

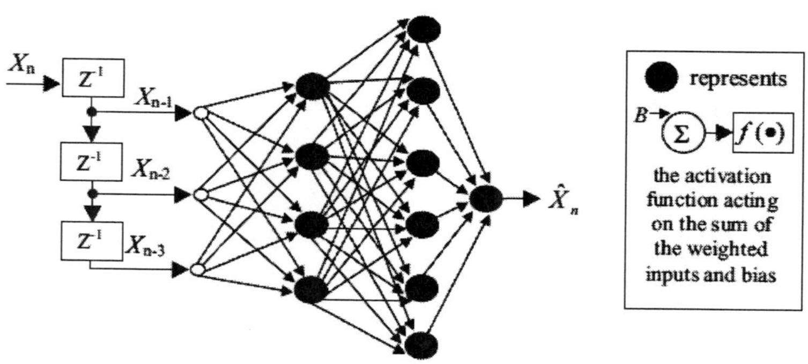

(c) MLFN

Figure 2. Continued.

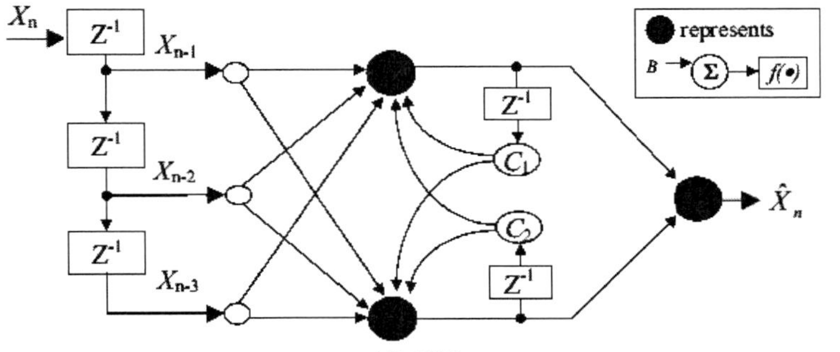

(d) EN

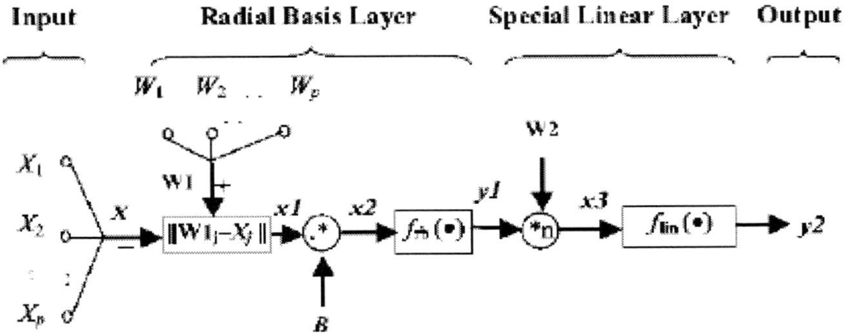

(e) GRNN

Figure 2. The ANN predictor architectures used in the experiments.

3.1. Rosenblatt Perceptron (RP)

The RP is a single-layer ANN model that uses feedforward propagation and the hardlimiter (step) activation function ($f(\bullet)$) for decision making. This type of network essentially handles only binary classification problems. A pth-order RP predictor would contain p input nodes to use the p past values to predict the current value. In recent years, the term "perceptron" has also been interchangeably used to refer to single- and multi-layer feedforward ANN using other non-step activation functions [9]-[10]. These are more popular and are described next.

3.2. Single-Layer Feedforward Network (SLFN)

Also known as the single-layer perceptron (SLP) [11], the SLFN is a dual layer network consisting of an input and output layer. Since the input nodes are not processing elements (PE), the input layer is not "counted", thus the architecture is known as a single layer network. For a more complete evaluation, three implementations of single-layer ANN are modeled in this work, namely, the RP and the two SLFN models below.

3.2.1. SLFN with lower-order functions (SLFNL)

The SLFNL is a straightforward implementation of the SLFN where the architecture used is similar to that of the RP. As with the RP, this approach uses only the past p input samples to determine the current sample. The SLFNL differs from the RP with regards to the learning and decision thresholds, and can be used for multi-level classification.

3.2.2. SLFN with higher-order functions (SLFNH)

An alternative network setup of the SLFN is the SLFNH, which not only uses the past p inputs, but also calculates and includes the higher order functions of the p past input as additional inputs, as shown in Figure 2. Thus, a 3rd-order SLFNH would have seven input nodes, i.e. one each for the three past samples and the remainder four nodes for the combinations of the product of those past samples. This is beneficial in situations where it is known that the combinations of the input values are meaningful and can contribute to more accurate prediction of the current value.

3.3. Multi-Layer Feedforward Network (MLFN)

The MLFN or multi-layer perceptron (MLP) [10] consists of more than two layers, with a number of middle / hidden layers between the input and output layers. As with the SLFN, the input layer is excluded in the naming convention, so an L-layer MLFN has the input layer, L-1 hidden layers and the output layer. The middle layers are generally feature detectors, but its numbers are kept low to minimize the complexity of the network. It has been shown in [12] that an MLFN with as few a one hidden layer is capable of arbitrarily accurate approximation to an arbitrary mapping. Most MLFN implementations rely only on a single hidden layer, as the costs of adding layers increases significantly in terms of processing power, execution time and storage requirements. The remaining ANN architectures tested in this work are multi-layer networks.

3.4. Elman Network (EN)

A recurrent ANN allows the output from one or more layers to be fed back into itself, or to one or more preceding layers. The advantage of such a network is that the results obtained from past iterations influence the current iteration. In this manner, the network is capable of adapting to changing signals and learning from past experience. Very often, this includes not only temporal information, but also takes the state of the system into consideration. The ANN architectures discussed above do not implicitly take such information into account, although past values of the data are presented to the ANN in the predictor implementation.

Popular recurrent ANN include the Hopfield network [13] and the Elman network (EN) [14], where the EN is more popular for realistic applications. The EN is a two-layer network with feedback from the output of the hidden layer back to itself. The delay incorporated in the hidden layer connection stores the network state information at time t-1 and sets a context for processing at time t. A special set of context units, each with the activation Cj as given by Eq. (1), are defined such that they receive feedback signals from the hidden layer.

$$Cj(t) = a \times Cj\ (t\text{-}1) + fj\ (t\text{-}1) \quad (1)$$

Where
Cj = activation of a context unit j
a = strength of self-connections (a<1)
fj = feedback received from non-context units

This ensures that past memory is kept within the EN, but gradually decays over time. The context can be described as temporal or spatial adjacency, and the information is implicitly and dynamically set in recurrent networks [15]. The ability to store information for future reference allows the EN to learn both temporal as well as spatial patterns. It can be trained to respond to and generate both kinds of stable state patterns. Furthermore, because the EN is an extension of the two-layer sigmoid / linear architecture, it inherits the ability to fit any input / output function with a finite number of discontinuities [16]. Each arc to the hidden and output PE (i.e. the inputs and the values from the context units) is acted upon by its own respective weight.

3.5. Generalised Regression Neural Network (GRNN)

The GRNN [17] is a two-layer radial basis (RB) network that is generally used for function approximation. A RB network is one that makes use of the normal or Gaussian function. The GRNN consists of an input layer, the hidden RB layer and a special linear layer. The RB first layer, when given a value x, returns $exp(-x^2)$. The weights of this layer are set to detect the difference between a sample value and its training input. As such, if the input vector $\boldsymbol{X}$ is used during training, the weights are initialised and set to $\boldsymbol{X}$. The biases are initialised and set to (0.8326 / *N*spread), where the constant *N*spread is supplied. Since the weights and biases are predetermined, these networks are set up very fast as epochs (iterations) of training is not required to set these values. The "distance" between the input and the weights is calculated by taking the dot product of the weight and the input vectors. This result is multiplied by the bias and is evaluated by the RB function.

The second layer is linear, where the weights are set to the target output. There are no biases in this layer. The characteristics of the GRNN allowed for three possible alternative implementations as the predictor in the proposed scheme, namely, prediction, approximation and estimation.

3.5.1. GRNN predictor (GRNNP)

This model is similar to the implementation of the other predictor schemes in which a number of past values are used to predict the current value. The training is done by selecting a training window for each block of data and setting the weights and biases accordingly.

3.5.2. GRNN approximator (GRNNA)

The GRNN is best suited for function approximation when given samples at significant points of a distribution. The training strategy for the GRNNA is to sample the entire block at regular intervals, I_{GRNNA}, as given by Eq. (2).

$$I_{GRNNA} = S_{Block} / p \tag{2}$$

where

I_{GRNNA} = interval between samples

S_{Block} = block size used

P = order of the predictor (= number of input nodes)

Regular intervals are chosen for fast and convenient implementation, without being hindered by the overheads of a distribution pattern analyzer. Training of the GRNNA is much faster than the predictor-based models as the number of training sets presented to the network is greatly reduced. For a typical predictor, $S_{Block} - p$ sets of training data are presented to train the network at each block, but only one set is presented to set up the GRNNA. This approximation technique requires that the entire block be buffered at the transmitting end. If the input data is not split into blocks, and is intended to be treated as a continuous single block (SB), the entire input stream would be buffered. The sampled values are sent to the GRNNA, which adjusts the weights and biases such that the network approximates the closest function that generates the input signal.

3.5.3. GRNN Estimator (GRNNE)

The GRNNA forces the whole block to be buffered during training. This is a very costly operation in terms of hardware and time constraints for real-time applications. As a compromise, the GRNNE is introduced. It has the similar structure as the GRNNA, but uses the interval I_{GRNNE} to approximate a function to the training set of the input using only the training window of a block (e.g. 20% of S_{Block}). The sampling intervals is given by Eq. (3).

$$I_{GRNNE} = 0.2\ S_{Block} / p \tag{3}$$

The trained pattern is then projected (stretched) to cover the entire prediction block (S_{Block}) by feeding the p number of input data values to to the GRNNE at $I_{Stretch}$ intervals given by Eq. (4).

$$I_{Stretch} = S_{Block} / p \tag{4}$$

4. Predictor Validation

Before testing for robustness, the ANN predictors developed need to be validated in terms of its performance on the test data. This is to ensure that the ANN predictors' robustness is not gained by sacrificing the expected prediction performance. Numerical analysis confirmed the losslessness of the prediction process. However, for a realistic comparison of performance, benchmarking against other known predictors was undertaken.

4.1. Benchmark Non-ANN Predictors

Linear reversible filters such as the finite impulse response (FIR) have long been used as predictors [18], and are the traditional benchmark. The filter uses a number of past values to predict the present value, where the past values are the coefficient parameters that may be fixed throughout the prediction process or be updated adaptively using algorithms such as the normalised least mean squares (NLMS) [19], during the prediction process. Both the FIR and NLMS are used in this work as benchmarks for evaluation of the proposed ANN predictors. For completeness, a high performance non-ANN adaptive predictor known as the Recursive Least Squares Lattice (RLSL) [19] is also included in the evaluation. This scheme incorporates both forward and backward feedback between layers (lattices) to improve prediction performance. Details of this architecture will not be described here, but can be found in [19].

4.2. Configuration

For the purposes of benchmarking, the best configurations of each predictor need to be used. The configuration of the various predictors used in this work, based on the best performance through experimental trials, are given in Table 1. The block size (S_{Block}) used for adaptive training is discussed below. As the non-adaptive FIR required no training, S_{Block} is set to its file size since the data stream is not split.

4.3. Block-adaptive Training

Typically, ANN are either dynamic or static, where the former adaptively learns to adjust to the input patterns during run-time while the latter is limited to the patterns learned during the training phase. However, fully adaptive training causes the system to suffer in terms of processing time and risk over-training a ANN, making it rigid, inhibiting it's generalization capabilities. In this work, a block-adaptive scheme is used, where the ANN is retrained at intervals, i.e. at the beginning of each block, as shown in Figure 3. The block size (S_{Block}) for each ANN was chosen empirically based on the best results achieved.

Note that the in the figure, the first prediction sequence starts at sample p until the end of the 20% of the second block. The p original samples are transmitted for the pth order predictor to enable the prediction to start at the receiver end. However, this transmission only happens after the ANN is trained and the state of the ANN is duplicated at the receiver, in order to ensure exact prediction at both ends. The second sequence of prediction in the figure uses the ANN trained with the first 20% of samples in the second block, and so on until the end of the input stream. Notice that the implementation capitalizes not just on prediction, but also in minimizing unnecessary transmission large numbers of original sample values (on the premise that the actual sample values are much larger that the transmitted residues), as this setup has been primed for lossless data compression.

Table 1. Configuration details of the Predictors

Predictor		No. neurons in *input-hidden-output* layers	Activation/ Transfer Function	Learning Rule	S_{Block}
ANN	RP	2-0-1	Hardlimiter/step	Perceptron learning	50
	SLFNL	4-0-1	Linear (output)	Widrow-Hoff	1500
	SLFNH	1-0-1			2000
	MLFN	2-1-1	Linear (hidden), Sigmoidal (output)	Backpropagation	1000
	EN	5-2-1			50
	GRNNP	5-45-1	Gaussian (hidden), Linear (output)	Backpropagation	50
	GRNNA	10-10-10			50
	GRNNE	5-5-5			50
Non-ANN	FIR	5th-order	1,-4,7,-7,4,-1	fixed	*v* (file size)
	NLMS	5th-order	variable coefficients	adaptive FIR	1
	RLSL	2-lattice		adaptive lattice	1

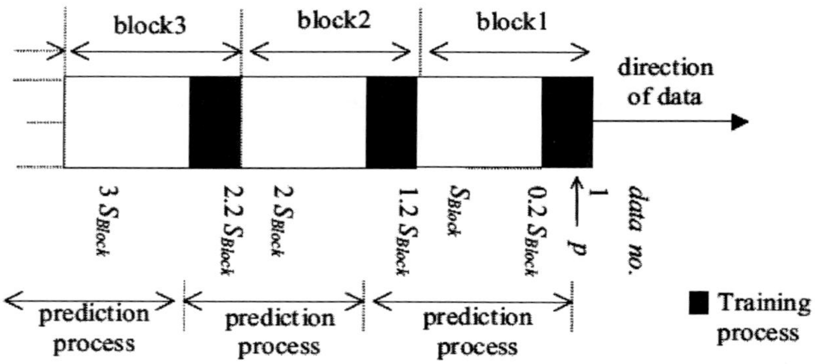

Figure 3. Real-time block-adaptive training scheme.

Table 2. The average Compression Ratio achieved by the Predictors

Predictor		Average C_R
ANN	RP	2.93
	SLFNL	8.76
	SLFNH	8.13
	MLFN	7.43
	EN	4.41
	GRNNP	6.81
	GRNNA	5.68
	GRNNE	5.67
Non-ANN	FIR	2.70
	NLMS	5.30
	RLSL	9.25

4.4. Compression Performance

For the purposes of testing the compression performance, six text files with varying distributions were used. The predictors were configured as given in Table 1. The compression performance in terms of compression ratio (C_R), i.e. the ratio of the original file size to the resulting encoded stream, is given in Table 2. The results show that the compression performance of the ANN predictors are comparable to those of the benchmark predictors.

As a more detailed analysis of the actual prediction, the average word size of the actual generated residues are examined. The results obtained in Table 3 once again show that the compression performance of the ANN predictors are comparable with the best benchmark predictor.

Table 3. Average word sizes (in bits) of the input and residue streams

Predictor		Word Size (bits)
Original		16.0
ANN	RP	9.8
	SLFNL	3.2
	SLFNH	2.7
	MLFN	3.2
	EN	2.8
	GRNNP	2.0
	GRNNA	2.5
	GRNNE	2.0
Non-ANN	FIR	10.7
	NLMS	9.3
	RLSL	3.8

Table 4. Average word sizes (in bits) of the input and residue streams

Predictor		Estimated time, in seconds ($T_{estimate}$)
ANN	RP	0.22
	SLFNL	0.22
	SLFNH	0.22
	MLFN	0.24
	EN	0.43
	GRNNP	0.33
	GRNNA	0.21
	GRNNE	0.21
Non-ANN	FIR	0.24
	NLMS	0.33
	RLSL	0.36

Another important aspect of performance is the processing speed or time taken for the prediction. The average values of each predictor on the test files are given in Table 4. The experiment was conducted on a Sun Ultra 10 machine for the benchmark predictors. However, as the ANN contains parallel architecture, but simulations were conducted using MATLAB [16], the results given in the table are careful estimates.

The calculations were done on a stream of 45207 samples (average of the total number of samples of all six test files). Each product (multiplication operation) was assumed to be equivalent to 32 summations (using the principle of a 32-bit full-adder), thus the estimated processing time was calculated using :

$$\begin{aligned} T_{\text{estimate}} &= \text{time taken for (training + execution + I/O)} \\ &= [c\,(s)\,(o_t + l)\,t_{\text{sum}}] + [(v - p)\,(o_e + l)\,t_{\text{sum}}] + t_{\text{I/O}} \end{aligned} \tag{5}$$

where

c = number of epochs (cycles) used for training
s = number of training sets used
o_t = number of training operations per sample (= per set)
l = number of layers with PE
t_{sum} = average time for each summation (i.e. t_{sum} » 5.755×10^{-9} seconds)
v = number of samples (i.e. v = 271242 / 6 = 45207 samples)
p = number of input nodes
o_e = number of execution operations per sample (= per set)
$t_{I/O}$ = total time taken for all I/O operations (i.e. $t_{I/O}$ » 0.19 seconds)

The time taken for each summation (t_{sum}) and for input/output operations ($t_{I/O}$) are assumed to be similar to those taken by the benchmark predictors, while the time taken for other miscellaneous operations (e.g. time to display results, transition time on system bus etc.) are already considered in both these values. From the estimated processing times in the table, it would appear that the ANN predictors generally perform faster that the benchmark predictors.

5. Evaluation of Robustness

The purpose of this chapter is to explore the robustness of the ANN predictors in the scheme. From the previous section, we are convinced that the ANN predictors have been set up correctly and are able to produce good performance in terms of compression and processing time. Thus, the experiment can now focus on subjecting the ANN and benchmark predictors to common problems expected in the normal application of such a scheme. The sample problems simulated in this section include noise, interference and hardware failure, in order to test the robustness of the ANN predictors to such influences.

5.1. Ambient Noise

The first test is to evaluate the performance of the predictors in the typical ambient noise/atmospheric radiation scenario. The popular simulation for this is the application of white gaussian noise to the data. For the purposes of this test, the ANN is trained using the clean signal as before and only the run-time data is "polluted" with the noise, simulated using the with Gaussian (0,1) (i.e. function with a mean of zero and variance of 1). The entire residue stream is subjected to this noise. The calculation of the errors is conducted between the actual sample values and the reconstructed ones.

The results obtained are plotted in Figure 4. The mean square error (MSE) results in Figure 4(a) confirm that in almost all situations, the ANN predictors managed to compensate for the noise better than the FIR and NLMS predictors. In general, the RP, MLFN and GRNN variations were more error tolerant to the Gaussian noise than even the RLSL. The EN's performance was relatively poor, although better than the FIR and NLMS, as it recurrently feeds (the noisy) values back into the network, thus influencing the consequent predicted values.

The SLFNH configuration was that of a first-order network. This makes it sensitive to each input value as it uses only one input value to predict the next. Even with this limitation of not having any PE except in the output, the SLFNH still managed to outperform the FIR, NLMS and RP. The lack of PE put the performance of the SLFNL at approximately the same level as the SLFNH.

The effect of the gaussian noise on the generated residue signal can be analysed via the signal-to-noise ratios (SNR) estimation, as given by the second plot in the Figure 4. The SNR is calculated as follows:

$$\text{SNR} = 10 \log_{10}(\text{MSV}_{residue} \,/\, \text{MSV}_{gaussian\_noise}) \tag{6}$$

where

$\text{MSV}_{residue}$ = mean square value of the residue stream

$\text{MSV}_{gaussian\_noise}$ = mean square value of the applied gaussian noise

Small SNR values indicate that the actual values are of small magnitude but heavily influenced by the applied noise. As it is the residue stream that is analysed, this indicates good compression but poor reconstruction. An SNR of 5 dB signifies the power of the noise is approximately a third of the signal power of the residue stream. From the plot, it is observed that different magnitudes of SNR are prevalent amongst the different test files, this indicating the differences in data patterns. This trend is also reflected by the MSE plot.

The mean average error (MAE) provides information on the average size of the errors encountered. The MAE values obtained are given in Figure 4(c), reflecting the findings of the MSE plot, with minor variations. Thus, the same conclusions may be drawn.

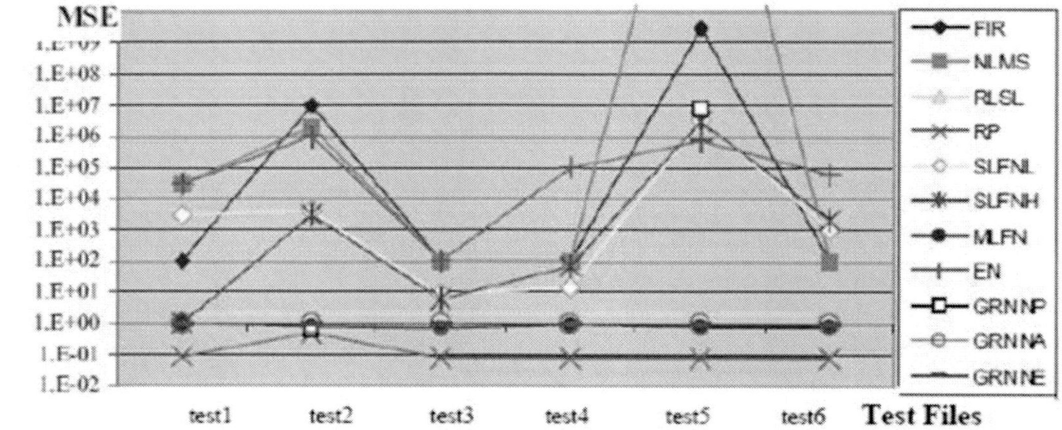

(a) Mean square error (MSE)

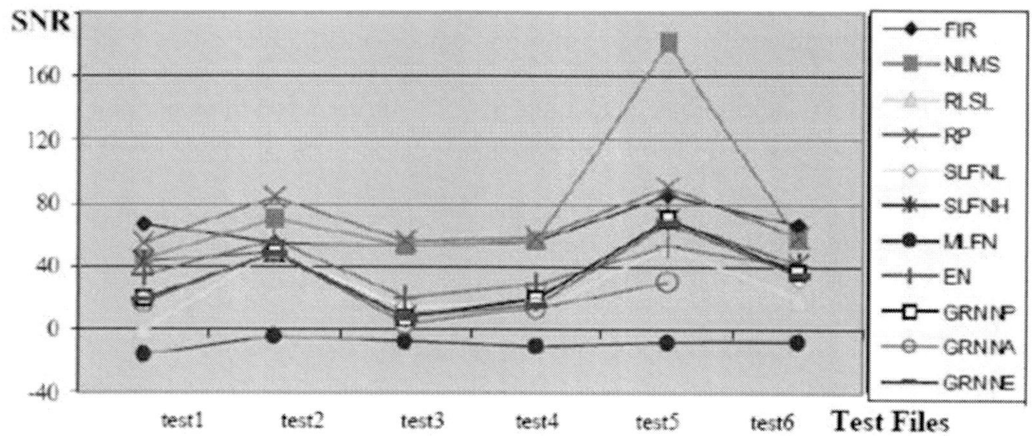

(b) Signal-to-noise ratio (SNR)

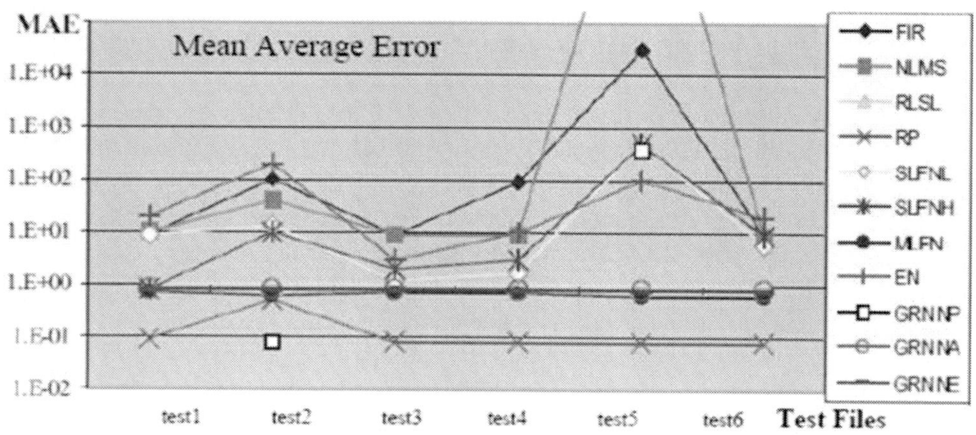

(c) Mean average error (MAE)

Figure 4. Results for Gaussian Noise Robustness Test.

5.2. Impulse Noise

Natural phenomena such as lightning, electromagnetic interference from solar flares / sun spots, and interference from nearby electrical and electronic devices are known to cause spikes in signals. This is especially prominent in places where there is a large presence of modernization and man-made influences such as cellular communications, wireless equipment etc. The impulse or burst noise simulation was designed to test the robustness of the predictors to the presence of such high frequency interference that occur in very short periods. The simulation consisted of two bursts of amplitude 5 that was comparable to the average amplitude of the residues, spanning a period of 5 samples. These impulses were applied at a quarter and three-quarters way through the transmission stream.

The results are measured in terms of MSE and MAE, as shown in Figure 5. A logarithmic y-axis is used to clearly show the trend in a compact manner. Unfortunately, as the logarithmic scale does not display the plots of 0, some points on the graphs are not shown. Overall, the trends observed for this simulation also reflects that in the previous test, with the GRNN and RP predictors having the lowest MSE values. As before, this was expected due their lesser sensitivity to the input values, as observed from the poorer compression performance in the previous section.

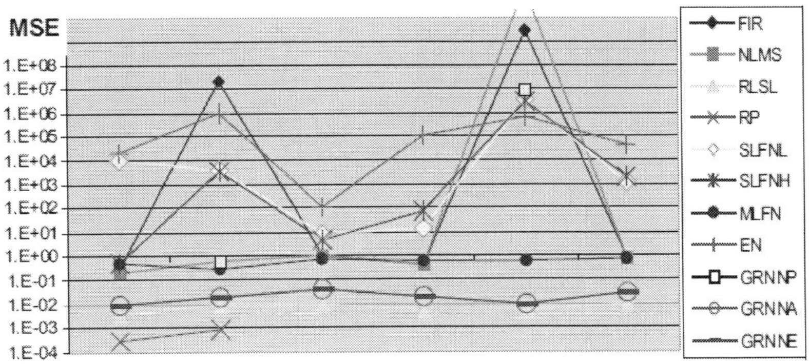

(a) Mean square error (MSE)

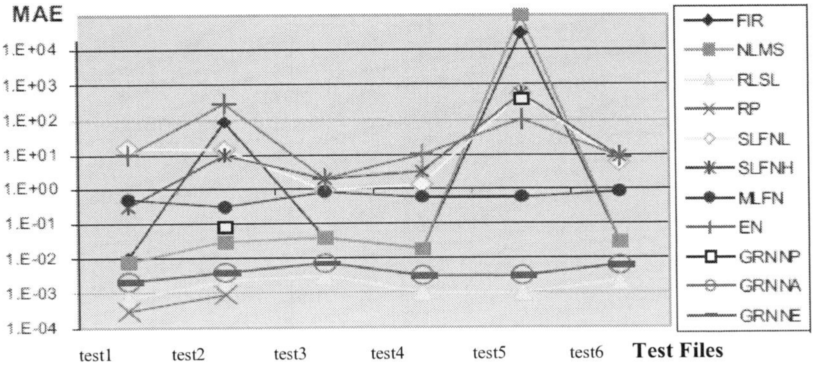

(b) Mean average error (MAE)

Figure 5. Results for Impulse Noise Robustness Test.

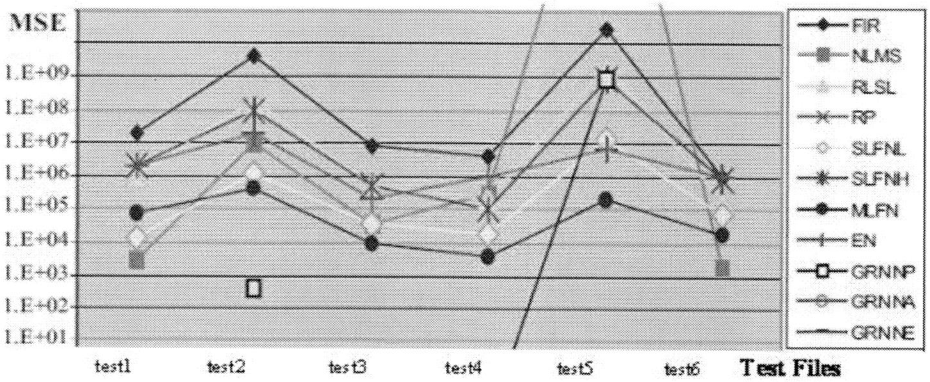

(a) Mean square error (MSE)

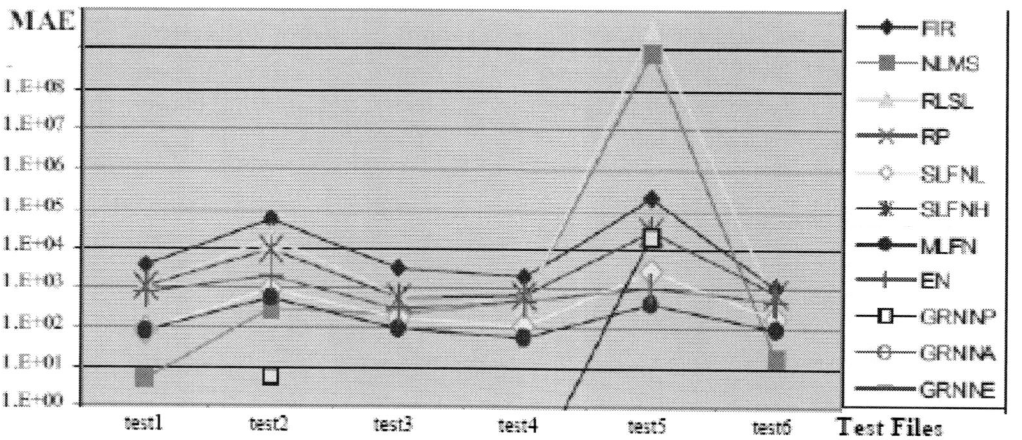

(b) Mean average error (MAE)

Figure 6. Results for Faulty Input Robustness Test.

5.3. Faulty Input

Most systems fail to produce the correct results when a component becomes faulty. One of the well known characteristics of the ANN architecture is hardware fault tolerance. Just like the brain, the ANN possesses the ability to reroute connections and information to compensate for a certain amount of network errors. In the presence of errors and faults, the well built ANN should fail gracefully.

To analyze the predictor robustness in this sense, a simple simulation of a failure of the input receptor was implemented by setting the first input receptor to 0. In this manner, no matter what the actual value is, the first input node value would be zero. In a large architecture, the failure of a single node, that too a non-PE, should be insignificant. However, as seen in Table 1, the architectures used in this work are relatively small and thus, the failure involves a relatively significant size of the entire predictor structure. This simulation produced interesting results, as observed in Figure 6.

From the figure, it is found that the ANN predictors generally performed more consistently than the classical predictors. Of particular interest is the fact that the 1st-order SLFNH was able to perform at a moderate level even though its only input node was faulty. As the trained GRNN approximators do not rely on the input stream for "predicting" the current values, they were not affected by the input failure. In relation to the 2nd-order RP, the processing was automatically adjusted such that it relied on only the second input node and not the first, thus its performance did not deteriorate noticeably.

5.4. Faulty Processing Element

In the previous test, an input node was subjected to failure, therefore eliminating the use of that input on any connections between that node and any subsequent nodes in the hidden and output layers. A further analysis would be to cause a coefficient (weight) of a PE to fail, such that the actual processing is affected in terms of the significance of an input to the evaluation of that PE. This implies internal architectural failure (as opposed the faulty input case which may also indicate failure external to the predictor).

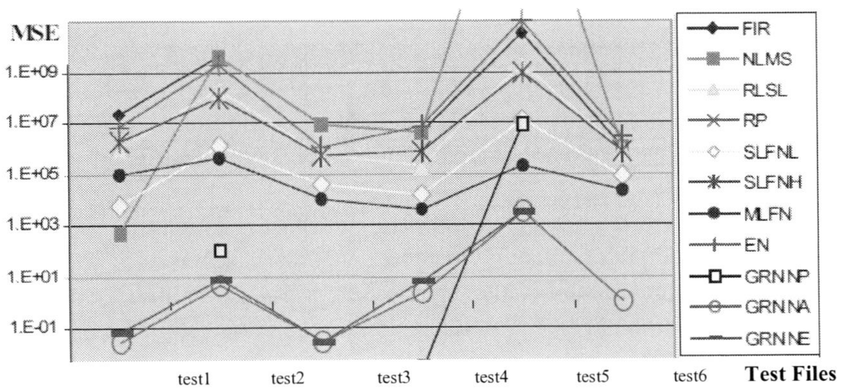

(a) Mean square error (MSE)

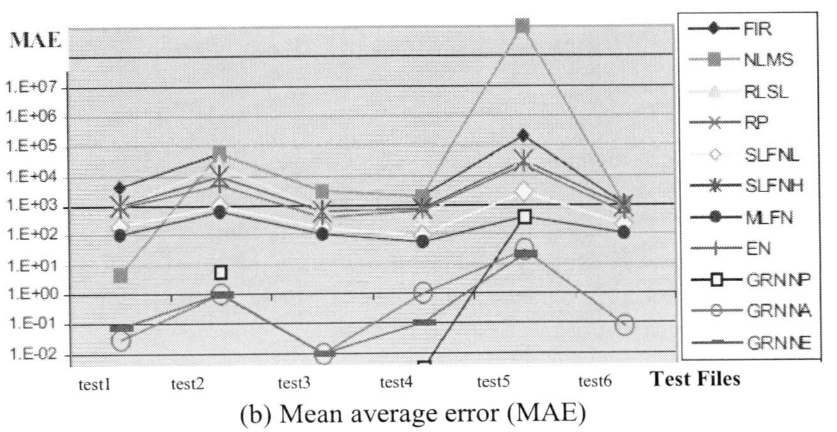

(b) Mean average error (MAE)

Figure 7. Results for Faulty Processing Element Robustness Test.

This robustness simulation involved setting the first weight coefficient to 0 to mimic hardware failure of the weight. It also implies a multiplier failure at the weight, thus although correct input is received, the node nullifies the input on that connection. Note that in a fully-connected architecture or one that involves several connections to the affected node, only one weighted input would be directly affected in this simulation, although the subsequent layers may suffer from the propagated error.

The results of this robustness simulation are given in Figure 7. The plots of some points for the RP, GRNNP and GRNNE are not shown as they were zero. The faulty weight and processing element affected the GRNN this time as it hindered its approximation abilities by erroneously simulating the wrong approximated function. The RP was not affected, which may imply that the first input was not a significant contributor to this predictor. As the SLFNH only has one input and one output, the failure of the weight had the same effect as the previous input failure experiment. The MLFN used also only had one node, thus displaying the same outcome. In general, the GRNN produced the best results for this test as it contained the most nodes and had better ability to compensate for the failure.

6. Discussions

The above results show that the ANN are indeed robust and fault tolerant with little adverse effect to the types of influences and faults simulated. To understand the basis of the robustness, this section examines an overall comparison of the schemes, in order to provide a better understanding of what and which contribute to the robustness of the ANNs used in this work.

From a comparison of the compression performance, it is found that the results achieved by the ANN were good, if not better than the benchmark schemes. The RLSL is a very sophisticated and computationally intensive scheme, thus its high performance is not surprising, and the performance of the ANN should not be underestimated based on the RLSL achievements. Based on the compression performance and size of the network, the SLFN had the highest C_R achievement and lowest number of interconnections, and as such, could be assumed to be the best choice for practical applications.

6.1. Network Size

Robustness, and accuracy, of an ANN generally improves with network size. Error tolerance is better in the presence of redundancy. Therefore, larger architectures are able to cope better with errors. This is especially true when the number of PE in the hidden layers is increased as the architecture would be able to recognize more complex patterns. However, relatively small networks were tested in this work, in the interest of reduced requirement of memory and processing resources.

The robustness results obtained by the ANN were significant although the architectural setup utilized a very small number of nodes (conventional implementations may use hundreds of nodes) and did not sufficiently exploit the strength of the ANN architectures. Even with this ‘handicap’, it has been shown that the ANN performs well, if not better than the traditional benchmark algorithms. Larger networks should display even better error and failure tolerance capabilities. In addition, with more PE, the larger networks would be able to store more "knowledge" and be better predictors by producing output (residues) with lower dynamic range, thus improving the compression performance.

6.2. Configuration

There are many possible configurations that may be optimized to improve the performance of the ANN. Some of these include detailed selection of the most appropriate type of ANN, choice of training algorithms, activation functions, inter-layer interconnections, S_{Block} etc. The latest trends allow for customized solutions by incorporating other high performance algorithms into hybrid ANN applications [20]. Wavelets, fractals, genetic algorithms (GA), particle swarm optimization (PSO) and fuzzy logic are just some of the technologies that have already been incorporated into modern ANN to solve complex problems.

The setup and evaluations undertaken in this work had to be trimmed down by intelligent sampling and selection based on existing knowledge, trends of results observed, recommendations, "rules of thumb", default values as well as limitations of the simulation package. The results described in this work were based on averages of multiple iterations of design and development. The values obtained were those for semi-optimum small networks without over-training the ANN nor through guided selection of samples. Mostly, values used were just those in the stream, without prior pattern analysis to optimize the training. Multiple distributions were used in order to develop a robust network. If the reader so desires, the proposed scheme could be fine-tuned to optimize the performance to particular trends in the data.

6.3. Nearest Integer Implementation (N_INT)

In order to reduce complexity and internal resource requirements during processing, the scheme was implemented to capitalize on integer arithmetic. When the input file consists of floating-point data, the data is shifted (e.g. by multiplying by 10^f such that the f is sufficient to enable precision behind the decimal points to be shifted before the decimal point and the floating-point value is represented by an integer. Of course, with the availability of multi-core systems nowadays, along with high-end graphic cards with dedicated graphics processing units (GPU), floating point arithmetic is now handled very efficiently and the N_INT implementation may no longer be necessary.

6.4. Other Considerations

All robustness evaluations conducted in the experiments were via software simulations using a combination of compiled C language programs for the non-ANN processing components, and MATLAB with the Neural Network Toolbox for the ANN customizations.

If available, the experiments could be conducted on dedicated ANN hardware, of on purpose-built ASIC (application specific integrated circuits), very large-scale integration (VLSI) technology or FPGA (field programmable gate arrays). A further suite of tests could also be conducted to test the system in terms of robustness to various other natural and man-made conditions that are likely to be of interest to the users of the ANN.

7. Conclusions

This chapter has provided some information and test results to verify that the ANN in the proposed purpose-built data compression scheme is indeed robust. It has been shown to be able to handle various distributions of data with good accuracy despite being subjected to various simulated faults such as noise and architectural failures. The ANN is a technology that has been widely used for several decades. Although newer and better technologies have evolved over the years, the use of ANN remains popular due to its ability to automatically learn and generalize patterns, and its robustness in withstanding a certain amount of faults, a capability which make it a very useful technology for remote systems.

This work provides empirical evidence of the robustness and fault tolerant characteristic of the ANN that is often taken for granted. It should be remembered that although intrinsically versatile, the properties of a technology may change when incorporated into a system or scheme. With this in mind, there should be verification of the desired properties of the system after integration. This chapter reports such work. The findings and discussions provided here may be useful in determining the suitability of ANN architectures in future implementations that require sustainable robustness to influences such noise and network failures in multi-stage algorithms employing the ANN.

References

[1] Rajesh, M. V., Archana, R., Unnikrishnan, A., Gopikakumari, R. & Jacob, J. (2008). Evaluation of the ANN Based Nonlinear System Models in the MSE and CRLB Senses. World Academy of Science, *Engineering and Technology*, *48*, 211-215.

[2] Gen, M., Ida, K. & Kobuchi, R. (1997). Neural Network Technique and Genetic Algorithm for Solving Nonlinear Integer Programming Problem. Proceedings of the Australia-Japan *Joint Workshop on Intelligent & Evolutionary Systems*, 95–105.

[3] Tchernev, E. B., Mulvaney, R. G. & Phatak, D. S. (2005). Investigating the Fault *Tolerance of Neural Networks. Neural Computation*, *17(7)*, 1646-1664.

[4] Zhou, Z. H. & Chen, S. F. (2004). Evolving Fault-Tolerant Neural Networks. Journal *Neural Computing & Applications*, *11(3-4)*,156-160.

[5] Salvo, M. A. G., Lee, J. & Lee, J. S. (2009). Fault Tolerance Based on Neural Networks for the Intelligent Distributed Framework. *Advances in Computational Intelligence*, *116*, 83-92.

[6] Logeswaran, R. (2002). A Prediction-Based Neural Network Scheme for Lossless Data Compression. IEEE Transactions on Systems, Man, and Cybernetics - Part C: *Applications and Reviews*, *32(4)*, 358-365.

[7] Logeswaran, R. (2004). Fast Two-Stage Lempel-Ziv Lossless Numeric Telemetry Data Compression Using A Neural Network Predictor. *Journal of Universal Computer Science, 10(9)*, 1199-1211.

[8] Stearns, S. (1995). Arithmetic Coding in Lossless Waveform Compression. *IEEE Transactions on Signal Processing, 43(8)*, 1874-1879.

[9] Dony, R. D. & Haykin, S. (1993). Optimally Integrated Adaptive Learning. Proceedings IEEE International Conference Acoustics, *Speech and Signal Processing, 1*, 609-612.

[10] Frate, F. D., Lichtenegger, J. & Solimini, D. (1999). Monitoring Urban Areas by Using ERS-SAR Data and Neural Networks Algorithms. *IEEE International Geoscience and Remote Sensing, 5*, 2696-2698.

[11] Widrow, B. & Lehr, M. A. (1990). 30 Years of Adaptive Neural Networks: Perceptron, Madaline and Backpropagation. *Proceedings of the IEEE, 78(9)*, 1415-1442.

[12] Hornik, K. M., Stinchcombe, M. & White, H. (1989). Multi-layer Feedforward Networks are Universal Approximators. *Neural Networks, 2*, 359-366.

[13] Hopfield, J. J. (1982). Neural Networks and Physical Systems with Emergent Collective Computational Abilities. *Proceedings of the National Academy of Sciences, 79*, 2554-2558.

[14] Elman, J. L. (1990). Finding Structure in Time. *Cognitive Science, 14*, 179-211.

[15] Fu, L. (1994). *Neural Networks in Computer Intelligence*. New York : McGraw Hill.

[16] Demuth, H. & Beale, M. (1996). *Neural Network Toolbox User's Guide*. Massachusetts The MathWorks, Inc.

[17] Wasserman, P. D. (1993). *Advanced Methods in Neural Computing*. New York : Van Nostrand Reinhold.

[18] McCoy, J. W., Magotra, N. & Stearns, S. (1994). Lossless Predictive Coding. IEEE *Midwest Symposium on Circuits and Systems*, 927-930.

[19] Haykin, S. (1996) *Adaptive Filter Theory (*3rd ed.). New Jersey : Prentice Hall.

[20] Przytula, K. W. & Prasanna, V. K. (1993). *Parallel Digital Implementations of Neural* Networks. New Jersey : Prentice-Hall.

In: Focus on Artificial Neural Networks
Editor: John A. Flores, pp. 299-322
ISBN: 978-1-61342-285-8

Chapter 14

INTELLIGENT INVERSE KINEMATICS SOLUTION FOR SERIAL MANIPULATORS PASSING THROUGH SINGULAR CONFIGURATIONS WITH PERFORMANCE PREDICTION NETWORK

***Ali T. Hasan*[1] *and H. M. A. A. Al-Assadi*[2]**
[1]Ministery of Higher Education and Scientific Research
Baghdad, Iraq
[2]Faculty of Mechanical Engineering, Universiti Teknologi MARA
(UiTM) 40450 Shah Alam, Selangor, Malaysia

ABSTRACT

This chapter is devoted to the application of Artificial Neural Networks (ANN) to the solution of the *Inverse Kinematics* (IK) problem for serial robot manipulators, in this chapter two networks were trained and compared to examine the effect of considering the Jacobian Matrix to the efficiency of the IK solution.

Offline smooth geometric paths in the joint space of the manipulator are obtained through trajectory planning process to give the desired trajectory of the end effector of the manipulator in a free of obstacles workspace. Some of the obtained data sets were used in the training phase while the remaining data sets were used in the testing phase.

Even though it is very difficult in practice, data used in this study were recorded experimentally from sensors fixed on robot's joints to overcome the effect of kinematics uncertainties presence in the real world such as ill-defined linkage parameters, links flexibility and backlashes in gear train

The generality and efficiency of the proposed algorithm are demonstrated through simulation of a general six DOF serial robot manipulator, finally the obtained results were verified experimentally.

1. INTRODUCTION

The most frequently attempted to be solved problem for serial robots, is the Inverse Kinematics (*IK*) task. The complexity in the solution arises from robots geometry and non-linear equations (trigonometric equations) occurring when transforming between Cartesian and joint spaces (multiple solutions and singularities exist), mathematical solutions for the problem may not always correspond to physical solutions and methods of solution depends on the robot configuration [1].

Researchers have used three approaches to the IK solutions; first approach is the analytical solution (close form solution) where all of the joints variables are solved analytically according to given configuration data [2], closed form solution is preferable because in many applications where the manipulator supports or is to be supported by a sensory system, the results from kinematics computations need to be supplied rapidly in order to have control actions, unfortunately, close form solution is only exist for robots with simple geometry.

In the second approach, all of the joint variables are obtained iterative computational procedures. There are four disadvantages in these (incorrect initial estimations before executing the inverse kinematics algorithms, convergence to the correct solution can not be guarantied, multiple solutions are not known, and, there is no solution if the Jacobian matrix is in singular configuration).

In the third approach, some of the joint variables are determined analytically in terms of two or three joints variables and these joint variables computed numerically. Disadvantage of numerical approaches to inverse kinematics problems is also heavy computational calculation and big computational time [3].

Techniques for the IK problem solution have been the subject of considerable research effort in the past years, as close-form analytical solution can only be found for manipulators having simple geometric structures, a number of techniques mainly based on the inversion of the Jacobian Matrix have been proposed for all these structures that cannot be solved in close-form [4].

One of the first techniques employed was the Resolved Motion Rate-Control method [5] which uses the pseudoinverse of the Jacobian matrix to obtain the joint velocities corresponding to a given end-effector velocity, an important drawback of this method was the singularity problem.

To overcome the problem of kinematics singularities, the use of a damped least squares inverse of the Jacobian matrix has been later proposed in lieu of the pseudoinverse [6,7].

Since in the above algorithmic methods the joint angles are obtained by numerical integration of the joint velocities, these and other related techniques suffer from errors due to both long-term numerical integration drift and incorrect initial joint angles.

To alleviate the difficulty, algorithms based on the feedback error correction are introduced [8]. However, it is assumed that the exact model of manipulator Jacobian matrix of the mapping from joint coordinate to Cartesian coordinate is exactly known. It is also not sure to what extent the uncertainty could be allowed. Therefore, most research on robot control has assumed that the exact kinematics and Jacobian matrix of the manipulator from joint space to Cartesian space are known. This assumption leads to several open problems in the development of robot control laws today [4].

In other words, it is not possible to formulate a mathematical model that has a clear mapping between Cartesian space and joint space for inverse kinematics problem, to overcome this problem, Artificial Neural Networks (ANN) uses samples to obtain the nonlinear model of such systems. Their ability to learn by example makes artificial neural networks very flexible and powerful when traditional model-based modeling techniques break down. Many researchers have experimented with this approach by applying it to several robot configurations [9-13].

Studying the IK of a serial manipulator by using ANNs has two problems, one of these is the selection of the appropriate type of network and the other is the generating of suitable training data set [14,15].

Researchers have applied different methods for gathering training data, some of them have used the kinematics equations [3,10], others have used the network inversion method [9,12], while the cubic trajectory planning [11] was used by others, a simulation program [16] has been also used for this purpose. However, there are always kinematics uncertainties presence in the real world such as ill-defined linkage parameters, links flexibility and backlashes in gear train, in this approach, although this is very difficult in practice [17].

Training and testing data used in this study were recorded experimentally from sensors fixed on each joint. Euler (RPY) representation was used to represent the orientation, as was recommended by Karilk and Aydin [10], (as they have used the robot model to get the training data and used the homogeneous transformation matrix representation to represent the orientation).

During training phase, the resulting network was compared to another network where the Jacobian Matrix was considered, the network that was showing better response in terms of precision and iteration was chosen to apply the testing data. Testing data was meant to pass through the singular configurations. Finally the approach was experimentally verified using a six DOF serial robot.

2. Inverse Kinematics for Serial Manipulators

It is known that the vector of Cartesian space coordinates (the end effector position and orientation) x of a robot manipulator is related to the joint coordinates q by:

$$x = f(q) \tag{1}$$

Where $f(\cdot)$ is a nonlinear differential function.
if the Cartesian coordinates x were given, joint coordinates q can be obtained as :

$$q = f^{-1}(x) \tag{2}$$

Solving the above equation, Denavit and Hertenberg [18] proposed a matrix method of systematically establishing a coordinate system to each link of an articulated chain as shown in Figure 1, to describe both translational and rotational relationships between adjacent links.

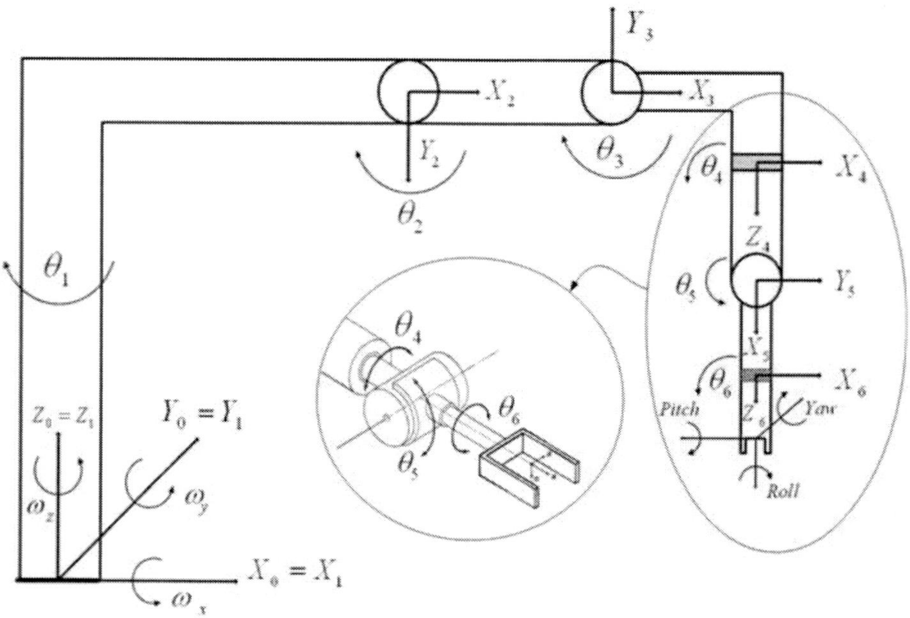

Figure 1. Schematic diagram for a general 6DOF serial robot showing the wrist mechanism.

In this method each of the manipulator links is modelled, this modelling describes the "*A*" homogeneous transformation matrix, which uses four link parameters [12,19].

Forward kinematics solution can be obtained as:

$$A_{END-EFFECTOR} = T_6 = A_1.A_2.A_3.A_4.A_5.A_6 = \begin{bmatrix} \begin{array}{c|c} Rotation & Position \\ matrix & vector \\ ---- & ---- \\ Perspective & \\ transformation & Scaling \end{array} \end{bmatrix} = \begin{bmatrix} n_x & s_x & a_x & p_x \\ n_y & s_y & a_y & p_y \\ n_z & s_z & a_z & p_z \\ 0 & 0 & 0 & 1 \end{bmatrix} \quad (3)$$

Where:

n : Normal vector of the hand. Assuming a parallel-jaw hand, it is orthogonal to the fingers of the robot arm.

s : Sliding vector of the hand. It is pointing in the direction of the finger motion as the gripper opens and closes.

a : Approach vector of the hand. It is pointing in the direction normal to the palm of the hand (i.e., normal to the tool mounting plate of the arm).

p :Position vector of the hand. It points from the origin of the base coordinate system to the origin of the hand coordinate system, which is usually located at the center point of the fully closed fingers.

The orientation of the hand is described according to the Euler (*RPY*) rotation as:

$$RPY(\phi_x,\phi_y,\phi_z) = Rot(Z_w,\phi_z).Rot(Y_w,\phi_y).Rot(X_w,\phi_x) \tag{4}$$

After T_6 matrix is solved:

$$\phi_z = ATAN2(n_y,n_x) \tag{5}$$

$$\phi_y = ATAN2(-n_z,n_x \cos\phi_z + n_y \sin\phi_z) \tag{6}$$

$$\phi_x = ATAN2(a_x \sin\phi_z - a_y \cos\phi_z, o_y \cos\phi_z - o_x \sin\phi_z) \tag{7}$$

These equations describe the orientation according to the *Euler* representation [10].

To find the IK solution, however, joints angels are found according to the manipulator's end position, described with respect to the world coordinate system.

IK solution can be shown as a function:

$$IK(X,Y,Z,\phi_x,\phi_y,\phi_z) = (\theta_1,\theta_2,\theta_3,\theta_4,\theta_5,\theta_6) \tag{8}$$

Model-based methods for solving the IK problem are inadequate if the structure of the robot is complex, therefore; techniques mainly based on inversion of the mapping established between the joint space and the task space of the manipulator's Jacobian matrix have been proposed for those structures that cannot be solved in closed form.

If a Cartesian linear velocity is denoted by *V*, the joint velocity vector $\dot{q}$ has the following relation:

$$V = J\dot{q} \tag{9}$$

Where J is the Jacobian matrix.

If V, is a desired Cartesian velocity vector which represents the linear velocity of the desired trajectory to be followed. Then, the joint velocity vector $\dot{q}$ can be resolved by:

$$\dot{q} = J^{-1}V \tag{10}$$

Inverting the Jacobian Matrix trying to solve the above equation normally results in the singularity problem.

The most commonly used techniques of coping with singularities are the following: avoiding singular configurations, robust inverse, normal form approach, extended Jacobian technique and channeling algorithms.

Singular configuration avoidance means keeping a current configuration away from a set of singular configurations. Unfortunately, it causes severe restrictions on the configuration space, as well as the workspace (task space), because the singular configurations split the configuration space into separate components [1].

To avoid ill conditioning of the Jacobian matrix, robust inverses are used instead inverting the original Jacobian matrix at singularity; a disturbed, well-conditioned Jacobian matrix is inverted. This method may force the robot to stop at singular configurations; also robust inverse methods increase errors in following a desired path [20].

The normal form technique, expresses original kinematics around singularity in the simplest, normal form. Both the kinematics are equivalent around singularity. Trajectory planning in the vicinity of singularity is significantly simpler with the use of the normal form kinematics than with the use of original kinematics. However, the transformation into the normal form is not computationally simple [21,22].

The extended Jacobian technique, supplements original kinematics with auxiliary functions. Then, extended Jacobian is formulated to be well-conditioned. Obviously, the extended Jacobian matrix has more rows than the original Jacobian matrix. Consequently, computational load of the inverse kinematics algorithm increases [20].

A channeling algorithm examines singular values of Jacobian matrix while approaching a singularity. As vanishing singular values detect a singular configuration, the algorithm forces to change signs of the singular values (the algorithm, contrary to the classical formulation, admits also negative singular values). The channeling algorithm works for any type of singularities but it is rather computationally involving as it is requires calling the singular values decomposition algorithm frequently [23].

In differential motion control, the desired trajectory is subdivided into sampling points separated by a time interval Δt between two terminal points of the path. Assuming that at time t_i the joint positions take on the value $q(t_i)$, the required q at time $(t_i + \Delta t)$ is conventionally updated by using:

$$q(t_i + \Delta t) = q(t_i) + \dot{q}\,\Delta t \tag{11}$$

Substituting Eqs. (2) and (10) into (11) yields:

$$q(t_i + \Delta t) = f^{-1}(x)(t_i) + J^{-1} V \Delta t \tag{12}$$

Equation (12) is a kinematics control law used to update the joint position q and is evaluated on each sampling interval. The resulting $q(t_i + \Delta t)$ is then sent to the individual joint motor servo-controllers, each of which will independently drive the motor so that the robotic manipulator can be maneuvered to follow the desired trajectory [9].

Therefore, to analyze the singular conditions of a manipulator and develop effective algorithms to resolve the inverse kinematics problem at or in the vicinity of singularities are of great importance.

3. THE ARTIFICIAL NEURAL NETWORKS

Artificial neural networks (ANNs) are collections of small individual interconnected processing units. Information is passed among these units along interconnections; the incoming connection has two values associated with it, an input value and a weight. The output of the unit is a function of the summed value.

The elementary nerve cell called a neuron, which is the fundamental building block of the biological neural network [24]. Its schematic diagram is shown in Figure 2.

A typical cell has three major regions: the cell body, which is also called the soma, the axon, and the dendrites. Dendrites form a dendritic tree, which is a very fine bush of thin fibbers around the neuron's body. Dendrites receive information from neurons through axons-Long fibbers that serve as transmission lines.

An axon is a long cylindrical connection that carries impulses from the neuron. The end part of an axon splits into a fine arborization. Each branch of it terminates in a small end bulb almost touching the dendrites of neighbouring neurons. The axon-dendrite contact organ is called a *synapse.* The synapse is where the neuron introduces its signal to the neighbouring neuron.

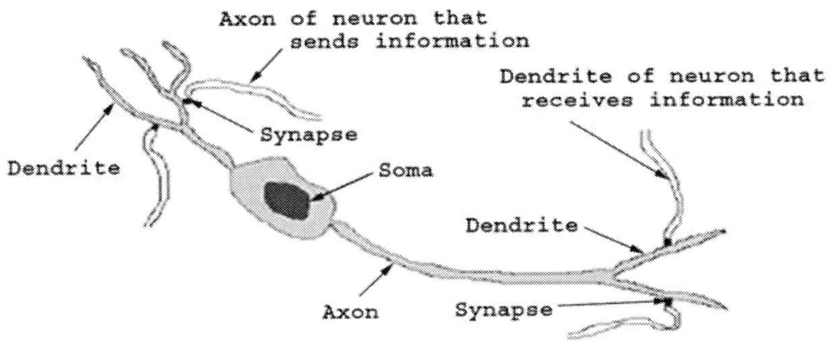

Figure 2. Schematic diagram for the biological neuron [24].

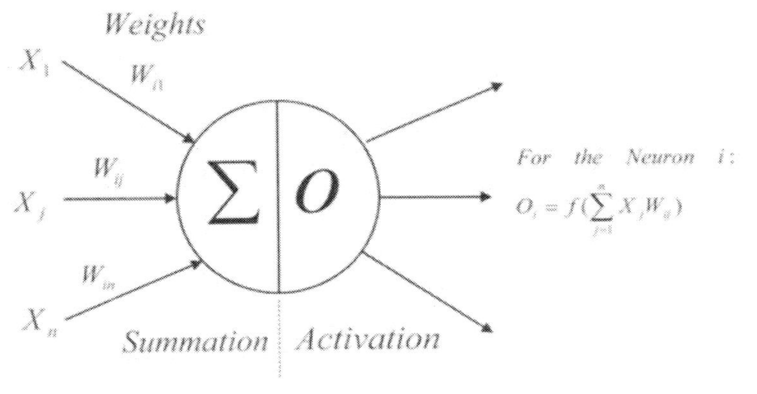

Figure 3. Information processing in the neural unit.

To stimulate some important aspects of the real biological neuron. An ANN is a group of interconnected artificial neurons usually referred to as “node” interacting with one another in a concerted manner, Figure 3 illustrates how information is processed through a single node.

The node receives weighted activation of other nodes through its incoming connections. First, these are added up (summation). The result is then passed through an activation function and the outcome is the activation of the node.

The activation function can be a threshold function that passes information only if the combined activity level reaches a certain value, or it could be a continues function of the combined input, the most common to use is a sigmoid function for this purpose. For each of the outgoing connections, this activation value is multiplied by the specific weight and transferred to the next node, An artificial neural network consists of many nods joined together usually organized in groups called ‘layers’, a typical network consists of a sequence of layers with full or random connections between successive layers, there are typically two layers with connection to the outside world as can be seen in Figure 4.

An input buffer where data is presented to the network, and an output buffer which holds the response of the network to a given input pattern, layers distinct from the input and output buffers called ‘hidden layer’, in principle there could be more than one hidden layer, In such a system, excitation is applied to the input layer of the network. Following some suitable operation, it results in a desired output. Knowledge is usually stored as a set of connecting weights (presumably corresponding to synapse efficiency in biological neural system) [24,25].

A neural network is a massively parallel-distributed processor that has a natural propensity for storing experiential knowledge and making it available for use. It resembles the human brain in two respects; the knowledge is acquired by the network through a learning process, and interneuron connection strengths known as synaptic weights are used to store the knowledge [26].

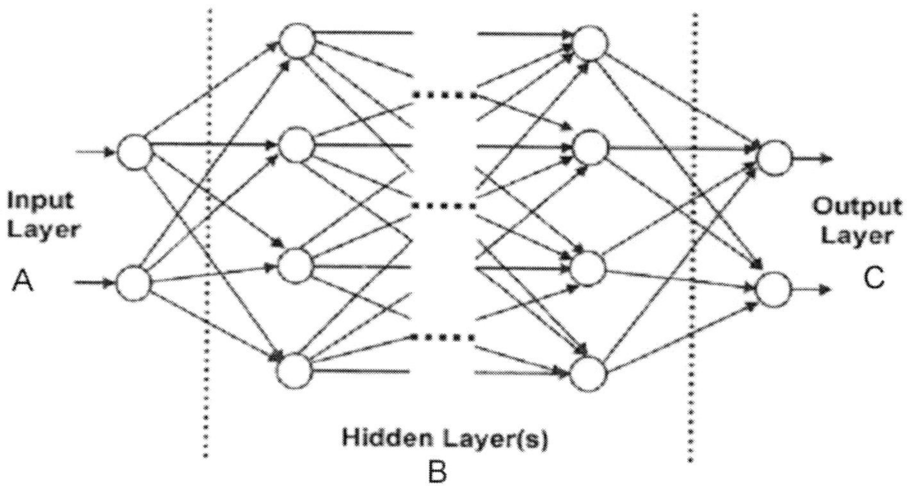

Figure 4. Typical architecture of an artificial neural network [24].

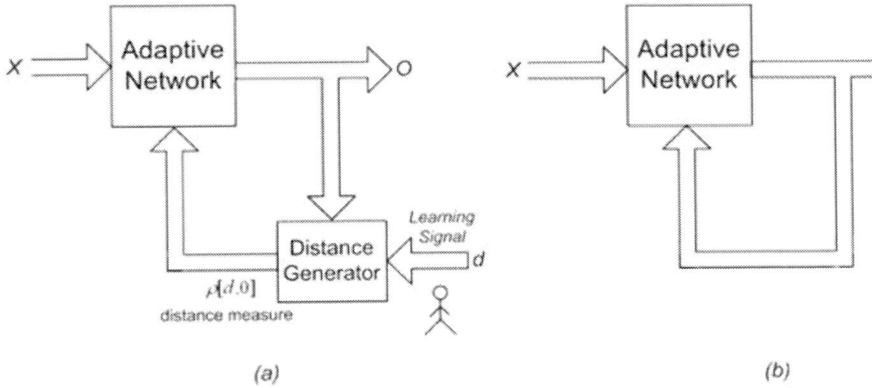

Figure 5. Basic learning modes [27].

Training is the process of modifying the connection weights in some orderly fashion using a suitable learning method. The network uses a learning mode, in which an input is presented to the network along with the desired output and the weights are adjusted so that the network attempts to produce the desired output. Weights after training contain meaningful information whereas before training they are random and have no meaning [24].

Two different types of learning can be distinguished: *supervised* and *unsupervised* learning, In *supervised learning* it is assumed that at each instant of time when the input is applied, the desired response d of the system is provided by the teacher. This is illustrated in Figure 5-a. The distance ρ [d,o] between the actual and the desired response serves as an error measure and is used to correct network parameters externally. Since adjustable weights have been assumed, the teacher may implement a reward and punishment scheme to adapt the network's weight. For instance, in learning classifications of input patterns or situations with known responses, the error can be used to modify weights so that the error decreases. This mode of learning is very pervasive. Also, it is used in many situations of learning. A set of input and output patterns called a *training set* is required for this learning mode [27].

Figure 5-b shows the block diagram of *unsupervised* learning. In *unsupervised learning,* the desired response is not known; thus, explicit error information cannot be used to improve network's behaviour. Since no information is available as to correctness or incorrectness of responses, learning must somehow be accomplished based on observations of responses to inputs that we have marginal or no knowledge about [27].

As it can be seen in Figure 4, the output of the units in layer ***A*** are multiplied by appropriate weights $\boldsymbol{W_{ij}}$ and these are fed as inputs to the hidden layer. Hence if $\mathbf{O}_i$ are the output of units in layer ***A,*** then the total input to the hidden layer, i.e., layer ***B*** is:

$$Sum_B = \sum_i O_i W_{ij} \tag{13}$$

And the output $\boldsymbol{O_j}$ of a unit in layer ***B*** is:

$$O_j = f(sum_B) \tag{14}$$

Where f is the non-linear activation function, it is a common practice to choose the sigmoid function given by: -

$$f(O_j) = \frac{1}{1 + e^{-O_j}} \tag{15}$$

As the nonlinear activation function.

However, any input-output function that possesses a bounded derivative can be used in place of the sigmoid function.

If there is a fixed, finite set of input-output pairs, the total error in the performance of the network with a particular set of weights can be computed by comparing the actual and the desired output vectors for each presentation of an input vector. The error at any output unit $\boldsymbol{e_K}$ in the layer **C** can be calculated by: -

$$e_K = d_K - O_K \tag{16}$$

Where $\boldsymbol{d_K}$ is the desired output for that unit in layer **C** and $\boldsymbol{O_K}$ is the actual output produced by the network .the total error $\boldsymbol{E}$ at the output can be calculated by: -

$$E = \frac{1}{2}\sum_K (d_K - O_K)^2 \tag{17}$$

Learning comprises changing weights so as to minimize the error function. To minimize E by the gradient descent method. It is necessary to compute the partial derivative of E with respect to each weight in the network. Equations (13) and (14) describe the forward pass through the network where units in each layer have there states determined by the inputs they received from units of lower layer. The backward pass through the network that involves "back propagation " of weight error derivatives from the output layer back to the input layer is more complicated.

For the sigmoid activation function given in equation (15), the so-called delta-rule for iterative convergence towards a solution my be stated in general as: -

$$\Delta W_{JK} = \eta \delta_K O_J \tag{18}$$

Where η is the learning rate parameter, and the error δ_K at an output layer unit K is given by: -

$$\delta_K = O_K(1 - O_K)(d_K - O_K) \tag{19}$$

And the error δ_J at a hidden layer unit is given by: -

$$\delta_J = O_J(1 - O_J)\sum_K \delta_K W_{JK} \tag{20}$$

Using the generalize delta rule to adjust weights leading to the hidden units is back propagating the error-adjustment, which allows for adjustment of weights leading to the hidden layer neurons in addition to the usual adjustments to the weights leading to the output layer neurons.

A back propagation network trains with tow step procedure as it is shown in Figure 6, the activity from the input pattern flows forward through the network and the error signal flows backwards to adjust the weights using the following equations: -

$$W_{IJ} = W_{IJ} + \eta\delta_J O_I \tag{21}$$

$$W_{JK} = W_{JK} + \eta\delta_K O_J \tag{22}$$

Until for each input vector the output vector produced by the network is the same as (or sufficiently close to) the desired output vector.

The fundamental idea underlying the design of the network is that the information entering the input layer is mapped as an internal representation in the units of the hidden layer(s) and the outputs are generated by this internal representation rather than by the input vector. Given that there are enough hidden neurons, input vectors can always be encoded in a form so that the appropriate output vector can be generated from any input vector [25]. ANNs while implemented on computers are not programmed to perform specific tasks. Instead, they are trained with respect to data sets until they learn the patterns presented to them. Once they are trained, new patterns may be presented to them for prediction or classification [24].

4. Experiment Design (Data Collection Procedure)

Trajectory planning was performed to generate the angular position and velocity for each joint, and then these generated data were fed to the robot's controller to generate the corresponding Cartesian position, orientation and linear velocity of the end-effector, which were recorded experimentally from sensors fixed on the robot joints.

In trajectory planning of a manipulator, it is interested in getting the robot from an initial position to a target position with free of obstacles path. Cubic trajectory planning method has been used in order to find a function for each joint between the initial position, θ_0, and final position, θ_f of each joint.

It is necessary to have at least four-limit value on the $\theta(t)$ function that belongs to each joint, where $\theta(t)$ denotes the angular position at time t.

Two limit values of the function are the initial and final position of the joint, where:

$$\theta(0) = \theta_0 \tag{23}$$

$$\theta(t_f) = \theta_f \tag{24}$$

Additional two limit values, the angular velocity will be zero at the beginning and the target position of the joint, where:

$$\dot{\theta}(0) = 0 \tag{25}$$

$$\dot{\theta}(t_f) = 0 \tag{26}$$

Based on the constrains of typical joint trajectory listed above, a third order polynomial function can be used to satisfy these four conditions; since a cubic polynomial has four coefficients.

These conditions can determine the cubic path, where a cubic trajectory equation can be written as:

$$\theta(t) = a_0 + a_1 t + a_2 t^2 + a_3 t^3 \tag{27}$$

The angular velocity and acceleration can be found by differentiation, as follows:

$$\dot{\theta}(t) = a_1 + 2a_2 t + 3a_3 t^2 \tag{28}$$

$$\ddot{\theta}(t) = 2a_2 + 6a_3 t \tag{29}$$

Substituting the constrains conditions in the above equations results in four equations with four unknowns:

$$\theta 0 = a_0, \tag{30}$$

$$\theta_f = a_0 + a_1 t_f + a_2 t_f^2 + a_3 t_f^3,$$

$$0 = a_0,$$

$$0 = a_1 + 2a_2 t_f + 3a_3 t_f^2$$

The coefficients are found by solving the above equations.

$$a_0 = \theta_0,$$
$$a_1 = 0,$$
$$a_2 = \frac{3}{t_f^2}(\theta_f - \theta_0),$$
$$a_3 = \frac{-2}{t_f^3}(\theta_f - \theta_0) \quad (31)$$

Angular position and velocity can be calculated by substituting the coefficients driven in Eq. (31) into the cubic trajectory Eqs. (27) and (28) respectively [11], which yield:

$$\theta_i(t) = \theta_i + \frac{3}{t_f^2}(\theta_{if} - \theta_{i0})t^2 - \frac{2}{t_f^3}(\theta_{if} - \theta_{i0})t^3, \quad (33)$$

$$\dot{\theta}_i(t) = \frac{6}{t_f^2}(\theta_{if} - \theta_{i0})t - \frac{6}{t_f^3}(\theta_{if} - \theta_{i0})t^2 \quad (33)$$

$$i = 1,2,.........., n$$

Where n is the joint number.

Joint angles generated ranged from amongst all the possible joint angles that do not exceed the physical limits of each joint.

Trajectory used for the training process has meant to be random trajectory rather than a common trajectory performed by the robot in order to cover as most space as possible of the robot's working cell.

As these joints are moving simultaneously with each other to complete the trajectory together. The interval of 1 second was used between a trajectory segment and another where the final position for one segment is going to be the initial position for the next segment and so on for every joint of the six joints of the robot.

After generating the joint angles and their corresponding angular velocities, these data are fed to the robot controller, which is provided with a sensor system that can detect the angular position and velocity on one hand and the Cartesian position, orientation and the linear velocity of the end-effector on the other hand; which are recorded to be used for the networks' training and testing.

5. Implementing the ANN

Two supervised feed forward ANN have been designed using *C programming language* to overcome the singularities and uncertainties in the arm configurations.

Both networks consist of input, output and one hidden layer, every neuron in each of the networks is fully connected with each other, sigmoid transfer function was chosen to be the activation function, generalized backpropagation delta learning rule (GDR) algorithm was used in the training process.

Off-line training was implemented; Trajectory planning was performed for 600 data set for every 1-second interval from amongst all the possible joint angles in the robot's workspace, then data sets were recorded experimentally from sensors fixed on the robot joints as was recommended by Karilk and Aydin [10], 400 set were used for the training while the other 200 sets were used for the testing the network.

All input and output values are usually scaled individually such that overall variance in data set is maximized, this is necessary as it leads to faster learning, all the vectors were scaled to reflect continuous values ranges from –1 to 1.

FANUC M-710i robot was used in this study, which is a serial robot manipulator consisting of axes and arms driven by servomotors. The place at which arm is connected is a joint, or an axis. This type of robot has three main axes; the basic configuration of the robot depends on whether each main axis functions as a linear axis or rotation axis. The wrist axes are used to move an end effecter (tool) mounted on the wrist flange. The wrist itself can be wagged about one wrist axis and the end effecter rotated about the other wrist axis, this highly non-linear structure makes this robot very useful in typical industrial applications such as the material handling, assembly of parts and painting.

5.1. Training Phase

In order to find the best network's configuration to solve the IK problem, and to make sure that for a certain trajectory the angular position of each joint will be the same as or sufficiently close to the desired when planning the trajectory for the robot, the ANN technique has been utilized where learning is only based on observation of the input–output relationship. Two networks have been trained and compared.

5.1.1. The First configuration (6-6 network configuration)

As was recommended by Karilk and Aydin [10], the input vector for the network consists of the position of the end effector of the robot along the X, Y and Z coordinates of the global coordinate system and the orientation according to the Euler representation (RPY), while the output vector was the angular position of each of the 6 joints as can be seen in Figure 6.

Number of neurons in the hidden layer has chosen to be 40 with a constant learning factor of 0.85 by trial and error. Table 1 shows the percentage of error of each of the 6 joints after the training was finished after 40,000 iteration, while Figure 7 shows the building knowledge curve for this network configuration.

Table 1. Overall error percentages for the training data for the 6-6 configuration

θ_1	θ_2	θ_3	θ_4	θ_5	θ_6
5.9%	2.487%	2.703%	6.73%	6.803%	3.658%

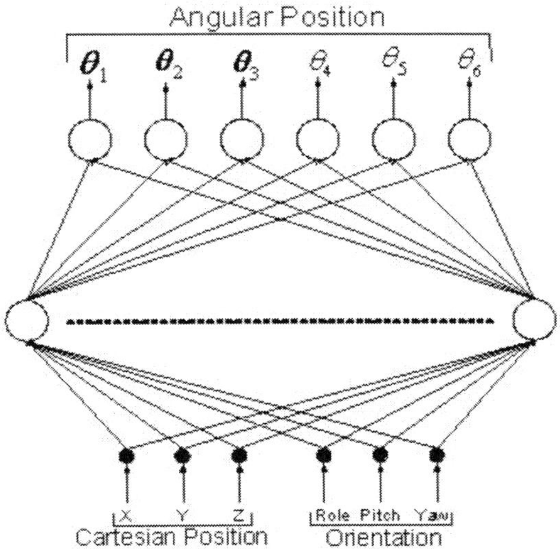

Figure 6. The topology of the 6-6 Network Configuration.

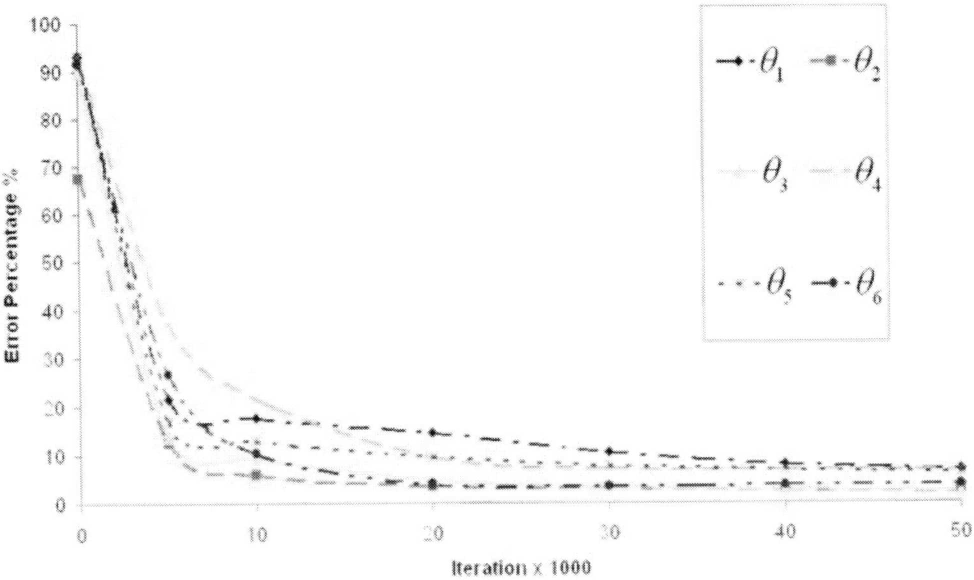

Figure 7. Building Knowledge curve for the 6-6 Network Configuration.

Even though one hidden layer was used in this study and the Euler representation (RPY) was used to represent the orientation, while the previous study of Karilk and Aydin [10] has used two hidden layers and the homogeneous transformation representation to represent the orientation; the results were almost similar (higher error percentage for the last three joints was obtained as compared to the second and third joints) even when different number of training patterns was used. The only difference here was that the error percentage of the first joint was almost similar to the percentages of the fourth and fifth joints.

5.1.2. The Second configuration (7-12 network configuration)

To examine the effect of considering the Jacobian matrix to the IK solution, another network has been designed, as in Figure 8, the new network consists of the Cartesian Velocity added to the input buffer and the angular velocity of each of the 6 joints added to the output buffer of the previous network.

Number of the neurons in the hidden layer was set to be 55 with constant learning factor of 0.9 by trial and error. Table 2 shows the percentage of error of each of the 6 joints after the training was finished after 40,000 iteration, while Figure 9 shows the building knowledge for this network configuration.

5.1.3. Networks' performance

To drive the robot to follow a desired trajectory, it will be necessary to divide the path into small portions, and to move the robot through all intermediate points. To accomplish this task, at each intermediate location, the robot's IK equations are solved, a set of joint variables is calculated, and the controller is directed to drive the robot to the next segment. When all segments are completed, the robot would be at the end point as desired.

Table 2. Overall error percentages for the training data for the 7-12 configuration

	Joint 1	Joint 2	Joint 3	Joint 4	Joint 5	Joint 6
Angular Position	4.32%	1.408%	1.403%	4.237%	5.165%	1.623%
Angular Velocity	2.627%	3.09%	2.33%	2.692%	2.803%	2.325%

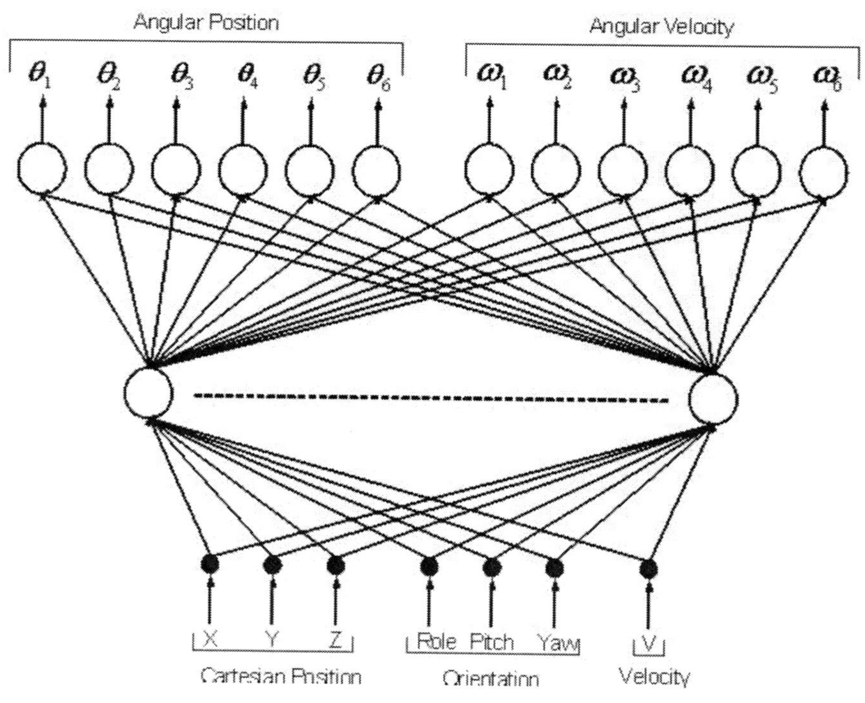

Figure 8. The topology of the 7-12 Network configuration.

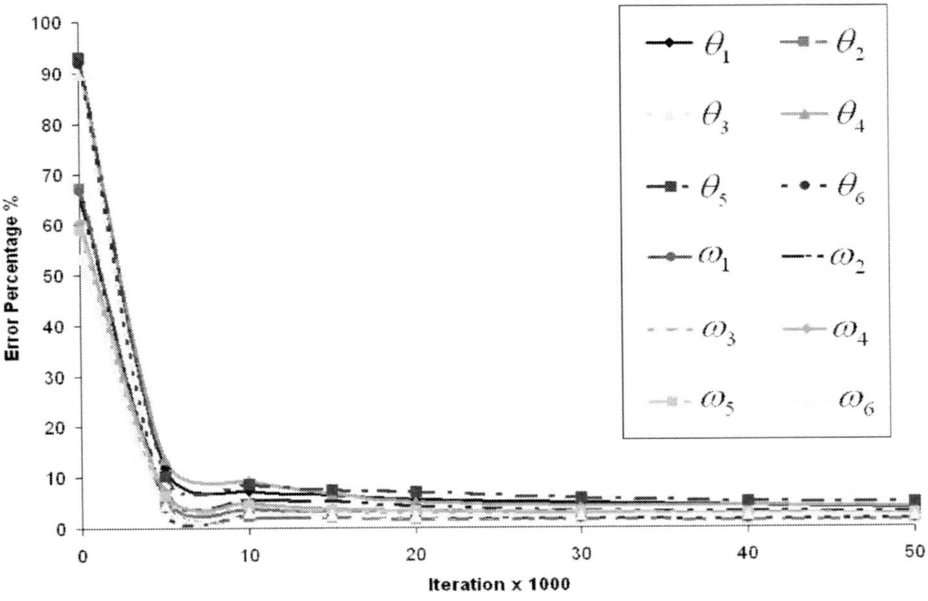

Figure 9. Building Knowledge curve for the 7-12 Network Configuration.

Figures 10 show the experimental trajectory tracking for the robot over the X Coordinates of the global coordinates system as an example, for both of the networks configurations compared to each other verses the desired trajectory.

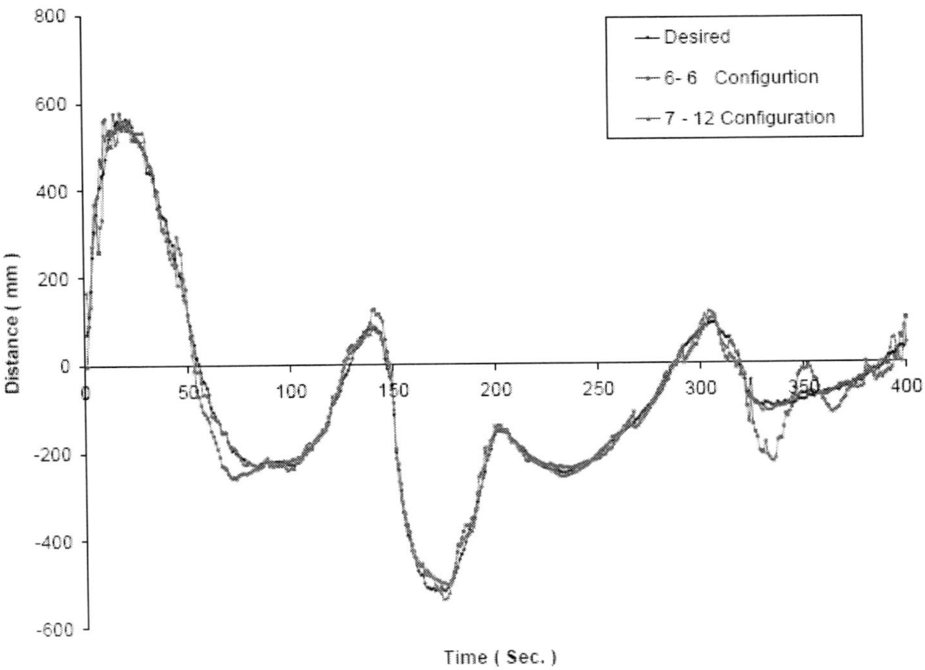

Figure 10. Experimental Trajectory tracking of the X coordinate for both Network Configurations compared to each other.

Table 3. Experimental trajectory tracking error percentages for the 6- 6 Network Configuration compared to the 7-12 Network Configuration

Network's Configuration	Cartesian Coordinates			Orientation Angles		
	X	Y	Z	Roll	Pitch	Yaw
6 - 6	14.6%	11%	2.24%	7.82%	13.68%	7.6%
7 - 12	3.65%	2.1%	1.32%	2.4%	4.74%	2.6%

The performance of the first network has improved when considering the Jacobian Matrix in the second configuration as can be seen in Table 3 which shows the experimental trajectory tracking error percentages for the 6-6 Network Configuration compared to the 7-12 Network Configuration.

5.2. Testing Phase

As the second configuration has shown better response than the first, it has been chosen to implement the testing data, new data that has never been introduced to the network before have been fed to the trained network in order to test its ability to make prediction and generalization to any set of data later overcoming problems resulting from applying the robot model.

Testing data were meant to pass through singular configurations (fourth and fifth joints); these configurations have been determined by setting the determinant of the Jacobian matrix to zero. Table 4 shows the percentages of error for the testing data set for each joint.

In order to verify the testing results, experiment has been performed to make sure that the output is the same or sufficiently close to the desired trajectory, and to show the combined effect of error, Figures 11 to 16 show the trajectory tracking of the X, Y, and Z coordinates with the Roll, Pitch and Yaw orientation angles respectively. The locus of which robot is passing through singular configurations are also shown. Through these figures, it can be seen that a good prediction has been achieved. The error percentages in the experimental data are shown in Table 5.

Table 4. Overall error percentages for the testing data

	Joint 1	Joint 2	Joint 3	Joint 4	Joint 5	Joint 6
Angular Position	2.515%	0.29%	1.745%	10.065%	9.755%	1.78%
Angular Velocity	4.2%	6.1%	3.22%	3.7%	6.5%	2.6%

Table 5. Overall experimental error percentages for the testing data

Cartesian Trajectory			Orientation Angles		
X	Y	Z	Roll	Pitch	Yaw
3.72%	3.06%	1.042%	3.562%	6.2%	4.964%

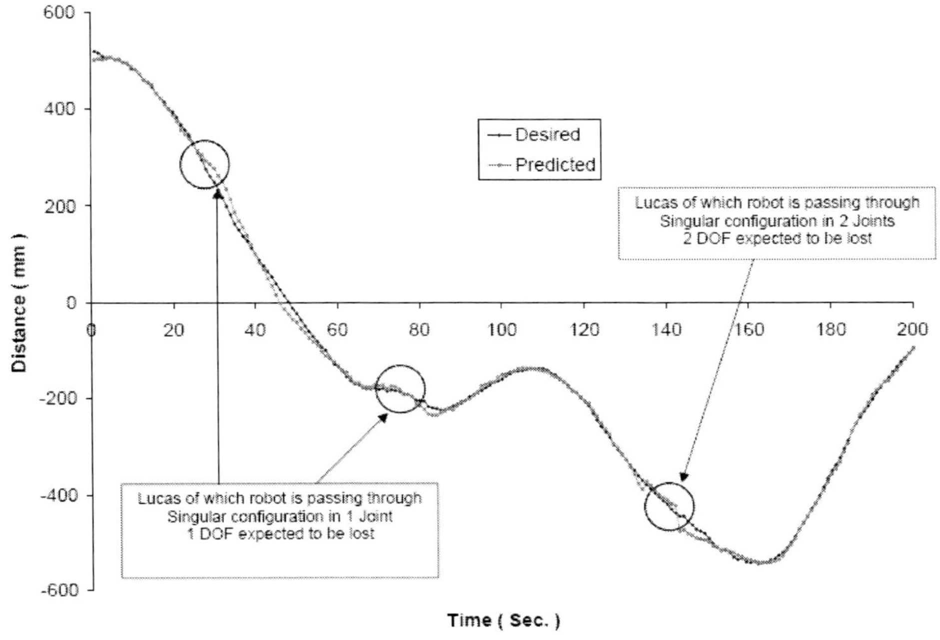

Figure 11. Experimental trajectory tracking of the X coordinate for the testing data.

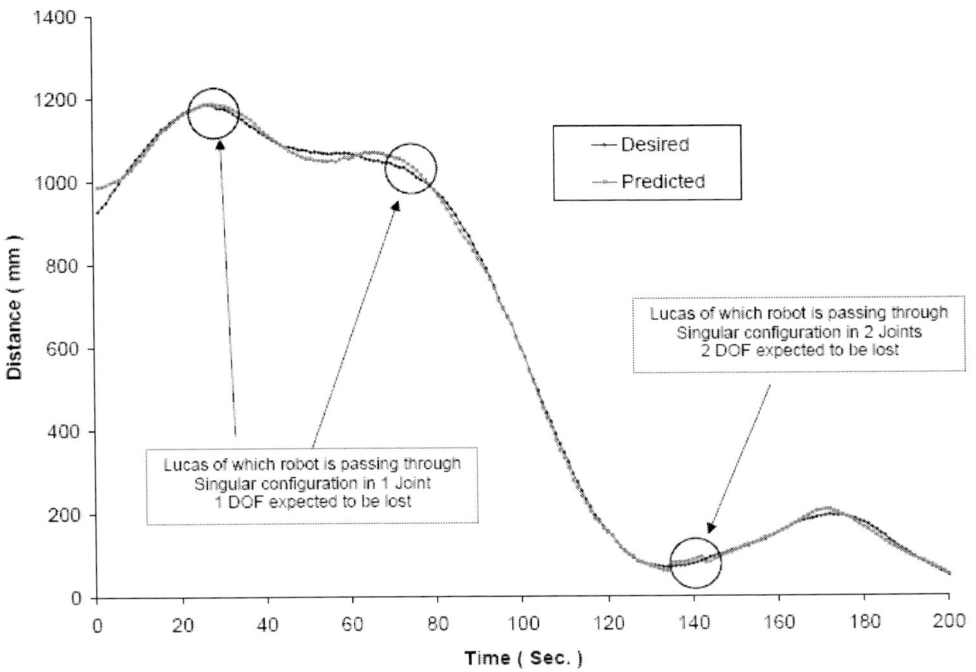

Figure 12. Experimental trajectory tracking of the Y coordinate for the testing data.

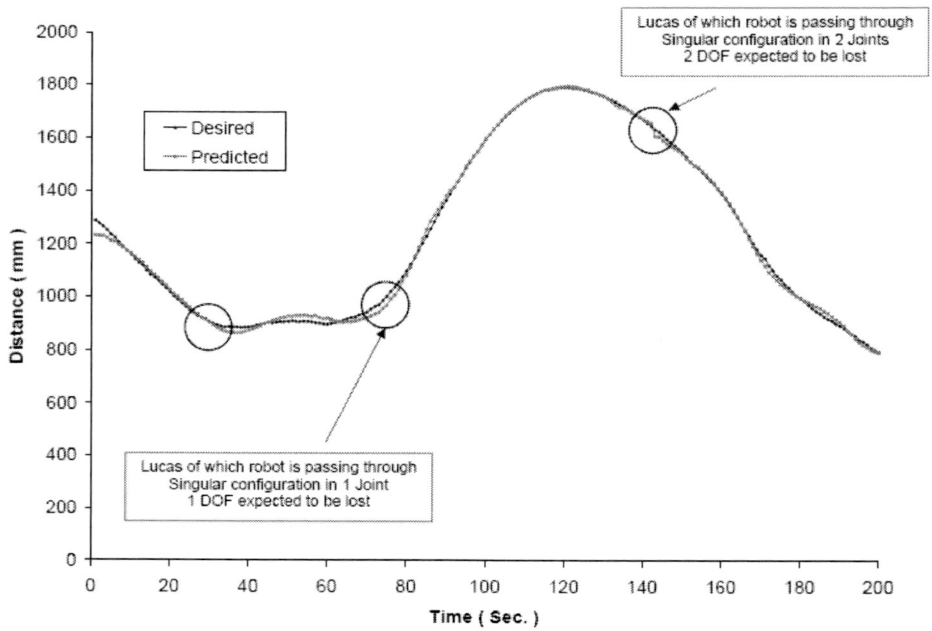

Figure 13. Experimental trajectory tracking of the Z coordinate for the testing data.

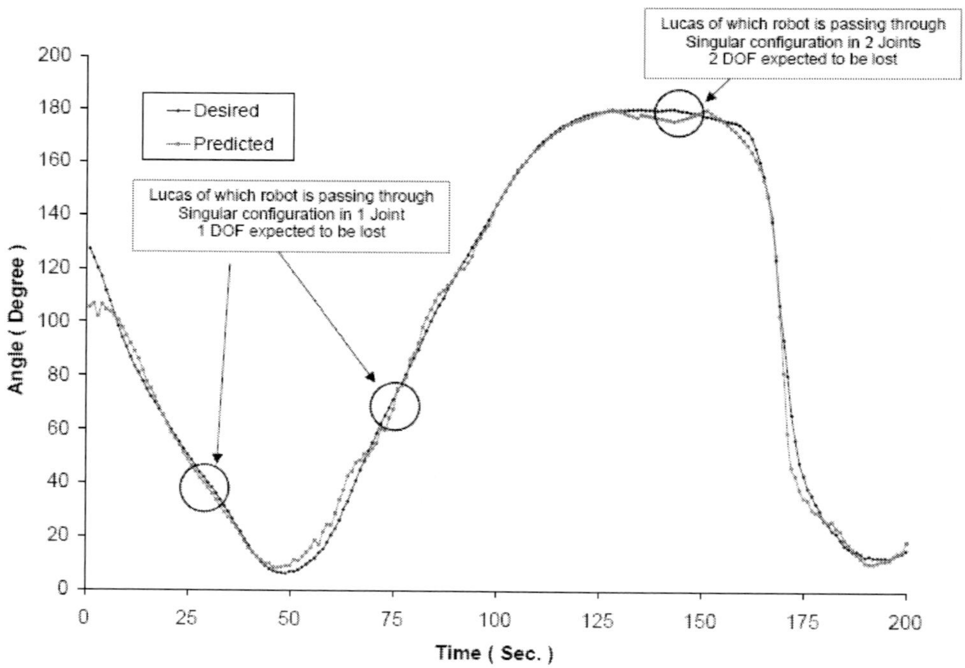

Figure 14. Experimental orientation tracking of the Roll angle for the testing data.

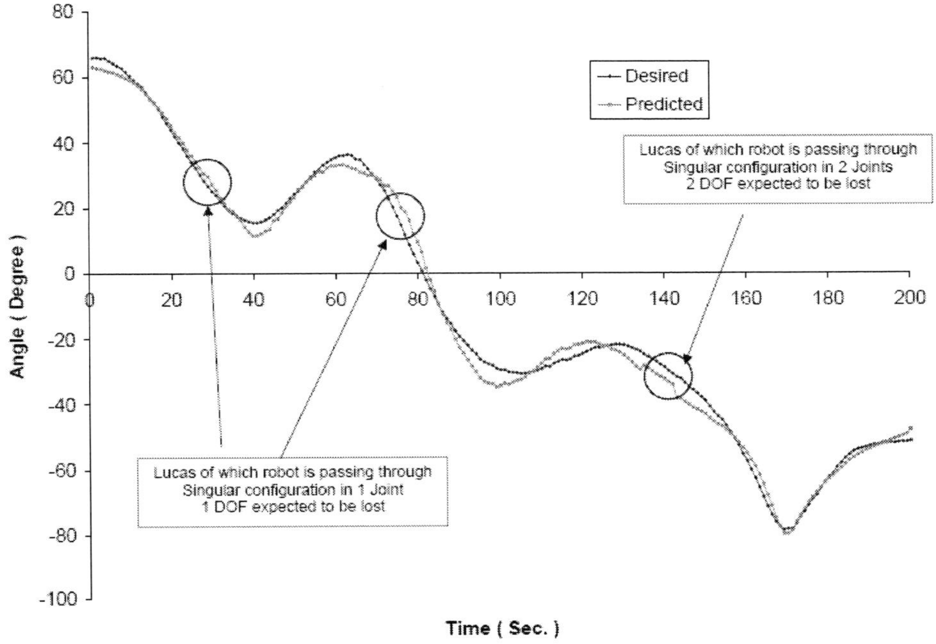

Figure 15. Experimental orientation tracking of the Pitch angle for the testing data.

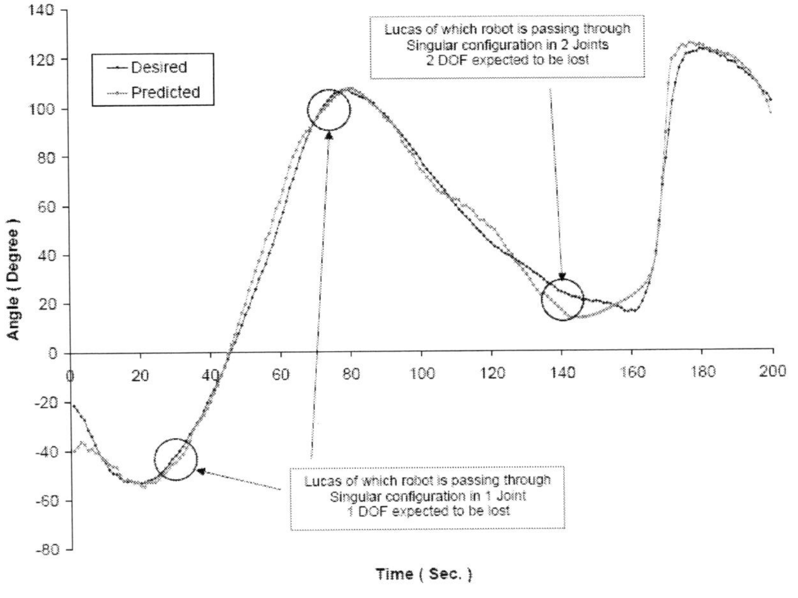

Figure 16. Experimental orientation tracking of the Yaw angle for the testing data.

6. Conclusions

This chapter was devoted to the implementation of the Neural Network technique to solve the inverse kinematics problem for serial robot manipulators, which are mainly singularities and uncertainties in arm configurations.

As demonstrated experimentally, the proposed technique can be effectively used to solve the problem.

Comparing the training results of the 6-6 Network Configuration to the literature has shown almost similar behavior, where the training error of the fourth and fifth joints were higher than the error of the rest of the joints despite the fact that different representation of the orientation and different number of training patterns have been used. The only difference in using this network was that the error of the first joint was almost similar to the error in the fourth and fifth joints.

Considering the effect of the Jacobian matrix to the solution of the Inverse Kinematics for the network configuration studied (6-6 Network Configuration) has shown that the networks where the Jacobian matrix effect is considered (7-12 Network Configuration) has given better results in terms of precision and iteration.

Using Artificial Neural Networks have shown the ability of the networks to remember the training patterns presented to them as well as making prediction for previously unknown trajectories overcoming drawbacks raised using other methods such as the Fuzzy Logic.

References

[1] Dulęba, I. & Sasiadek, J. Z. (2002). Redundant manipulators motion through singularities based on modified Jacobian method. Third International Workshop on *Robot Motion and Control*, November, 9-11, 331-336.

[2] Duffy, J. (1980). *Analysis of mechanism and robot manipulators*, John Wiley, New York.USA.

[3] Bingual, Z., Ertunc, H. M. & Oysu, C. (2005). *Comparison of Inverse Kinematics Solutions Using Neural Network for 6R Robot Manipulator with Offset. ICSC congress on Computational Intelligence.*

[4] Antonelli, G., Chiaverini, S. & Fusco, G. (2003). A new on-line algorithm for inverse kinematics of robot manipulators ensuring path-tracking capability under joint limits. IEEE *Transaction on Robotics and Automation*, *19(1)*, 162-167.

[5] Whitney, E. (1969). Resolved motion rate control of manipulators and human prostheses, *IEEE Transaction Man–Mach. Systems*, *10*, 47–53.

[6] Nakamura, Y. & Hanafusa, H. (1986). Inverse kinematic solutions with singularity robustness for robot manipulator control. *Journal of Dynamic Systems Measurements Control*, *108*, 163–171.

[7] Wampler, C. W. (1986). Manipulator inverse kinematic solutions based on vector formulations and damped least-squares methods. *IEEE Transaction Syst., Man, Cybernetics*, *16*, 93-101.

[8] Wampler, C. W. & Leifer, L. J. (1988). Applications of damped least-squares methods to resolved-rate and resolved-acceleration control of manipulators, *Journal of Dynamic Systems Measurements Control*, *110*, 31-38.

[9] Kuroe, Y., Nakai, Y. & Mori, T. (1994). A new Neural Network Learning on Inverse Kinematics of Robot Manipulators. International Conference on Neural Networks, *IEEE world congress on computational Intelligence, 5,* 2819-2824.

[10] Karilk, B. & Aydin, S. (2000). An improved approach to the solution of inverse kinematics problems for robot manipulators. *Engineering applications of artificial intelligence*, *13*, 159-164.

[11] Köker, R., Öz, C., Çakar, T. & Ekiz, H. (2004). A study of neural network based inverse kinematics solution for a three-joint robot. *Robotics and Autonomous Systems*, *49*, 227–234.

[12] Köker, R. (2005). Reliability-based approach to the inverse kinematics solution of robots using Elman's networks. *Engineering applications of artificial intelligence*, *18*, 685-693.

[13] Hasan, A. T., Hamouda, A. M. S., Ismail, N. & Al-Assadi, H. M. A. A. (2006). An adaptive-learning algorithm to solve the inverse kinematics problem of a 6 D.O.F serial robot manipulator. *Journal of Advances in Engineering Software*, *37*, 432-438.

[14] Funahashi, K. I. (1998). On the approximate realization of continuous mapping by neural networks. *Neural Networks*, *2(3)*, 183-192.

[15] Hasan, A. T., Hamouda, A. M. S., Ismail, N. & Al-Assadi, H. M. A. A. (2007). A new adaptive learning algorithm for robot manipulator control. Proceeding of the I Mech E, Part I: *Journal of System and Control Engineering, 221(4)*, 663-672.

[16] Driscoll, J. A. (2000). Comparison of neural network architectures for the modeling of robot inverse kinematics. Proceedings of the IEEE, *south astcon*, 44-51.

[17] Hornik, K. (1991). Approximation capabilities of multi-layer feed forward networks. IEEE *Trans.Neural Networks*, *4(2)*, 251-257.

[18] Denavit, J. & Hertenberg, R. S. A. (1955). Kinematics Notation for lower Pair Mechanism Based on Matrices. *Applied mechanics*, *77*, 215-221.

[19] Fu, K. S., Gonzalez, R. C. & Lee, C. S. G. (1987). *Robotics Control, Vision, and Intelligence.* McGraw-Hill book Co. Singapore, international edition.

[20] Nakamura, Y. (1991). *Advanced Robotics: Redundancy and Optimization*, Addison Wesley, New York, USA.

[21] Tchoń, K. & Muszyński, R. (1998). Singular Inverse Kinematic Problem for Robotic Manipulators: A Normal Form Approach, IEEE *Trans. on Robotics and Automation*, *14(1)*, 93-104.

[22] Muszyński, R. & Tchoń, K. (1997). Singularities of non-redundant robot kinematics. *Robotic Research*, *1997*, *16(1)*, 60-76.

[23] Dulęba, I. (2000). Channel algorithm of transversal passing through singularities for non-redundant robot manipulators. *IEEE conference on Robotics and Automation*, San Francisco, CA.USA, 1302-1307.

[24] Soteris, A. (2001). Kalogirou. Artificial Neural Networks In Renewable Energy Systems Applications: a review. *Renewable and Sustainable Energy Reviews*, *5*, 373-401.

[25] Santosh, A. & Devendra, P. (1993). Garg. Training back propagation and CMAC neural networks for control of a SCARA robot. *Engineering Applications of Artificial Intelligence*, *6(2)*, 105-115.

[26] Haykin, S. (1994). *Neural Networks. A Comprehensive Foundation*. New York: Macmillan.

[27] Jacek, M. (1992). *Zurada. Introduction to Artificial Neural Systems*. West Publishing Company. Singapore.

In: Focus on Artificial Neural Networks
Editor: John A. Flores, pp. 325-340

ISBN: 978-1-61324-285-8

Chapter 15

USING ARTIFICIAL NEURAL NETWORKS FOR CONTINUOUSLY DECREASING TIME SERIES DATA FORECASTING

Mebruk Mohammed[1], Kunio Watanabe[1] and Shinji Takeuchi[2]
[1]Geosphere research Institute of Saitama University, Shimo Okubo 255, sakura-ku, Saitama 338-8570, Japan
[2]Tono Geoscience Center, Japan Atomic Energy Agency (JAEA), 1-64, Yamanonichi, Akeyo, Mizunami, Gifu 509-6132, Japan

ABSTRACT

Data preprocessing is an issue that is often recommended to create more uniform data to facilitate ANN learning, meet transfer function requirements, and avoid computation problems. In ANN typical transfer functions, such as the sigmoid logistic function, or the hyperbolic tangent function, cannot distinguish between two very large values, because both yield identical threshold output values of 1.0. It is then necessary to normalize (preprocess) the inputs and outputs of a network. Usually normalization is carried out using the minimum and maximum values obtained in the in-sample (calibration) data. Such a network will result in absurd output, if the out-of-sample (test) data contain values that are beyond the in-sample data range. This ultimately limits the application of ANN in forecasting continuously increasing or decreasing time series data.

This study will present a novel and successful application of ANN, which is trained by the error backpropagation algorithm, in the context of forecasting beyond in-sample data range. The emphasis here is on continuously decreasing hydraulic pressure data forecasting that are observed at Mizunami underground research laboratory construction site, Japan. The ANN utilizes the sigmoid logistic function in its hidden and output layers.

1. INTRODUCTION

The common approach to modeling of a phenomenon involves applying physical principles and results from full-scale experiments. The power of such modeling approach is predicated on its ability to assign via its equations spatial and/or temporal variability of input parameters and conditions. Natural, random, real world conditions like heterogeneity and anisotropy, which exert significant effect on a system behavior, can be incorporated directly into the simulation model. Almost paradoxically, its capability for representing natural real world complexity underscores the uncertainty of model input values and the corresponding output predictions.

Seeking to circumvent parameter uncertainty and simplifying mathematical and physical assumptions inherent to such models, alternative Black-box approaches for the simulation of a system behavior have been sought. A Black-box model is a mathematical model describing relations between input and output data for a given process. Artificial neural networks (ANN), is one of such Black-box model types.

A major implementation advantage of the empirically based approaches, like ANN, is that it typically uses input variables that are fundamentally more accessible and less uncertain than input parameters required by physical based models.

A major disadvantage of ANN model is that it can only directly predict a system behavior at specific conditions for which it has been developed with a corresponding set of field data. ANN developed with exhaustive data, can achieve excellent predictive accuracy. Exhaustive data here implies paired input-output data that contain the possible future maximum and minimum data values. In cases where activities bring about continuously increasing or decreasing change in a system, it would be difficult to acquire the necessary exhaustive data. Such continuous system would ultimately require the need to predict beyond the observed conditions (in sample data). ANN model is intolerant to such conditions where prediction involves out of observed conditions.

This chapter thus will show, with practical example, a successful effort to model continuously increasing or decreasing system variations using ANN. The ANN is trained by the backpropagation algorithm (backpropagation neural network (BPN)) having the sigmoid logistic activation function in its hidden and output layers.

2. EXISTING FEEDFORWARD ARTIFICIAL NEURAL NETWORK MODEL

Feedforward artificial neural networks trained by the backpropagation algorithm (BPN) are currently used in variety of applications with great success. The reason behind this widespread adoption can be found in two very important abilities that they exhibit. BPN can be trained to learn through examples (memorization ability) and can respond to cases that are similar but not identical to the ones that they have been trained with (generalization ability).

The building block of a BPN is called a neuron; its mathematical model is as shown in Figure 1. Each neuron, receives inputs in the form of weighted signals, sums them along with a bias term and applies a function f, called activation function (usually non-linear), to determine its own output signal, denoted by y. A typical BPN is composed of several such neurons, which are arranged in layers can be seen in Figure 2.

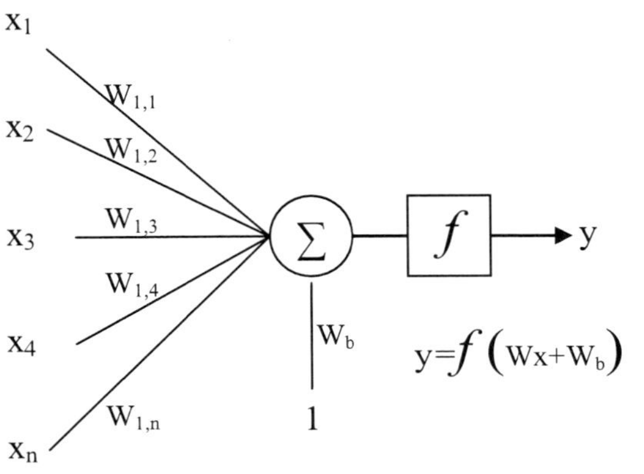

Figure 1. Mathematical model of a typical neuron.

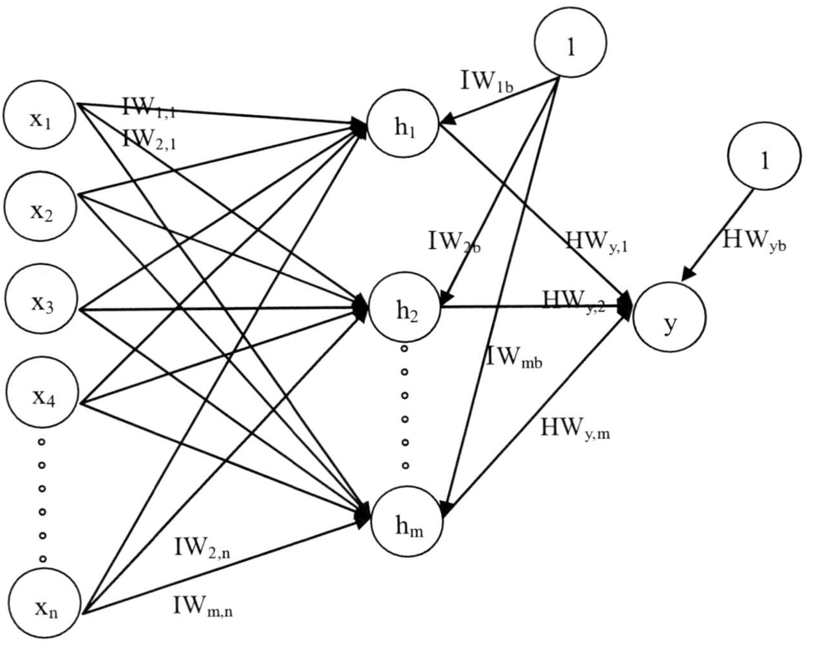

Figure 2. Architecture of a typical BPN.

The mathematical notations in Figure 1 and Figure 2 are described as:

- x_i is the output signal of the ith neuron in the input layer
- h_j is the output signal of the jth neuron in the hidden layer
- y is ANN's output signal
- $IW_{j,i}$ is weight coefficient between the ith input neuron and the jth hidden neuron
- $HW_{y,j}$ is weight coefficient between the jth hidden neuron and the output neuron
- IW_{jb} is the bias weight coefficient of the jth hidden neuron
- HW_{yb} is the bias weight coefficient of the output neuron
- n is the number of neurons in the input layer
- m is the number of neurons in hidden layer

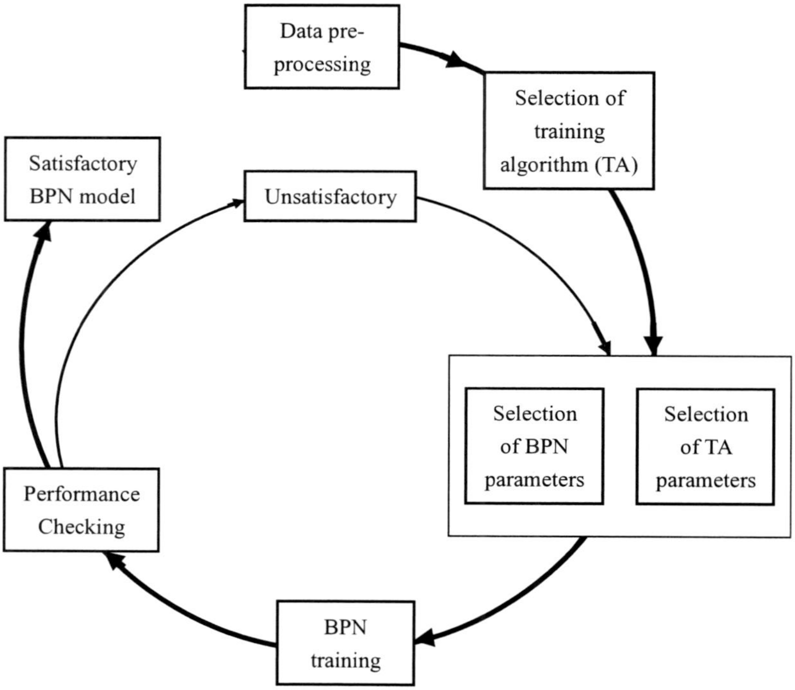

Figure 3. Development of a BPN.

2.1. Development of a BPN

The development of a BPN involves several stages, from pre-processing the necessary data to creating a satisfactory model (Figure 3). The most important aspect of this process is the selection of certain parameters that are crucial for the model's performance.

2.2. Data Pre-Processing

2.2.1. Data partitioning

Before building a BPN model, partition the data using a partition utility is carried out. Partitioning yields mutually exclusive datasets: a training dataset, a validation dataset and a test dataset. In ANN model, the training dataset is used to obtain the network weights.

Once a model is built on training data, the accuracy of the model on unseen data shall be carried out. For this, the model should be used on a dataset that was not used in the training process i.e. a dataset where the actual value of the target variable is known. The discrepancy between the actual value and the predicted value of the target variable is the error in prediction. If the training data itself were used to compute the accuracy of the model fit, an overly optimistic estimate of the accuracy of the model would be obtained. This is because the training or model fitting process ensures that the accuracy of the model for the training data is as high as possible; the model is specifically suited to the training data. To get a more realistic estimate of how the model would perform with unseen data, a part of the original data should be set aside and not use it in the training process. This dataset is known as the validation dataset. After fitting the model on the training dataset, its performance should be tested on the validation dataset.

The validation dataset is often used to fine-tune models. For example, a number of neural network models with various architectures might be used and their accuracy on the validation dataset is compared to choose among the competing architectures. In such a case, when a model is finally chosen, its accuracy with the validation dataset is still an optimistic estimate of how it would perform with unseen data. This is because the final model has come out as the winner among the competing models based on the fact that its accuracy with the validation dataset is highest. Thus, yet another portion of data which is used neither in training nor in validation, is set aside. This set is known as the test dataset. The accuracy of the model on the test data gives a realistic estimate of the performance of the model on completely unseen data.

There are two ways of data partitioning: random partitioning and user-defined partitioning. In simple random partitioning, every observation in the main dataset has equal probability of being selected for the partition dataset. For example, if 60% is specified for the training dataset, then 60% of the total observations would be randomly selected and would comprise the training dataset. In other words, each observation has a 60% chance of being selected for the training partition. In user-defined partitioning, the partition variable that is used to partition the dataset should be specified. This is useful when the observations that should be used in training, validation, and/or test dataset are already predetermined.

In most monitoring and management problems since the test dataset represent the future (planned) conditions, 'time' is an important partition variable. Thus the test dataset is partitioned in user-defined way. For example in Figure 4, hydraulic pressure variation caused by continuous vertical shafts construction is depicted. To model such hydraulic pressure variation using BPN, the partition variable should be time (shaft construction schedule).

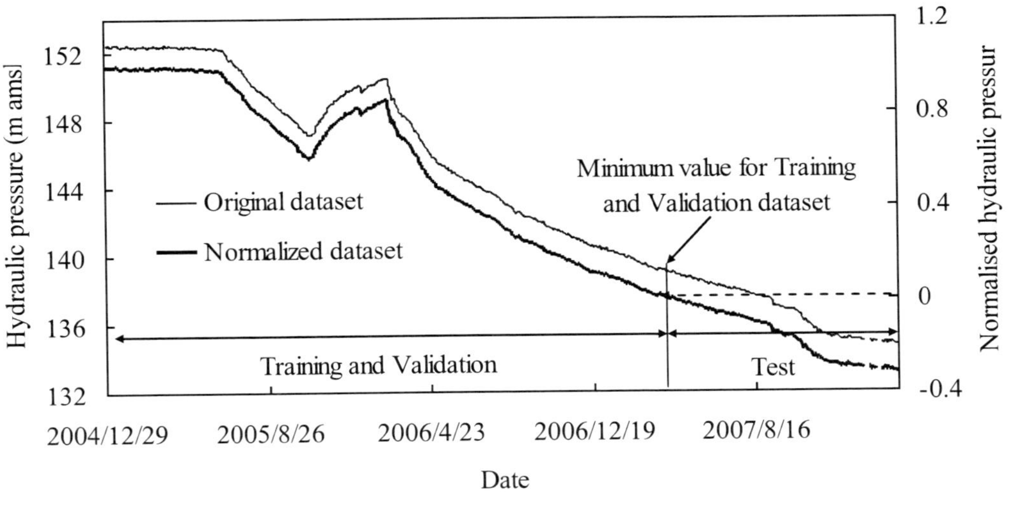

Figure 4. Data partitioning and normalization.

2.2.2. Normalization

The performance of a BPN depends on the effective way of presentation of data. If the input and the output variables are not of the same order of magnitude, some variables may appear to have more significance than they actually do. The training algorithm has to compensate for order-of-magnitude differences by adjusting the network weights, which is not very effective in many of the training algorithms such as backpropagation algorithm. For example, if one input variable has a value of thousands and other input variable has a value in tens, the assigned weight for the second variable entering a node of the first hidden layer must be much greater than that for the first. In addition, typical transfer functions, such as a sigmoid function, or a hyperbolic tangent function, cannot distinguish between two values when both are very large, because both yield identical threshold output values of 1.0. It is then necessary to normalize the input and output in the same order of magnitude. In general the equation for normalization is written as:

$$X_{Norm} = emi + \frac{X - X_{Min}}{X_{Max} - X_{Min}} \times (ema - emi) \tag{1}$$

Where, X_{Min} and X_{Max} are the minimum and maximum values of a given data set X. *emi* and *ema* are the minimum and maximum values of normalization ranges, respectively.

Whenever normalization of a dataset is carried out, determination of minimum and maximum value of the given dataset is needed. The problem is that these maximum and minimum values restrict the operating range of the network (Welstead 1994). Usually normalization is carried out using the minimum and maximum values obtained in the training and validation datasets. If such normalization is carried out for the hydraulic pressure trend shown in Figure 4, then the test dataset usually result in values below the expected normalization range (negative in this case).

For BPN, depending on the activation functions used in hidden and output layer nodes, different normalization ranges can be used. Chaturvedi (2008), has pointed out that if the normalization range for sigmoid and hyperbolic tangent function are in the range of 0.1 to 0.9 and -0.9 to +0.9, respectively, then the BPN model will take least number of epochs.

2.2.3. Sequence of data presentation

BPN training performance is very much dependent on the way the data is presented to it. If clustered data are presented to BPN, then it will learn more efficiently and quickly (Chaturvedi 2008). For effective training of BPN, beside normalization of the training and validation data, it is equally important that how the sequences of these data for learning are presented. This is especially important when on-line training is adopted.

2.3. Selection of Training Algorithm

Feedforward ANN has been applied successfully to solve some difficult and diverse problems by training them in a supervised manner with a highly popular algorithm known as the error backpropagation algorithm. This algorithm is based on the error-correction learning rule. Backpropagation learning may be implemented in one of two basic ways; (1) Sequential mode (also referred to as the on-line mode or stochastic mode) in which, adjustments are made to the weights of the network on an example-by-example basis. (2) Batch mode: in which adjustments are made to the weights of the network on an epoch-by-epoch basis, where each epoch consists of the entire set of training examples.

The backpropagation algorithm involves two phases: the forward phase and the backward phase (Rumelhart et al. 1986). In forward phase, the weights ($IW_{j,i}$, IW_{jb}, HW_{yb} and $HW_{y,j}$ in Figure 2) of the network are fixed, and the input signal is propagated through the network layer-by-layer. The function signal appearing at the hidden neuron *j* and at the output neuron *y* is computed, respectively as:

$$h_j(p) = f(u_j(p)), \quad u_j(p) = IW_{jb} + \sum_{i=1}^{n}\left[IW_{j,i} \times x_i(p)\right] \tag{2a}$$

$$y(p) = f(v(p)), \quad v(p) = HW_{yb} + \sum_{i=1}^{m}\left[HW_{y,i} \times h_i(p)\right] \tag{2b}$$

The forward phase finishes with the computation of an error signal for each pattern *p* in the training phase:

$$error(p) = d(p) - y(p) \tag{3}$$

where *d(p)* is the desired response and *y(p)* is the actual output produced by the network in response to the *n* inputs $x_i(p)$.

If *N* is the total number of patterns in the training set, the average squared error E_{av} over all the patterns is given by

$$E_{av} = \frac{1}{N}\sum_{p=1}^{N}\left[error(p)\right]^2 = \frac{1}{N}\sum_{p=1}^{N}\left[d(p) - y(p)\right]^2 \tag{4}$$

If E_{av} is within the acceptable limit the process is terminated otherwise a backward phase is carried out for updating the synaptic weights by using the backpropagation equation. In the backward phase, the error signals computed are passed leftward through the network (Figure 2), layer-by-layer and recursively computing the local gradient δ for each neuron as:

$$\delta(p) = error(p) \times f'(p) \tag{5}$$

where $f'(p)$ is the derivative of the activation function.

The computation of δ for each neuron of the architecture in Figure 2 requires the derivative of the activation function, *f* (.) associated with that neuron. For this derivative to exist, the function *f* (.) needs to be continuous. In basic terms, differentiability is the only criterion that an activation function would have to satisfy (Chaturvedi 2008). It has been observed that a non-linear activation function with maximum variation in the mid-values gives stability to the learning process. Such an activation commonly used is the sigmoid activation, whose derivative attains maximum at mid-value. Thus the sigmoid logistic function (Eq. 6) was adopted as an activation function in both hidden- and output layers' nodes.

$$f(u_j(p)) = \frac{1}{1+e^{-u_j(p)}} \tag{6a}$$

$$f(v(p)) = \frac{1}{1+e^{-v(p)}} \tag{6b}$$

where $v\ (p)$ and $u_j\ (p)$ are as defined in Eq. 2.

For sigmoid logistic function the weight updating is computed by Eq. 7 or Eq. 8, depending on whether the neuron is in the output layer or hidden layer(s), respectively (Rojas 1996). The recursive computation is continued layer-by-layer, by propagating the changes to all synaptic weights from output layer to input layer in sequential mode.

$$\begin{gathered} HW_{y,j} = HW_{y,j} + \eta \times \sigma(p) \times f(u_j(p)) + \alpha \times \lfloor \Delta HW_{y,j} \rfloor \\ \sigma(p) = y(p) \times (d(p) - y(p)) \times (1 - y(p)) \end{gathered} \tag{7}$$

$$\begin{gathered} IW_{i,j} = IW_{i,j} + \eta \times \beta_j(p) \times x_j(p) + \alpha \times \lfloor \Delta IW_{i,j} \rfloor \\ \beta_j(p) = f(u_j(p)) \times \sigma(p) \times (1 - f(u_j(p))) \times HW_{y,j} \end{gathered} \tag{8}$$

where $\Delta HW_{y,j}$ and $\Delta IW_{i,j}$ are the previous pattern's weight change of $HW_{y,j}$ and $IW_{i,j}$, respectively; η is the learning rate and α is the momentum factor. See Rojas (1996) and Chaturvedi (2008) for detail explanation of these parameters.

2.4. Selection of BPN Parameters

Parameter selection for BPN is mainly dependant on the experience of the modeler regarding the ANN, backpropagation and the studied phenomenon. Selection of parameters for BPN includes initialization of the network synaptic weights, number of hidden layers, number of neurons in the hidden layers, the learning rate, the momentum factor, maximum number of epochs and/or the error tolerance. The backpropagation algorithm provides an "approximation" to the trajectory in the error-parameter space computed by the method of steepest descent. According to the method of steepest descent, the synaptic weights are adjusted in an iterative fashion along the error surface with an aim of moving them progressively toward the optimum solution (Rojas 1996).

2.4.1. Initial synaptic weights

Every training procedure starts by initializing the synaptic weight coefficients, i.e. by assigning values to them. The training goal is to find weight values that minimize the networks error function. Since the initial values of the weights define the starting point of the training algorithm on the error function, they affect both the training speed and the achieved training error. This depends on whether this point is close to the global minimum or located in an area with many local minima. The most common method used to initialize weights is to randomly select values from pre-defined value field (usually centered on zero).

2.4.2. BPN Architecture

BPN architecture is directly related to the complexity of the solution space that it represents. A network that is fairly simple might not be able to learn the interactions underlying the training data, while a very complex network will memorize them to such extent that it will no longer be able to respond to unknown data (Chaturvedi 2008). Obtaining the right architecture is the most crucial stage in the development of a BPN model. Given that there is no theory as to what this architecture is or how to obtain it; it is one of the most difficult stages to perform.

Number of layers

While developing BPN model, two layers are fixed, namely input layer and output layer. Generally, at the input layer, the inputs are distributed to other neurons in the next layer and no processing takes place at this layer. Unlike the input layer, at output layer processing is done. In literatures it is mentioned that three layer network is a universal approximator and could handle most of the problems (Hornik et al. 1989). Then also for complex problems, it is difficult to train ANN with three layers network structure. Hence, most of the time the ANN developer uses trial and error method to select the number of layers in the ANN structure. In this study we have adopted a three layered architecture.

Number of neurons in each layer

The number of input variables determines the total number of input nodes. Most of the time trial-and-error method is used to select the number of neurons in a hidden layer. There are two ways to deal with this problem. Firstly, one can start with small number of neurons and then during training the number of neurons may be increased till the satisfactory performance is obtained. Second method to handle this situation is, one could begin with large number of neurons and then start deleting a neuron, till the ANN size is optimal.

In the present study, the number of hidden nodes was selected starting from the minimum and then increased by one consecutively, efficiency of the ANN model checked, and the best number of hidden nodes was selected.

2.4.3. Learning rate

The learning rate, η decides the scaling of the gradient of the error surface to be used for the synaptic weights adjustment. The smaller the learning rate value, the smaller will be the changes to synaptic weights in the network from one iteration to the other and the smoother will be the trajectory in the error space. This improvement is achieved at the cost of a slower learning. If the learning rate is too large, so as to speed up the rate of learning, the resulting large changes in the synaptic weight may make the trajectory in the error space oscillatory and unstable.

2.4.4. Momentum factor

Another simple method of increasing the rate of learning, and yet avoiding the danger of instability, is to include a momentum factor,α. The incorporation of momentum factor in the backpropagation algorithm represents a minor modification to the synaptic weight update and yet it can have highly beneficial effects on learning behavior of the algorithm. The momentum term also helps in preventing the learning process from trapping in local minima (Chaturvedi 2008).

2.4.5. Maximum number of epochs and error tolerance

There are several stopping criteria, each with its own practical merit, which may be used to terminate the synaptic weight adjustments. The logical thing to do is to think in terms of the tolerable error. In the two phases of the backpropagation algorithm, all patterns in the training set is used in the forward phase to compute the average square error if this error is less than the error of tolerance the synaptic weights that bring about this error are used for prediction. If the average square error is not less than the error of tolerance the backward phase of the algorithm is carried out. Therefore the total number of epochs determines how much such iterations required to reach the desired goal (error tolerance).

3. APPLICATION OF EXISTING BPN FOR CONTINUOUSLY DECREASING TIME SERIES DATA

In this study BPN application was tested for a time series data (Figure 5) obtained in MSB-1 borehole drilled at Mizunami underground research laboratory site, here after this site is called MIU. The observed hydraulic pressure variation in Figure 5 is mainly caused by the construction of two vertical shafts at MIU. As the figure indicates continuous decrease in the observed hydraulic pressure is seen except for the pressure sensor at top most elevation. The observation period was from January 1, 2005 to March 7, 2008. For details of the activities at MIU please see Mohammed et al. (2009; 2010), JNC (2003), Kumazaki et al. (2003), Goto et al. (2002) and Takeuchi et al. (2007).

Here we will see two examples; the first (Example 1) will involve the use of observed hydraulic pressure at some other locations to predict the pressure at a different location. For example the hydraulic pressure at 56.8 m above mean sea level (amsl) can be modeled by taking the observed pressures at 72.5 m, 117.3 m and 132.6 m amsl as inputs to BPN. The second example (Example 2) will use a hybrid approach that uses three finite element model results for the sensor at 56.8 m amsl as inputs to BPN to model the observed hydraulic pressures (see Figure 6). For the details of the finite element method (FEM) and the hybrid approach please refer to Mohammed et al. (2009; 2010).

The analysis period in the case study (from January 1, 2005 to March 7, 2008) was divided in 4646 patterns of 6 hour interval. After ignoring some faulty measured hydraulic pressure values a total of 4560 were used in this study. Of these patterns 50%, 20% and 30% were used for Training, Validation and Test sets, respectively. While discussing the results of the BPN both Training and Validation sets are combined and called it as Calibration set. Thus 70% of the total data, 3192 patterns, were used for Calibration phase and the rest 30% (1368 patterns) were used as Test phase.

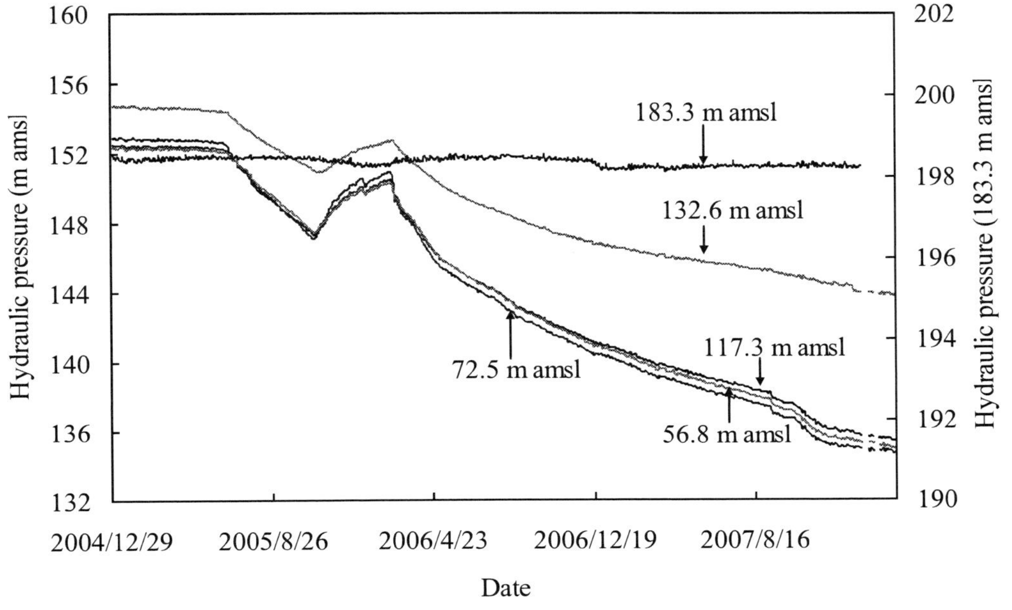

Figure 5. The five hydraulic pressure values recorded in MSB-1.

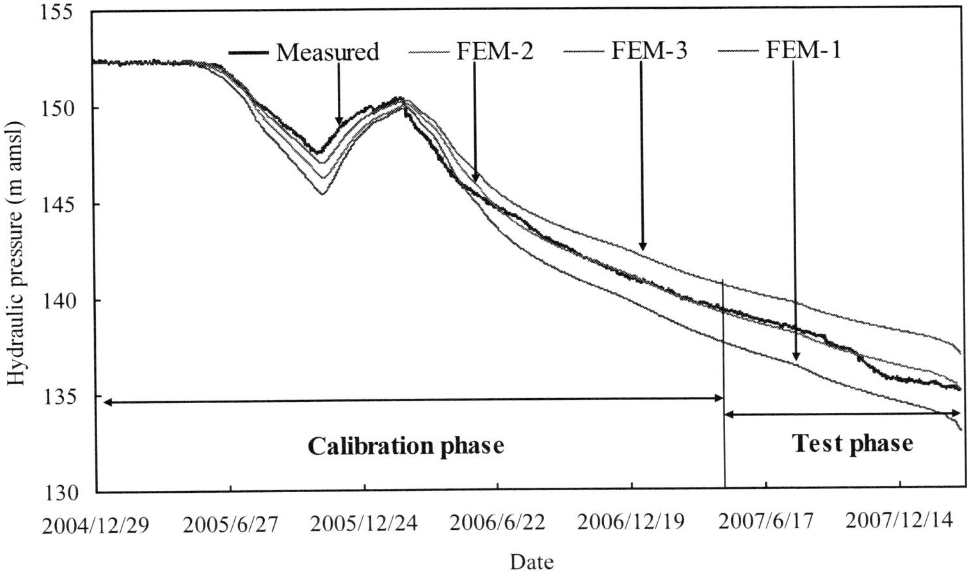

Figure 6. The three FEM and measured hydraulic pressure values at 56.8 m amsl.

The method typically used to quantity model error is to compute the residual which is the difference between predicted and observed values. In addition to the residual, three types of measures of the goodness of fit were used to check the performance of the proposed Grey-box model: (1) coefficient of efficiency (CE), (2) coefficient of determination (CD) and (3) root mean square error (RMSE). The formulae adopted for CD, CE and RMSE (Loague and Green 1991) were as follows:

$$CD = \frac{\sum_{i=1}^{N} (O_i - \bar{o})^2}{\sum_{i=1}^{N} (P_i - \bar{o})^2} \tag{9}$$

$$CE = 1 - \frac{\sum_{i=1}^{N} (P_i - O_i)^2}{\sum_{i=1}^{N} (O_i - \bar{o})^2} \tag{10}$$

$$RMSE = \left[\frac{\sum_{i=1}^{N} (P_i - O_i)^2}{N} \right]^{1/2} \tag{11}$$

where O is observed value, P is predicted value, $\bar{o}$ is mean of observed value and N is total number of data.

For perfect prediction CD and CE tend to one, while RMSE tends to zero. The lower limit for RMSE and CD is zero. CE can become negative. If CE is negative the model predicted values are worse than simply using the observed mean. The CD is a measure of the proportion of the total variance of observed data explained by the predicted data.

In addition to the above statistical displays of a model performance, Graphical displays can also be useful for showing trends, types of errors and distribution patterns (Loague and Green 1991). Several types of graphical display are possible. In this study a graphical display that shows the comparison of observed and predicted hydraulic pressure profiles will be used. Such graphical display can be used to judge the quality of the model performance at specific time.

The degree of model error that is acceptable depends on several factors (Spitz and Moreno 1996): (1) degree of natural heterogeneity (2) location, number and accuracy of measurements (3) purpose for which the model has been developed. Shamseldin (1997) claims that predictive models having CE values above 0.9 are very satisfactory, between 0.8-0.9 are fairly good and below 0.8 are unsatisfactory.

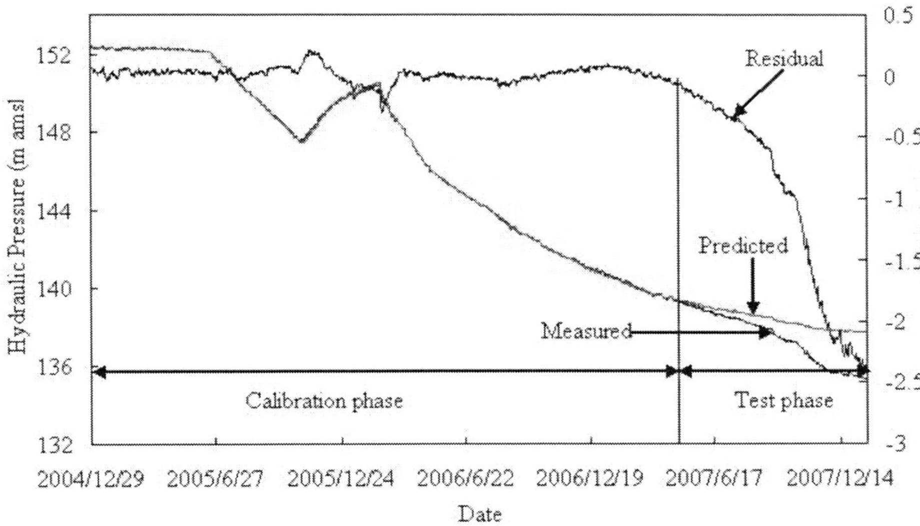

Figure 7. Existing BPN result for Example 1.

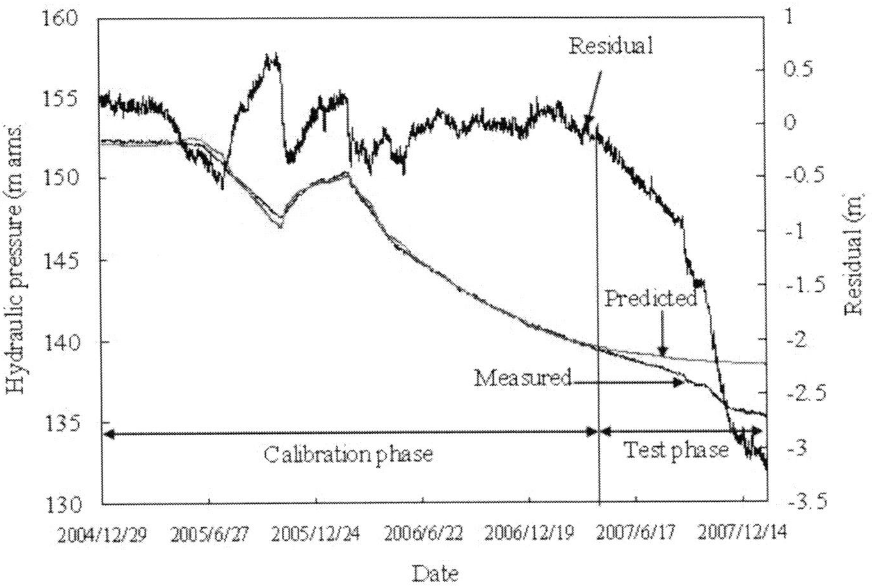

Figure 8. Existing BPN result for Example 2.

3.1. Existing BPN Model Results

The BPN parameters used during calibration is shown in table 1 for both examples. In calibration phase, the value of RMSE, CE, and CD can be seen in Table 2 for both examples. Models having CE greater than or equal to 0.9 are very satisfactory, according to this criterion the results in Table 2 indicate the models were calibrated successfully.

Table 1. Parameters used in the calibration of the BPN

Parameter	Example 1	Example 2
Initial synaptic weight	±1.0	±1.0
Number of hidden layers	1	1
Number of nodes in hidden layer	3	3
Learning rate	0.001	0.001
Momentum factor	0.9	0.9
Maximum epoch number	60000	60000
Tolerable error	0.001	0.001

Table 2. Goodness of fit indicators obtained during calibration

Parameter	Example 1	Example 2
Maximum magnitude of residual	0.297 m	0.677 m
Coefficient of efficiency (CE)	0.999	0.997
Coefficient of determination (CD)	0.997	0.998
Root mean square error (RMSE)	0.064 m	0.230 m

The entire model results are shown in Figure 7 and 8 for example 1 and 2, respectively. From these figures it is evident that for continuously decreasing time series data adopting the usual BPN approach would bring about inaccurate test phase results. In both figures the residual after subtracting the BPN model result from the measured one is also depicted. The residual obtained in the test phase is gradually increasing, which indicates the prediction shorter period predictions are better. This is mainly due to the lower limit of normalization range (*emi*) used was 0.1, that allow some normalized values between 0 and 0.1 to be used in the BPN model.

3.2. Modified Preprocessing Approach for Continuously Increasing and/or Decreasing Time Series Data

In the previous section, the performance aspect of BPN has been discussed in detail. One of the main problems of the existing BPN model was its inability to predict beyond its calibration range. To overcome this problem and improve its training and testing performance, a modified preprocessing approach is developed in this section.

In the common BPN with the sigmoid logistic activation function all the inputs and outputs are normalized in the range of 0 and 1. This normalization would make the outputs and inputs in the same order of magnitude. Moreover the normalized output shall also be compared with the BPN result in order to compute the mean square error, which ultimately used in adjusting the synaptic weights.

The output of a BPN, with sigmoid logistic activation function in its output layer nodes, will have a value in the range of 0 and 1. However, for continuously decreasing time series data (Figure 4) the normalized output values in the test period could be negative. Likewise for continuously increasing time series data the normalized output values could be greater than 1.

In such cases the BPN result would have different result from the normalized outputs, the former being between 0 and 1, while the later will be beyond the range of 0 and 1. In order to use the existing BPN for solving such problems a modified form of input and output data preprocessing is carried out. In this modified preprocessing approach, the basic character of BPN, i.e. its ability to respond to cases that are similar but not identical to the ones that it has been trained with (generalization ability) comes in to being. In other words though the test phase normalized input-output pairs are beyond the range of the activation function used in the output layer node, they are assumed to be similar to the input-output pairs that have been used in the training process. Thus the major point here is that how do we form this similarity? More specifically, how one can represent input-output data pairs, which are beyond the required normalization range by similar input-output pairs that can be recognized by the trained BPN?

In this study case the similarity was set by taking 'equidistant similarity' from *emi* and *ema* for continuously decreasing and increasing time series trends, respectively as shown in Figure 9. Here *emi* and *ema* are as defined in Eq. 1. In the figure the equation of a sigmoid logistic function is $y = \frac{1}{1+e^{-x}}$. In Figure 9 the brocken double arrow shows the similarity taken for normalized values above *ema*, while the solid double arrow is for normalized values below *emi*. A normalized value which is, say, *s* distance below *emi* is assumed to be similar to a value that is obtained at *s* distance above *emi*. In similar maner a normalized value that is *s* distance above *ema* is assumed to be similar to a value obtained at *s* distance below *ema*. In this way any value that is beyond the normalization range will be transferred in to the range required by the activation function. This similarity can be expressed mathematicaly as:

$$\begin{aligned} zmo &= 2 \times emi - z \quad for \;\; z < emi \\ zmo &= 2 \times ema - z \quad for \;\; z > ema \end{aligned} \tag{12}$$

where *zmo* and *z* are the modified and actual normalized values, respectively.

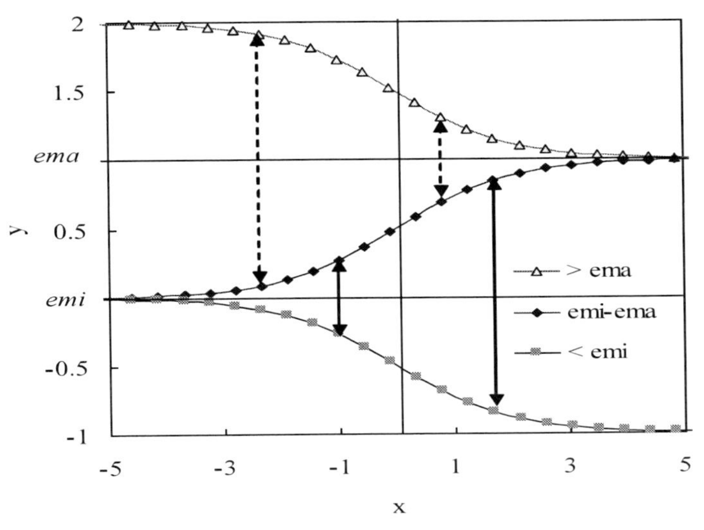

Figure 9. Similarity formation.

It is wise to note that such similarity will work as long as the normalized value is between (*emi* - *ema*) and (*emi* + *ema*). For the modified values (Eq. 12) the BPN result would be in the range of the activation function used in the output layer node. Thus the BPN result shall also be modified as per Eq. 13. The de-normalization process will use the modified BPN results (*bmo*) to get the final inferred result.

$$\begin{aligned} bmo &= 2 \times emi - b \quad for \;\; z < emi \\ bmo &= 2 \times ema - b \quad for \;\; z > ema \end{aligned} \tag{13}$$

where *bmo* and *b* are the modified and actual BPN results, respectively.

Adopting such modification in the formation of similarity between the test phase and calibration phases in the two examples has bringabout the results shown in Figure 10 and Figure 11. In both figures the residual after subtracting BPN model result from the measured value is also depicted. In both examples, the maximum magnitude residual found was 0.276 m for Example 1 and 1.476 m for Example 2.

The comparison of the two BPN models (with and without the proposed preprocessing approach) can be seen from Table 3. The over all performance of the BPN in both examples in the test period indicate the successful application of the proposed preprocessing method.

Table 3. Test phase RMSE, CD and CE values obtained for the two examples

Examples		RMSE	CD	CE
Example 1	Without modification	1.451	1.471	0.111
	With modification	0.122	1.076	0.994
Example 2	Without modification	1.560	1.346	-0.182
	With modification	0.713	1.844	0.765

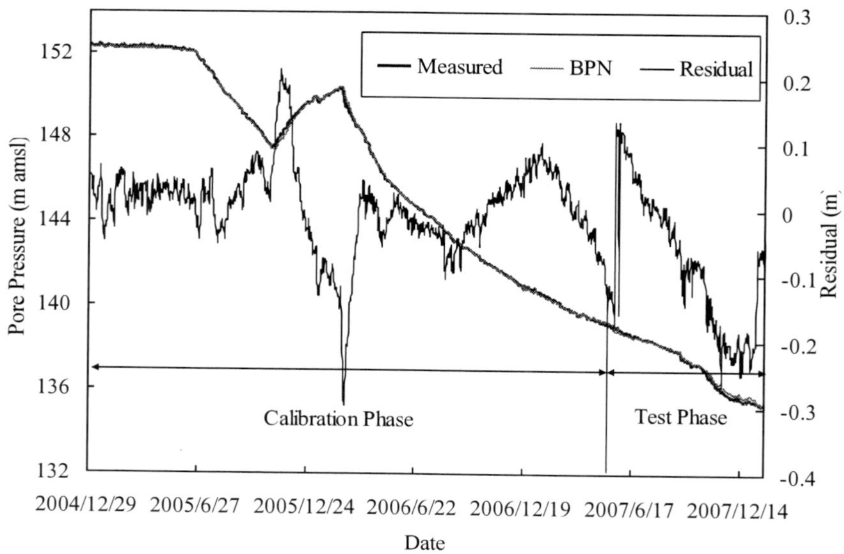

Figure 10. Modified BPN result for Example 1.

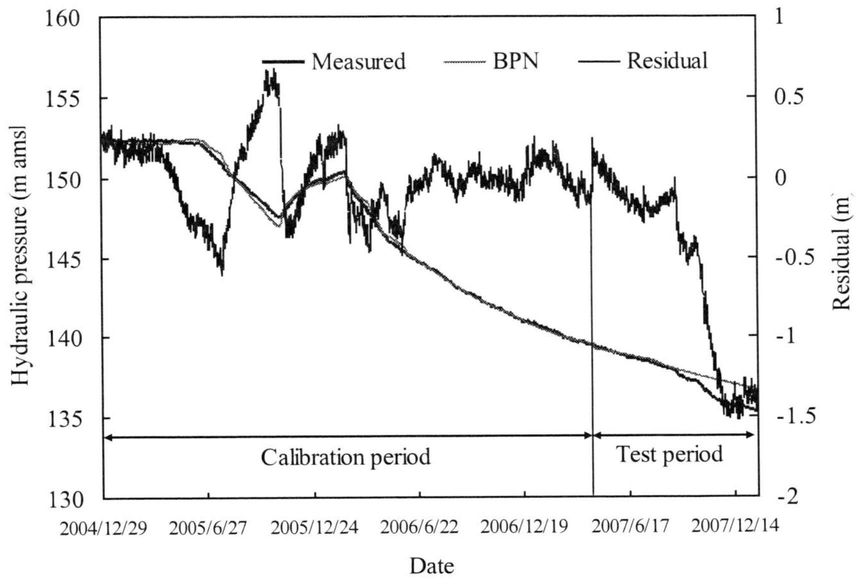

Figure 11. Modified BPN result for Example 2.

4. CONCLUSION

In this study, a modified preprocessing approach is proposed for improving the existing BPN problem of inability to work beyond their calibration range. As shown for the two examples in Mizunami underground research laboratory case, by finding a similar pattern that lies within the range of the activation function, the prediction ability of the BPN beyond its calibration range was dramatically improved. The measures of goodness of fit (CD, CE, and RMSE) values obtained for Calibration and Test phases are very good. This can clearly indicate the successful application of the proposed preprocessing approach.

Although the dynamicity of the groundwater flow pattern in MIU project area is complex due to construction of shafts, the proposed preprocessing have shown very good result. Therefore, this preprocessing approach would also have good results if it is applied in case of analyzing continuously increasing or decreasing time series data.

While encouraging, the results and implications of this study must be regarded with caution. A BPN with sigmoid logistic activation function in its hidden and output layers was used to illustrate the mathematical and modeling approach. More research is needed to investigate the potential strengths and limitations of this proposed preprocessing methodology. Only through rigorous testing under a variety of ANN models with different activation functions will the potential merits of the proposed preprocessing method be established. Still, based upon its theoretical foundation and the results obtained in the two examples, the method may hold promise as a powerful and flexible preprocessing approach and warrants future investigation.

REFERENCES

Chaturvedi, DK. *Soft Computing: Techniques and its application in electrical engineering,* Springer-Verlag, Berlin, 2008.

Goto, J; Ikeda, K; Kumazaki, N; Mukai, K; Iwatsuki, T; Hama, K. Working Program for Shallow Borehole Investigations, *JNC,* Tono Geoscience Center, JNC TN7400 2002-005, 2002. Available via http://www.jaea.go.jp/04/tono/miu_e/publ/tn74002002-005.pdf, accessed April, 2010.

Hornik, K; Stinchcombe, M; White, H. Multilayer feedforward networks are universal approximators, *Neural Networks*, 1989, Vol.2, 359–366.

JNC, Master Plan of the Mizunami Underground Research Laboratory Project, *JNC Technical Report,* JNC TN7410 2003-001, 2003. Via http://www.jaea.go.jp/04/ tono/miu_e/ publ/tn74102003-001.pdf, accessed April, 2010.

Kumazaki, N; Ikeda, K; Goto, J; Mukai, K; Iwatsuki, T; Furue, R. Synthesis of the Shallow Borehole Investigations at the MIU Construction Site, Japan, JNC Development Institute, Tono Geoscience Centre JNC TN7400 2003-005, 2003. Via http://www.jaea.g o.jp/04/tono/miu_e/publ/jnctn74002003005/msbreport.html, accessed April, 2010.

Loague, K; Green, RF. Statistical and Graphical methods for evaluating solute transport models: Overview and application, *J. Contam. Hydrol.*, 1991, Vol 7, 51-73.

Mohammed, M; Watanabe, K; Takeuchi, S. Grey model for prediction of pore pressure change, Environ. Earth Sci., 2009, In press, DOI:10.1007/s12665-009-0287-y.

Mohammed, M; Watanabe, K; Takeuchi, S. Adaptive Neuro Fuzzy Inference System approach for prediction of hydraulic pressure change. *Ann. J. hydraulic engineering*, 2010, 54, 43-48.

Reed, RD; Marks II, RJ. *Neural Smithing: Supervised Learning in Feedforward Artificial Neural Network*, the MIT Press, Cambridge, MA, 1999.

Rumelhart, DE. Parallel distributed processing: explorations in micro structures of cognition; M$_{c}$Clelland, J. L; Ed; MIT: *Foundations Cambridge*, MA, 1986, Vol. 1.

Rojas, R. Neural Networks a systematic introduction, Springer-Verlag, *Berlin*, 1996.

Shamseldin, AY. Application of a neural network technique to rainfall runoff modeling, *J. Hydrol.*, 1997, 199, 272-294.

Spitz, K; Moreno, JA. *practical guide to groundwater and solute transport modeling,* John Wiley & Sons. Inc, New York, 1996.

Takeuchi, S; Takeuchi R; Salden W; Saegusa, H; Arai T; Matsuki K. Hydro-geological conceptual model determined from baseline and construction phase groundwater pressure and surface tilt meter data at the Mizunami Underground Research Laboratory, Japan, Proceeding 11th International Conference on *environmental remediation and radioactive waste management, ASME,* 2007, Bruges, Belgium.

Welstead, ST. *Neural Networks and Fuzzy Applications in C/C++,* Wiley, New York, 1994.

In: Focus on Artificial Neural Networks
Editor: John A. Flores, pp. 341-353
ISBN: 978-1-61324-285-8

Chapter 16

Application of Artificial Neural Networks in Enzyme Technology

Mohd Basyaruddin Abdul Rahman[1,2*], Naz Chaibakhsh[1,2], Mahiran Basri[1,3] and Abu Bakar Salleh[3,4]

[1]Department of Chemistry, Faculty of Science, 43400 UPM Serdang, Universiti Putra Malaysia, Selangor, Malaysia
[2]Structural Biology Research Center, Malaysia Genome Institute, Bangi, 43600 Bangi, Selangor, Malaysia
[3]Department of Biochemistry, Faculty of Biotechnology & Biomolecular Sciences, 43400 UPM Serdang, Universiti Putra Malaysia, Selangor, Malaysia
[4]Laboratory of Industrial Biotechnology, Institute of Bioscience, 43400 UPM Serdang, Universiti Putra Malaysia, Selangor, Malaysia

Introduction

Enzymes are protein molecules that speed up biochemical reactions without being consumed, so act as biocatalysts that help make or break the covalent bonds (Alberts, 1998). Enzyme technology is the technological concepts that enable application of enzymes in production processes to achieve sustainable industrial development. This field is leading to discovery, development and purification of enzymes, and their application in different industry sectors (van Beilen and Li, 2002). Custom design of enzyme activity for desired industrial applications, process control and bioparameter estimation are major goals in enzymatic process development.

Mathematical modeling and simulation is a powerful approach for understanding the complexity and nonlinear behavior of biological systems and identifying natural laws describing their behavior (Meng et al. 2004). Computational Intelligence (CI) techniques have been successfully applied to solve problems in the identification and control of biological systems (do Carmo Nicoletti and Jain, 2009). Artificial Neural Networks (ANNs), in

* Corresponding author: Email : basya@science.upm.edu.my

particular, provide an adequate approach in estimating variables from incomplete information and handling nonlinear dynamic systems like enzymatic processes.

One of the major problems of ANNs is the cost of model development due to requiring relatively extensive training data (Montague and Morris, 1994). It is also difficult to interpret the network, and convergence to a solution is slow and depends on the network's structure (do Carmo Nicoletti and Jain, 2009). In order to overcome these limitations, Design of Experiments (DOE) has been introduced as a better methodology than the common trial and error techniques for generating the ANN's training data (Balestrassi et al., 2009).

This chapter reviews some applications of ANNs in enzyme technology. Some practical considerations including utilization of DOE for training the neural networks in enzymatic processes have also been introduced.

Neural Network Contributions in Enzyme Technology

Various types of network architecture and processing functions have been reported in the literature, which were applied to enzymatic processes from enzyme production to synthesis reaction yield prediction.

Prediction of Reaction Yield

Predicting the reaction yield using intelligent methods helps to save time and money since the most promising conditions can simply be verified in the laboratory instead of performing a large number of experiments to gain the same information.

ANN has been applied by Kirk et al. (1990) for predicting the yield of lipase-catalyzed synthesis of ethyl 6-O-dodecanoyl D-glucopyranoside. The ANN, trained by back propagation algorithm, consisted of three layers, an input layer of five neurons representing the process parameters (temperature, pressure, agitation speed, enzyme amount, and substrate molar ratio), a hidden layer of four neurons, and an output layer of two neurons. The neurons in the output layer represented the 6-O-monoester and 2,6-O-diester formed after 22 hours. A very good correlation was observed between the prediction values and the experimental data. A test data set was generated consisting of all the possible combinations of parameters within the selected range. The conditions which were predicted to obtain a yield of more than 85% of 6-O-monoester and not more than 5% 2,6-O-diester, were selected. The results showed that ANN can be used to predict and optimize the reaction conditions.

In another work, Abdul Rahman et al. (2008) introduced ANN as an adequate tool for the prediction yield of an esterification reaction. In this study, a multilayer feedforward neural network trained by backpropagation algorithm, Levenberg– Marquardt, was employed to characterize the essential behavior of immobilized *Candida antarctica* lipase B-catalyzed synthesis of dioctyl adipate ester. After evaluating various ANN configurations, the best network was found to consist of seven hidden nodes using hyperbolic tangent sigmoid transfer function. The R^2 and mean absolute error (MAE) values between the actual data and predictions made by the network were determined as 0.9998 and 0.0966 for training set. A simulation test with a testing dataset showed that the MAE was low and R^2 was close to 1.

The results indicated the good generalization of the developed ANN model and its capability to predict the esterification yield. Comparison of the performance of radial basis function network (RBF), using the same training and testing data, with the developed models showed RBF model was more accurate than backpropagation network for training the ANN with R^2 0.9999 and MAE 0.06857. However, with respect to testing data, the RBF model showed a poorer performance and generalization capability with R^2 0.3844 and MAE 6.8576.

ANN has also been used, with considerable good performance, for predicting the reaction yield in enzyme-catalyzed esterification of isoamyl alcohol and isovaleric acid (Chowdary et al., 2002), esterification of anthranilic acid with methanol (Manohar and Divakar, 2005), transesterification of palm oil and oleyl alcohol (Basri et al., 2007), transesterification of triolein and palmitic acid (Ciftci et al., 2008), and esterification of betulinic acid and phthalic anhydride (Ghaffari Moghaddam et al., 2010).

Prediction of Reaction Rate and Kinetic Parameters

The rate of an enzymatic reaction is measured to determine the biocatalyst activity (Aehle, 2007). Several researchers have used ANN to predict the rate of an enzyme-catalyzed reaction (Bas et al., 2007; Szaleniec et al., 2006). Modeling of the penicillin G acylase (PGA)-catalyzed synthesis of amoxicillin through the reaction of *p*-hydroxyphenylglycine methyl ester and 6-aminopenicillanic acid was reported by Silva et al. (2008). A hybrid model, with two feedforward neural networks, that coupled the ANNs to mass-balance equations were trained and used to predict the rates of amoxicillin and *p*-hydroxyphenylglycine net production.

Hybrid models combine the mechanistic and empirical modeling strategies with "black-box" techniques and use all the available information to produce dynamic models with much better accuracy (do Carmo Nicoletti and Jain, 2009). Regarding bioprocess systems, usually mechanistic models include three types of equations, which are mass and energy-balances, rate equations, and temperature dependant equations (Zorzetto et al., 2000). Hybrid models can be developed in different manners. One approach is having the mathematical model as the basis, and the ANN calculates unknown parameters. The network can also be used to learn the deviation between the mathematical model output and the target output. Another possible way is applying the deterministic model as reinforcement for mapping the relationship between input and output data (Aguiar and Filho, 2001). The approach considered in this study uses a mass-balance model as reinforcement for the relation between input and output data.

The input variables included substrate and product concentration and the output variable was the rate of reaction. The number of hidden layers and neurons was chosen by trial and error. The network was trained by backpropagation algorithm with a random-search method. According to the authors, the hybrid model provided highly accurate results. The optimal number of nodes was 15 for both rates of reaction, and the number of iterations was 10,000 and 65,000 for the amoxicillin and *p*-hydroxyphenylglycine net production.

Mazutti et al. (2010) described a hybrid neural network approach to estimate model parameters for the production of inulinase enzyme by the yeast *Kluyveromyces marxianus*, in a batch bioreactor. Inulinase is an enzyme that acts on 2,1-beta-d-fructoside links in inulin, releasing d-fructose. It is especially used in the food industry for the synthesis of fructooligosaccharides. Inulinase production has shown a complex dynamic kinetic behavior. The authors investigated a method for overcoming the main drawback of the mathematical modeling of bioprocesses, which was the parameter estimation procedure. A feedforward neural network was employed to predict the kinetic parameters of the mathematical model based on mass-balance equations. The four input variables were molasses initial concentration, corn steep liquor, total reducing sugars and fermentation time. The six outputs were the parameters of the proposed mathematical model: maximum specific microbial growth μ_{max}, Contois saturation constant k_s, yield coefficient for limiting substrate $Y_{x/s}$, growth-associated constant α, non-growth associated constant β and deactivation coefficient λ. In this study, the weights and bias of the ANN were optimized by a simulated annealing algorithm combined with the Nelder–Mead Simplex (NMS) method. The number of hidden neurons was changed from 1 to 12 and the best topology for the ANN was found to be 4-8-6. Using the hyperbolic-tangent transfer function increased the network learning rate and performance. The hybrid model exhibited good performance, and was able to estimate the model parameters with high values of correlation coefficient, more than 0.85.

Prediction of Enantioselectivity

Considerable efforts have been made to understand the effect of steric factors and electronic properties of substrates and enzymes in determining the enantioselectivity of biocatalysts. Different computational tools such as partial least squares (PLS) regression methods, with the help of conformational-dependent chirality code (CDCC) or classification and regression trees (CART) techniques have been used for the prediction of enantiomeric excess (%ee) using molecular descriptors (Mazurek et al., 2007; Caetano et al., 2005). ANN is another approach applied by several researchers for this purpose (Zhang and Aires-de-Sousa, 2006). Mazurek et al. (2007) have reported on the use of the counter propagation artificial neural network (CP-ANN) for modeling and prediction the enantioselectivity of artificial metalloenzymes. The aim of their work was to develop a model capable of predicting %ee for a new ligand with a structure similar to the ligands used in the study, and to set up a methodology which can be applied for different sets of proteins, ligands and catalysts. In order to construct a general descriptor-based model, 18 biotinylated rhodium-diphosphine ligands with 20 proteins forming a total of 360 ligand–protein catalysts, were combined. Incorporation of a biotinylated organometallic catalyst into a host protein (here, streptavidin) results in versatile artificial metalloenzymes for enantioselective catalytic applications. In this work, the models of the artificial metalloenzyme structures were obtained by docking simulations. By docking procedure, the position of the ligands inside streptavidin can be predicted. The molecular descriptors for the docked ligand structures and three amino acid residues (T111-Ser112X-G113; representing the amino acid mutated at the position 112 and its two closest neighbors, threonine and glycine in positions 111 and 113, respectively) as well as the Rh-C_α distances (where C_α is the asymmetric carbon of each amino acid residue in the host protein; distance descriptors computed from docking simulations) were used as inputs of the ANN and the output was the enantiomeric excess, %ee obtained for the hydrogenation of acetamidoacrylic acid to produce N-acetamidoalanine. Counter-propagation neural networks (CP-ANN) usually consisted of two layers. One layer is the input or Kohonen layer and the other is the output layer. The input layer performs the mapping of the input data into lower dimensional array by use of competitive learning. In CP-ANN, the vectors with the N input variables are compared only with the weights of the neurons in the Kohonen layer. Once the central (or winning) neuron is found among the neurons in the Kohonen layer, the weights of Kohonen layer is adjusted and the position of the central neuron is transferred from the input to the output layer and the weights in the output layer are corrected according to the given target value, so the network is trained (Kuzmanovski and Novic, 2008). After training the network, the complete set of the training vectors is once more run through the ANN. In the last run, the labeling of the neurons excited by the input vectors is made into a table called "top map" (Zupan et al., 1997). In this study, from the 360 ligand-protein complexes, 114 samples were chosen for validation of the model and the remaining 246 samples were used as the training set. After testing over 2,300 networks, the lowest root mean square error (8.8 %ee) and a high correlation coefficient value (R=0.953) was found for the network with 21×21 neurons, trained for 50 learning cycles with the maximum correction factor of 0.25. The top map of predicted enantioselectivity shows the positive and negative enantiomeric excess (%ee) corresponded to R-, S- and non selective ligand–protein combinations. The obtained results demonstrate the potential use of ANN as a tool for finding new efficient, highly selective ligand-protein catalysts.

Production of Enzyme

ANN has been successfully applied as a prediction tool for production of various enzymes by submerged and solid-state fermentation. Composition of the fermentation medium is very important in enzyme production due to effect of medium components on the cell's metabolism and also on operating costs (Pandey et al., 2006). Application of ANN for prediction of cellulase and xylanase production by *Trichoderma reesei* under solid-state fermentation has been described by Singh et al. (2008). The three input variables were wheat bran-sugarcane bagasse content, water to substrate ratio, and incubation time. A 3-10-10-1 ANN was used to predict the enzyme activity (the output). A correlation coefficient (>0.8) and root mean square error (<0.4) for the validation data indicated the good predictability of the developed model. In another study, Jin et al. (2007) have used an ANN pattern recognition (ANNPR) model-based on-line adaptive control strategy for the fed-batch phytase production process with recombinant *Pichia pastoris*, to demonstrate its universal ability, and operational simplicity. Phytase enzyme is an essential additive to animal feed that can break down the indigestible phytic acid part found in grains and oil seeds. Phytase can be efficiently produced by attaining high cell density in a short time and low-cost on-line process monitoring and control. On-line statistical process control is a simple useful tool for improving the process performance and product reproducibility monitoring. In this study, a substrate feeding rate control strategy in fed-batch cultivation process was applied to culture the *P. pastoris* to a high cell density level with glucose as the carbon source. The cells grew in batch mode until glucose was completely consumed. At this stage, either the traditional DO-Stat method or the ANNPR control strategies were initiated. In cultivation phase, the ANNPR was used to assure the cells growth at the highest rate by controlling glucose concentration.

The glucose was fed by a pump with on–off mode. The pump was automatically switched on when dissolved oxygen (DO) exceeded a pre-determined level. Through on-line collecting DO/pH series data and recognizing the DO and pH's changing patterns of a fermentation process, the two physiological states of "substrate starvation" and "substrate in excess" were concluded. Finally, based on the recognition results, the feeding speed of a programmable peristaltic pump was regulated by a PC driven D/A converter to add the glucose. In this work, the substrate feeding control system performance based on DO-Stat method and the ANNPR model was compared. The time series data of pH and DO measurements were input to each ANNPR model for training. A three hidden layer back-propagation network using sigmoid transfer function was adopted. The results showed that the final phytase activity and yield achieved with the ANNPR control system was much higher than that with DO-Stat method. The cultivation time also could be shortened for about 30% with the ANNPR model-based control strategy.

Successful application of ANN to enzyme production has been reported in the literatures for production of other enzymes such as lipase, glucoamylase, β-galactosidase, endonuclease and α-amylase (Linko et al., 1999; Günay et al., 2008).

Enzyme Biosensor

Biosensor is a device that incorporates a biological sensing element, with high selectivity to substances, with a transducer. The enzyme biosensor can measure substrate conversion or product formation based on potentiometric or amperometric methods (Webster, 2001). A number of ANN-based application studies using enzyme biosensors for quantitative analysis have been published (Zhang et al., 2009; Baronas et al., 2004; Ferreira et al., 2003). Torrecilla et al. (2007) have used a glucose oxidase amperometric biosensor based on a colloidal gold-cysteamine-gold disk electrode for the analysis of a mixture of glucose, ascorbic acid and uric acid. Feedforward back-propagation ANN was trained with the data obtained from cyclic voltammograms, generated by the biosensor, and was used to process signals from the biosensor. The current signal was input into the ANN (11 input nodes including voltammograms composed of current intensity values) and the output layer consisted of three neurons for the output variables, which were the glucose, ascorbic acid and uric acid concentrations in the range of 0.1–1.0 mM. Bayesian regularization back-propagation algorithm (TRAINBR) provided the best results. TRAINBR is a training function that updates the weight and bias values according to Levenberg-Marquardt optimization. It minimizes a combination of squared errors and weights, and finds the correct combination to generate a network that generalizes well (Aggarwal et al., 2005). The optimum number of hidden neurons was found to be 13. For the validation datasets, the R^2 was higher than 0.99, and the MPE (mean prediction error) was less than 0.017. The authors concluded that the developed ANN was a powerful tool to quantify the analytes separately and solve the interference effect on the analysis of glucose, ascorbic and uric acids.

Prediction of Enzyme Thermostability

Thermal processes are of significant importance in the food and pharmaceutical industries and hence enzymes possessing higher stability have received much attention in this context (Nath et al., 1997). Enzyme structures are stabilized with various non-covalent interactions such as hydrophobic, electrostatic, van der Waals and hydrogen bonds (Gromiha, 2007). It has been found that intra-helical salt bridges and the amino acid composition on protein surface may be important factors in the stability of thermophilic proteins. Several methods have been proposed for predicting the protein stability which are mainly based on distance and torsion potentials, structural environment-dependent amino acid substitution and propensity tables, empirical energy functions, multiple regression techniques, contact potentials, support vector machines, decision trees, average assignment, classification and regression tools and backbone flexibility (Ebrahimi et al., 2009; Gromiha, 2007). Ebrahimi et al. (2009) applied the neural network to define the most important features contributing to xylanase enzyme thermostability and to find a suitable tool to predict the best probable enzymes' mutants at high temperatures. From UniProtKB/Swiss-Prot database, 74 protein attributes or features of all available xylanase enzymes were extracted and used as the input to the feed forward back-propagation ANN. Optimum temperature was set as the output. The input features were classified as continuous variables. Feature selection algorithm was applied to find out the most important enzymes' features based on the *p*-value and analysis of variance. The output then was converted to a flag variable of true or false. Five different algorithms including Quick, Dynamic, Multiple, Prune, Exhaustive Prune and RBF (radial basis function) were used to generate the models in order to compare their prediction accuracy. The dataset was divided into three groups (training, testing and validation with 80%, 10% and 10% ratio, respectively). The most complicated network was generated by using Multiple method containing 200 neurons in network layers; while the simplest networks were obtained by using Exhaustive Prune and Prune methods with a total of 7 neurons in network layers. Estimated accuracy of developed ANNs varied from 80% to 90%. The best estimated accuracy (90.638%) was obtained in Multiple and the weakest accuracy (80.560%) found in Dynamic method. The results showed that frequency of amino acids Met and Lys had the weakest correlation with thermostability; and frequency of Gln was the most important feature contributing to xylanase thermostability. The best model based on Multiple algorithm was used for testing thermostability in 7030 possible mutants of xylanase from *Bacillus halodurans*, that were virtually generated. According to Ebrahimi et al. (2009), in some of the mutants up to 10°C thermal stability improvement was observed.

Prediction of Enzyme Active Sites

Enzymes perform a variety of catalytic functions by interacting with substrates at specific locations called "active sites". The active site consists of the amino acid residues that play important role in the catalysis (Tsai, 2007). Identification of active sites and catalytic residues in enzymes is extremely important for understanding their catalytic mechanism, and exploring their applications (Tang et al., 2008). Finding the location of the protein functional sites is also helpful for the study of targeted mutants and structure-based drug design. Traditional molecular biology techniques such as mutagenesis, pH dependence and chemical labeling that are usually used for finding the functional sites are mostly time-consuming and dependant on some prior knowledge of the protein function (Gutteridge et al., 2003). *In silico* technologies can be helpful for finding the functional sites in novel protein molecules (Rao, 2007). Techniques for finding functional sites such as evolutionary trace (ET), generally focus on searching for three-dimensional clusters of conserved residues. Gutteridge et al. (2003) described a new method for prediction of active sites of enzymes using neural network instead of searching for clusters of conserved residues. A data set contains 159 proteins comprising nearly 55,000 non-catalytic residues and 550 catalytic residues were used for training the ANN. The parameters used for discriminating catalytic and non-catalytic residues were conservation, relative solvent accessibility, secondary structure, cleft and depth. All the parameters were coded before presentation to the network. The performance of different networks in predicting catalytic residues and locating active sites was compared, based on the analysis of the structure and sequence. A single-layer feed forward neural network trained by a scaled conjugate gradients algorithm was used in this study. To improve estimation of the network generalization performance, a ten-fold cross validation resampling was carried out. The performance of the ANN models was measured by Matthews correlation coefficient (MCC). Ranking and clustering the residues improves the prediction and locating the active sites. The catalytic residues in each structure were ranked by network score. All residues that scored above a cut-off value were used in the clustering algorithm. A pair of residues, which their atoms laid within 4Å distance of each other, was clustered together, and this cluster was defined as a sphere. In order to check whether the prediction was correct, the overlap between the predicted sphere and the closest known active site was calculated. The prediction was correct when the overlap was larger than 50% of the volume of the known active site. In more than 69% of the test samples the active site was correctly predicted. The MCC for different networks in predicting catalytic residues was compared. When a combination of sequence and structural information was used for training the network, the clustering algorithm showed better performance. In this study, the authors also applied the ANN to identify the active sites in five novel enzymes with unknown structures. In most cases, the functional residues were successfully identified.

Other researchers also have reported application of ANN for prediction of active sites of enzymes such as Pande et al. (2007) who used an ANN trained by scaled conjugate gradient to predict the catalytic sites of the enzyme phosphoenolpyruvate mutase based on sequence-derived information; and Stahl et al. (2000) who used a self-organizing neural network for prediction of active sites and class of enzymes including adenosine deaminase, carbonic anhydrase, aldolase, phosphatase, β-lactamase, carboxypeptidases, Cu/Zn superoxide dismutase, adamalysin, matrixins, and other metalloproteinases.

Practical Considerations

Although an artificial neural network is immensely efficient and powerful problem-solving strategy and lots of learning and estimation methods are available that can be utilized, for practical applications a number of issues should be considered.

In data driven techniques such as ANNs, selection of an appropriate set of input parameters is essential to develop a reliable model. Reducing the complexity of input and output variables results in simpler and more accurate models. Too many input parameters can extremely slow down the learning process and too few parameters can provide insufficient information and decrease the generalization capabilities of the network (Rafiq et al., 2001). Several techniques can be employed to reduce the number of selected input variables such as wrapper and filter methodologies, principal components analysis, Brute force, Mutual information techniques and genetic algorithm (da Costa Couto, 2009).

Selection of an adequate number of training data is also an important issue. In fact, one of the limitations of ANNs consists of requiring relatively extensive training data hence high cost and long computation time. There is no general rule for determining the number of training data (Rafiq et al., 2001). Design of Experiments (DOE) approach can be useful to reduce the number of training data in some applications of ANN. DOE is an organized and structured method for determining the relationship between factors influencing a process and the output of the process (Chakraborty, 2009). Hypercube design with the data selected from the corner of the cube, within the cube, and mid sides of the cube faces is usually sufficient for training the network, especially for small scale problems (Jenkins, 1997). In a case study, Alam et al. (2004) investigated the effect of DOE on the performance of ANN. The neural network metamodels were developed using six different experimental design including full factorial design (FFD), random sampling design (RSD), central composite design (CCD), modified Latin Hypercube design (LHD), and modified Latin Hypercube design supplemented with domain knowledge (LHD + DK). The authors found that a modified-Latin Hypercube design, supplemented by domain knowledge, could be an effective and robust method for the development of neural network simulation metamodels with better predictive accuracy than conventional and random sampling designs. Validating the model with experimental data, prior to online implementation, is very important for practical applications. Validation set must be independent of the data used for training. About 20% of the data obtained by DOE can be used to validate the model.

Several researches have used DOE (such as full factorial design, central composite rotatable design, and crossed mixture design) methods to generate ANN training data sets for predicting the enzymatic reaction yield or improving the enzyme production (Didier et al., 2009; Riveros et al., 2009; Ebrahimpour et al., 2008). The authors stated that by using DOE, the advantages of ANN over classical statistical methods would become more evident.

Conclusion

In this chapter, some applications of ANNs in enzyme technology have been briefly reviewed. Artificial neural networks proved to be a powerful approach to handle the highly complicated behaviors of nonlinear systems. Significant improvement may be obtained by combining other approaches such as genetic algorithms (GA), partial least squares (PLS), principal component analysis (PCA) and quantitative structure-activity relationship (QSAR). The application of these useful tools provides valuable solutions to the problems of enzyme engineering and technology, and it seems that there is still a potential area of research in this field to fully exploit its capabilities.

REFERENCES

Abdul Rahman, MB; Chaibakhsh, N; Basri, M; Salleh, AB; Rahman, RNZRA. Application of artificial neural network for yield prediction of lipase-catalyzed synthesis of dioctyl adipate. *Applied Biochemistry and Biotechnology,* 2009, 158, 722–735.

Aggarwal, KK; Singh, Y; Chandra, P; Puri, M. Bayesian regularization in a neural network model to estimate lines of code using function points. *Journal of Computer Sciences,* 2005, 1, 505-509.

Aguiar, HC; Filho, RM. Neural network and hybrid model: a discussion about different modeling techniques to predict pulping degree with industrial data. Chemical *Engineering Science,* 2001, 56, 565-570.

Aehle W. *Enzymes in industry, production and applications.* Wiley-VCH Verlag GmbH & Co. Weinheim. 2004

Alam, FM; McNaught, KR; Ringrose, TJ. A comparison of experimental designs in the development of a neural network simulation metamodel. *Simulation Modelling Practice and Theory,* 2004, 12, 559-578.

Alberts, B; Bray, D; Johnson, A; Lewis, J; Raff, M; Roberts, K; Walter, P. (1998). Essential cell biology, an introduction to the molecular biology of the cell. *Garland Publishing Inc*, New York.

Baronas, R; Ivanauskas, F; Maslovskis, R; Vaitkus, P. An Analysis of Mixtures Using Amperometric Biosensors and Artificial Neural Networks. *Journal of Mathematical Chemistry,* 2004, 36, 281-297.

Bas, D; Dudak, FC; Boyaci, IH. Modeling and optimization III, Reaction rate estimation using artificial neural network (ANN) without a kinetic model. *Journal of Food Engineering,* 2007, 79, 622-628.

Basri, M; Rahman, RNZRA; Ebrahimpour, A; Salleh, AB; Gunawan, ER; Abdul Rahman, MB. Comparison of estimation capabilities of response surface methodology (RSM) with artificial neural network (ANN) in lipase-catalyzed synthesis of palm-based wax ester. *BMC Biotechnology*, 2007, 7, 53-63.

Caetano, S; Aires-de-Sousab, J; Daszykowski, M; Vander Heyden, Y. Prediction of enantioselectivity using chirality codes and Classification and Regression Trees. *Analytica Chimica Acta,* 2005, 544, 315–326.

Chakraborty, UK. *Computational Intelligence in Flow Shop and Job Shop Scheduling.* Springer-Verlag, Berlin, Heidelberg, 2009.

Chowdary, GV; Ramesh, MN; Prapulla, SG. Application of artificial neural network for the maximization of isoamyl isovalerate ester yields in n-hexane. *Asian Journal of Microbiology,* Biotechnology and Environmental Sciences, 2002, 4, 561-566.

Ciftci, ON; Fadiloglu, S; Gogus, F; Guven, A. Prediction of a model enzymatic acidolysis system using neural networks. *Grasas Y Aceites*, 2008, 59, 375-382.

da Costa Couto, MP. Review of input determination techniques for neural network models based on mutual information and genetic algorithms. *Neural Computing & Applications,* 2009, 18, 891-901.

do Carmo Nicoletti, M; Jain, LC. 2009. *Computational Intelligence Techniques for Bioprocess Modelling, Supervision and Control.* Springer-Verlag Heidelberg.

Ebrahimi, M; Ebrahimie, E; Ebrahimi, M; Deihimi, T; Delavari, A; Mohammadi-dehcheshmeh, M. Application of neural networks methods to define the most important features contributing to xylanase enzyme thermostability. *IEEE Congress on Evolutionary Computation (CEC 2009)*

Ebrahimpour, A; Rahman, RNZRA; Chng, DHE; Basri, M; Salleh, AB. A modeling study by response surface methodology and artificial neural network on culture parameters optimization for thermostable lipase production from a newly isolated thermophilic *Geobacillus sp.* strain ARM. *BMC Biotechnology*, 2008, 8, 96-111.

Ferreira, LS; Souza Jr., MB; Trierweiler, JO; Hitzmann, B; Folly, ROM. Analysis of experimental biosensor/FIA lactose measurements. *Brazilian Journal of Chemical Engineering,* 2003, 20, 7-13.

Didiera, C; Forno, G; Etcheverrigaray, M; Kratje, R; Goicoechea, H. Novel chemometric strategy based on the application of artificial neural networks to crossed mixture design for the improvement of recombinant protein production in continuous culture. *Analytica Chimica Acta*, 2009, 650, 167-174.

Ghaffari Moghaddam, M; Ahmad, FBH; Basri, M; Abdul Rahman, AB; Artificial neural network modeling studies to predict the yield of enzymatic synthesis of betulinic acid ester. Electronic *Journal of Biotechnology,* 2010.

http, //www.ejbiotechnology.info/content/next/abstract/a17.html#.

Günay, ME; I. Nikerel, E; Oner, ET; Kirdar, B; Yildirim, R. Simultaneous modeling of enzyme production and biomass growth in recombinant *Escherichia coli* using artificial neural networks. *Biochemical Engineering Journal*, 2008, 42, 329–335.

Gutteridge, A; Bartlett, GJ; Thornton, JM. Using A neural network and spatial clustering to predict the location of active sites in enzymes. *Journal of Molecular Biology*, 2003, 330, 719-734.

Jenkins, WM. An introduction to neural computing for the structural engineer. *The struct Engng*, 1997, 75, 38-41.

Jin, H; Zheng, Z; Gao, M; Duan, Z; Shi, Z; Wang, Z; Jin, J. Effective induction of phytase in *Pichia pastoris* fed-batch culture using an ANN pattern recognition model-based on-line adaptive control strategy. *Biochemical Engineering Journal*, 2007, 37, 26-33.

Kirk, O; Barfoed, M; Bjfrkling, F. Application of a neural network in the optimization of an enzymatic synthesis. *Tetrahedron Computer Methodology*, 1990, 3, 239- 243.

Kuzmanovski, I; Novic, M. Counter-propagation neural networks in Matlab. Chemometrics and Intelligent *Laboratory Systems,* 2008, 90, 84-91.

Linko, S; Zhu, YH; Linko, P. Applying neural networks as software sensors for enzyme engineering. *Trends in Biotechnology*, 1999, 17, 155-162.

Manohar, B; Divakar, S. An artificial neural network analysis of porcine pancreas lipase catalysed esterification of anthranilic acid with methanol. *Process Biochemistry*, 2005, 40, 3372-3376.

Mazurek, S; Ward, TR; Novic, M. Counter propagation artificial neural networks modeling of an enantioselectivity of artificial metalloenzymes. *Molecular Diversity*, 2007, 11, 141-152.

Mazutti, MA; Corazza, ML; Maugeri, F; Rodrigues, MI; Oliveira, JV; Treichel, H; Corazza, FC. Hybrid modeling of inulinase bio-production process. Journal of Chemical *Technology and Biotechnology*, 2010, 85, 512-519.

Meng, TC; Somani, S; Dhar, P. Modeling and simulation of biological systems with stochasticity. *In Silico Biology*, 2004, 4, 24.

Montague, G; Morris, J. Neural-network contributions in biotechnology. *Trend in Biotechnology*, 1994, 12, 312-324.

Nath, S; Satpathy, GR; Mantri, R; Deep, S; Ahluwalia, JC. Evaluation of enzyme thermostability by enzyme assay and differential scanning calorimetry, A study of alcohol dehydrogenase. *Journal of the Chemical Society*, Faraday Transactions articles, 1997, 93, 3351-3354.

Pande, S; Raheja, A; Livesay, DR; Prediction of enzyme catalytic sites from sequence using neural networks. Proceedings of the 2007 IEEE Symposium on Computational Intelligence in *Bioinformatics and Computational Biology*, (CIBCB 2007).

Pandey, A; Webb, C; Soccol, CR; Larroche, C. (2006). *Enzyme technology*. Springer-Verlag, New York.

Rafiq, MY; Bugmann, G; Easterbrook, DJ. Neural network design for engineering applications. *Computers and Structures*, 2001, 79, 1541-1552.

Rao, AA. *Computer-aided prediction of active sites in enzymes*. http, //allamapparao.org/en/papers/paper26.pdf, 2007.

Riveros, TA; Porcasi, L; Muliadi, S; Hanrahan, G; Gomez, FA. Application of artificial neural networks in the prediction of product distribution in electrophoretically mediated microanalysis. *Electrophoresis*, 2009, 30, 2385-2389.

Singh, A; Tatewar, D; Shastri, PN; Pandharipande, SL. Application of ANN for prediction of cellulase and xylanase production by *Trichoderma reesei* under SSF condition. Indian *Journal of Chemical Technology*, 2008, 15, 53-58.

Stahl, M; Taroni, C; Schneider, G. Mapping of protein surface cavities and prediction of enzyme class by a self-organizing neural network. *Protein Engineering*, 2000, 13, 83-88.

Szaleniec, M; Witko, M; Tadeusiewicz, R; Goclon, J. Application of artificial neural networks and DFT-based parameters for prediction of reaction kinetics of ethylbenzene dehydrogenase. *Journal of Computer-Aided Molecular Design,* 2006, 20, 145-157.

Tang, YR; Sheng, ZY; Chen, YZ; Zhang, Z. An improved prediction of catalytic residues in enzyme structures. Protein Engineering, *Design & Selection*, 2008, 21, 295-302.

Torrecilla, JS; Mena, ML; Yanez-Sedeno, P; Garcia, J. A neural network approach based on gold-nanoparticle enzyme biosensor. *Journal of Chemometrics*, 2008, 22 46-53.

Tsai, CS. (2007). *Biomacromolecules, introduction to structure, function and informatics.* John Wiley & Sons, Inc, NJ.

van Beilen, JB; Li, Z. Enzyme technology, an overview. Current *Opinion in Biotechnology,* 2002, 13, 338–344.

Webster, GJ. *Minimally invasive medical technology*. IOP Publishing Ltd. 2001.

Zhang, Y; Li, WW; Zeng, GM; Tang, L; Feng, CL; Huang, DL; Li, YP. *Environmental Engineering Science*, 2009, 26, 1063-1070.

Zhang, IY; Aires-de-Sousa, J. Physicochemical stereodescriptors of atomic chiral centers. *Journal of Chemical Information and Modeling,* 2006, 46, 2278–2287.

Zorzetto, LFM; Maciel Filho, R; Wolf-Maciel, MR. Process modelling development through artificial neural networks and hybrid models. *Computers and Chemical Engineering*, 2000, 24, 1355-1360.

Zupan, J; Novic, M; Ruisanchez, I. Kohonen and counterpropagation artificial neural networks in analytical chemistry. *Chemometrics and Intelligent Laboratory Systems,* 1997, 38, 1-23.

In: Focus on Artificial Neural Networks
Editor: John A. Flores, pp. 355-374
ISBN: 978-1-61324-285-8

Chapter 17

DEVELOPMENT OF AN ANN MODEL FOR RUNOFF PREDICTION

A. Bandyopadhyay** and A. Bhadra
Department of Agricultural Engineering, North Eastern Regional Institute of Science and Technology, Nirjuli, Itanagar, Arunachal Pradesh, India

ABSTRACT

Over the years, several hydrological models ranging from empirical relationships to physically based models have been developed for prediction of runoff. The physically based models are better as they relate physical processes but at the same time their data requirement is also high. Therefore, there is a need to look for alternative methods for prediction of runoff using readily available information such as rainfall. Artificial Neural Network (ANN) is an information processing system that is composed of many nonlinear and densely interconnected processing elements or neurons. Feed forward multilayer neural networks are widely used as predictors in several fields of applications. The purpose of this study is to demonstrate development of an ANN model using both steepest descent and Levenberg-Marquardt optimization training algorithms and to investigate its potential for accurate runoff estimation. Different ANN networks were trained and tested to predict the daily runoff for Kangsabati reservoir catchment. The networks were selected using one, two, and three hidden layers. The network models were trained for seven years data and tested for one year data for different sizes of architecture. Training was conducted using both steepest descent and Levenberg-Marquardt Back Propagation where the input and output were presented to the neural network as a series of learning patterns. Results indicated that the neural networks trained with Levenberg-Marquardt Back Propagation converged much faster than simple steepest descent back propagation. Further, the ANN models performance improved with increase in number of hidden neurons as well as with increase in number of hidden layers up to a certain point 15-20-20-1 (best network architecture), after which the performance deteriorated.

* Corresponding author: Email: arnabbandyo@yahoo.co.in

Keywords: Rainfall-runoff modeling, Artificial neural network, Levenberg-Marquardt.

INTRODUCTION

The study of the human brain is thousands of years old. With the advent of modern electronics, it was only natural to try to harness this thinking process. Inspired by a desire to understand the human brain and emulate its functioning, the development of Artificial Neural Networks (ANN) began approximately 70 years ago when McCulloch and Pitts (1943) modeled a simple neural network with electrical circuits.

After that it has experienced a huge resurgence due to the development of more sophisticated algorithms and emergence of more powerful computers. Extensive research has been devoted to investigate the potential of ANNs as computational tool that acquire, represent, and compute a mapping from one multivariate input space to another. The ability to identify a relationship from given patterns makes it possible for ANNs to solve large scale complex problems like pattern recognition, nonlinear modeling, classification, etc. A tremendous growth in the field of ANNs has occurred since Rumelhart and McClelland (1986) discovered a mathematically rigorous back propagation algorithm. Consequently, ANNs have found application in diverse disciplines ranging almost all branches of engineering and science.

ANN is described as an information processing system that is composed of many nonlinear and densely interconnected processing elements or neurons. ANN acquires the knowledge through a learning process that involves finding an optimal set of weight for the connection and threshold value for the nodes. ANNs have been proven to provide better solution when applied to complex systems that are difficult to understand; problems that deal with noise or involve pattern recognition; and where input is incomplete or ambiguous by nature. These are calibrated using automatic calibration techniques, thus eliminating the lengthy calibration cycle. These properties suggest that ANNs may be well suited to the problem of estimation and prediction in hydraulics and hydrology which is complicated because of the nonlinearity of physical processes and uncertainty in parameter estimation. After appropriate training, ANNs are found to generate satisfactory results for many prediction problems. In the present study its ability in accurate estimation of runoff is investigated.

The application of ANNs in water resource and hydrologic modeling was introduced to the water resource community by Daniell (1991), who used ANNs to predict monthly water consumption and to estimate flood occurrence. Since then, ANNs have been used for a variety of water resource applications. These include time-series prediction for rainfall forecasting, reservoir inflow time series, river salinity, and rainfall-runoff processes. The ASCE Task Committee on Application of Artificial Neural Networks in Hydrology (2000a and b) and Dawson and Wilby (2001) give good state-of-the-art reviews on ANN modeling in hydrology. A number of researchers (e.g., Zhu et al., 1994; Dawson and Wibly, 1998; Tokar and Johnson, 1999; Coulibaly et al., 2000) have investigated the potential of using neural networks in modeling watershed runoff based on rainfall inputs. Majority of studies have proved that ANNs are capable of performing quite satisfactorily in rainfall-runoff modeling (Minns and Hall, 1996; Shamseldin, 1997; Sajikumar and Thandaveswara, 1999; Tokar and

Markus, 2000; Jain and Srinivasulu, 2004; Rajurkar et al., 2004; de Vos and Rientjes, 2005; Ahmad and Simonovic, 2005; Tayfur and Singh, 2006).

It is clear that the rainfall-runoff relationship is an extremely complex and difficult problem involving many variables, which are interconnected in a very complicated way. Most of the models work best when data on the physical characteristics of the watershed are available at the model grid scale. This kind of data is rarely available, even in heavily instrumented research watersheds. This complexity and non-linearity makes it attractive to try the neural network approach, which is inherently suited to problems that are mathematically difficult to describe. Therefore, the ANN approach does provide a viable and effective alternative to conceptual models for developing input-output simulation and forecasting models in situations that do not require modeling of the internal structure of the watershed. However, because the ANN approach does not have physically realistic components and parameters, it is by no means a substitute for conceptual watershed modeling (Hsu et al., 1995).

ANN approach is called a "model" as it has many features in common with other modeling approaches in hydrology. The choice of appropriate network architecture can be considered equivalent to the process of model selection. Similarly, network training and testing can be identified with model calibration and validation. Unlike regression-based approaches in hydrology, ANNs do not require specification of a mathematical form. Otherwise ANNs are similar to regression-based approaches in many respects since both the approaches are from black box group of models and develop a relationship between the inputs and outputs of the system only without describing the underlying physics of the process. In addition, ANNs are more versatile than other hydrological models because of the freedom available with the choice of the number of hidden layers and the nodes associated with each of these layers (ASCE Task Committee on application of ANNs in Hydrology, 2000a). An attractive feature of ANN is their ability to extract the relation between the inputs and outputs of a process, without the physics being explicitly provided to them. ANNs are able to provide a mapping from one multivariate space to another, given a set of data representing that mapping.

Coulibaly et al. (1999) reported that about 90% of the neural network applications in hydrology extensively make use of the multilayer feed-forward neural networks trained by the standard steepest decent back propagation (BP) algorithm. BP training is widely used for runoff prediction because it is capable of handling very large learning problems for implicit and complicated modeling (Neuralware, 1993). This algorithm is also a popular learning method capable of very large learning problems. However, its convergence tends to be very slow, and it often yields sub-optimal solutions (Baldi and Hornik, 1989; Mühlenbein, 1990; Sima, 1996). An efficient back propagation training process requires some numerical optimization technique (Hagan et. al., 1996). Levenberg-Marquardt optimization technique (Levenberg, 1944; Marquardt, 1963) can be incorporated into back propagation algorithm to make training faster and efficient to find better optima for a variety of problems (Hagan and Menhaj, 1994; Masters, 1995; Coulibaly et al., 2000; Raghuwanshi et al., 2006; Nayebi et al., 2006).

Given the reported success in applications of ANNs in pattern recognition and simulation of non-linear systems, an attempt has been made in this study to develop and test an ANN model for prediction of runoff using rainfall as input. The feed forward back-propagation algorithm (Werbos, 1974; Rumelhart and McClelland, 1986) was selected for runoff modeling because it is considered as a good choice for implicit and complicated modeling. This algorithm is a popular learning method capable of very large learning problems. However, due to the reasons stated above, the Levenberg-Marquardt algorithm, a standard second-order nonlinear least squares technique, based on the back propagation process to increase the speed and efficiency of the training was used.

The specific objectives of the study were as follows:

1. To develop ANN models for predicting daily runoff, and
2. To test the developed model with data from a catchment.

THEORETICAL CONSIDERATIONS

Neural Network Concept

The neural-network approach, also referred to as connectionism or paralleled distributed processing, adopts a brain metaphor of information processing. Information processing in a neural network occurs through interactions involving large number of simulated neurons.

This simulated neuron, or unit, has four important components:

1. Input connections (synapses), through which the unit receives activation from other units;
2. Summation function that combines the various input activations into a single activation;
3. Threshold function that converts this summation of input activation into output activation; and
4. The output connections (axonal paths) by which a unit's output activation arrives as input activation at other units in the system.

Artificial neural networks (ANNs) are massively parallel systems composed of many processing elements connected by links of variable weights. The back propagation network is the most popular of many ANN paradigms. The network consists of layers of neurons, with each layer being fully connected to the proceeding layer by inter connection strengths or weights W. Figure 1 illustrates a three-layer neural network consisting of input layer (L_I), hidden layer (L_H), and the output layer (L_O) with the inter-connection weights W_{ih} and W_{ho} between layers of neurons. Initial estimated weight values (i.e., W_{ih} and W_{ho}) are progressively corrected during a training process that compares predicted outputs to known outputs and back propagates any errors to determine the appropriate weight adjustments necessary to minimize the errors. In modeling any process, the input neurons in the input layer (L_I) and output layer (L_O) consists of the input and output of the process, respectively. However, in the subsequent section for generalization, the output of hidden neurons in the hidden layer is denoted by H_{oj} for j^{th} neuron and output of the output neuron in the output layer L_O is denoted by O_j for j^{th} neuron. L, M and N are the number of neurons in input, hidden, and output layers, respectively.

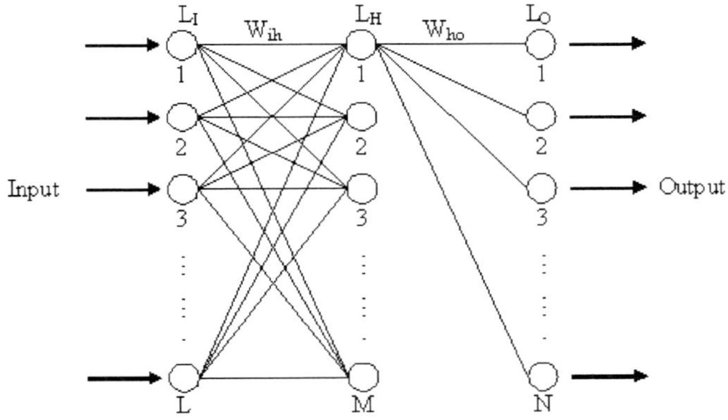

Figure 1. Configuration of three-layer neural network.

The total input H_{ij} to hidden units j is a linear function of outputs x_i of the units that are connected to j and of the weights w_{ij} on these connections:

$$H_{ij} = \sum_{i} x_i \, w_{ij} \tag{1}$$

A hidden unit has a real-value output $H_{oj,}$ which is a non-linear function of its total input. Units can be given biases (θ_j) by introducing on extra input to each unit which always has a value of 1:

$$H_{oj} = \frac{1}{1 + e^{-(H_{ij} + \theta_j)}} \tag{2}$$

The use of a linear function for combining the inputs to a unit before applying the non-linearity greatly simplifies the learning procedure. The aim is to find a set of weights that ensure that for each input vector, the output vector produced by the network is the same as (or sufficiently close to) the desired output vector. If there is a fixed, finite set of input-output cases, the total error in the performance of the network with a particular set of weights can be computed by comparing the actual and desired output vectors for every case. The total error E is defined as:

$$E = \frac{1}{2} \sum_{C} \sum_{f} (O_{j,c} - T_{j,c})^2 \tag{3}$$

where,

c = index over cases (input-output pairs)
j = index over output units
C and f = two maximum input and output pairs
O = actual state of an output neuron in the output layer
T = targeted state

Neural Network Architecture Parameters

Determination of appropriate neural network architecture is one of the most important tasks in the model-building process. Various types of neural networks are analyzed to find the most appropriate architecture of a particular problem. Multilayer feed forward network are found to outperform all the others. Although multilayer feed forward networks are one of the most fundamental models, they are the most popular type of ANN structure suited for practical applications.

Neural Network Models

Neural network models are specified by the net topology, node characteristics, training and learning rules. These rules specify an initial set of weights and indicate how weights should be adopted to improve performance. Both design procedures and training algorithms are the topics of current research.

Training Algorithms

A major concern in the development of a neural network is determining an appropriate set of weights that make it perform the desired function. There are many ways that this can be done, the most popular class of these algorithms is based on supervised training. Supervised training starts with a network comprising an arbitrary number of hidden neurons, a fixed topology of connections, and randomly selected values for weights. The network is then presented with a set of training patterns, each comprising an example of the problem to be solved (the inputs) and its corresponding solution (the targeted output). Each problem is input into the network in turn, and the resultant output is compared to the targeted solution providing a measure of total error in the network for the set of training patterns. Properly trained back propagation network give reasonable results when presented with new input during validation.

In the process of model development several network architectures with different number of input neurons in input layer with varying number of hidden neurons are considered to select the optimal architecture of the network. A trial and error procedure based on the minimum error during validation is used to select the best network architecture.

Number of Hidden Layers

There is no fixed rule for selection of hidden layers of a network. Therefore trial and error method was used for selection of number of hidden layers. Hecht-Nielsen (1990) provided a proof that even one hidden layer of neuron (operating sigmoid activation function) can also be sufficient to model any solution surface of practical interest.

Number of Hidden Neurons

The ability of the ANN to generalize data not included in training depends on selection of sufficient number of hidden neurons to provide a means for storing higher order relationships necessary for adequately abstracting the process. There is no direct and precise way of determining the most appropriate number of neurons to include in hidden layer, and this problem becomes more complicated as number of hidden layer increases. It is observed that more number of hidden layer neurons provide a greater potential for developing a solution surface that fits closely to that implied by training patterns. In practice, however, a large number of hidden neurons can lead to a solution surface that while fitting the training surface exactly, deviate significantly from the trend of the surface at intermediate points or provide too literal interpretation of the training points which is called over fitting. In addition, large number of neurons also slows down the operation of network both during training and during use. Conversely, an accurate model of some or all features in the solution surface may not be achieved if too few hidden neurons are included in the network. To solve this problem, several neural networks with different number of hidden neurons are used for calibration and one with best performance together with compact structure is accepted.

Training Data

The distribution of the training pattern within the problem domain can have a significant effect on the learning and generalization performance of a network. A major limitation of ANN is that they are not usually able to extrapolate; therefore, training pattern should go at least to the edge of the problem domain in all dimensions. It is also essential to have training pattern evenly distributed within this region. If not then trained network performs poorly for the region where the density of training data is less.

Training data set significantly influence a network's ability to learn and generalize the patterns. Increasing the number of training patterns provide more information about the shape of the solution surface, and thus increase the potential level of accuracy that can be achieved by the network. While it is essential that the information contained in the input set is sufficient to determine the network output, a large training set, however, can overwhelm training algorithm, thus increasing the probability to stick in the local error minima. Moreover, there is a limit to the amount of information that can be modeled by network with fixed number of neurons. Also, collection or creation of data is a time and money consuming activity and generally fairly accurate results are envisaged within a definite error boundary. Therefore, a balance should be made between the accuracy of the results achieved against the cost and time required to collect or produce the training pattern.

Levenberg-Marquardt Training Algorithm

Levenberg-Marquardt algorithm works by making the assumption that the underlying function being modeled by the neural network is linear. Based on this calculation, the minimum can be determined exactly in a single step. The calculated minimum is tested, and if the error there is lower, the algorithm moves the weights to the new point. This process is repeated iteratively on each generation. Since the linear assumption is ill-founded, it can easily lead Levenberg-Marquardt algorithm to test a point that is inferior (perhaps even wildly inferior) to the current one. The clever aspect of Levenberg-Marquardt algorithm is that the determination of the new point is actually a compromise between a step in the direction of steepest descent and the above-mentioned leap. Successful steps are accepted and lead to a strengthening of the linearity assumption (which is approximately true near to a minimum). Unsuccessful steps are rejected and lead to a more cautious downhill step. Thus, Levenberg-Marquardt algorithm continuously switches its approach and can make very rapid progress.

The Levenberg-Marquardt algorithm uses the approximate Hessian matrix (second derivatives of error matrix, E) in the weight update procedure as follows:

$$\Delta W_{ji} = -[H + \mu I]^{-1} J^T r \tag{4}$$

where,

r = residual error vector

μ = variable small scalar that controls the learning process

I = identity matrix

J = Jacobian matrix

= ∇E

$$= \begin{bmatrix} \partial r_1(w)/\partial w_1 & \partial r_1(w)/\partial w_2 & . \; . \; . & \partial r_1(w)/\partial w_n \\ \partial r_2(w)/\partial w_1 & \partial r_2(w)/\partial w_2 & . \; . \; . & \partial r_2(w)/\partial w_n \\ . & . & . \; . \; . & . \\ . & . & . \; . \; . & . \\ . & . & . \; . \; . & . \\ \partial r_n(w)/\partial w_1 & \partial r_n(w)/\partial w_2 & . \; . \; . & \partial r_n(w)/\partial w_n \end{bmatrix}$$

J^T = transpose of the J

H = Hessian matrix

= $\nabla^2 E$

= $J^T J$ approximately

The flowchart of the Levenberg-Marquardt algorithm is shown in Figure 2. The Maximum Cycle (MaxCycle) and Error Tolerance ($E_{tolerance}$) are the inputs for terminating the program. In practice, Levenberg-Marquardt algorithm is faster and finds better optima for a variety of problems than do the other usual methods.

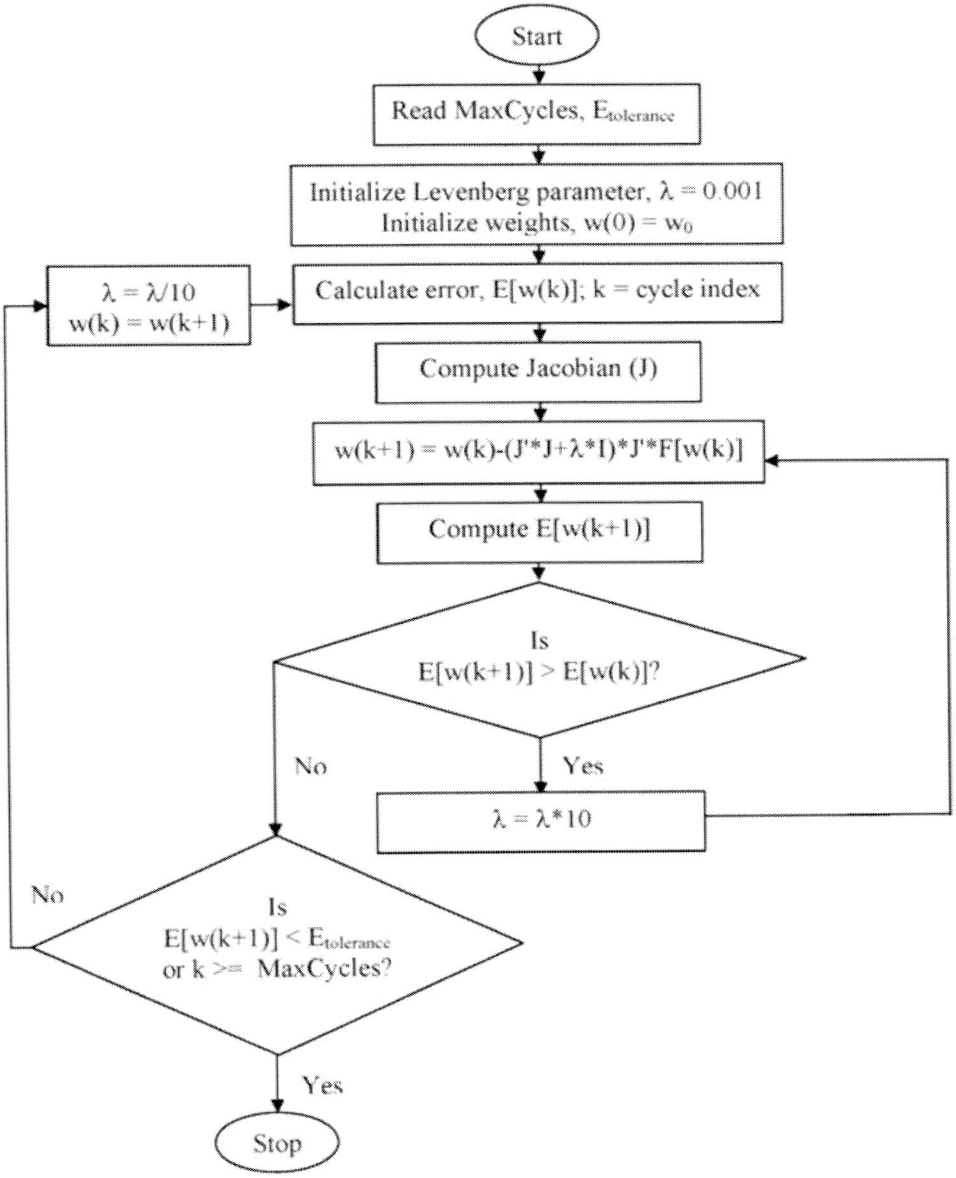

Figure 2. Flowchart of Levenberg-Marquardt algorithm.

DEVELOPMENT OF SOFTWARE

The software for training and testing the networks was developed in the current study using the back-propagation algorithm with steepest descent and Levenberg-Marquardt training optimization. The entire code was written in C language. This code was linked to a Graphical User Interface (GUI) developed in Visual Basic 6.0.

Coding in C

Based on the back-propagation algorithm with steepest descent and Levenberg-Marquardt training optimization techniques described earlier, the executable code was developed in C language for training and testing different neural networks for rainfall-runoff modeling. The flow chart of the developed software is shown in Figure 3.

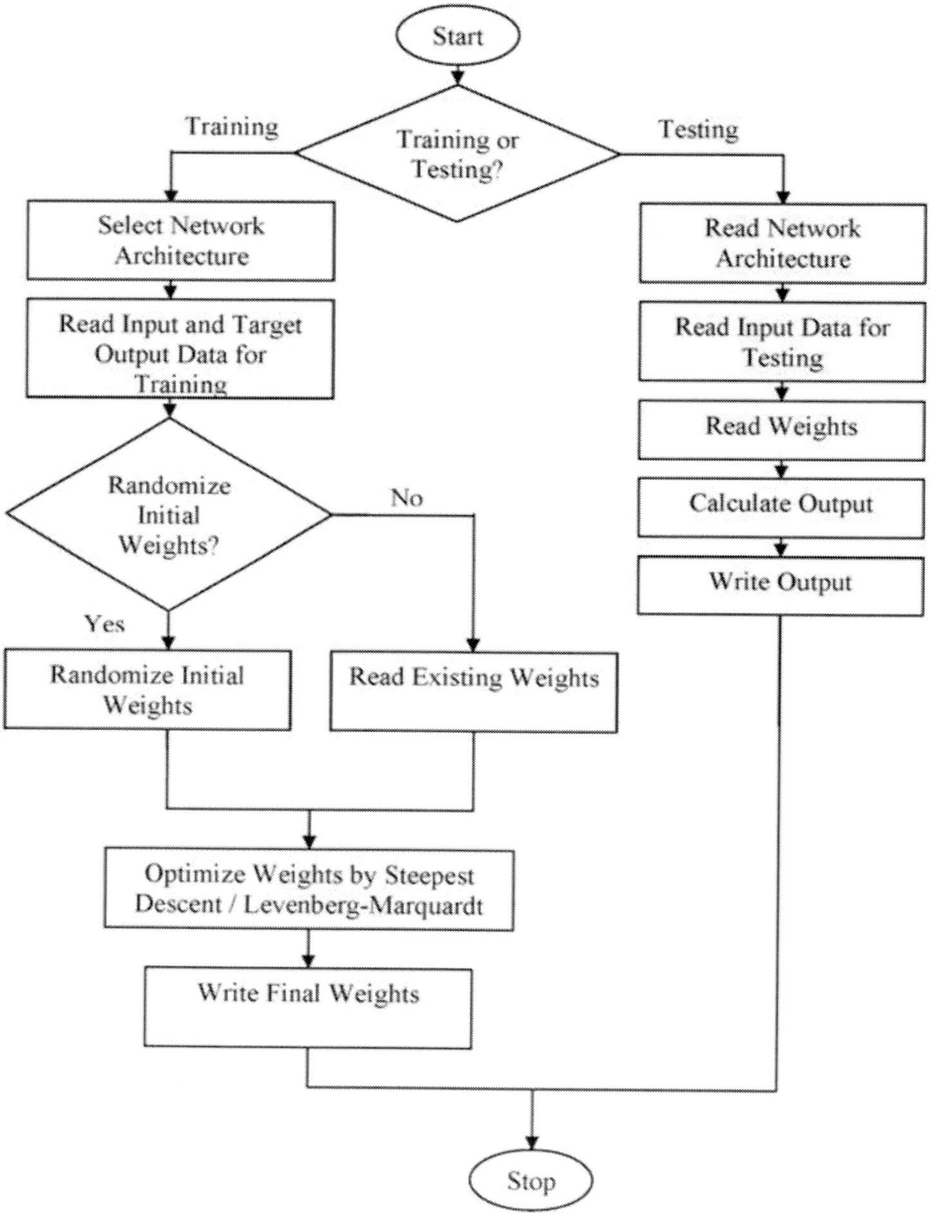

Figure 3. Flowchart of the developed software.

Development of GUI

To make the model user friendly, graphical user interface (GUI) was developed using Visual Basic 6.0. Several windows were developed for entering network information and training parameters, and generating and displaying graphical outputs along with statistical parameters.

The *Training* window (Figure 4) of the GUI asks from user information regarding network architecture, general training parameters, e.g., error tolerance, maximum cycles, and training algorithm specific parameters, e.g., learning rate parameter (beta), Levenberg parameter (lambda), etc., and the training patterns. The weights can either be randomly initialized or can be read from an existing weight file from previous training session. All these information can be saved in a file and later retrieved. The network is trained on clicking the button *Start Training*. At the end of the training, the result of training is displayed in the text box for RMSE.

The *Testing* window (Figure 5) opens on clicking the *Testing* tab of the GUI. In this window the user can provide the input patterns only can also be saved in a file and later retrieved. On clicking the *Start Testing* button, predicted outputs are obtained as the testing results on the right side of the window. For statistical analysis of the testing results the *Statistical Analysis* command button of the *Testing* window needs to be clicked. A new window gets opened in which observed runoff can be entered and results of statistical analysis (Figure 6) along with a plot between predicted and observed runoff can be viewed.

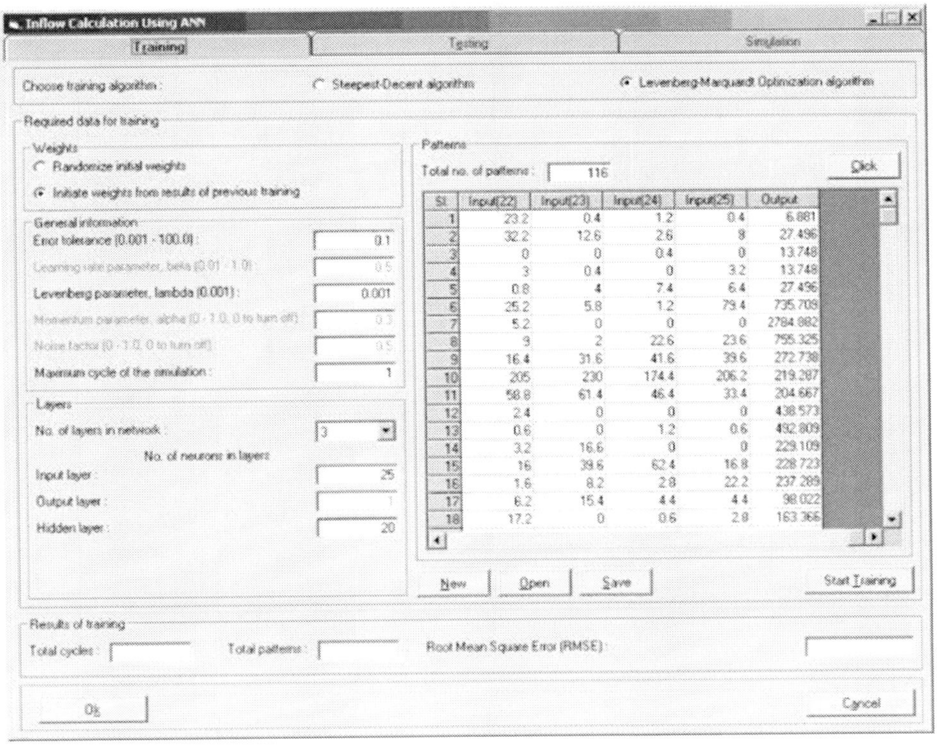

Figure 4. Training window of the developed GUI.

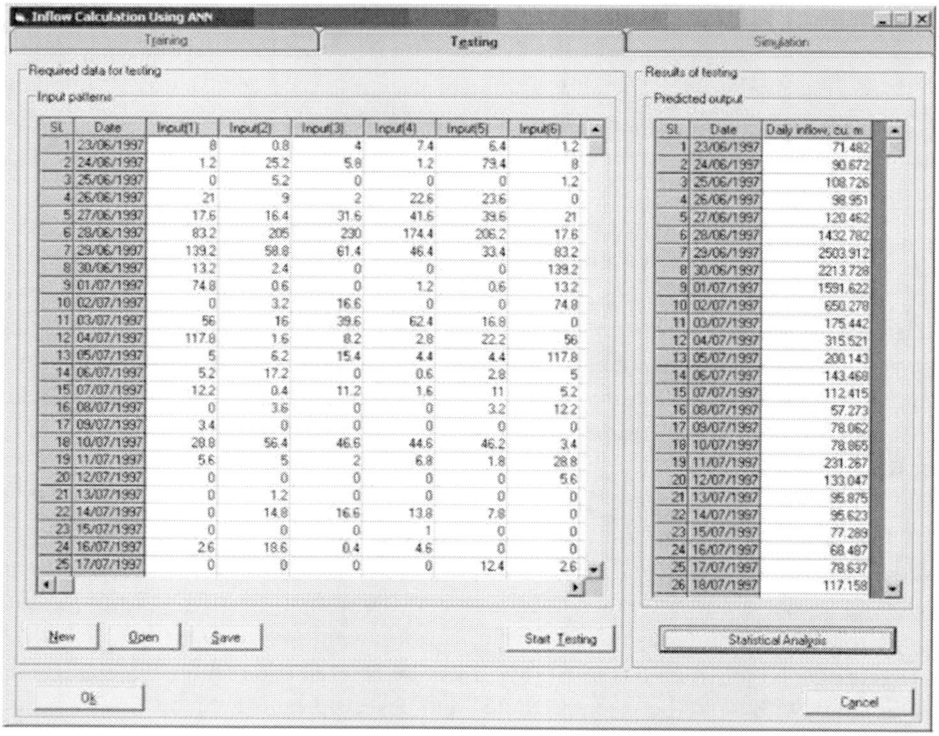

Figure 5. Testing window of the developed GUI.

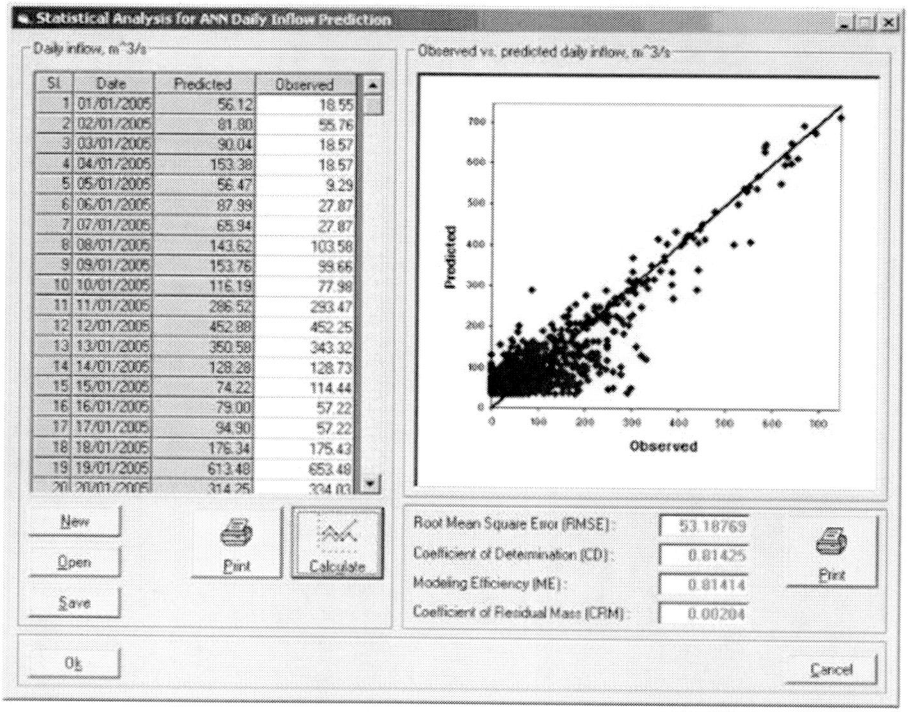

Figure 6. Statistical analysis window of the developed GUI.

Linking of C-Code With GUI

Developed C-code of ANN model was linked with GUI through *Shellexecute* function. Required input files are created by the GUI and saved in a particular project folder. During execution, the executable program (EXE file made from C-code) is called using *Shellexecute* function. The program first finds its required input files in the particular project folder. If these files exist, the program reads those files and executes.

After successful execution of the program, output files are created and saved in the same folder by the EXE file. Then, from the output files, data are extracted and displayed in the tabular and graphical forms through GUI output windows.

DATA AND METHODOLOGY

Study Area

Kangsabati Irrigation Project, situated in the western part of West Bengal, India, was chosen as the study area. Kangsabati river and its main tributaries originate from Chhotanagpur hills. After flowing south-easternly, Kangsabati river falls in the Hoogly river. Kangsabati dam, built just above the confluence of Kangsabati and its tributaries Kumari, is located at 22° 57' 30" N latitude and 86° 45' 30" E longitude. Kangsabati dam was constructed in two phases. The dam was first constructed on Kangsabati river in 1965. Subsequently, the dam over tributary Kumari was constructed in 1973 and both dams were connected to form a single reservoir, viz., Kangsabati reservoir. Figure 7 shows the catchment of Kangsabati reservoir including the raingauge and river network within its catchment. Total catchment area of Kangsabati reservoir is about 3626 sq. km and was considered for estimation of reservoir inflow in this study.

Data

Twelve years (1986-97) daily rainfall data measured in five rainguage stations in Kangsabati reservoir catchment, namely, Kangsabati dam site, Rangagora, Khariduar, Tusama, and Simulia, were collected from Central Water Commission, Ministry of Water Resources, Govt. of India. These daily rainfall data were used as input to the networks. Daily reservoir inflow data were also collected for same twelve years (1986-97) from Irrigation and Waterways Department, Bankura, Govt. of West Bengal. These inflow data were used as target output.

Out of these twelve years (1986-97) rainfall-runoff data, two different training and testing data sets were prepared using only the monsoon period data (July to October). Seven years (1986-91 and 1993) monsoon data were used for training and one year (1997) monsoon data were used for testing the networks. Other four years (1992 and 1994-96) data were discontinuous, resulted in poorer training performance when included and hence were discarded.

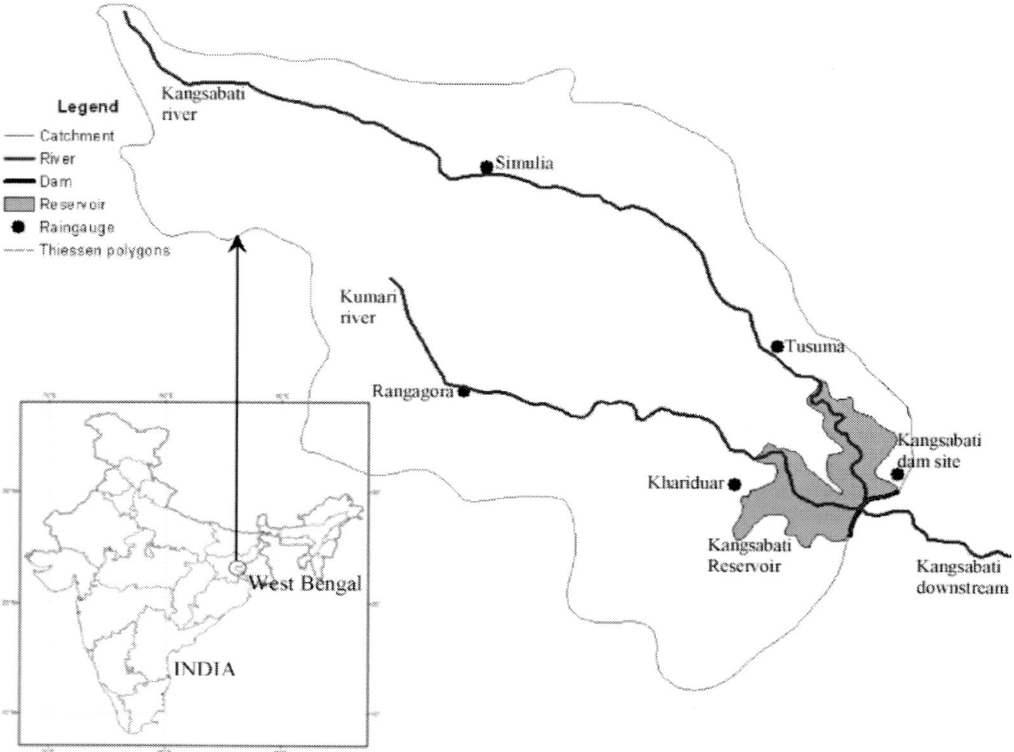

Figure 7. Kangsabati reservoir catchment.

The daily data used for training and testing the networks were normalized such that the data lie between 0 and 1 because the sigmoid activation function used for training the networks has lower and upper limits of 0 and 1, respectively. The normalization process also removes the cyclicity of the data. The following procedure was adopted for normalizing the input and output data of each data set. Each variable, $X_{i,}$ in the data set was normalized ($X_{i, norm}$) between 0 and 1 by dividing its value by the upper limit of the data set, $X_{i, max}$. The resulting normalized data were then used for mapping as:

$$X_{i, norm} = X_i / X_{i, max} \tag{5}$$

Development of ANN Architecture

The selection of training data that represents the characteristics of meteorological pattern is extremely important in modeling. The training data should be large enough to contain the characteristics of the catchment and to accommodate the requirements of the ANN architecture. In this study, the networks were trained for different number of input neurons (from 5 to 30, step 5) to consider daily rainfall data from five different rain gauge stations and previous days rainfall data for taking into account the effect of antecedent moisture content. It was found that the networks having 15 neurons (rainfall data of five rain gauge stations for a particular day and past two days) in the input layer performed better. Hence, all the networks studied further consisted of 15 neurons in input layer, one neuron in output layer (considering single outlet point of the catchment) and one, two, and three hidden layers with different combinations of hidden neurons. No specific method was available in the literature to determine the number of neurons in the hidden layer(s). So the networks were trained with different number of neurons (from 10 to 30, step 5) in hidden layer(s).

Performance Indicators

Hydrologic models are used most frequently to simulate or predict flow either on a continuous basis or for a particular event. In all cases the moel computed value is compared with the measured value. A variety of verification criteria which could be used for evaluation and inter comparison of different models were proposed by World Meteorological Organization (WMO). The graphical performance criteria coefficient of determination (r^2) and the statistical performance criteria root mean square error (RMSE) has been used in this study for estimating quantitative performance of the developed ANN model.

A linear scale plot is drawn between the observed and ANN predicted runoff for training phase. The coefficient of determination (r^2) is selected to measure the correlation between the actual and the predicted values. The r^2 value close to unity and the closeness of the best fit line with the 1:1 line indicate good agreement between the observed and predicted values.

Of the several numerical indicators, the one numerical indicator selected for the present study was root mean square error (RMSE). RMSE can give a quantitative indication of the model error in terms of a dimensional quantity. RMSE equal to zero indicates a perfect match between the observed and predicted values. Based on RMSE values, the best networks were selected among all the trained networks. One best network from each one, two, and three hidden layer networks was chosen. The selected trained networks were then validated (tested) using the observed daily rainfall data for the monsoon season of 1997. Thereafter, the observed daily runoff for the monsoon season of 1997 was compared with its predicted counterparts for evaluation of networks validation performance.

Results and Discussion

Training of Networks

The networks were trained with one, two, and three hidden layers and with different combinations of hidden neurons to get the daily runoff with only input of daily rainfall. It was observed that the RMSE values in general decreased with increase in hidden nodes as well as with number of hidden layers when the number of training cycles remains constant. A single network architecture corresponding to the minimum RMSE was selected for each input and hidden layer conditions. Networks 15-30-1, 15-20-20-1 and 15-20-20-20-1 gave minimum RMSE values (m^3/s) of 42.247, 44.067, and 40.771, respectively. The peak of the predicted runoff matched consistently well with the measured runoff in all the years for the training period. The network predicted high runoff events quite accurately but slightly under predicted the other events. Continued training could reduce the RMSE further but made the network too case specific by over-fitting to the training data set and resulted in poorer performance in testing.

Regression analysis was performed between the observed and predicted daily runoff values for one, two, and three hidden layer networks. The values of coefficient of determination (r^2) for one, two, and three hidden layer networks were found to be 0.897, 0.888, and 0.891, respectively. Example time series and scatter plots of the observed and predicted daily runoff for one hidden layer network are shown in Figures 8 and 9. These high r^2 values indicate a close match between the observed and predicted daily runoff.

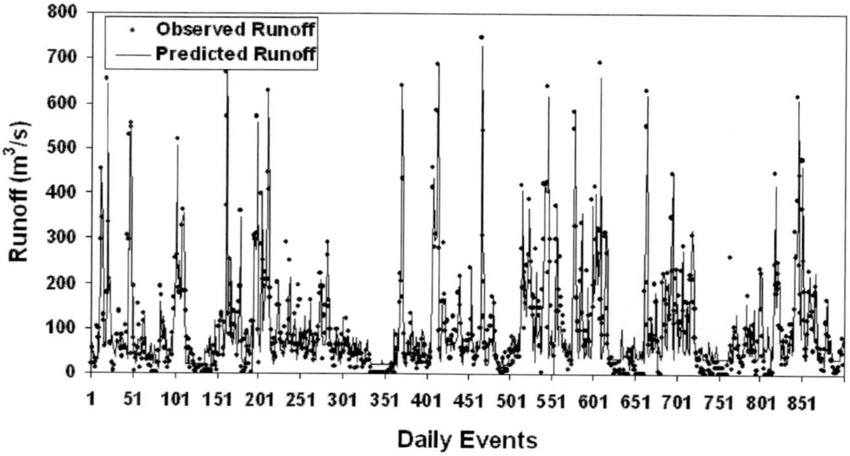

Figure 8. Time series plot of observed and predicted runoff (15-30-1 network training).

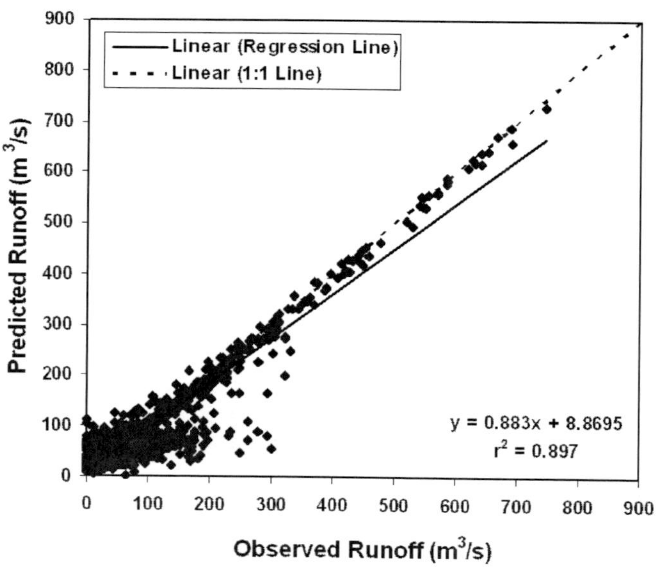

Figure 9. Scatter plot between observed and predicted runoff (15-30-1 network training).

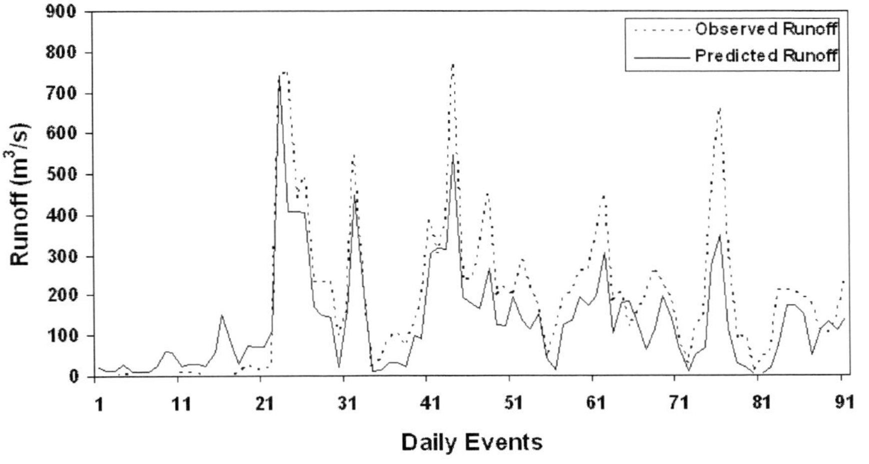

Figure 10. Time series plot of observed and predicted runoff (15-20-20-1 network testing).

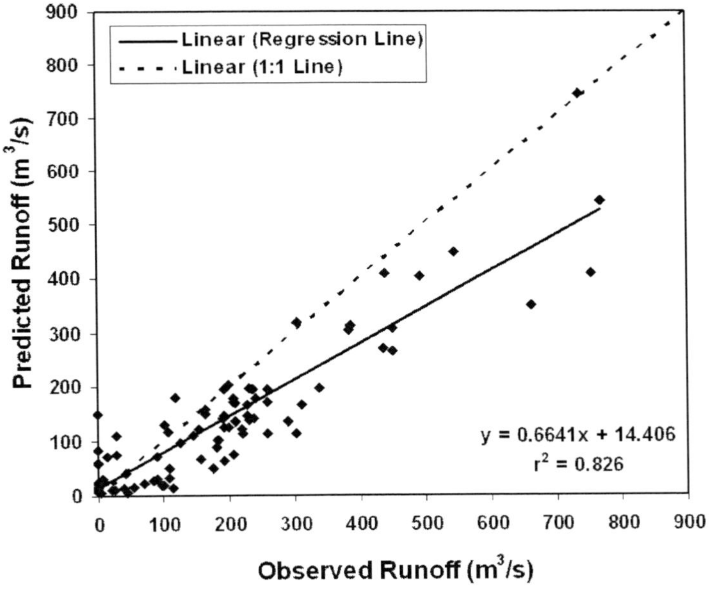

Figure 11. Scatter plot between observed and predicted runoff (15-20-20-1 network testing).

Comparison of Steepest Descent and Levenberg-Marquardt Algorithms

As evident from Table 1, Levenberg-Marquardt needed much less number of cycles compared to steepest descent for obtaining same order of RMSE values. The weight files generated by Levenberg-Marquardt also produced much better results during validation as compared to weight files generated by steepest descent for similar training RMSE values. This may be accounted to the fact that steepest descent often converges to local minima. Therefore, the validation results in this study were obtained using the Levenberg-Marquardt generated weight files.

Table 1. Number of cycles required in training

Algorithms	Network architecture		
	15-30-1	15-20-20-1	15-20-20-20-1
Steepest Descent	1520	842	765
Levenberg-Marquardt	32	49	70

Testing of Networks Using Levenberg-Marquardt Algorithm

For the three selected Levenberg-Marquardt trained networks for each one, two, and three hidden layer networks (15-30-1, 15-20-20-1, and 15-20-20-20-1), r^2 values obtained were 0.779, 0.826, and 0.794, respectively, during testing. The RMSE values for the same were 92.445, 90.841, and 91.673 m^3/s, respectively. These high values of r^2 indicated a close relationship between the observed and predicted daily runoff values. The 15-20-20-1 network was therefore identified as the best network for the rainfall-runoff problem envisaged in this study due to its highest r^2 value in testing. The time series and scatter plots of the daily observed and predicted runoff are shown in Figures 10 and 11.

Conclusion

A user-friendly, easy to operate multilayer feed forward back propagation ANN model was developed in this study. Two different training algorithms were provided in the model, viz., steepest descent and Levenberg-Marquardt. In this study, the developed ANN model was used for rainfall-runoff mapping of a scantily gauged catchment. However, efforts were made to make the model as generalized as possible and was provided with a user-friendly and flexible GUI (graphical user interface). It can, therefore, be used to model any type of input-output relationships by changing the network structure suitably.

The neural networks trained with Levenberg-Marquardt Back Propagation converged very fast in comparison to the simple steepest descent back propagation. Further the ANN models convergence performance improved with increase in number of hidden neurons as well as with increase in number of hidden layers up to a certain point (best network architecture), after which the performance deteriorated. For the problem studied, 15-20-20-1 network architecture was found to be the best with an r^2 value of 0.888 in training and 0.826 in testing.

References

Ahmad, S. & Simonovic, S. P. (2005). An artificial neural network model for generating hydrograph from hydro-meteorological parameters. *J. Hydrol. 315(1-4),* 236-251.

ASCE Task Committee on Artificial Neural Networks in Hydrology. 2000a. Artificial neural networks in hydrology I: Preliminary concepts. *J. Hydrol. Engrg., 5(2)*, 115-123.

ASCE Task Committee on Artificial Neural Networks in Hydrology. 2000b. Artificial neural networks in hydrology II: Hydrologic applications. *J. Hydrol. Engrg., 5(2)*, 124-137.

Baldi, P. & Hornik, K. (1989). Neural networks and principal components analysis: Learning from examples without local minima. *Neural Networks, 2*, 53-58.

Coulibaly, P., Anctil, F. & Bobée, B. (1999). Prévision hydrologique par réseaux de neurones artificiels: État de l'art. *Revue canadienne de génie civil, 26(3)*, 293-304.

Coulibaly, P., Anctil, F. & Bobee, B. (2000). Daily reservoir inflow forecasting using artificial neural networks with stopped training approach. *J. Hydrol., 230*, 244-257.

Daniell, T. M. (1991). Neural networks – Applications in hydrology and water resources engineering. *Proc. Int. Hydrol. and Water Symp.,* Nat. Conf. Publ. 91/22, Inst. of Eng., Australia, Barton, ACT, Australia, *3*, 797-802.

Dawson, C. W. & Wilby, R. L. (1998). An artificial neural network approach to rainfall-runoff modeling. *Hydrol. Sci., 43(1)*, 47-66.

Dawson, C. W. & Wilby, R. L. (2001). Hydrological modeling using artificial neural networks. *Progress in Physical Geography, 25*, 80-108.

de Vos, N. J. & Rientjes, T. H. M. (2005). Constraints of artificial neural networks for rainfall-runoff modeling: Trade-offs in hydrological state representation and model evaluation. *Hydrol. Earth Sy. Sci., 9(1)*, 111-126.

Hagan, M. T., Demuth, H. B. & Beale, M. (1996). *Neural network design.* PWS Publishing Company, Boston, USA.

Hagan, M. T. & Menhaj, M. B. (1994). Training feed forward networks with the Marquardt algorithm. IEEE Trans. *Neural Networks, 5(6)*, 989-993.

Hecht-Nielsen, R. (1990). *Neurocomputing,* Addison-Wesley, ISBN 0-201-09255-3.

Hsu, K., Gupta, H. V. & Sorooshian, S. (1995). Artificial neural network modeling of the rainfall-runoff process. *Water Resour. Res., 31(10)*, 2517-2530.

Jain, A. & Srinivasulu, S. (2004). Development of effective and efficient rainfall-runoff models using integration of deterministic, real-coded genetic algorithms and artificial neural network techniques. *Water Resour. Res., 40(4)*.

Levenberg, K. (1944). A method for the solution of certain problems in least squares. *Quart. Appl. Mat., 2(2)*, 164-168.

Marquardt, D. (1963). An algorithm for least-squares estimation of nonlinear parameters. Soc. Industr. *Appl. Mat. J. Appl. Math., 11*, 431-441.

Masters, T. (1995). *Advanced algorithms for neural networks: A C++ sourcebook.* Wiley, NY, USA.

McCulloch, W. S. & Pitts, W. H. (1943). A Logical Calculus of the Ideas Immanent in Neural Nets. *Bull. Mathematical Biophysics, 5.*

Minns, A. W. & Hall, M. J. (1996). Artificial neural networks as rainfall runoff models. *Hydrologic. Sci. J., 41(3)*, 399-417.

Mühlenbein, H. (1990). Limitations of multi-layer perceptron networks - Steps towards genetic neural networks. *Parallel Compg., 14*, 249-260.

Nayebi, M., Khalili, D., Amin, S. & Zand-Parsa, Sh. (2006). Daily stream flow prediction capability of artificial neural networks as influenced by minimum air temperature data. *Biosys. Engrg., 95(4)*, 557-567.

Neuralware. (1993). *Neural computing: A technology hand book for NeuralWorks Professional II/plus and NeuralWorks Explorer.* Pittsburgh, pa: Neural ware, Inc.

Raghuwanshi, N. S., Singh, R. & Reddy, L. S. 2006. Runoff and sediment yield modeling using artificial neural networks: Upper Siwane river, India. *J. Hydrolog. Engrg., ASCE, 11(1)*, 71-79.

Rajurkar, M. P., Kothyari, U. C. & Chaube, U. C. (2004). Modeling of the daily rainfall-runoff relationship with artificial neural network. *J. Hydrol., 285*, 96-113.

Rumelhart, D. E. & McClelland, J. L. (Eds). (1986). Parallel distributed processing: *Explorations in microstructure of Cognition, Vol. I*, MIT Press, Cambridge, Mass.

Sajikumar, N. & Thandaveswara, B. S. (1999). A non-linear rainfall-runoff model using artificial neural networks. *J. Hydrol., 216(1-2)*, 32-55.

Shamseldin, A. Y. (1997). Application of a neural network technique to rainfall-runoff modeling. *J. Hydrol.* (Amsterdam), *199(3-4)*, 272-294.

Sima, S. (1996). Back-propagation is not efficient. *Neural Networks, 9(6)*, 1017-1023.

Tayfur, G. & Singh, V. P. (2006). ANN and fuzzy logic models for simulating event-based rainfall-runoff. *J. Hyd. Engrg.*, ASCE, *132(12)*, 1321-1330.

Tokar, A. S. & Jhonson, P. A. (1999). Rainfall-runoff modeling using artificial neural networks. *J. Hydrol. Engrg., 4(3)*, 232-239.

Tokar, A. S. & Markus, M. (2000). Precipitation-runoff modeling using artificial neural networks and conceptual models. *J. Hydrol. Eng., 5(2)*, 156-160.

Werbos, P. J. (1974). Beyond regression: New tools for prediction and analysis in behavior sciences. *Ph.D Thesis*, Harward University, Cambridge, Mass.

Zhu, M., Fujita, M. & Hashimoto, N. (1994). Application of neural networks to runoff prediction. In: Hipel, K.W. (Ed.). *Stochastic and statistical method in hydrology and environmental engineering, Vol. 3*, 205-216.

In: Focus on Artificial Neural Networks
Editor: John A. Flores, pp.375-397
ISBN: 978-1-61324-285-8

Chapter 18

ARTIFICIAL NEURAL NETWORKS CONCEPT: TOOLS TO SIMULATE, PREDICT AND CONTROL PROCESSES

Abdoul-Fatah Kanta*
Service de Science des Matériaux, Université de Mons, Mons, Belgique

ABSTRACT

Artificial neural network (ANN) is a powerful statistical procedure permitting to relate the parameter of a given problem to its desired result by considering a complex network of artificial neurons. This concept is based on model which offers the possibility to develop such a global and integrated approach, without providing any physical explanation for the relationships that have to be validated from a physical point of view.

The design of neural networks structures is an important problem for ANN applications which is difficult to solve theoretically. The definition of optimal network architecture for any particular problem is quite difficult and remains an open problem. The contribution of this paper is the description and the implementation of a formal neural networks concept.

1. INTRODUCTION

The first version of the formal neuron was introduced by W.S. Mc Culloch and W. Pitts. Hebb introduced the first empirical rule for the modification of the synaptic weights [1]. The concept of artificial neural networks (ANN) introduced by Mc Culloch and Pitts [2] is a concept which is based on mathematical and data-processing models, assemblies of calculating units called formal neurons, and whose original inspiration was a model of the human nervous cell [3-4].

* Corresponding author: Tel. (0032) 65 37 44 47, Fax. (0032) 65 37 44 16, E-mail: abdoul.kanta@umons.ac.be

The neurons are inter-connected by one-way networks called "connections" and realize an algebraic function of its inputs [5]. Each calculating unit has only one output connection which can be duplicated in as many specimens as wished, the duplicates transmitting the same signal.

Feed forward ANN was considered in this part. Signals are received at the input layer, pass through the hidden layer and reach to the output layer. The learning process primarily involves the determination of connection weights and patterns of connections.

Figure 1 presents the ANN architecture. It can be divided in three sets [3-4] following:

- input neurons unit which receives the input data in the shape of a scalar vector values, the input vector represents the parameters of the problem. These values are communicated to the neurons via their external input values. Thus, they influence their activation, and by extension, the neural networks behavior;
- output neurons layer whose activations constitute the output vector. They are collectively interpreted like the neural networks result. The same neuron can be the input and the output of the neural networks at the same time; therefore input unit and output unit are not necessarily disjoint;
- hidden neurons layer which connecting the input unit and output unit represents the correlations encoded by the system. In general, the presence of hidden neurons in neural networks reduces its computing power, and allows it to tackle more complex problems.

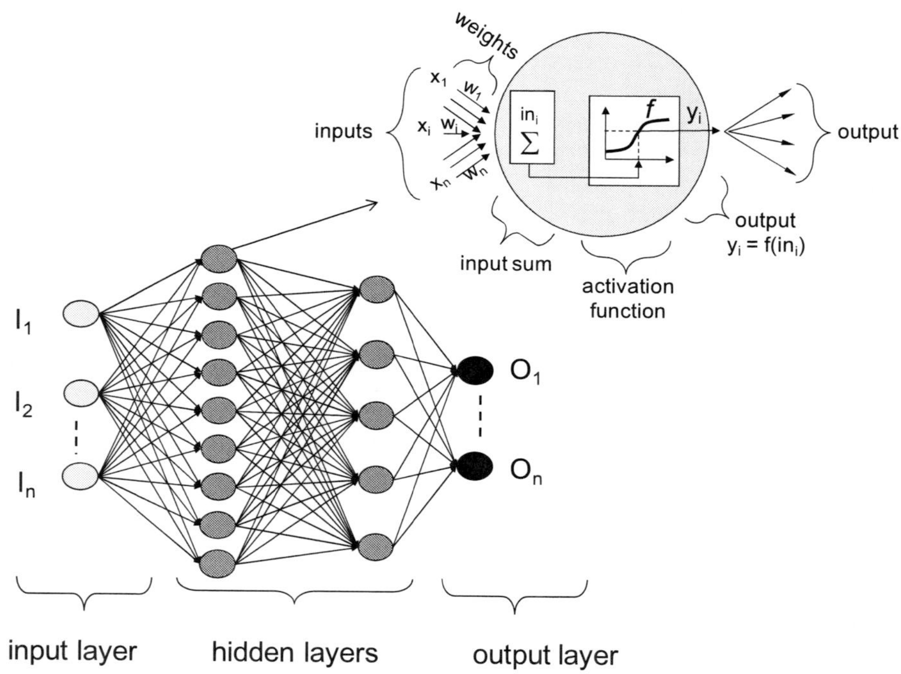

Figure 1. Feed forward artificial neural networks model.

The neural networks represent nonlinear families of function modular, able to approximate any function [5-6]. The computational power of artificial neural networks derives from their parallel distributed structure and their inherent ability to adapt to specific problems, learn and generalize. These characteristics provide ANN the ability to solve problems that are too complex for conventional technologies; problems that do not have an algorithmic solution or for which an algorithmic solution is too complex to be defined.

Able to adapt automatically, the neural networks find their application to a large variety of fields such as (i) patter recognition: it is one of the fields where the neural networks are more powerful; (ii) competition with the statistical methods: the neural networks are used in the fields calling upon the processing of many statistical data, marketing, economy, etc.

2. Models

Model is an abstract object which intended to reproduce a certain reality in the simulation phase [6]. Simulation permits global observation for the model whose employment when the relations of input/output cannot be described simply.

In the case of a structural model, two problems must be solved: propagate information through the model structure and to interpret the behaviors of its components. Build a structural model consists to choose its structure and the behavior of its components in order to answer certain specifications [6-7]. Hence, the development of the neural networks model consists to calculate its parameters so that the output of the model predicts the process output. In statistical terms, it means to estimate the parameters in such way that the model output is as near as possible to the expectation of the process output. This expectation, which is an unknown function of the input, is called regression. The parameters estimation (training) is realized with database constituted by couples {inputs values; outputs values measured on the process} and training algorithm destined to find the values of the parameters which minimize a function of the error [5, 8-9] (difference between the process output and the ANN output).

2.1. Construction Methodes

The construction of a structural model requires to define the modeling paradigm, Figure 2. The choices are realized automatically with algorithmic rules or heuristic (which lead to the result in the majority of the cases, but can fail sometimes) [6].

The problem of prediction consists to estimate the future state of process from its behavior history and contiguous environmental variables. In control, it permits to estimate passively the future state and to intervene in order to achieve a given objective [3].

2.2. Formal Neuron

The formal neuron (elementary processor of ANN) is an independent automate whose state, scalar value, names activation or activity [3].

The neuron architecture is characterized by [4]:

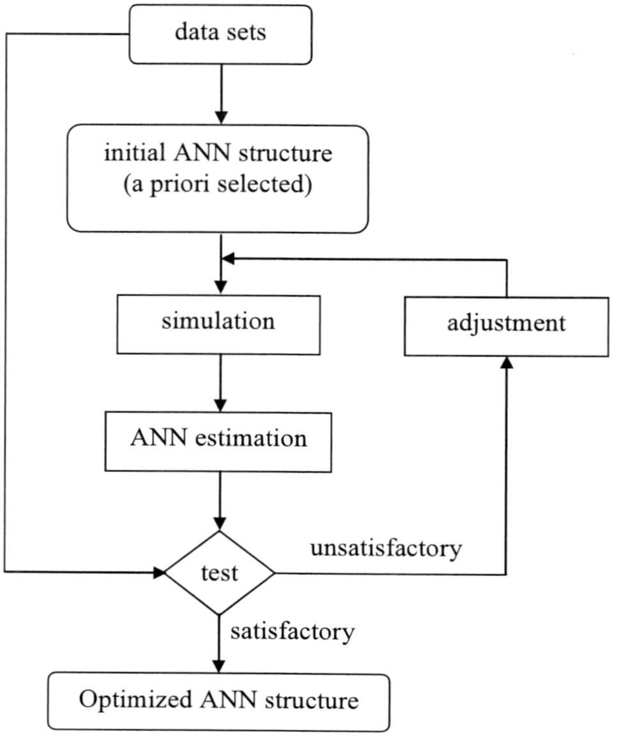

Figure 2. Heuristic general method of structural method construction.

- the input being the sum of flow coming from the other neurons connected upstream, Figure 3;
- the activation (transfer) function making the input nonlinear [10] and animating the neuron by determining its activation. The activation of the current neuron is propagated along synaptic bonds towards the downstream neurons, and gradually, through the network. This function is generally nonlinear (sigmoid, hyperbolic, etc.) 5, 11-12]. The non-linearity is necessary so that the hidden layers of the neural networks are useful [3]. The principal role of this function is to encoder the non-linearity of the problem on the scale of the neurons;
- the output resulting from the transformation allows to supply the neurons connected downstream.

Mathematically, a formal neuron is an algebraic operator which summed the inputs, called potential. Its output (Equation 1) is a function of this potential [5]:

$$y = f(v) = f\left(\theta_0 + \sum_{i=1}^{n} \theta_i x_i\right) \qquad \text{Equation 1}$$

where {x} are the n neurons inputs, v represents the potential and y is the ANN output. The parameters $\{\theta_i\}$ are called weights or synaptic coefficients, θ_0 is the biais and f is the activation function.

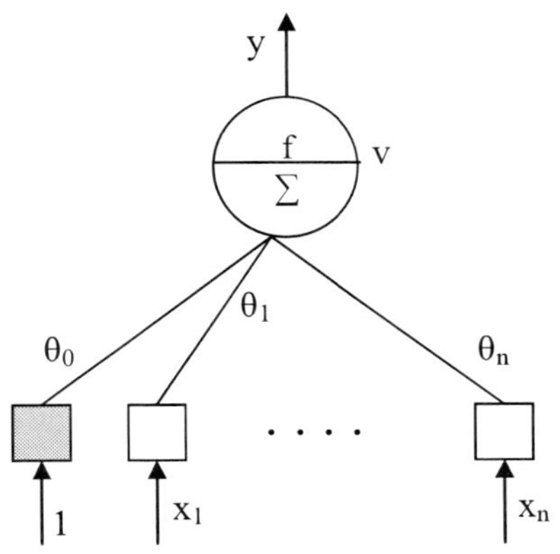

Figure 3. Formal neuron.

2.3. Interconnection Structure

The connections between the neurons which compose the neural networks describe the model topology. It can be unspecified, but generally it is possible to distinguish regularity [13]:

- multi-layer networks : the neurons are arranged by layer. There is no connection between the neurons of the same layer and the connections are only made with the neurons of the downstream layers. Usually, each neuron of a layer is connected to all neurons of the following layer and to this only. This permits to introduce the concept of "direction course" of the information (activation) within the neural networks and thus, to define the concepts of inputs neurons and outputs neurons;
- local connections: it is a multi-layer structure which preserving a singular topology. Each neuron maintains relationships to reduced and localized number of neurons to upstream layer;
- recurrent connections: in the kind of this neural networks, connections bring back information compared to the propagation direction defined in a multi-layer neural networks. These connections are generally locals;
- complete connections: the structure of interconnection is most general. Each neuron is connected to all neurons of the neural networks and with itself.

3. Artificial Neural Networks Principle

A multilayer perceptron (MLP) consists of three layers: an input layer, one or more hidden layers and an output layer [8]. Each layer is composed of a predefined number of neurons. The neurons in the input layer only act as buffers for distributing the input signals to neurons in the hidden layer. Each neuron in the hidden layer sums up its input signals after weighing them with the strengths of the respective connections from the input layer, and computes its output.

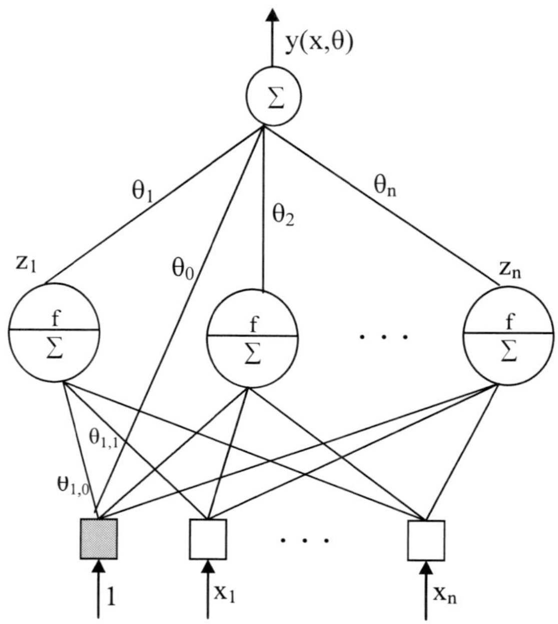

Figure 4. Artificial neural networks generic structure.

A coding mechanism which is external with the neural networks transforms the data into a sequence of inputs vectors [3, 14]. In the same way, a decoding mechanism recovers the output vectors and interprets to produce usable values [3].

The criteria of convergence in training is based on minimizing the error level, where a satisfactory agreement is found with the training set results of the network result. According to Figure 4, the output is given by:

$$y(x,\theta) = \theta_0 + \sum_{1}^{n} \theta_i z_i \qquad \text{Equation 2}$$

$$y(x,\theta) = \theta_0 + \sum_{i=1}^{n} \theta_i f\left(\theta_{i0} + \sum_{j=1}^{n} \theta_{ij} x_j \right) \qquad \text{Equation 3}$$

where $\{x_j\}$ are the n neurons inputs, $\{z_i\}$ represent the outputs of the hidden layer, y is the ANN output, $\{\theta_i\}$ are weights between hidden layer and output layer, $\{\theta_{ij}\}$ represent weights between input layer and hidden layer, θ_0 is the biais and f is the activation function.

However, it can retain that the upstream neurons have an influence which is related to their own activation and the weight (θ) connecting them to the current neuron. The operating mode of the neurons imposes the adjustment of the parameters to optimize the neural networks structure. These parameters can be assembled in several categories [15] like the connection mode, the training algorithm kind, the layers definition.

3.1. Corpus Constitution

The corpus of training is proportional to the number of neural networks weights. An important number of weights generally require long training on larger corpus [3]. When the neural networks have many neurons, the time is too large during calculation.

For technical reasons and especially economic, it is not always possible to create a database comprising the necessary case numbers. The database was extended to characterize the system flexibility around a point into the space of the parameters [4]. The most known technique to increase artificially the data number of the training corpus consists to create artificial data by adding a random noise to the real data. The noise corresponds to the variability of the results compared to the average measured.

To ensure that each input variable provides an equal contribution in the ANN, the input and output data were normalized to a common interval of [0, 1] [7, 14], according Equation 4:

$$\hat{x} = \frac{\widetilde{x} - \overline{x}}{\sigma_x} \qquad \text{Equation 4}$$

where $\hat{x}$ is the formatted expression, $\widetilde{x}$ the experimental value, $\overline{x}$ the average value of the database and σ_x represent the standard deviation.

Database was divided by random selection in categories, Figure 5.

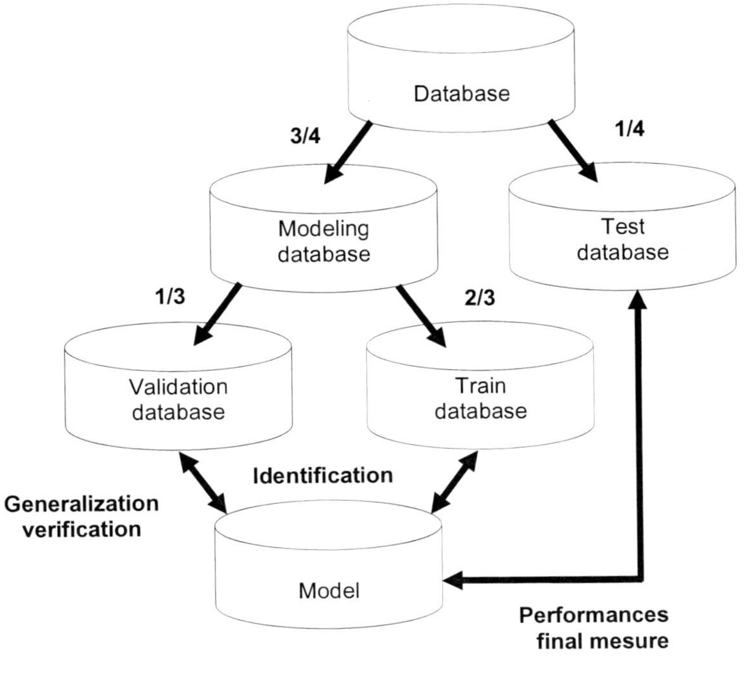

Figure 5. Database constitution.

3.2. Neural Networks Architecture

Two important factors must be considered in order to ensure a successful application of MLP. These factors concerned the number of hidden layers and the number of neurons. Determining the optimal number of hidden layers is a very crucial step in designing the MLP. However, such a determination is not an easy task since it depends largely on a designer's experience. The choice of the hidden layer size is a dependent problem and do influences the quality of the training pattern.

ANN architecture optimization approach can be based on network pruning, where from large network architecture with an adequate performance, the network is pruned by removing non-useful connections during the training. One approach to pruning is based on information from Hessian matrix of the error surface, and it is used here with the utilization of the so-called optimal brain surgeon (OBS) [17].

The basic idea of this method is to start a neural network with excessive numbers of neurons inside the hidden layer during the training step and to cut all weights that have slight influence on error [18-19]. The OBS strategy considers that if a network has the smallest training error, then it is an optimal network. The pruning technique indeed reduces the complexity of neural networks, and improves their ability of prediction because it avoids the use of over parameterized models [20]. The implementation of the optimization process is to retrain each of the intermediate networks and to calculate the training error each time when a weight has been eliminated, then select the network with the smallest training error as the final network [21-22].

Therefore, the ANN is retrained fitting the training errors by a quadratic function, to warrant a possible minimum [23-24]. It is difficult to decide how often the network should be retrained during the pruning session. The suggestion consists in retraining each time 5% of the weights have been eliminated [23].

The most used transfer function is the sigmoid function. This function is characterized by the fact that its slope must approach zero as the input gets large. This causes a problem when using steepest descent to train a MLP network with sigmoid function, since the gradient can have a very small magnitude; and therefore, cause small changes in the weights and biases, even though the weights and biases are far from their optimal values.

3.3. Training Procedure

Data are encoded in the vector shape called input pattern, which is communicated to the inputs neurons. The response of the neural networks is interpreted from the activation value of output neuron, whose vector is called output pattern. The training consists of four steps:

- weights initialization: the weights found by the ANN at the end of the training depend partly on the initial weights values. The current practice is to initialize the weights at random values at the beginning of training;
- input data presentation and activation propagation;
- error calculation (standard deviation between ANN output and real output) : for each neuron, an error value is calculated from its activation and the activation of the

neurons which are connected to it. Within the supervised training, the error takes also account of the difference between activation of the outputs neurons and the reference pattern;

- correction vector calculation : from the error values, the correction to the weights connections and the thresholds of the neurons are determined. The effective correction of the weights can be done after each pattern presentation. The pattern number to be presented at the ANN before correction is called batch size (update).

Multiple problems can occur during training, such as:

- local minima: the training by gradient descent can lead to a solution under optimal. Really, this type of problem is particularly difficult to prevent because the form of the error surface is generally unknown;
- overfitting: when the training is prolonged, the weights choice reflects excessively the corpus characteristics to the detriment of the real task. This kind of problem can be avoided by using cross validation [27];
- if the pattern number is not enough large, the error cannot correctly estimate. Thus, they cannot trust the value of the cost function after training. In this case, better estimators such as the average quadratic error on data which are independent to the training data and the average quadratic errors obtained by cross validation are used. These estimators are better than the cost at end of training because they take into account the errors of the model on examples which did not used to estimate the parameters. They descript better the aptitude for model generalization. When the best model was selected, it is essential to calculate the estimation by interval, i.e. to construct a confidence interval for the regression E(Yp), which is the expectation of the process output [5].

3.3.1. Neural network parameters initialization

The potential v_i=0 of each hidden neuron i defines a hyper plan in the inputs space:

$$v_i = \theta_{i0} + \sum_{j=1}^{N} \theta_{ij} x_j = 0 \qquad \text{Equation 5}$$

If the training parameters {θij} are initialized to zero, the retropropagated values of hidden neurons are zero at the first iteration. It results from this that the parameters of the hidden neurons will never be modified. Thus, to start the algorithm, it is necessary that these parameters are initially different to zero. In addition, to start good training in good conditions, a practical and effective solution consists to initialize at zero the parameters of the hidden neurons relative to the constant inputs (biais). Thus, the initial hyper plans pass by the origin. Another precaution must be taken for the initialization of the parameters values: it is necessary that they are not too large, thus, all potentials of the hidden neurons $\{v_i\}$ are in the quasi linear part of the transfer function.

3.3.2. Neuronal parameters estimation

Figure 4 shows architecture of neural networks models. It is constituted with n inputs and q parameters. This network defines parameterized functions noted ξ= {g(**x**, θ), **x** pertain to R^n, θ pertaining to R^q, $\{\mathbf{x}^k, y^k_p\}_{k=1,N}$ the training database with N examples.

To obtain the parameters θ values, a function of the quadratic cost is minimized.

$$J(\theta)=\frac{1}{2}\sum_{k=1}^{N}\left(y_p^k-g\left(\mathbf{x}^k,\theta\right)\right)^2 \quad \text{Equation 6}$$

The output being nonlinear compared to the parameters, the minimum of the cost function search requires to use an iterative algorithm of minimization. But these algorithms do not guarantee an absolute minimum, thus it is necessary to carry out several minimizations for several values of the initial parameter vector. In practice, several networks models are training and their performance is used in order to select the best [5]. The performance of ANN is generally defined by the average quadratic error (i.e. expectation of quadratic error).

The total cost states:

$$J(\theta)=\sum_{k=1}^{N}J^k(\theta)=\frac{1}{2}\sum_{k=1}^{N}\left(y_p^k-g\left(\mathbf{x}^k,\theta\right)\right)^2 \quad \text{Equation 7}$$

where $J^k(\theta)$ represent partial cost relative to example k, y_p^k is the data measured on the process (experimental data), $g\left(\mathbf{x}^k,\theta\right)$ function calculated by ANN, $\mathbf{x}^\mathbf{k}$ parameter x value to example k.

The J minimization cannot be carried out directly with a matrix calculation because the model output g(x^k, θ) is not linear. Moreover, J (θ) is not quadratic; it may be that there are several minima. To find θ value which corresponds to a minimum J(θ), iterative or recursive algorithm acting by successive θ modifications are used.

The neural networks parameters estimation always comprises the following phases [5]:

- the cost function definition;
- the training algorithm choice;
- parameters initialization.

At the presentation of one pattern example for recursive algorithm or all patterns examples for iterative algorithm, following required modules are necessary [5]:

- cost calculation;
- cost gradient calculation;
- descent direction calculation;
- calculation of the step progression in the descent direction;
- parameters modification.

To define these parameters, ANN constitutes by hidden neurons, potential and derivable activation function is considered. The only adjustable parameters are the weights of connection: they constitute the vector θ.

For the iterative algorithms, the descent direction is a linear transformation of the total cost gradient. It is given by:

$$\theta(iter) = \theta(iter-1) + \mu(iter)d(iter)$$ Equation 8

$$\theta(iter) = \theta(iter-1) + \mu(iter)m(iter)\frac{\partial J}{\partial \theta}\Big|_{\theta(iter-1)}$$ Equation 9

where $\mu(iter) > 0$, $m(iter)$ positive square matrix, $d(iter)$ descent direction calculated at iteration "iter" with the value $\theta(iter-1)$ of θ obtained in the preceding iteration and $\mu(iter)$ is the step progression.

The total cost (Equation 7) is the sum of N partials costs. The gradient of the total cost at iteration "iter" is obtained by calculating successively the gradients of the N partial costs and by summing them:

$$\frac{\partial J}{\partial \theta}\Big|_{\theta(iter-1)} = \sum_{k=1}^{N}\frac{\partial J^k}{\partial \theta}\Big|_{\theta(iter-1)}$$ Equation 10

Recursive algorithm modifies the parameters values when each example is presented. It uses the gradient of partial cost $\frac{\partial J^k}{\partial \theta}\Big|_{\theta(k-1)}$ relative to the example k which calculated with the available parameters θ(k-1).

That, the algorithm is iterative (founded on total cost) or recursive (founded on partial cost) it is necessary to calculate the gradient of the partial cost.

4. Algoritms of Minimization

In the case of total cost minimization, it is necessary to define rules which stopped the algorithm implemented. These rules relate to the magnitudes which, when they are very low, translate the fact that the algorithm reached an absolute or relative minimum, such as:

- euclidian norm of the total cost gradient;
- euclidian norm of the total cost gradient variation between two iterations;
- euclidian norm of the vector parameters variation between two iterations.

Moreover, it is necessary to define a stop criteria based on maximum iterations number if the preceding rules don't stop the training.

4.1. Algorithms Implementation Elements

4.1.1. Gradient vector of the cost

When the neural networks has one output, the gradient of the quadratic cost J(θ) evaluated at the point $\theta=\theta_0$ is a vector of q dimension noted $\frac{\partial J}{\partial \theta}\Big|_{\theta_0}$

The expression of the $j^{ième}$ component of the gradient is [5]:

$$\frac{\partial J}{\partial \theta}\Big|_{\theta_0} = -\sum_{k=1}^{N}\left(y_p^k - y^k(\theta_0)\right)\frac{\partial y^k}{\partial \theta_j}\Big|_{\theta_0} = -\sum_{k=1}^{N} e^k(\theta_0)\frac{\partial y^k}{\partial \theta_j}\Big|_{\theta_0} \qquad \text{Equation 11}$$

Thus, there is:

$$\frac{\partial J}{\partial \theta}\Big|_{\theta_0} = -\begin{bmatrix} \frac{\partial y^1}{\partial \theta_1}e^1 + \ldots + \frac{\partial y^N}{\partial \theta_1}e^N \\ . \\ . \\ . \\ \frac{\partial y^1}{\partial \theta_q}e^1 + \ldots + \frac{\partial y^N}{\partial \theta_q}e^N \end{bmatrix} \qquad \text{Equation 12}$$

The network Jacobien at point θ_0 is defined by:

$$Z(\theta_0) = \frac{\partial y}{\partial \theta^T}\Big|_{\theta_0} = -\frac{\partial e}{\partial \theta^T}\Big|_{\theta_0} = \begin{bmatrix} \frac{\partial y^1}{\partial \theta_1}e^1 + \ldots + \frac{\partial y^N}{\partial \theta_1}e^N \\ . \\ . \\ . \\ \frac{\partial y^1}{\partial \theta_q}e^1 + \ldots + \frac{\partial y^N}{\partial \theta_q}e^N \end{bmatrix}_{\theta=\theta_0} \qquad \text{Equation 13}$$

The N*q elements of the jacobien matrix $Z=Z(\theta_0)$ are calculated by evaluated the gradient in the direct direction.

Generally, while omitting to indicate the evaluation point, they obtain:

$$\frac{\partial J}{\partial \theta} = \sum_{k=1}^{N}\frac{\partial J^k}{\partial \theta} = \sum_{k=1}^{N} e^k\frac{\partial e^k}{\partial \theta} = -\sum_{k=1}^{N} e^k\frac{\partial y^k}{\partial \theta} = -Z^k e \qquad \text{Equation 14}$$

When the network has N_S outputs $(y_1, y_2, \ldots, y_{ns})$, the cost is:

$$J = \sum_{k=1}^{N} J^k = \sum_{k=1}^{N}\sum_{r=1}^{n_s} J_r^k$$ Equation 15

The component (ij) of the gradient is:

$$\frac{\partial J}{\partial \theta_{ij}} = \sum_{k=1}^{N} \frac{\partial J^k}{\partial \theta_{ij}} = \sum_{k=1}^{N}\sum_{r=1}^{n_s} \frac{\partial J_r^k}{\partial \theta_{ij}}$$ Equation 16

Thus, it results that:

$$\frac{\partial J}{\partial} = \sum_{r=1}^{n_s} Z_r^T e_r$$ Equation 17

Where $$Z_r = \frac{\partial \mathbf{y}_r}{\partial \theta^T}$$ Equation 18

4.1.2. Cost hessian

The second derivate matrix of the scalar J is called Hessian or hessian matrix. It is a symmetrical square matrix of q order. The Hessian evaluated at $\theta=\theta_0$ is:

$$H(\theta_0) = \frac{\partial^2 J}{\partial \theta \partial \theta^T}\Big|_{\theta_0}$$ Equation 19

In a general, the element (ij) of the Hessian is:

$$[H]_{ij} = \frac{\partial^2 J}{\partial \theta_i \partial \theta_j} = \sum_{k=1}^{N}\left(\frac{\partial e^k}{\partial \theta_i}\frac{\partial e^k}{\partial \theta_j} + \frac{\partial^2 e^k}{\partial \theta_i \partial \theta_j} e^k \right)$$ for i, j=1, q Equation 20

When the model is linear, all partial derivates $\partial^2 e^k / \partial \theta_i \partial \theta_j$ are zero and:

$$[H]_{ij} = \sum_{k=1}^{N} \frac{\partial^2 e^k}{\partial \theta_i \partial \theta_j} = \sum_{k=1}^{N} x_i^k x_j^k$$ Equation 21

Thus: $H = X^T X$ Equation 22

The second term of Equation 20 is neglect because it is small if surface is near to quadratic. When ANN has one output, the Hessian approximate expression is:

$$\widetilde{H} = Z^T Z \qquad \text{Equation 23}$$

For n_s output, there is

$$J = \sum_{k=1}^{N} J^k = \sum_{k=1}^{N}\sum_{r=1}^{n_s} J_r^k \qquad \text{Equation 24}$$

thus:

$$\frac{\partial^2 J}{\partial\theta_i \partial\theta_j} = \sum_{k=1}^{N} \frac{\partial^2 J^k}{\partial\theta_i \partial\theta_j} = \sum_{k=1}^{N}\sum_{r=1}^{n_s} \frac{\partial^2 J_r^k}{\partial\theta_i \partial\theta_j} \qquad \text{Equation 25}$$

and finally:

$$H = \sum_{r=1}^{n_s} H_r \qquad \text{Equation 26}$$

for the output r, if $\partial^2 J_r^k / \partial\theta_i \partial\theta_j$ is approximates by $\left(\partial^2 e_r^k / \partial\theta_i\right)\left(\partial^2 e_r^k / \partial\theta_j\right)$, the relation becomes:

$$\widetilde{H} = \sum_{r=1}^{n_s} \widetilde{H}_r = \sum_{r=1}^{n_s} Z_r^T Z_r \qquad \text{Equation 27}$$

4.1.3. Jacobien matrix

The Hessian approximate expression requires to calculate the n_s outputs of matrix Zr. It is thus convenient to calculate line by line, a line is relative to an example. That can be carried out with retropropagation algorithm, in the direct direction.

4.1.3.1. Error retropropagation

The partial cost J_r^k relative to the output r and the example k is:

$$J_r^K(\theta) = \frac{1}{2}\left(e_r^k(\theta)\right)^2 = \frac{1}{2}\left(y_{pr}^k - g_r\left(\mathbf{x}^k, \theta\right)\right)^2 = \frac{1}{2}\left(y_{pr}^k - y_r^k(\theta)\right)^2 \qquad \text{Equation 28}$$

The component of the partial cost gradient J_r^k relative to the parameter θ_{ij} of the connection which goes from the unit j (network input or neuron output) towards neuron i is:

$$\frac{\partial J_r^k}{\partial \theta_{ij}} = \frac{\partial J_r^k}{\partial e_r^k}\frac{\partial e_r^k}{\partial \theta_{ij}} = e_r^k \frac{\partial e_r^k}{\partial \theta_{ij}}$$ Equation 29

The retropropagation algorithm in its usual form gives:

$$\frac{\partial J_r^k}{\partial \theta_{ij}} = \frac{\partial J_r^k}{\partial v_i}\frac{\partial v_i}{\partial \theta_{ij}} = -q_i^k x_j^k$$ Equation 30

The element (k,ij) of the matrix Zr is:

$$[Z_r]_{k,ij} = -\frac{\partial e_r^k}{\partial \theta_{ij}} = \frac{\partial y_r^k}{\partial \theta_{ij}} = \frac{q_i^k x_j^k}{e_r^k}$$ Equation 31

For the example k, it is necessary to carry out n_s retropropagation to obtain the lines of the n_s matrix Zr by dividing the components of the gradient by the corresponding error.

4.1.3.2. Direct direction

Always it is mean to calculate the partials derivates:

$$\frac{\partial y_1^k}{\partial \theta_{ij}} \cdots \frac{\partial y_r^k}{\partial \theta_{ij}} \cdots \frac{\partial y_{n_s}^k}{\partial \theta_{ij}}$$ Equation 32

for the q parameters θ_{ij} of the ANN.

4.2. Iterative Algorithms

The iterative algorithms consist to modify the parameters at each iteration in the descent direction d(i) which is a linear transformation of the gradient at the current point:

$$\theta(i) = \theta(i-1) + \mu(i)d(i) = \theta(i-1) - \mu(i)M(i)\frac{\partial J}{\partial \theta}\Big|_{\theta(i-1)}$$ Equation 33

where M(i) is positive, the scalar $\mu(i) > 0$ is the progression step in the descent direction.

4.2.1. Gradient algorithm

The algorithm of the gradient consists to modify the parameters θ at each iteration according to the direction of the gradient and in the opposite sense:

$$\theta(i) = \theta(i-1) - \mu(i)\frac{\partial J}{\partial \theta}\Big|_{\theta(i-1)} \text{ where } \mu(i) > 0 \qquad \text{Equation 34}$$

The direction of descent is thus the opposite of the gradient direction: it is the direction according which the cost function decreases quickly. μ is the step progression in the descent direction.

If the ANN has one output, the algorithm is:

$$\theta(i) = \theta(i-1) + \mu(i)Z(i)^T e(i) \qquad \text{Equation 35}$$

where Z(i) is the output matrix of the partial derivates calculated at iteration i with the parameters θ(i-1).

If the ANN has n_s outputs, by using the relation of the gradient, the algorithm becomes:

$$\theta(i) = \theta(i-1) + \mu(i)\sum_{r=1}^{n_s} Z_r(i)^T e_r(i) \qquad \text{Equation 36}$$

$$\text{where } Z_r(i) = \frac{\partial \mathbf{y}_r}{\partial \theta^T}\Big|_{\theta(i-1)} \qquad \text{Equation 37}$$

The algorithm of the gradient is very effective far a minimum when the gradient has a great value. It is a simple algorithm which shows the following characteristics:

- it is generally effective far from a minimum of the cost;
- in near minimum, the gradient tends towards zero, therefore the parameters evolution becomes very slow.

4.2.2. Newton algorithm

The algorithm of Newton consists to modify the parameters at each iteration by:

$$\theta(i) = \theta(i-1) + d(i) = \theta(i-1) - \left(H\big|_{\theta(i-1)}\right)^{-1} \frac{\partial J}{\partial \theta}\Big|_{\theta(i-1)} \qquad \text{Equation 38}$$

This can write:

$$\theta(i) = \theta(i-1) + H(i)^{-1} Z(i)^T e(i) \qquad \text{Equation 39}$$

In practice, at iteration i, the Hessien square matrix H(i) are calculated with the parameters θ(i-1) and θ(i) is expressed as the solution of a linear system. If the ANN has only one output, the system of unknown factors θ(i) is:

$$H(i)\theta(i) = H(i)\theta(i-1) + Z(i)^T e(i) \qquad \text{Equation 40}$$

The direction d(i) is a descent direction if the hesien matrix is positive.

The value of θ(i) calculated by the algorithm of Newton corresponds to the minimum of the osculatory paraboloid P(θ) at the surface J(θ) to the point θ(i-1). The osculatory paraboloid has for equation:

$$P(\theta) = a + \left(\frac{\partial J}{\partial \theta}\Big|_{\theta(i-1)} - H\Big|_{\theta(i-1)}\theta(i-1) \right)^T \theta + \frac{1}{2}\theta^T H\Big|_{\theta(i-1)}\theta \quad \text{Equation 41}$$

where a is a scalar independent of θ.

If H is positive, the vector θ_{min} corresponding to the minimum of the paraboloid P(θ) is obtained for:

$$\frac{\partial P}{\partial \theta} = \frac{\partial J}{\partial \theta} - H\theta(i-1) + H\theta_{min} = 0 \quad \text{Equation 42}$$

$$\text{thus: } \theta_{min} = \theta(i-1) - H^{-1}\frac{\partial J}{\partial \theta} \quad \text{Equation 43}$$

4.2.3. Levenberg-marquardt algorithm

Levenberg [28] and Marquardt [29] studies resulted in adding matrix Z^TZ to the algorithm of Newton in order to reduce calculations. The Levenberg-Marquardt algorithm is deduced by:

$$\theta(i) = \theta(i-1) + \left(Z(i)^T Z(i) + \lambda(i)I_q\right)^{-1} Z(i)^T e(i) \quad \text{Equation 44}$$

Thus, ∀ λ>0, the matrix of the linear system to solve at each iteration is conditioned better than Z^TZ.

The increment $\delta(i) = \theta(i) - \theta(i-1)$ of this algorithm is the solution of least squares relating to the equation:

$$A\delta = b \quad \text{Equation 45}$$

where: $A = \begin{bmatrix} -Z \\ \sqrt{\lambda}I_q \end{bmatrix}$ and $b = \begin{bmatrix} -e \\ 0 \end{bmatrix}$

Thus the normal equation is:

$$A^T A\delta = A^T b \quad \text{Equation 46}$$

At each iteration, the solution of the normal equation must thus calculate.

4.2.4. Hesian approximation calculation

If the ANN has only one output, Hessien H of the cost is approximated by Z^TZ. When the example k is presented, the kth line of Z noted $(\mathbf{z}^k)^T$ is calculated and the external product $z^k(z^k)^T$ is added to the sum of the contributions which are already calculated:

$$\sum_{j=1}^{k-1} \mathbf{z}^j \left(\mathbf{z}^j\right)^T \qquad \text{Equation 47}$$

Thus, it is not necessary to store the matrix Z(N,q), but a square matrix of q degree.

If the neural networks offer n_s outputs, the Hessien is approximated by:

$$\sum_{r=1}^{n_s} (Z_r)^T Z_r \qquad \text{Equation 48}$$

When the example k is presented, the kth line of n_s matrices Zr is calculated and the external product is carries out:

$$\sum_{r=1}^{n_s} \mathbf{z}_r^k \left(\mathbf{z}_r^k\right)^T \qquad \text{Equation 49}$$

which is added with the matrix:

$$\sum_{j=1}^{k-1} \sum_{r=1}^{n_s} \mathbf{z}_r^j \left(\mathbf{z}_r^j\right)^T \qquad \text{Equation 50}$$

4.3. Recursive Algorithms

The iterative algorithms considering previously minimize the total cost. The recursive algorithms carry out the parameters modifications of the partial cost gradient relative to the presented example according:

$$\theta(k) = \theta(k-1) - \mu(k) m(k) \frac{\partial J^k}{\partial \theta}\Big|_{\theta(k-1)} \qquad \text{Equation 51}$$

where θ(k) is the value of the vector parameters after the presentation of the example k and the corresponding modification.

A recursive algorithm carries out the modifications on the N examples than it is necessary so that a criterion of bearing stop on the total cost is satisfied [30].

The iterative algorithms carry out at each iteration only one modification of the parameters corresponding to the presentation of the complete training database. Recursive algorithms carry out a modification of the parameters after the presentation of each example of training database. The recursive algorithm of the gradient presents the advantage of the simplicity of its implementation. Often, a term of inertia (momentum term) is however added [9, 31]:

$$\Delta\theta(k) = \mu_1 \Delta\theta(k-1) - \mu_2 \frac{\partial J^k}{\partial\theta}\Big|_{\theta(k-1)} \qquad \text{Equation 52}$$

where $0 < \mu_1 < 1$ and $\mu_2 > 0$.

This technique which tends to attenuate possible oscillations of the cost function can increase the convergence velocity if $\mu_1$et μ_2 are judiciously selected. But, this is slow comparing to other algorithms.

4.3.1. Recursive algorithm of gradient (partial cost gradient)

The direction of the descent is the direction of the partial gradient relative to the example k:

$$\theta(k) = \theta(k-1) - \mu(k)\frac{\partial J^k}{\partial\theta}\Big|_{\theta(k-1)} \qquad \text{Equation 53}$$

$$\theta(k) = \theta(k-1) + \mu(k)\frac{\partial J^k}{\partial\theta}\Big|_{\theta(k-1)} e^k \qquad \text{Equation 54}$$

In the case of the linear model, the relation is defined following:

$$\begin{cases} y^k = \left(\mathbf{x}^k\right)^T \theta(k-1) \\ \dfrac{\partial J^k}{\partial\theta}\Big|_{\theta(k-1)} = \mathbf{x}^k \end{cases} \qquad \text{Equation 55}$$

Thus the algorithm is written by:

$$\theta(k) = \theta(k-1) + \mu(k) e^k \mathbf{x}^k \qquad \text{Equation 56}$$

4.3.2. Gauss-newton algorithms

The disadvantage of the Newton algorithm is to require at each iteration the calculation of Hessien H, matrix of degree q and the resolution of a linear system of dimension q.

A quasi-Newton algorithm consists to replace the reverse of Hessien H by the estimation M(i) put at each iteration.

5. Test Procedure

After the networks training, the model is tested by the use of the rest data sets to verify whether the network acquires source dynamics. If the network output for each test pattern is almost close to the respective target, the network is considered to have acquired the underlying dynamics from the training patterns.

The test procedure permits to validate the results of the training process [32]. The test error is an important quantity since it can be viewed as an estimate of the generalization error. This should not be too large compared to training error, in which case one must suspect that the network is over-fitting the training data [33].

Optimization of a neural network is an important task of neural network-based studies and there are some methods applied. The performance of the neural network model was evaluated with the root mean square error (RMSE) and the mean absolute percentage error (MAPE) (Equation (57 and Equation 58).

$$RMSE = \sqrt{\frac{1}{T}\sum_{k=1}^{T}\left(y_{net}(k) - y(k)\right)^2} \qquad \text{Equation 57}$$

When the RMSE is at the minimum, a model can be judged as very good [34].

$$MAPE = \left[\frac{1}{T}\sum_{k=1}^{T}\frac{\left|y_{net}(k) - y(k)\right|}{y(k)}\right] * 100\% \qquad \text{Equation 58}$$

where y_{net} represents ANN output, y is the experimental data and T is the maximum number of trading days in the corresponding (training or test) period.

RMSE and MAPE are one of the most widely used measurements. They represent the fit between the neural network predictions and the actual targets. If the network outputs for each test pattern are relatively close to the respective targets, RMSE and MAPE will have small values. The network is considered to have acquired the underlying dynamics from the training patterns.

6. Validation Procedure

The path from validation to training is due to the criterion having local minima. The main role of this step is to estimate the average generalization error by final prediction error (FPE) [35].

The main potential error sources can be [35-36]:

- over fitting error: it appears when the training adjustment is excessive. There are unnecessary degrees of freedom, meaning that more hidden neurons are used than required. An ANN with this feature has lost its generalization ability [37-38].

- bias error: it is the opposite parameter. The degrees of freedom are reduced, and hence, the generalization ability is considered to have been excessively favored. The validation of the ANN makes a considerable error.

7. Conclusion

The application of the neural networks is summarized in a problem of optimization. ANN has a fundamental property which distinguishes it from the traditional techniques of processing data. To obtain a precise nonlinear model, an ANN needs less adjustable parameters than the traditional methods of regression (for example polynomial regression). Consequently, an ANN requires less data than a method of regression traditional to carry out a nonlinear model of equivalent precision; an ANN permits to elaborate from the same data base, a more precise nonlinear model.

References

[1] Pelissier, A. & Tête, A. (1995). *Sciences cognitives : textes fondateurs* (1943-1950), Paris, Presse Universitaire Française.

[2] Mc Culloch, W. C. & Pitts, W. H. (1943). A logical calculus ideas imanent in nervous activity, *Bulletin of Mathematical Biophysics*, *vol. 5*, 115-133.

[3] Jodouin, J. F. (1994). *Les réseaux neuromimétiques*, Paris, Edition Hermès.

[4] Nelson, M. M. & Illingworth, W. T. (1991). *A practical guide to neural nets*, New York, Addison-Wesley, 3rd edition.

[5] Personnaz, L. & Rivals, I. (2003). *Réseaux de neurones formels pour la modélisation, la commande et la classificatio*n, Paris, Collection sciences et techniques de l'ingénieur, CNRS Edition.

[6] Sarzeaud, O. (1995). *Les réseaux de neurones : Contribution à une théorie,* Ouest Editions.

[7] Pham, D. T. & Xing, L. (1995). *Neural networks for identification, prediction and control,* London, 2nd printing, Springer.

[8] Zhan, J. & Li, F. (1994). *A New Approch in multilayer perceptron-forward propagation*, Proceeding of IEEE International Conference on Neural Networks, Piscataway, New Jerse, USA, 88-93.

[9] Arisawa, M. & Watada, J. (1994). Enhanced back-propagation learning and its application to business evaluation, Proceeding of IEEE International Conference on Neural Networks, Piscataway, *New Jerse,* USA, 155-160.

[10] Tetko, I. V., Tanchuk, V. Y. & Luik, A. I. (1994). Simple heuristic methods for input parameters' estimation in neural networks, Proceeding of IEEE International *Conference on Neural Networks,* Piscataway, New Jerse, USA, pp. 386-390.

[11] Werbos, P. J. (1988). Generalization of back propagation with application to recurrent gas market model, *Neural Netw.*, *vol. 1* (No. 4), 339-356.

[12] Holmstrom, L. & Koistinen, P. (1992). Using additive noise in back-propagation training, IEEE Trans. *Neural Netw.*, *vol. 3*, 24-38.

[13] Brightwell, G., Kenyon, C. & Paugam-Moisy, H. (1997). Multilayer neural network: one or two hidden layers, NeuroCOLT Technical Report Series, NC-TR-97-001, University of London.

[14] Guessasma, S., Montavon, G., Gougeon, P. & Coddet, C. (2003). Designing expert system using neural computation in view of the control of plasma spray processes, *Mater. Des., vol. 24,* (No. 7), 497-502.

[15] Guessasma, S., Montavon, G. & Coddet, C. (2004). Neural computation to predict in flight particle characteristic dependences from processing parameters in the APS process, *J. Therm. Spray Technol., vol. 13* (No. 4), 570-585.

[16] Jim, K. C., Giles, C. L. & Horne, B. G. (1996). An analysis of noise in recurrent neural networks: convergence and generalization, *IEEE Trans. Neural Netw., vol. 7,* 1424-1438.

[17] Sozen, A. & Arcaklioglu, E. (2005). Solar potential in Turkey, *Applied Energ., vol.* 80, 35-45.

[18] Kasabov, N. K. (1998). *Foundations of neural networks, fuzzy systems and knowledge engineering,* MIT Press, Cambridge.

[19] Xia, P. Q. (2003). An inverse model of MR damper using optimal neural network and system identification, *J. Sound Vibr., vol. 266,* (No.5), pp. 1009-1023.

[20] Hassibi, B. & Stork, D. G. (1993). Second order derivatives for network pruning: optimal brain surgeon, *Advances in Neural Information Processing Systems,* Morgan Kaufmann, San Mateo, CA, *vol. 5,* 164-171.

[21] Hansen, L. K. & Larsen, J. (1996). Linear unlearning for cross-validation, *Adv. Comput. Math., vol. 5,* 269-280.

[22] Zhu, Y., Lu, Y. & Li, Q. (2006). MW-OBS: An improved pruning method for topology design of neural networks. *Tsinghua Sci. Technol., vol. 11,* (No. 3), pp. 307-312.

[23] Heitz, J., Krogh, A. & Palmer, R. G. (1991). Introduction to the theory of neural computation, *Addison-Wesley,* New York.

[24] Haykin, S. (1994). *Neural networks, A comprehensive foundation,* Macmillan College Publishing, New York.

[25] umelhart, D. E., Hinton, R. G. E. & Williams, R. J. (1986). Learning representations by back-propagating errors, *Nature, vol. 323,* 533-536.

[26] Karayiannis, N. B. (1994). Accelerating the training of feed-forward neural networks using generalized Hebbian rules for initializing the internal representations, *Proceeding of IEEE International Conference on Neural Networks, Piscataway,* New Jerse, USA, 32-37).

[27] Pados, D. A. & Papantoni-Kazakos, P. (1994). A note on estimation of the generalization error and the prevention of overfitting, *Proceeding of IEEE International Conference on Neural Networks,* Piscataway, New Jerse, USA, 321-326.

[28] Levenberg, K. (1994). A method for the solution of certain non-linear problems in least squares, *Q. J. Appl. Math., Vol. 2,* (No. 2), 164-168.

[29] Marquardt, D. W. (1963). An algorithm for least squares estimation of non-linear parameters. *J Soc. Ind. Appl. Math., vol. 11,* (No. 2), 431-441.

[30] Piché, S. W. (1944). The second derivative of reccurent Network, *Proceeding of IEEE International Conference on Neural Networks, Piscataway,* New Jerse, USA, 245-250.

[31] Bishop, M. (1995). *Neural networks for pattern recognition,* Clarendon Press, Oxford.

[32] Reich, Y. & Barai, S. V. (1999). Evaluating machine learning models for engineering problems, *Artif Intell Eng*., *vol*. 13, (No. 3), 257-272.

[33] Reich, Y. & Barai, S. V. (2000). A methodology for building neural networks models from empirical engineering data, *Eng Appl Artif Intell*., *vol*. 13, (No. 6), 685-694.

[34] Pados, D. A. & Papantoni-Kazakos, P. (1994). A note on estimation of the generalization error and the prevention of overfitting, *Proceeding of IEEE International Conference on Neural Networks*, Piscataway, New Jerse, USA, 321-326.

[35] Guessasma, S., Trifa, F. I., Montavon, G. & Coddet, C. (2004). Al_2O_3–13% weight TiO_2 deposit profiles as a function of the atmospheric plasma spraying processing parameters. *Mater. Des., vol. 25*, (No. 4), 307-315.

[36] Gareta, R., Romeo, L. M. & Gil, A. (2006). Forecasting of electricity prices with neural networks, *Energy Conv. Manag*., *vol. 47*, (No. 13-14), 1770-1778.

[37] Reich, Y. & Barai, S. V. (2000). A methodology for building neural networks models from empirical engineering data, *Eng Appl. Artif. Intell*., *vol*. 13, (No. 6), 685-694.

[38] Bishop, C. M. (1994). Neural networks and their applications, *Rev Sci Instrum*, *vol. 65*, 1803-1832.

INDEX

2

20th century, 91, 94

A

accelerometers, 118
accessibility, 348
accounting, 77, 89
accurate models, 349
acid, 5, 10, 20, 25, 343, 344, 345, 346, 347, 351
active site, 347, 348, 351, 352
activity level, 306
actual output, 40, 214, 308, 329
actuators, 124
adaptability, xii, 57, 277
adaptation, 99, 107, 240, 248
adenosine, 348
adjustment, 249, 309, 331, 380, 394
adverse conditions, 152
aerosols, 27
age, 141
aggregation, 100
air quality, 71
air temperature, xi, 61, 74, 87, 209, 210, 219, 222, 223, 224, 225, 227, 228, 373
air temperature fluctuations, 228
alcohols, 5, 19, 21
aldehydes, 266
aldolase, 348
aluminium, 53, 199
amino, 192, 205, 344, 347
amino acid, 192, 205, 344, 347
amino acids, 192, 347
amplitude, 199, 201, 204, 235, 279, 291
amylase, 346
anatomy, ix, 129, 130
anhydrase, 348
anisotropy, 324
annealing, 47, 53, 344
aqueous solutions, 73
Argentina, viii, 75, 86, 91, 93, 94
arithmetic, 108, 176, 279, 295
arson, 220
arteries, 137
artificial intelligence, xii, 2, 4, 100, 153, 154, 167, 258, 278, 321
Artificial Neural Networks, i, iii, v, vi, xii, xiii, 1, 2, 25, 26, 27, 55, 56, 62, 83, 100, 129, 171, 173, 174, 188, 226, 229, 256, 257, 259, 277, 299, 301, 305, 320, 321, 323, 341, 350, 356, 372, 373, 375, 379
ascorbic acid, 346
Asia, 53
ASS, 110, 111, 122
assessment, 4, 8, 65, 82, 91, 228
atmosphere, 61, 78
ATO, 5
atoms, 195, 348
automata, 194
automate, 377
Automobile, 170
avoidance, 304
axons, 305

B

background noise, 200
bankruptcy, 30, 51, 52, 53
base, 69, 78, 100, 169, 193, 235, 259, 302, 395
behaviors, 349, 377
benchmarking, 284, 285
benchmarks, 285
beneficial effect, 332
benefits, 176, 177, 264
benign, 62, 130, 148
benzene, 264, 265
bias, 56, 102, 105, 173, 174, 175, 214, 218, 279, 283, 324, 325, 344, 346, 395
bioavailability, 5
biocatalysts, xiii, 341, 344

biochemical processes, 64
biodiesel, 64, 71
biodiversity, 60, 62
biological processes, viii, xi, 55, 56, 64, 209, 210
biological sciences, 83
biological systems, xiii, 341, 352
biomass, 68, 71, 88, 351
biomass growth, 351
biosensors, 346
biotechnology, 352
birefringence, xi, 191, 192, 196, 204
bounds, 47, 48, 49, 149
brain, ix, 56, 83, 129, 130, 154, 155, 172, 173, 176, 194, 195, 205, 206, 207, 278, 292, 356, 358, 382, 396
breakdown, 207

C

C++, 23, 340, 373
CAD, ix, 97, 98, 142, 143, 144, 146, 147, 148, 150
cadmium, 268, 269
calculus, 395
calibration, xii, xiii, 71, 220, 257, 259, 266, 323, 335, 336, 338, 339, 356, 357, 361
CAM, ix, 97, 98
cancer, 139, 149, 150
cancer death, 139
candidates, 144, 154
carbon, viii, xi, 22, 75, 77, 80, 81, 82, 83, 85, 89, 94, 169, 191, 192, 195, 207, 344, 345
carbon atoms, 22
carbon nanotubes, xi, 191, 192, 195, 207
carcinoma, 149
cardiologist, 138, 139
case study, 69, 225, 273, 332, 349
castor oil, 5, 10, 13, 14, 16, 17
catastrophic failure, 176, 181
catchments, 60, 226
cell biology, 350
cell body, 172, 174, 305
cell division, 192
cellular communications, 291
central nervous system, 100
changing environment, 154
chaos, 207
chemical, vii, xii, 1, 5, 6, 7, 8, 10, 21, 27, 62, 63, 65, 67, 68, 69, 73, 88, 89, 153, 226, 257, 258, 259, 260, 264, 265, 272, 348
chemical characteristics, 8, 153
chemical industry, 63
chemical properties, vii, 1, 5, 6, 7, 8, 67, 258
chemical reactions, 27
chemicals, 268
chemometric techniques, 266
chemometrics, 64, 258
chest radiography, 139, 147
Chicago, 129, 147
China, 227, 275
chiral center, 352
chirality, 344, 350
chromatography, 259, 265
classes, 57, 105, 106, 130, 175, 182, 192
classification, ix, 9, 23, 53, 77, 87, 93, 106, 129, 130, 131, 133, 144, 182, 188, 273, 281, 282, 309, 344, 347, 356, 395
clavicle, 140
climate, viii, 62, 72, 75, 76, 77, 78, 79, 81, 82, 84, 86, 87, 88, 89, 90, 93, 94
climate change, 62, 72
climates, 93
clustering, 3, 15, 18, 69, 70, 348, 351
clusters, 2, 60, 348
coding, 183, 184, 185, 380
cognition, 212, 340
coherence, 206
color, iv, 68
combined effect, 316
commercial, 14, 268
communication, xi, 191, 204
communication systems, xi, 191
community, 70, 356
compaction, 28, 259, 261
comparative analysis, 73
compensation, 99, 124, 147
competition, 377
complex interactions, 6, 83
complexity, viii, xiii, 3, 4, 22, 31, 55, 56, 61, 83, 85, 106, 153, 183, 184, 216, 242, 254, 260, 282, 295, 300, 324, 331, 341, 349, 357, 382
composites, 168
composition, 4, 6, 7, 11, 19, 21, 23, 24, 62, 71, 81, 140, 141, 142, 269, 347
compounds, 8, 258, 264, 265, 266
compression, 3, 259, 278, 285, 287, 288, 289, 291, 294, 295, 296
computation, xiii, 2, 30, 183, 184, 194, 206, 243, 274, 323, 329, 330, 349, 396
computed tomography, 148
computer, ix, 3, 23, 27, 110, 129, 130, 139, 148, 150, 154, 155, 176, 183, 205, 206, 252, 253, 279
computer simulations, 252
computing, xii, 26, 27, 124, 172, 176, 184, 236, 242, 243, 257, 329, 351, 373, 376
conceptual model, 71, 340, 357, 374
condensation, 195, 207

conditioning, 304
conference, 189, 274, 321
configuration, 57, 63, 64, 259, 264, 285, 289, 300, 304, 312, 314, 316, 320
congress, 320, 321
conjugate gradient method, 263
connectivity, 172, 176
consciousness, 206
consensus, 70
conservation, 72, 348
constituents, vii, xi, 1, 5, 6, 153, 191, 192
Constitution, 381
construction, xiii, 6, 14, 19, 23, 85, 86, 210, 255, 260, 272, 323, 327, 332, 339, 340, 377, 378
consumption, 59, 356
convergence, xiii, 3, 39, 47, 50, 85, 99, 114, 216, 264, 300, 308, 342, 357, 372, 380, 393, 396
cooling, 63, 74, 114, 157, 163
cooling process, 63
correlation, 15, 78, 80, 86, 93, 185, 203, 211, 220, 228, 258, 270, 271, 342, 344, 345, 347, 348, 369
correlation coefficient, 15, 185, 211, 220, 270, 271, 344, 345, 348
correlations, 81, 82, 376
corruption, 176
cost, xiii, 23, 24, 64, 216, 254, 331, 342, 345, 349, 361, 383, 384, 385, 386, 388, 390, 392, 393
cost minimization, 385
cotton, 82, 92, 93
covalent bond, xiii, 341
covering, 56, 58, 67, 156, 262
CPU, 46, 48
critical period, 76, 78, 79
critical value, 9, 10
Croatia, 231
crop, viii, 75, 76, 78, 79, 81, 82, 83, 84, 85, 86, 87, 88, 90, 91, 92, 94, 95
crop production, 87, 94, 95
crops, ix, 61, 76, 90, 92, 93
cross-validation, 52, 57, 396
crystalline, 10, 192
crystals, 193
CT scan, 150
cultivars, 79, 91
cultivation, 93, 345
culture, 345, 351
cumulative percentage, 4
cycles, 8, 23, 288, 345, 365, 369, 371, 372
cycling, 88
cyclodextrins, 264
cytoplasm, 194
cytoskeleton, xi, 191, 192, 206, 207

D

data analysis, 222
data collection, 93
data generation, 155
data mining, 30, 154, 178
data processing, xii, 257, 258, 266
data set, xii, 10, 15, 17, 22, 26, 44, 57, 89, 107, 157, 158, 159, 160, 177, 178, 180, 182, 185, 222, 224, 257, 258, 259, 270, 273, 299, 301, 309, 312, 316, 328, 348, 349, 361, 367, 368, 370, 394
database, 59, 185, 347, 377, 381, 384, 392
decision trees, 347
decoding, 266, 380
decomposition, x, 140, 141, 149, 150, 151, 196, 260, 304
decomposition temperature, x, 151
defects, 205
deficit, 76, 79, 87
deformation, x, 151, 153
degradation, 11, 88
dendrites, 172, 174, 305
denoising, 261
dependent variable, 181
depolymerization, 192
depth, ix, xi, 76, 80, 81, 82, 89, 94, 97, 98, 99, 108, 113, 114, 115, 187, 192, 209, 210, 348
derivatives, 264, 308, 362, 396
detection, ix, 3, 129, 130, 143, 144, 146, 147, 148, 149, 150, 196, 265
deviation, 15, 85, 93, 108, 115, 116, 236, 343
DFT, 259, 352
diastole, 139
differential equations, 234
differential scanning, 23, 352
differential scanning calorimetry, 23, 352
dilation, 261
dimensionality, xii, 148, 257, 264
dipole moments, 194
directionality, 46, 48
discrete data, 107
discretization, 52
discrimination, 72
diseases, 86, 139
dispersion, 61, 73, 269
displacement, 110, 111, 118, 123
dissolved oxygen, 346
distilled water, 19, 23, 200, 201
distribution, x, xii, 9, 14, 30, 62, 66, 143, 144, 145, 151, 153, 155, 157, 173, 182, 193, 194, 197, 257, 283, 284, 334, 352, 361
diversity, viii, 55
DOI, 51, 340

dosage, 2, 10, 26
double bonds, 265
drainage, 76, 92
drought, 79, 88, 91, 93
drug delivery, vii, 1, 5, 6, 7, 8, 19, 25, 27
drug design, 347
drug interaction, 11
drug release, 4, 24, 27
drugs, 5, 10, 27
drying, 63, 65, 71
DSC, 23, 25, 254
dyes, 265
dynamical systems, 100

E

early warning, 61
ecosystem, 88
editors, 205, 207
electric charge, 193
electric field, 196, 198, 199, 200, 202, 204
electrical conductivity, 8
electricity, 255, 397
electrochemical impedance, 258
electrodes, 199
electromagnetic, xi, 191, 194, 195, 196, 197, 199, 203, 204, 232, 236, 237, 238, 251, 291
electromagnetic fields, xi, 191, 196, 204
electromagnetic waves, 195
electron, 192, 193, 205
empirical methods, 87
employment, 9, 377
energy, x, xi, 61, 64, 140, 142, 149, 151, 193, 194, 195, 197, 206, 231, 343, 347
energy transfer, 194, 206
engineering, 56, 62, 63, 65, 66, 68, 69, 72, 340, 349, 351, 352, 356, 373, 374, 396, 397
environment, 30, 47, 92, 154, 172, 173, 176, 178, 181, 194, 227, 347
environmental change, 176, 178
environmental conditions, 253
environmental quality, 88
environmental variables, 87, 377
enzyme, xiii, 341, 342, 343, 345, 346, 347, 348, 349, 351, 352
enzymes, xiii, 341, 344, 345, 346, 347, 348, 351, 352
epidemic, 84
equilibrium, 210, 225
equipment, 65, 67, 140, 291
ester, 265, 342, 343, 350, 351
estimation process, 251
ethanol, 5, 68, 265
ethylene, 5, 19, 27
ethylene glycol, 5, 19, 27
eukaryotic, xi, 191, 192
eukaryotic cell, xi, 191, 192
evapotranspiration, viii, 59, 61, 72, 75, 77, 78, 79, 83, 85, 86, 88, 93
evidence, 201, 296
evolution, x, 4, 51, 151, 192, 390
exchange rate, 185, 186, 189
excitation, 110, 121, 306
execution, 3, 282, 288, 367
experimental design, xii, 26, 257, 349, 350
explosives, 63, 65
external environment, 100
extraction, ix, 129, 130, 258, 259, 265, 267, 273

F

fabrication, 254
factor analysis, 186, 187
false positive, 143, 144, 146, 147, 148, 150
families, 261, 377
Faraday effect, 196, 199, 200, 201
farmers, 90
fatty acids, 5
fault diagnosis, 265
fault tolerance, xii, 69, 176, 177, 181, 277, 292
feature detectors, 282
feature selection, xii, 257, 258
FEM, 332, 333
fermentation, 66, 71, 73, 344, 345
fertility, 80, 82, 89, 91
film formation, 169
filters, x, 121, 129, 130, 131, 133, 136, 142, 146, 147, 197, 285
filtration, 63, 226
financial, 30
finite element method, 332
flexibility, vii, xiii, 1, 254, 299, 301, 347, 381
flexible manufacturing, 30
flowering period, 76, 79
fluidized bed, 27
food, 65, 344, 347
food industry, 344
force, x, 98, 99, 111, 122, 124, 145, 151, 304, 349
forecasting, viii, xi, xiii, 30, 60, 61, 72, 75, 82, 83, 87, 93, 181, 185, 186, 187, 188, 189, 209, 220, 225, 226, 228, 229, 323, 356, 357, 373
formation, 5, 6, 7, 8, 11, 17, 19, 25, 26, 61, 169, 337, 338, 346
formula, 78, 93, 185
freedom, 99, 121, 155, 188, 357, 394, 395
friction, x, 99, 124, 151, 152, 153, 156, 167, 168, 169, 170, 234

fructose, 344
fuzzy sets, 185, 212

G

Gaussian membership, 212
GDP, 193
genes, 23, 192
genetic programming, 4
geometry, xi, 8, 108, 209, 210, 211, 219, 265, 300
Germany, 169, 188, 189
glass transition, x, 151
glass transition temperature, x, 151
glucoamylase, 346
glucose, 345, 346
glucose oxidase, 346
glycol, 5, 20, 21, 25, 27
grades, 212
graph, 172, 173, 174, 175, 176, 182, 188, 269, 271
graphite, 266
gravity, 153
Greece, 82, 93
groundwater, 94, 226, 229, 339, 340

H

halogen, 265
heat transfer, 68
heavy metals, 266, 270
height, 65, 192
helical polymer, 193
heterogeneity, 324, 334
hexagonal lattice, 19, 192, 193
hexane, 350
historical data, 66, 185
human, vii, 1, 30, 56, 61, 62, 88, 124, 154, 212, 226, 228, 259, 306, 320, 356, 375
human actions, 88
human brain, vii, 1, 56, 154, 259, 306, 356
human resources, 30
humidity, 61, 65, 74
hybrid, 31, 39, 51, 121, 186, 188, 189, 213, 226, 229, 295, 332, 343, 350, 353
hydrocarbons, 265
hydrogen, 64, 68, 193, 347
hydrogen bonds, 347
hydrogen peroxide, 64
hydrogenation, 344
hydrophobic properties, 8
hypothesis, 194, 278
hysteresis, 240, 241

I

ideal, 19, 88, 134, 269, 272
identification, xi, xiii, 70, 155, 157, 212, 229, 231, 235, 245, 246, 248, 249, 250, 273, 341, 395, 396
identification problem, 273
identity, 174, 175, 192, 269, 362
image, ix, 129, 130, 131, 132, 133, 134, 135, 136, 137, 138, 140, 141, 142, 143, 144, 145, 146, 147
image analysis, ix, 129, 130, 142
imagery, 62
images, ix, 62, 129, 130, 132, 134, 136, 137, 138, 139, 140, 141, 142, 143, 144, 145, 146, 147, 148
improvements, 188
impulses, 291, 305
in vitro, 4, 194, 196, 205
in vivo, 194
incubation time, 345
independence, vii, 1
independent variable, 77, 80, 82, 83, 86, 87, 258
India, 29, 355, 367, 374
indirect effect, 81
indium, 268
induction, xi, 231, 232, 233, 234, 235, 236, 237, 238, 239, 240, 245, 247, 248, 251, 252, 253, 254, 255, 351
industries, 347
industry, xi, xiii, 2, 64, 68, 72, 152, 195, 231, 232, 235, 255, 264, 341, 350
inertia, 156, 234, 393
inferences, 50, 210
information processing, xiv, 56, 154, 195, 206, 355, 356, 358
infrared spectroscopy, 72, 259
integrated circuits, 296
integration, 80, 85, 227, 240, 253, 254, 296, 300, 373
integrity, 11, 279
intelligence, 27, 154
intelligent systems, 169
intensity values, 346
interface, x, 5, 151, 152, 153, 154, 155, 156, 157, 159, 160, 162, 163, 164, 165, 166, 167, 168, 254, 365, 372
interference, 105, 258, 268, 288, 291, 346
interneuron, 306
intervention, 258
inversion, 71, 300, 301, 303
iodine, 226, 265
ions, 193
Iran, 265, 266
iron, 152, 233, 251
irrigation, 59, 61, 87
Islam, 227

isolation, ix, 97
isoniazid, 10
issues, 65, 70, 93, 155, 176, 228, 348
Italy, 92, 191
iteration, 47, 114, 179, 180, 216, 282, 301, 312, 314, 320, 331, 383, 385, 389, 390, 391, 392, 393

J

Japan, xiii, 124, 139, 149, 168, 296, 323, 340
job flows, 53
joints, xiii, 299, 300, 303, 309, 311, 312, 313, 314, 316, 320

K

ketones, 266
kinetic model, 350
kinetic parameters, 344
kinetics, 65, 352
Korea, 186

L

labeling, 143, 150, 345, 348
lactose, 351
lattices, 206, 285
laws, 56, 300
layered architecture, 331
lead, viii, xii, 29, 51, 60, 62, 155, 186, 187, 188, 195, 196, 203, 210, 257, 269, 361, 362, 377, 383
learning behavior, 332
learning process, 23, 57, 83, 106, 155, 172, 173, 306, 330, 332, 349, 356, 362, 376
lecithin, 5, 19, 25
left ventricle, 130, 139
lens, 198
lesions, ix, 129, 130, 142, 143, 146, 149
ligand, 266, 344
light, 174, 196, 197, 198, 207, 251
linear function, 56, 134, 178, 213, 240, 359
linear model, 66, 393
linear systems, 169, 357
liquid crystals, 6, 8, 25
lung cancer, 139, 142, 148, 149
Luo, 98, 123

M

machine learning, x, 129, 131, 226, 397
magnetic field, 196, 198, 199, 200, 201, 202, 203, 204
magnetic resonance, 147
magnitude, 33, 196, 236, 242, 249, 289, 328, 336, 338, 382
majority, 187, 253, 377
Malaysia, 277, 299, 341
man, 62, 291, 296
management, 30, 86, 87, 93, 210, 227, 327, 340
manufacturing, 2, 53, 153, 170
map unit, 106, 107, 122
mapping, vii, ix, 1, 61, 62, 105, 107, 129, 130, 132, 134, 177, 178, 181, 182, 282, 300, 301, 303, 321, 343, 345, 356, 357, 368, 372
marketing, 377
Maryland, 129
mass, 6, 8, 14, 25, 65, 68, 99, 110, 111, 112, 121, 122, 194, 258, 259, 343, 344
mass spectrometry, 259
materials, 65, 76, 140, 153, 168, 169, 192, 195, 196, 207
mathematical methods, 154
mathematics, 188
matrix, 4, 26, 33, 34, 35, 37, 38, 99, 130, 142, 168, 175, 179, 181, 268, 300, 301, 303, 304, 314, 316, 320, 362, 382, 384, 385, 386, 387, 388, 389, 390, 391, 392, 393
matter, iv, 76, 82, 89, 204, 229, 292
measurement, ix, 67, 129, 130, 137, 195, 197, 235
measurements, 8, 26, 78, 201, 268, 334, 346, 351, 394
media, 27, 139, 196, 265
medical, ix, 27, 129, 130, 131, 137, 142, 143, 146, 147, 150, 352
medicine, ix, 56, 129, 130
Mediterranean, 87
membership, 212, 213
membrane separation processes, 66
membranes, 5, 63, 70, 73
memory, 100, 207, 242, 253, 254, 283, 294
memory capacity, 207
metabolism, 345
metalloenzymes, 344, 351
metals, 266, 268, 269, 270
metaphor, 358
meter, 111, 340
methanol, 343, 351
methodology, ix, xiii, 11, 24, 27, 51, 53, 63, 64, 66, 67, 70, 71, 76, 85, 90, 98, 156, 157, 158, 167, 259, 339, 342, 344, 350, 351, 397
Mexico, 257
microemulsion, vii, 1, 6, 7, 8, 10, 14, 16, 17, 18, 20, 21, 23, 24, 25, 27

micrometer, 114
microscope, 205
microscopy, 8, 205
microstructure, 5, 23, 374
microwaves, 196
Missouri, 94
misunderstanding, 186
misuse, 185
mixing, 197, 210
modelling, viii, x, 15, 55, 56, 59, 60, 61, 63, 65, 69, 70, 72, 73, 98, 99, 152, 153, 154, 169, 170, 228, 302, 353
modernization, 291
modifications, 384, 392
MODIS, 87
modules, 384
moisture, 65, 91, 368
moisture content, 65, 368
molasses, 344
molecular biology, 348, 350
molecular structure, 21, 65, 265
molecular weight, 5
molecules, xiii, 5, 7, 8, 19, 72, 193, 194, 195, 341, 348
momentum, 17, 104, 216, 251, 330, 332, 393
monomers, 193
Montenegro, 1
motion control, 254, 304
motor control, 245, 254
multidimensional, 258, 262, 263, 267
multiple regression, viii, 64, 75, 84, 86, 87, 209, 210, 219, 222, 223, 225, 347
multiple regression analysis, 64, 209, 210, 219, 222, 223, 225
multiplication, 288
multiplier, 180, 294
multivariate calibration, 259, 266
multivariate modeling, 273
mutagenesis, 348
mutation, 4, 47, 219
mutation rate, 219

N

naming, 282
nanomaterials, 195
nanometers, 196
nanoscale materials, 195
nanotube, 207
natural laws, xiii, 341
near infrared spectroscopy, 265
negative effects, 88
neglect, 387
nematode, 194
nerve, 172, 305
nervous system, 100
neural function, 64
Neural Network Model, v, vi, 151, 154, 169, 209, 324, 360
neurons, xiv, 2, 8, 11, 19, 21, 23, 33, 40, 56, 83, 85, 100, 101, 102, 107, 113, 122, 155, 158, 159, 160, 172, 173, 174, 194, 214, 215, 218, 237, 244, 252, 253, 254, 259, 263, 264, 267, 269, 270, 271, 279, 286, 305, 306, 309, 312, 314, 324, 325, 330, 331, 342, 343, 344, 345, 346, 347, 355, 356, 358, 360, 361, 368, 369, 372, 375, 376, 378, 379, 380, 381, 382, 383, 385, 394
neuroscience, 100
neutral, 193, 236, 242
next generation, 195
NIR, 71, 72
NIR-spectroscopy, 72
nitrogen, 72, 73, 92, 94, 197, 265
nitrogen dioxide, 73
nodes, vii, 1, 15, 29, 32, 33, 40, 41, 43, 44, 57, 66, 100, 103, 172, 173, 174, 175, 176, 178, 179, 181, 185, 186, 213, 214, 215, 216, 218, 237, 244, 250, 279, 281, 282, 284, 288, 293, 294, 306, 328, 330, 331, 336, 342, 343, 346, 356, 357, 369
nodules, 130, 139, 142, 143, 144, 145, 146, 147, 148, 149, 150
nonionic surfactants, vii, 2, 8
nonlinear dynamic systems, xiii, 342
non-linear equations, 300
nonlinear systems, 349
nuclear magnetic resonance, 259
numerical analysis, 121
nutrient, 68, 76, 77
nutrient concentrations, 68
nutrients, 82, 210

O

obstacles, xii, 299, 309
oil, vii, 1, 4, 6, 7, 8, 10, 14, 16, 17, 19, 25, 345
oleic acid, 5
omentum, 123
one dimension, 66
opacity, 144
operating costs, 345
operating range, 328
operations, ix, 65, 67, 97, 98, 99, 108, 150, 212, 261, 288
opportunities, 60, 183
optimization, vii, viii, xii, xiv, 1, 11, 14, 18, 26, 27, 29, 30, 31, 32, 33, 51, 53, 57, 60, 63, 65, 71, 211,

219, 226, 227, 251, 257, 295, 346, 350, 351, 355, 357, 363, 364, 382, 395
organ, ix, 129, 130, 137, 305, 344
organelles, 192
organic fibers, 169
organic matter, 76, 80, 82, 89, 91, 92, 93

P

Pacific, 53
palm oil, 343
pancreas, 351
parallel, xii, 67, 94, 154, 155, 157, 172, 173, 174, 176, 177, 178, 181, 182, 193, 194, 264, 277, 278, 288, 302, 306, 358, 377
parallel processing, 154
parallelism, 30
parameter estimation, 344, 356
Partial Least Squares, 259
partition, 85, 326, 327
pastures, 76
pattern recognition, xii, 30, 52, 58, 257, 345, 351, 356, 357, 396
PCA, 68, 259, 265, 349
PCR, 259
penalties, 38
penicillin, 343
peripheral nervous system, 100
permeation, 24
permission, 139, 140, 141, 143, 145
permit, 235
personal communication, 78
pH, 68, 193, 197, 205, 268, 346, 348
pharmaceutical, vii, 1, 2, 4, 5, 8, 26, 27, 347
pharmaceuticals, 27
phase diagram, vii, 1, 6, 9, 11, 13, 14, 18, 20, 21, 22, 23, 25
phenolic compounds, 266
Philadelphia, 227, 274, 275
phosphate, 265
phosphates, 193
phosphoenolpyruvate, 348
phosphorus, 72, 265
photons, 194
photosynthesis, 87
physical characteristics, 357
physical fields, 153
physical phenomena, 59
physical properties, 195, 196, 278
physicochemical properties, 18
physics, 195, 207, 357
physiology, 94
pitch, 99
plants, 64, 68, 76
PLS, 64, 68, 69, 95, 259, 266, 344, 349
polarity, 193, 265
polarization, 196, 197, 199
polarization planes, 197
polarized light microscopy, 8, 21
policy, 72
pollutants, 61
pollution, 61, 64, 266
polycyclic aromatic hydrocarbon, 265
polymer, 169
polymerization, 8, 68, 71, 192, 193
polymerization process, 68, 71
polymerization processes, 68
polymers, xi, 191, 192, 193
polyp, 147
polypeptide, 193
polyps, 130, 146, 148, 150
poor performance, 118
population, 4, 23, 37, 43, 46, 47, 49, 59, 61, 219, 225
population size, 37, 47, 219
Portugal, 55, 206
potassium, 70
potential benefits, 70
predictability, 345
prediction models, 82, 98, 99, 121, 259
predictive accuracy, 324, 349
predictor variables, 219
preparation, iv, 5, 7, 14, 19, 197
present value, 285
prevention, 149, 396, 397
principal component analysis, 69, 349
Principal Components Analysis, 259
principles, x, 129, 131, 248, 258, 324
prior knowledge, 154, 348
probability, 219, 327, 361
problem-solving, 348
process control, xiii, 65, 265, 341, 345
processing stages, 273
productive capacity, 88
profilometer, 114
programming, 4, 98, 100, 123, 185, 311
project, 192, 193, 273, 339, 367
propagation, viii, xi, xiv, 3, 8, 14, 18, 20, 21, 29, 31, 40, 46, 51, 83, 85, 100, 104, 105, 135, 149, 209, 210, 213, 214, 216, 226, 228, 264, 273, 281, 308, 309, 322, 342, 344, 346, 347, 351, 355, 356, 357, 358, 360, 363, 364, 372, 374, 379, 382, 395
propylene, 5, 27
prostheses, 320
protein folding, 204
proteins, 193, 199, 202, 204, 205, 344, 347, 348
prototype, 106, 107

prototypes, 106
pruning, 66, 382, 396
punishment, 307
purification, xiii, 341
purity, 67
pyrolysis, x, 151

Q

QED, 194, 206
quality improvement, 147
quality standards, 67
quantification, 88, 266, 269
quantization, 106
quantum dot, 195
quantum dots, 195
quantum field theory, 194
quantum gravity, 194
quantum state, 194
Queensland, 28

R

radial distance, 217
radiation, xi, 61, 76, 78, 92, 136, 137, 140, 196, 209, 210, 219, 222, 224, 225, 289
radio, xi, 191, 196, 207
radioactive waste, 340
radiography, 139, 140
radius, 106, 156
rainfall, viii, xiv, 59, 71, 73, 75, 76, 77, 78, 79, 81, 83, 85, 86, 88, 227, 228, 340, 355, 356, 357, 364, 367, 368, 369, 372, 373, 374
random numbers, 42
random walk, 185
reactions, xiii, 64, 73, 341
real numbers, 262
recognition, 3, 98, 123, 124, 226, 346, 377
recommendations, iv, 84, 295
reconstruction, 130, 289
recovery, xii, 153, 157, 160, 163, 169, 170, 257, 259, 272, 273
recycling, 64
reducing sugars, 344
redundancy, 294
reference frame, 234, 237, 238, 240, 244, 247, 248
refraction index, 196
regions of the world, 149
regression, viii, 3, 4, 15, 26, 61, 64, 69, 71, 73, 75, 77, 82, 83, 85, 86, 87, 90, 94, 104, 134, 154, 178, 181, 219, 259, 266, 269, 270, 271, 273, 344, 347, 357, 374, 377, 383, 395
regression analysis, 3
regression equation, 181
regression method, 77, 344
regression model, 64, 82, 83, 87, 154, 219
regulations, 2, 64
reinforcement, 343
rejection, 63, 65
relaxation, 195
relevance, viii, 5, 8, 55, 69
reliability, 61, 269
relief, 76
remediation, 94, 340
renewable energy, xi, 231
requirements, xiii, 92, 152, 229, 232, 248, 282, 295, 323, 368
researchers, vii, x, xii, 1, 4, 30, 47, 171, 178, 182, 183, 184, 185, 186, 187, 188, 277, 278, 301, 343, 344, 348, 356
residual error, 362
residuals, 178
residues, 193, 285, 287, 291, 295, 344, 347, 352
resilience, 277
resistance, 57, 70, 232, 240, 245, 246, 248, 249, 251
resolution, xii, 140, 141, 192, 227, 257, 259, 265, 266, 273, 393
resources, 59, 294
response, 15, 16, 44, 60, 62, 64, 66, 71, 82, 83, 86, 87, 92, 94, 98, 103, 146, 150, 172, 196, 201, 202, 204, 210, 222, 226, 234, 237, 258, 259, 264, 265, 266, 285, 301, 306, 307, 316, 329, 350, 351, 382
restrictions, 175, 304
rhodium, 344
rights, iv
risk, 62, 65, 73, 184, 285
risk assessment, 62
room temperature, 14, 194, 197
root, 15, 22, 80, 219, 220, 334, 345, 369, 394
rotation axis, 312
roughness, ix, 97, 98, 99, 107, 108, 109, 113, 114, 115, 116, 118, 121, 122, 123, 124, 125, 126
Royal Society, 206
rules, 70, 154, 212, 213, 253, 295, 360, 377, 385, 396
runoff, xiv, 59, 60, 71, 227, 340, 355, 356, 357, 358, 364, 365, 367, 369, 370, 371, 372, 373, 374

S

saturation, xi, 222, 231, 344
scaling, 76, 260, 262, 263, 267, 331
scatter, 370, 372
scatter plot, 370, 372
scattering, 200, 201

schema, 198
schemata, 195
scholarship, 273
science, xii, 72, 94, 258, 341, 356
scientific papers, 255
scope, 48, 258
sea level, 332
sediment, 225, 227, 374
selectivity, 258, 346
self-organization, 206
semiconductors, 195
sensing, 71, 73, 346
sensitivity, x, 63, 72, 142, 143, 146, 151, 153, 169, 200, 206, 223, 225, 240, 258, 291
sensors, viii, xii, xiii, 55, 56, 67, 69, 258, 273, 277, 299, 301, 309, 312, 351
sequencing, 30, 31, 33, 35, 36, 51, 52, 53, 72
Serbia, 1, 151
services, iv
set theory, 212
shape, 212, 361, 376, 382
showing, xii, 26, 62, 66, 84, 172, 258, 264, 301, 302, 334
signals, xi, 57, 60, 65, 100, 104, 122, 172, 191, 196, 213, 214, 215, 235, 240, 249, 253, 254, 258, 259, 260, 265, 266, 268, 273, 282, 291, 324, 329, 346, 379
signal-to-noise ratio, 265, 289
signs, 304
silicon, 195
simulation, xiii, 15, 27, 72, 94, 110, 204, 225, 289, 291, 292, 294, 295, 299, 301, 324, 341, 342, 349, 350, 352, 357, 377
simulations, 26, 62, 98, 110, 228, 288, 295, 344
Singapore, 321, 322
skin, 5
sludge, 72
smoothing, 15, 104, 131, 132, 199, 201, 202, 203
social sciences, 56
society, 64
sodium, 4, 28
software, viii, 8, 10, 17, 19, 21, 55, 56, 67, 69, 92, 185, 199, 242, 253, 254, 295, 351, 363, 364
solubility, 10, 26, 265
solution, vii, viii, xii, xiii, 10, 23, 29, 31, 34, 36, 37, 39, 40, 42, 43, 46, 47, 48, 49, 50, 51, 124, 183, 193, 196, 197, 199, 200, 202, 204, 211, 258, 268, 299, 300, 302, 303, 308, 314, 320, 321, 330, 331, 342, 356, 360, 361, 373, 377, 383, 390, 391, 396
solution space, 331
solvents, 265
sowing, 78, 79
space-time, 206
Spain, 82, 91, 93, 257
spatial frequency, 141
species, 62, 71, 72, 73, 76, 194, 259, 266, 272, 273
species richness, 62, 71
specifications, 197, 377
spectroscopy, 71, 73, 258, 259
stability, 5, 11, 67, 91, 152, 153, 154, 330, 347
stabilization, 11, 19
standard deviation, 41, 50, 136, 144, 217, 218, 272, 381, 382
standardization, 70
starvation, 346
state, 30, 52, 67, 173, 174, 175, 232, 234, 235, 237, 242, 247, 248, 251, 282, 283, 285, 345, 356, 359, 373, 377
states, 173, 194, 240, 242, 243, 247, 308, 346, 384
statistics, 76, 77, 188, 220
steel, 98, 114, 123
stenosis, 149
stimulus, 172, 174
stock exchange, 189
stock price, 52, 181, 189
storage, 60, 81, 89, 93, 282
streptokinase, 68
stress, 87, 88, 90, 92
string theory, 194
structuring, 222
substitution, 141, 347
substrate, 68, 342, 343, 344, 345, 346
substrates, 344, 347
subtraction, 140, 141, 149
success rate, 19
sugarcane, 345
Sun, 4, 27, 71, 288
suppression, ix, 129, 130, 131, 140, 143, 144, 146, 147, 148
surface area, 156
surfactant, 4, 6, 7, 8, 10, 14, 16, 17, 19, 20, 21, 22, 23, 24, 25, 27
surfactants, 5, 7, 8, 10, 19, 265
suspensions, 99, 124
Sweden, 71
symmetry, 186, 195, 207
synapse, 305, 306
synthesis, 260, 342, 343, 344, 350, 351

T

Taiwan, 185
target, 19, 44, 104, 105, 106, 155, 214, 268, 283, 309, 310, 326, 343, 345, 367, 394
techniques, viii, xii, xiii, 3, 18, 26, 55, 56, 58, 59, 60, 61, 62, 64, 65, 69, 72, 75, 80, 82, 83, 86, 87, 90,

93, 106, 140, 141, 152, 210, 222, 226, 227, 232, 234, 235, 240, 257, 258, 259, 265, 300, 301, 303, 341, 342, 343, 344, 347, 348, 349, 350, 356, 364, 373, 395
technologies, 170, 295, 296, 348
technology, xiii, 4, 27, 153, 195, 253, 254, 296, 341, 342, 349, 352, 373
temperature, vii, x, xi, 1, 5, 6, 8, 47, 49, 65, 66, 68, 70, 76, 77, 78, 79, 87, 93, 151, 152, 153, 154, 155, 156, 157, 158, 159, 160, 162, 163, 164, 165, 166, 167, 168, 169, 195, 197, 209, 210, 212, 219, 220, 221, 222, 223, 224, 225, 227, 228, 231, 240, 264, 265, 342, 343, 347
test data, 15, 17, 19, 158, 159, 223, 280, 284, 326, 327, 328, 342
test procedure, 394
testing, xii, 8, 15, 19, 21, 65, 85, 93, 99, 100, 113, 114, 126, 155, 156, 157, 158, 159, 160, 211, 220, 224, 269, 270, 272, 278, 284, 287, 299, 301, 311, 312, 316, 317, 318, 319, 336, 339, 342, 345, 347, 357, 363, 364, 365, 367, 368, 370, 371, 372
texture, 76, 80, 81, 82, 89, 124
thallium, 268
therapy, ix, 129, 130
thermal degradation, x, 151, 228
thermal energy, 153
thermal stability, 347
thermodynamics, 153
thermograms, 23
thermostability, 347, 351, 352
threonine, 344
time series, xi, xiii, 3, 60, 61, 69, 72, 187, 188, 189, 209, 220, 323, 332, 336, 337, 339, 346, 356, 370, 372
tissue, 140, 141, 142, 144
TMC, 114
toluene, 68, 73
topology, viii, 55, 57, 63, 64, 65, 66, 69, 70, 106, 172, 263, 267, 313, 314, 344, 360, 379, 396
torsion, 347
364, 365, 367, 368, 369, 370, 371, 372, 373, 377, 380, 381, 382, 383, 384, 385, 392, 394, 395, 396
training speed, 331
traits, 91
trajectory, xii, 65, 299, 301, 303, 304, 309, 310, 311, 312, 314, 315, 316, 317, 318, 330, 331
transducer, 111, 346
transesterification, 343
transformation, 2, 59, 154, 301, 302, 304, 313, 378, 385, 389
transformation matrix, 301, 302
translation, 260, 261, 262, 263
transmission, 152, 278, 279, 285, 291, 305
transparency, 4, 194
transpiration, 79
transport, 94, 192, 195, 228, 340
transportation, 64
treatment, xii, 53, 61, 63, 64, 68, 73, 257, 265, 273
trial, xiii, 6, 63, 65, 70, 148, 158, 237, 240, 312, 314, 331, 342, 343, 360
triglycerides, 5, 10
tropical forests, 71
tuberculosis, 139
turbulence, 229
Turkey, 97, 121, 171, 396
two-dimensional space, 62

U

U.S. Geological Survey, 220, 226, 228
UK, 60, 91, 92, 189, 255, 350
unconditioned, 269
uniform, xiii, 85, 114, 219, 323
unique features, 174
United, 92, 139, 169
United Kingdom, 139, 169
United States, 92, 139
universe, 185
updating, 176, 329, 330
urban, 227
uric acid, 346
USA, 10, 14, 21, 26, 93, 129, 168, 169, 170, 185, 188, 189, 197, 205, 209, 226, 228, 229, 274, 320, 321, 373, 395, 396, 397
UV, 64, 68, 70, 73, 265
UV light, 64

V

validation, 9, 17, 21, 22, 57, 60, 69, 83, 85, 87, 157, 158, 180, 220, 269, 270, 326, 327, 328, 345, 346, 347, 348, 357, 360, 369, 371, 383, 394, 395
valve, 111
variables, viii, xiii, 4, 6, 8, 14, 56, 59, 61, 66, 67, 69, 75, 78, 79, 81, 82, 83, 85, 86, 88, 89, 92, 98, 111, 123, 176, 185, 186, 187, 210, 212, 213, 222, 224, 228, 235, 236, 258, 259, 300, 314, 324, 328, 331, 342, 343, 344, 345, 346, 347, 349, 357
variations, ix, 87, 93, 97, 99, 119, 158, 240, 289, 324
varieties, 68, 72
vector, 19, 33, 41, 44, 57, 65, 69, 71, 72, 73, 100, 104, 105, 106, 107, 122, 134, 179, 180, 212, 216, 217, 232, 234, 235, 236, 237, 239, 240, 242, 243, 244, 245, 247, 248, 249, 253, 262, 263, 267, 268,

283, 301, 302, 303, 308, 309, 312, 320, 347, 359, 362, 376,뫰382, 383, 384, 385, 386, 391, 392
vegetation, 76, 87
vehicles, vii, ix, x, 1, 5, 6, 18, 97, 151, 152
velocity, 300, 303, 309, 310, 311, 314, 393
ventricle, 139
versatility, 130
vessels, 137, 142, 143, 144, 148, 149
vibration, ix, 97, 99, 117, 118, 124
visualization, 58, 69

W

Washington, 189
wastewater, 64, 68, 73, 227
water, vii, viii, xi, 1, 4, 6, 7, 8, 10, 13, 14, 16, 17, 18, 19, 21, 23, 25, 59, 60, 61, 62, 68, 71, 75, 76, 79, 80, 81, 82, 83, 85, 86, 87, 88, 89, 90, 92, 93, 94, 194, 195, 200, 209, 210, 219, 220, 222, 223, 225, 226, 227, 228, 278, 345, 356, 373
water quality, 210, 220, 227, 228
water resources, 60, 61, 219, 228, 373
watershed, 59, 356, 357
wavelengths, 196
wavelet, 72, 149, 228, 259, 260, 261, 262, 263, 264, 266, 267, 273, 274
wavelet analysis, 228
wavelet neural network, 274
wear, 98, 99, 153, 168, 169, 170
websites, 255
weight changes, 3
weight ratio, 19, 22
wind turbines, xi, 231
windows, 200, 365, 367
WNN, 259, 260, 262, 263, 264, 265, 266, 267, 268, 269, 270, 271, 272, 273
worldwide, viii, 75, 76, 82, 87, 139

Y

yeast, 66, 73, 343
yield, viii, xiii, 75, 76, 77, 79, 81, 82, 83, 84, 85, 86, 87, 88, 89, 90, 91, 92, 93, 94, 95, 187, 252, 254, 269, 272, 311, 323, 328, 342, 343, 344, 346, 349, 350, 351, 374

Z

zinc, 192